ENVIRONMENTAL SCIENCE

Fourth Edition

Stanley H. Anderson
University of Wyoming

•

Ronald E. Beiswenger
University of Wyoming

•

P. Walton Purdom

Macmillan Publishing Company
New York

Maxwell Macmillan Canada
Toronto

Maxwell Macmillan International
New York Oxford Singapore Sydney

Editor: Robert Pirtle
Production Supervisor: Dora Rizzuto
Production Manager: Roger Vergnes
Text Designer: Blake Logan
Cover Designer: Blake Logan

This book was set in Janson by Carlisle Communications.
Printed and bound by R.R. Donnelly & Sons, Crawfordsville.
The cover was printed by New England Book Components, Inc.

Printed in the United States of America

Macmillan Publishing Company
866 Third Avenue, New York, New York 10022

Macmillan Publishing Company is part
of the Maxwell Communication Group of Companies.

Maxwell Macmillan Canada, Inc.
1200 Eglinton Avenue East
Suite 200
Don Mills, Ontario M3C 3N1

Library of Congress Cataloging-in-Publication Data

Anderson, Stanley H.
Environmental science/ Stanley H. Anderson, Ronald E. Beiswenger
P. Walton Purdom. – 4th ed.
p. cm.
Includes bibliographical references and index.
ISBN 0-02-303191-3
1. Human ecology. 2. Environmental protection. 3. Environmental policy. 4. Ecology. I. Beiswenger, Ronald E. II. Purdom, P. Walton, 1917– . III. Title.
GF41.A48 1993
363.7 – dc20 92-18925
CIP

Printing: 1 2 3 4 5 6 7 8 Year: 3 4 5 6 7 8 9 0 1 2

PREFACE

In the mid-1990s more and more people have come to recognize the impact of human population on the environment. National debts run high as nations seek to borrow on their decreasing supplies of natural resources. World tensions rise as populated nations seek to gain territory, natural resources, and prestige. Growth-oriented individuals express the hope of increasing productivity as populations increase. There is anticipation of slower population growth as third world countries realize the necessity of dealing with their own overpopulation and malnutrition problems, rather than expecting help from the rest of the world.

Natural disasters compound these problems. While the United States was recovering from the effects of Mt. St. Helen's eruption in 1980, several major earthquakes occurred. In 1985, a devastating earthquake destroyed parts of Mexico City, and in 1989 the San Francisco Bay Area suffered another major earthquake. Hurricanes and tornadoes ravaged other parts of the world. As thousands of people succumbed to these natural disasters, people began to reflect on the fragile nature of their existence and to question the survival of the species.

The answer to this question lies in how we manage and use the environment. Can people live as a component of the natural system? If we do well, the quality of life will improve; if we do not, disaster awaits.

In this fourth edition, we present environmental science as a living subject. We have followed the suggestions of reviewers and students and arranged the material and topics to be most convenient. The first chapter discusses environmental science and relates it to other disciplines. Planning is introduced as a key component in management. We introduce the concept of natural systems in the next section and go on to explain how people are a component of the environment. We have expanded the information on population to include population biology, population genetics, and causes of extinction. A discussion of global climate change follows. Three chapters are now devoted to the management of our natural resources, including forests, grasslands, fish, wildlife, and biodiversity. We emphasize the importance of planning in managing these resources. A more comprehensive view of air, water, and soils is presented throughout the book. We also discuss the methods used to regulate our natural resources—laws, court decisions, and local actions such as zoning in each chapter. Throughout the book, we introduce sound management principles and examine them as the bases for maintaining our survival. Each chapter contains applicable case studies, clear illustrations, a summary with conclusions, and study questions. To keep abreast of natural as well as human-induced environmental changes, the fourth edition has been updated to include a number of current events. Our systematic approach incorporates the use of scientific hypotheses in the political arena to accomplish society's environmental goals.

In Part one we examine living systems and their interactions with the environment. We will discuss the amount of energy used for food and the methods and limitations of natural systems, their sustainability, along with the dynamics of populations. Part One also includes discussion of the geophysical and hydrological systems of the earth. Although these are dynamic systems, they are difficult to change by human means. Consequently, we discuss how best to live with the systems by showing how we alter through extraction of energy and minerals or adjust to their hazardous aspects.

Part Two deals with human interactions and the environment, especially atmospheric and hydrologic systems. We will see how the environment influences human health. Special hazards include storms and air and water pollution. We will also discuss pollution as th[illegible] result of the disruption of homeosta[illegible] how the symbiotic relationship [illegible] the environment can be reestabl[illegible] in this part we show how global clim[illegible] changing as the result of alteration of [illegible]on dioxide in the atmosphere.

Part Three is c[illegible]erned with managing the environment, e[illegible]asizing human use of the land and its re[illegible]rces. We will talk about forests and grassla[illegible] as systems we use to provide for fish, wil[illegible], food, and energy. Biodiversity is consid[illegible] as a concept and a tool to maintain a ba[illegible] in natural systems.

[illegible]vironmental toxins are discussed in Part [illegible]our, which evaluates pesticides, their effects on ecosystems and human health, and methods of pest control. This section also considers waste disposal, including toxic and radioactive wastes.

We will study human communities from the standpoint of protecting the environment and providing a healthy and desirable quality of life in Part Five. The goals we set and where we need to go to protect this earth are addressed.

Despite our best efforts, we do not have all the answers. We present general background information that will stimulate you to examine some unanswered questions about our environment, and we encourage you to seek answers from the many disciplines that contribute to the study of environmental science.

The purpose of this book is to create an understanding of (1) all facets of the environment that affect ecosystems and human life; (2) the impacts of human activity on various aspects of environmental quality; and (3) the environmental, economic, and cultural factors that shape urban development. We show how society and individuals can make rational decisions for achieving the quality of environment they de[illegible]. We make specific suggestions to encourage [illegible]he reader to adopt an active role in protecting our environment.

Many people helped with the preparation of the Fourth Edition. We are indebted to the reviewers of this edition for their perceptive comments and helpful suggestion. The staff of Macmillian Publishing provided extremely valuable guidance throughout the production process. We appreciate the assistance of Margaret Comaskey and Dora Rizzuto. In particular, we would like to thank Chris Waters and Linda Ohler for helping with the typing and editing of the manuscript.

Stanley H. Anderson
Ronald E. Beiswenger

CONTENTS

Chapter 1
ENVIRONMENTAL SCIENCE 1

What Is Environmental Science? 1
An Approach to Environmental Science 6
Systematic Environmental Management 8

PART ONE
NATURAL SYSTEMS 18

Chapter 2
ECOSYSTEMS 21

What Is an Ecosystem? 21
How Energy Moves Through an Ecosystem 23
Mineral Cycles 30
Ecosystem Succession 41
Physical Factors That Limit Distribution 45
Community Organization 48

Chapter 3
POPULATIONS 53

Population Dynamics 53
Population Distribution 60
Population Genetics 62
Evolution 68
Genetic Engineering 72
Extinction 73

Chapter 4
GEOPHYSICAL SYSTEMS 78

The History and Structure of the Earth 78
Continents on the Move 82
Volcanoes and Earthquakes 86
Geological Processes and Human Activities 94
Seashores 107

Chapter 5
THE HYDROLOGICAL SYSTEM 115

Water in Circulation 115
Lakes 123
Influences on the Water Cycle 124
Runoff 127
Water Supply 132

PART TWO
POLLUTION 140

Chapter 6
WATER QUALITY AND POLLUTION 143

Water Pollution: Its Sources and Effects 143
Drinking Water Quality 153
Supply and Disposal 157
Water Treatment 159

Chapter 7
THE ATMOSPHERE AND GLOBAL CLIMATE CHANGE 171

The Earth's Atmosphere 171
Human Activities and Climate 176
Air Circulation and Weather Patterns 182
Storms as Hazards 185

Chapter 8
AIR POLLUTION 191

Types of Air Pollution 191
Acid Deposition 210
Noise Pollution 214

PART THREE
ENVIRONMENTAL MANAGEMENT 220

Chapter 9
FORESTS AND GRASSLANDS 223

Land and Resources 223
Forest Ecosystems 225
Forestry 231
Forestry and the Future 237
Grassland 239
Rangeland Ecosystems 240
Range Management 241

Chapter 10
WILDLIFE AND FISHERIES 247

Wildlife Management 247
Species Management 259
Fisheries 266

Chapter 11
BIODIVERSITY 274

What Is Biodiversity? 274
Biomes 278
Loss of Biodiversity 283
Natural Biodiversity 285
Single-Species Management 287
Managing for Biodiversity 290
Tools Used to Manage for Biodiversity 292
National and Worldwide Considerations 294

Chapter 12
FOOD 299

History of Food Production 299
World Food Supply 305
Agricultural Land 306
Food from the Sea 311
Energy and Food 314
Nutrition Requirements 315
Prospects for the Future 320

Chapter 13
ENERGY 324

What Is Energy? 324
Energy Demands and Production 325
Electricity 329
Primary Sources of Energy 331
Secondary Energy Sources 345
Energy and the Environment 355
Energy Conservation 356

PART FOUR
TOXINS IN THE ENVIRONMENT 361

Chapter 14
PESTICIDES 363

What Is a Pest? 363
Pesticides 364
Impact on the Environment 366
Pesticides and Human Health 371
Economic Considerations 376
Alternative Methods of Pest Control 377

Chapter 15
WASTES 386

Solid Waste 386
Disposal 390
Toxic Wastes 399
Toxic Metals 402
Radiation 402

Sources of Radiation 411
Radioactive Waste Disposal 414

PART FIVE
PEOPLE AND THE ENVIRONMENT 419

Chapter 16
HUMAN POPULATION 421

Cultural Evolution 421
History of Population Growth 423
Human Demography 426
Stress and Distress in Modern Society 432
Reproduction 435
Population Control 437
Attitudes Toward Birth Control 443
Future Population Growth Patterns 444

Chapter 17
WHERE ARE WE, AND WHERE DO WE GO? 448

Development of Cities 448
Economic Growth 449
Factors Influencing Location 451
Environmental Quality 453
Interaction Between Urban Areas and the Natural System 455
Looking Ahead 458

APPENDIXES 459
1. The Metric System and Conversion Factors 460

2. Selected Environmental Periodicals 462

3. Environmental Organizations 464

GLOSSARY 467

INDEX 480

1

ENVIRONMENTAL SCIENCE

One of the unique characteristics of the species Homo sapiens *is unparalleled curiosity—about everything. Curiosity, together with extraordinary intelligence and motor coordination, beckon us to investigate everything—particularly those things that affect us directly. Other traits shared with animal life in general are the drives to reproduce and to survive. Over many centuries, we have learned much about ourselves and the physical systems around us. One of the results of our ability to apply this vast knowledge is a 1992 world population of more than 5 billion people. Our analytical minds tell us that by 2020 there will be more than 8 billion people. From this information we can deduce that we do not need to be concerned with extinction from a lack of progeny.*

Survival, however, is another issue. Air and water pollution, diminishing natural resources, and increasing food shortages are some of the physical problems we face. Less tangible, but nevertheless uncomfortable, are the emotional effects of overcrowding. These include stress due to an increase in crime, psychosomatic illness, and drug and alcohol abuse.

Our concern about such changes prompts us to ask questions and set goals. What quality of environment do we desire? What environmental conditions are necessary to the kind of life we want? What options do we have? Can we preserve the natural system? Can or should we shape the environment to our needs and desires? How can we accomplish our goals? Today as never before, we have the power to alter our internal and external environments. Through modern technology we can prevent or cure many diseases and live comfortably in inhospitable climates.

If we have the power, what else is needed to achieve the environmental goals essential to a satisfying quality of life, preservation of ecosystems, and our own survival? Knowledge itself is power. But one scientist may make a statement that contradicts another, stimulating emotionally charged debate. Fears are translated into "facts," and in other cases the facts are disregarded, denied, or deemphasized in favor of myth. How are we to distinguish between truth and fiction? This text provides the basic knowledge for members of society to make thoughtful choices.

WHAT IS ENVIRONMENTAL SCIENCE?

Environmental science is the application of knowledge from many disciplines to the study and management of the environment (Figure 1.1). It is an examination of the conditions, circumstances, and influences that affect all life and how life in turn responds.

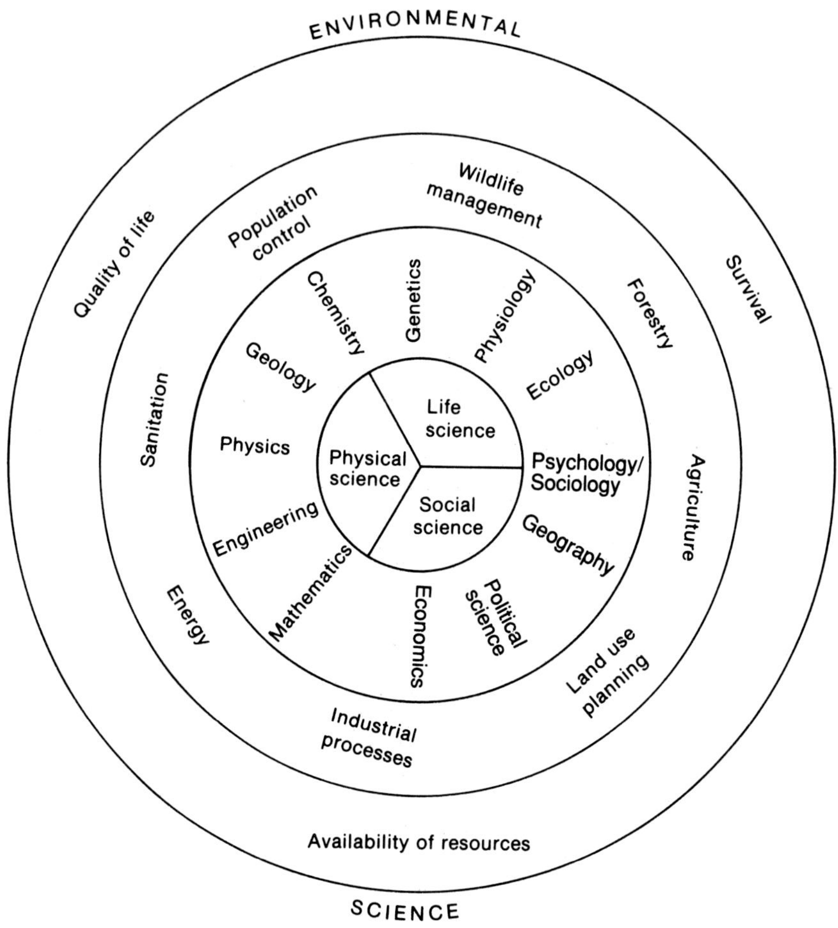

Figure 1.1
Environmental science is a multidisciplinary field.

Life

The meaning of life evolved with the passing of time and society's subsequent acquisition of knowledge. During ancient times, people viewed life with superstition. To fulfill a cognizant need, they developed myths of gods and rituals to explain life and death processes. But as civilizations advanced, people learned more about different aspects of life. Scientists discovered that all living organisms were made up of specific elements and compounds, primarily carbon and water. By the mid-1900s they had identified deoxyribonucleic acid (DNA) and ribonucleic acid (RNA) as the carriers of genetic information, thereby providing insight as to how life reproduced. While acquiring this gradual knowledge of life, however, scientists observed that one basic characteristic endured: Life, in whatever form, requires the correct balance of environmental conditions to survive.

Organization of Life

All material in the universe is made up of different elements that combine in different ways.

Living material is likewise composed of elements, including carbon, the basis of most living organisms, hydrogen, nitrogen, oxygen, sodium, phosphate, calcium and others, that combine in different ways.

There are ninety-two naturally occurring elements, not including the radioactive materials that decay. Of these ninety-two elements, carbon, hydrogen, nitrogen, oxygen, phosphorus, and sulfur make up 99 percent of all living material. Rocks, soil, air, and water are all composed of different combinations of elements. Soil is composed of a variety of different elements depending on its location. How the different elements combine determines the type of substance.

The most generally accepted scientific explanation of the formation of the solar system is the big bang theory. According to this theory, about 16 billion years ago, all matter was at a single point when an explosion vaporized the material and scattered it through space. This matter subsequently condensed in local areas in space to form the billions of stars, planets, comets, other objects, and our own galaxy, which contains about 100 billion stars and our sun and planets. The chemical particles in the condensing matter collided with each other to release energy that created heat, raising the temperature until thermonuclear reactions occurred.

The current hypothesis as to the beginning of life is offered by a Russian biochemist, A. I. Oparin. He believes a process of chemical evolution occurred before life began. This evolution took place in a primitive atmosphere and in early seas, which contained a variety of different chemicals. There was probably little or no free oxygen at the time, but a primitive water atmosphere and an abundance of hydrogen, oxygen, carbon, and nitrogen. (Today these elements make up 90 percent of living tissues.) Temperatures undoubtedly fluctuated greatly and climatic conditions were rather violent. The relatively thin atmosphere allowed a great deal of ultraviolet radiation to reach the primitive earth. Lightning was also common, providing a great deal of energy. Oparin believes that under these conditions chemicals combined to form organic molecules that collected in a thin soup in the earth's seas and lakes. Because there was no free oxygen to react and degrade these molecules to simple substances such as carbon dioxide, they tended to remain around, and as lakes and water systems dried, some molecules were able to concentrate. When these dried basins became moist again, another cycle of chemical change and concentration resulted in formation of new organic substances.

Laboratory tests in the 1950s using enclosed glass containers with electrical discharge tend to support Oparin's hypothesis that a variety of elements could combine into some basic precursors of life. Because these precursors of life persisted, it is believed that a boundary membrane formed around the early combinations of elements and separated them from the external environment. This process continued over millions of years, permitting early cells to form. The earliest fossils of some organisms similar to today's bacteria have been dated at about 3.5 billion years, or about 1.1 billion years after the formation of the earth.

When early life reached a stage at which it could divide to form identical or similar combinations of elements from existing units, we had the beginning of life as we now understand it.

From these early beginnings, changes on earth caused changes in the existing form of life (Figure 1.2). Conversely, the life that developed caused changes on the earth which in turn brought about a different environment. As living organisms began to interact with one another, a variety of events occurred. A transfer of energy from both the outer atmosphere, including the sun, and other physical elements provided a basis for living organisms, which were influenced by the surrounding environment. This situation continues today—changes are occurring all the time, and these changes in turn influence how life is found on earth.

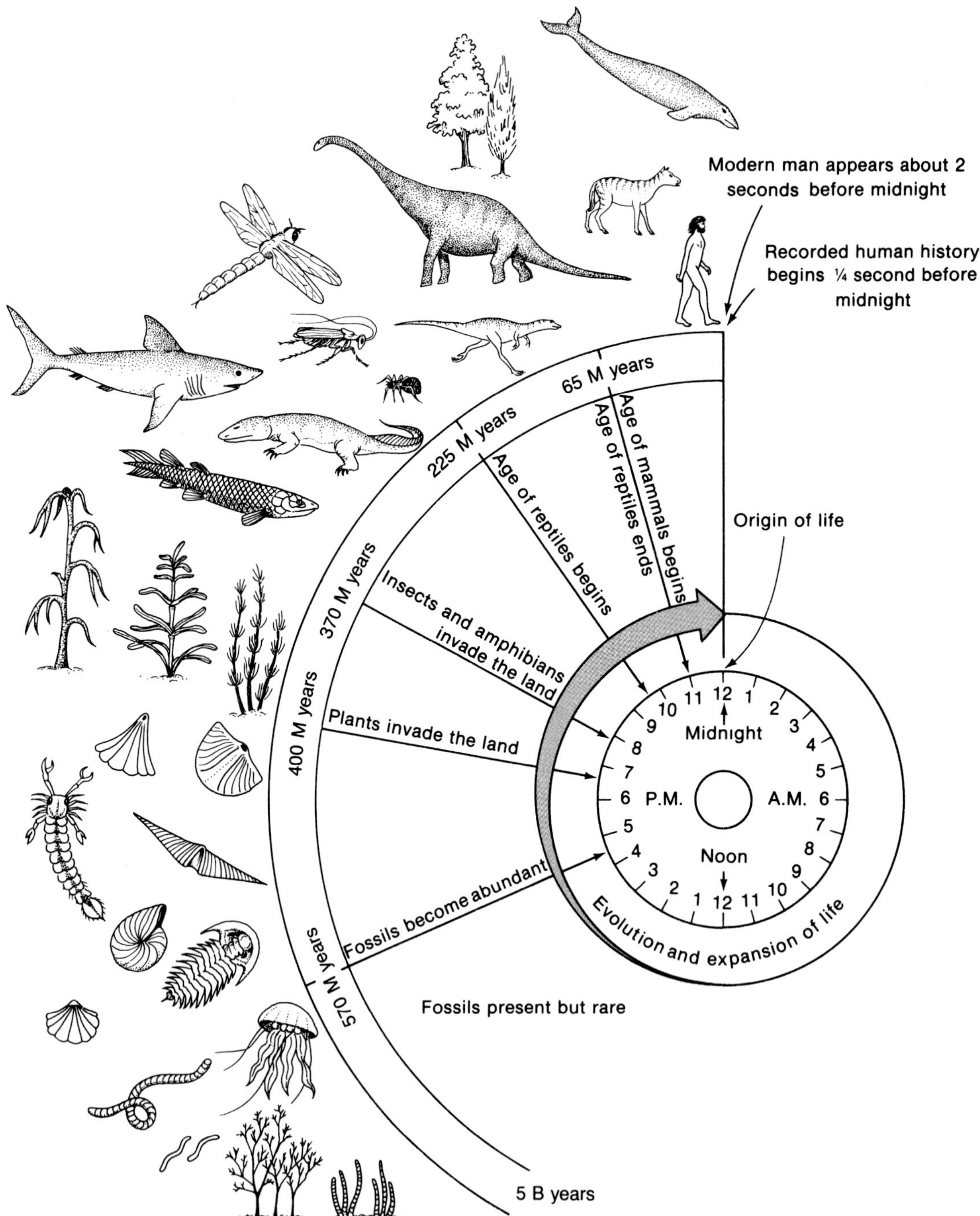

Figure 1.2
The history of life condensed to a 24-hour scale. (From *Life: An Introduction to Biology,* 2nd ed., by George Gaylord Simpson and William S. Beck, © 1965 by Harcourt Brace Jovanovich, Inc. Reprinted and reproduced by permission of the publisher.)

Perception of the Environment

The way a person perceives the environment reflects one's previous experience, education, lifestyle, and interests. For example, a special interest in its physical aspects may lead an earth scientist to the study of geology, meteorology, and hydrology. To completely understand these areas, however, scientists must consider how biological events influence their area of primary concern. A chemist might develop an initial interest in chemicals in the environment. This pursuit could lead to a discovery of the fate of pollutants discharged into the environment. An ecologist might investigate what happens when organisms consume the pollutants and how, or if, the concentration increases as one species eats another.

Worldwide different cultures look at the environment differently. Some remove "valuable" components for themselves. In poor countries people take all they can for survival. People in the United States often have difficulty in relating to methods by which other cultures perceive the environment.

Ecology and Environmental Science

Many people view ecology and environmental science as synonymous; thus environmentalists are frequently considered ecologists. Actually, anyone interested in and concerned about the environment is an environmentalist. Ecology is one of the disciplines constituting the core of environmental science. Within ecology there are many subdisciplines; thus, an aquatic ecologist has different advanced training than a radiation ecologist.

The word *ecology* is derived from the Greek word *oikos*, meaning a place to live. Most ecologists therefore define ecology as the study of the relationship of an organism or group of organisms to their environment. The ecologist Eugene Odum points out that ecologists are concerned with the biology of groups of organisms and with functional processes. He defines ecology as "the study of the structure and the function of nature."

While other disciplines also study the interactions of life and the environment, ecologists are concerned about what limits life, how living things use resources such as minerals and energy, and how living things interact. Ecology is the study of these processes; environmental science is the application of this knowledge to managing the environment.

Divisions Within the Environment

The focus of our study is on places where life is found—where all the elements needed to support life are present. This relatively thin layer at the earth's surface is called the **biosphere.** By nature the scope of the biosphere is restricted. Few living things are found above 3100 meters (10,000 feet) or in the depths of the earth. Although life exists in all parts of the oceans, it is dependent entirely on the food producers within the areas of sunlight penetration.

Within the biosphere are functional units or **ecosystems,** consisting of all living organisms plus nonliving components. Deserts, forests, and lakes are examples of these units. For the sake of convenience, we study these areas individually or in small parts. They are not isolated from the surrounding biosphere. Thus, interactions occur at the interface—between pond and shore, forest and field, or ocean and land.

Environmental Modification

Within recent history, people have modified the biosphere to such an extent that major changes have begun. Massive cities, variations in the atmosphere's mineral content, new nutrient balances in water systems, and major alterations in the earth's vegetation patterns have all combined to alter the environment. People have begun to realize that these transitions are causing life itself to change. As the structure of natural communities alters, we can expect these long-term modifications to result in different species of life.

We can already create artificial surroundings that mimic natural environments. These surroundings enable people to live in formerly adverse locations, such as the ocean and outer space. By controlling pressure and providing oxygen, we can descend to the ocean depths in submarines and bathospheres. Someday people may even live in enclosed cities on the ocean floor. In space travel, full pressure suits create an artificial balance of environmental conditions around astronauts so they can carry their life-supporting environment with them.

In recent years there has been an increasing emphasis on monitoring environmental modification on a global level and on recommending strategies for improving the well-being of people and sustaining the ecosystems on which they depend. It is now generally recognized that ecologically sound environmental management and economic development, particularly in developing countries, are closely linked. In 1980, the International Union for the Conservation of Nature and Natural Resources (IUCN), the U.N. Environmental Program (UNEP) and the World Wildlife Fund (WWF) collaborated to develop a World Conservation Strategy. The Strategy outlines goals and actions to achieve three major objectives:

1. To maintain essential ecological processes and life support systems
2. To preserve genetic diversity
3. To ensure that the utilization of living resources, and the ecosystems in which they are found, is sustainable

These objectives are based on the recognition that successful economic development and an improved quality of life take into account the capabilities and limitations of the natural environment, as well as the needs of future generations.

More recently these ideas have evolved into the idea of **sustainable development.** This is development that meets the needs and aspirations of the present without conflicting with natural resource conservation, environmental protection, and sustainable ecosystems. To understand sustainable development, we must examine all the processes that are occurring in our environment.

AN APPROACH TO ENVIRONMENTAL SCIENCE

In this book, we attempt to follow a path of scientific inquiry into how natural systems function, explore the past influences, present problems, and future prospects for people and the environment, and develop rational thought processes for evaluating environmental issues. Our goal is to propose a systematic approach to managing the environment with a view to maintaining our ecosystems.

Environmental science encompasses many disciplines, each with its own principles and concepts. These concepts unify scientific inquiry into a holistic understanding of the environment. The major concepts we use as guidelines are homeostasis, energy, limits, symbiosis, systems, and models.

Homeostasis

Because environmental science involves understanding and analysis of many complex interactions, we find a variety of approaches used in its study. **Cybernetics,** the science of controls, is one approach. How is the environment controlled? How are all the interactions within and among the physical and biological systems regulated? The control of such systems requires information in the form of feedback. **Feedback** is the return of output, or part of the output, to a system as input. That is, it gives the system information that will cause it to change so as to maintain a particular state. An example of a cybernetic control by means of feedback is the thermostat in your home. When you set a thermostat to a desired temperature, the information is fed to a temperature-sensing device. If the room temperature is lower than the temperature you indicate, the furnace starts up; if it is higher, the furnace cuts off. Room temperature is continually monitored as signals are sent to the furnace to regulate its operation.

Two forms of feedback occur in this example. When the temperature falls, more heat is supplied. This negative or reverse relationship between input of information and response is called *negative feedback.* When a change in the system in one direction is converted into a command to change the system in the same direction, *positive feedback* occurs. Generally, negative feedback keeps a system in equilibrium and positive feedback disrupts equilibrium and causes the system to become unstable.

Living systems, including groups of organisms living together in the same environment as well as individual organisms, have cybernetic or self-regulating feedback mechanisms that maintain their equilibrium. This tendency for biological systems to resist change and remain in a state of equilibrium is called **homeostasis.** Physiologists study many forms of homeostatic control. Regulation of body temperature and blood chemical content are examples of homeostatic control in individuals. Equilibrium between organisms and the environment can also be maintained by feedback mechanisms. Such processes are important in the balance of nature. Only recently have people begun to analyze the interplay of energy and materials in sustaining life.

The homeostatic mechanisms in living organisms generally operate in a common manner. The initial information input, or *stimulus,* activates a sensing device. The information is transmitted over a sensory pathway to a response selector that transmits a signal to effectors which, in turn, initiate a *response.* Thus, an animal that sees danger runs away or prepares to fight.

Response involving whole groups of organisms is much slower. Without predatory organisms, a prey population increases; and predator populations increase in response to a large number of prey animals.

Living systems interact in many ways. We can study the interactions of two individuals or two populations to determine which mechanisms influence their relationship. Likewise, we can examine different levels of biological organization until we encompass the whole universe. Each new bit of information supplies another piece in our puzzle of understanding life and the complicated interactions that take place in the natural system.

Throughout this book, consider how environmental actions influence the established equilibrium. What forms of feedback operate to maintain equilibrium? What are the consequences of disruption?

Energy

Physicists define **energy** as the ability to do work. Work can mean the movement of a car, growth and reproduction of a plant, or explosion of a bomb. Virtually all human actions require energy. Wasteful use of energy, such as needlessly discarding food or excessive use of cars, converts energy into less usable forms.

We introduce energy as a concept because it has an impact on almost all our actions. Since it actually imposes structure on living systems, energy supplies the driving force and determines the limitations of life on earth. For each topic in this book, consider the type of energy involved in the concept and the limits energy imposes.

Limits

The concept of a **limit** is more controversial than the other concepts we have described. Let us consider a small pond on which water lilies grow. In this pond, the lilies double in number each day. The size of the pond is such that this rate of growth will cause the pond to be completely covered in thirty days. When is the last day the pond can be cleared of lilies before it will be completely covered? On the twenty-ninth day the pond will be half covered, and the next day it will be completely covered.

It is because of this concept that some scientists speculate on the capacity of the earth to maintain an ever-increasing human population with its food, energy, and mineral consumption and waste generation. While the increase in human population is not as dramatic as the growth rate of the water lilies, a small increase every day will ultimately reach the capacity of the environment to support consumption or waste assimilation, whatever that capacity is. The area of speculation is not so much whether such a capacity exists, but when it will be reached. How much time do we have for making decisions?

Symbiosis

Symbiosis refers to dissimilar organisms living together. It is an important interaction maintained by homeostatic mechanisms. Various kinds of environmental disruptions result in a breakdown of many symbiotic relationships. We will discuss symbiosis and its ecological importance in Chapter 2.

Systems

The concept of a **system** is used to describe many things: a transportation system, the health care system, a school system, a highway system, an ecosystem. *Closed systems* have virtually no input from the outside. A space capsule, a submarine, or the earth as a whole are nearly closed systems. In contrast, forests, estuaries, air and river basins, and urban communities are more *open systems.* In general, a number of related interactions are necessary to the functioning of a system. A change in one part has repercussions in the rest. If a mechanic fails to repair a bus properly, it might not run. The bus system could be thrown off schedule as a result. If one link in a highway system is missing, there might be traffic jams on the interconnecting roads.

A system's boundaries are selected for convenience. A total transportation system would include not only the bus and highway systems, but the materials that supply power as well. Trucks and miscellaneous delivery vehicles might also be essential parts. Gasoline stations for private autos and the electric generating stations that supply a rail or trolley system could be added. However, a line has to be drawn somewhere; so intercity transportation might be excluded from a consideration of the transportation system of a local community. Even so, the intercity connections influence the intracity system and have to be considered.

In a living system, interactions are numerous and complex. Species interact not only with each other, but also with the environment. The movement of energy between the sun and groups of organisms and the role of minerals in the interrelationships between the living and nonliving environment are examples of such interactions.

SYSTEMATIC ENVIRONMENTAL MANAGEMENT

A major theme of this book is systematic environmental management. The principles and concepts described in the preceding sections help us develop a holistic understanding of the environment. Understanding brings the possibility of controlling use of the environment so that we sustain and not degrade it while we benefit from its resources.

What Are Resources?

Minerals, land, timber, and wildlife are all resources. Lakes, streams, oceans, and underground water make up our water resources. A **resource,** then, is anything that satisfies the needs or wants of civilization. Early societies used wood rather than oil for fuel, so oil was not one of their resources. Historically, people have increasingly taxed the extent of the earth's resources, turning to new ones to fill their needs when the old ones were used up. In Chapter 2 we will examine how the sun is the driving force of life and study the facts about mineral cycles. Thus, we can view the entire earth as a biosphere in a dynamic or changing state. We can influence this dynamic state through our actions. While we cannot significantly increase the amount of sunlight entering the earth's atmosphere, we *can* change the state of some resources so that they are no longer available for use. For example, we can upset the earth's heat balance by producing gases such as carbon dioxide that leave heat in the atmosphere. This leads to global warming.

Preservation, Conservation, and Management

Some people think that a segment or a large portion of any resource should be set aside and preserved for future use. Thus, they might protect a section of forest land from being disturbed or perhaps save an oil reserve. Although **preservation** of physical resources such as forests or oil might save these commodities for future generations, natural systems are dynamic and cannot be kept from changing. *Natural succession* means that preserved areas may change and no longer be the same as when first set aside. Mineral cycles can also vary as other natural systems change. Since we cannot prevent change, how can we best sustain our environment while using its resources?

People frequently speak of preservation when they mean **conservation.** The word *conservation* is derived from two Latin words: *con*, meaning together, and *servare*, meaning to keep or guard. So conservation literally means *to keep together.* The word itself was coined by Pinchot shortly after the 1908 White House conference. Today we think of conservation as using our natural resources wisely rather than keeping things together in the status quo. This idea incorporates concepts of manipulation and decision making in regard to resources, but we must first know what our resources are and what part they play in the ecosystem. Aldo Leopold, a noted conservationist, pointed out that people must understand ecological processes to practice conservation. He further indicated that people trained in this field must have a basic knowledge of geography, botany, zoology, history, and economics. Conservation of some areas of our wilderness and river systems is now mandated by laws such as the National Wilderness Preservation System and the National Wild and Scenic Rivers System.

We manage or manipulate the environment in countless ways—some helping and sustaining ecosystems and some detrimental to them. Resource **management** is the process of directing or controlling the production or use of resources. To do this properly, it is important to see all contributing influences and predict the effects of habitat manipulation. We must remember that people decide what species of plants or animals are desirable; in the natural system there are no good or bad species.

Planning

The planning process is used to solve conservation problems or to arrange orderly use of natural resources. It can also involve the approach to a particular problem. For example, why is the deer population increasing in Pennsylvania? A sound, relevant approach would consist of securing knowledge of the historical deer population trends from Pennsylvania Game Commission data to determine what changes are occurring in deer populations in the state and contacting experts in the field. Comparisons could be made and a research plan developed and implemented to determine the exact reason for the change.

Another, perhaps more common, use of planning is directed toward the long-term use of a natural resource. While rotational grazing, mineral extraction, and forest harvesting of national lands can all be part of a master plan, their simultaneous recreational use for sportsmen, campers, and backpackers has to be considered.

To successfully sustain our environment, we must understand how planning operates. Management really consists of planning, organizing, and controlling. The organizing and controlling processes carry out decisions made during the planning process. Part of the study of environmental science involves planning so we can control our environment. If we allow air and water pollution to increase, we harm the health of all living things. Deforestation in the tropics destroys the habitats of many plants and animals as well as affecting the world balance of air chemicals and water. Thus, we *must* plan and thereby manage to control pollution and the use of our natural resources.

Planning is a systems approach. It can be conceptualized through a series of questions:

1. Where are we?
2. Where do we want to be?
3. How do we get there?
4. Did we make it?

Where are we? To answer this question, we must know what we have. We could have an inventory of animals, nuclear power plants, supplies, and so on, depending on the planning being done. We must decide what needs to be inventoried and the level of details required. This concerns time and dollars.

Where do we want to be? To answer this question we need to formulate goals, objectives, and strategies. A goal is a statement of mission or policy; many are very general (i.e., "To reduce the level of all pollution," or "To conserve our wildlife resources"). Objectives define measurable results (i.e., "Decrease air pollution by a specific amount," or "Provide so many days of deer hunting"). Strategies are methods that can be used to achieve objectives or overcome problems that prevent us from reaching our objectives. Development of a number of strategies may be necessary to achieve an objective. Reducing the number of days people can drive their cars and enforcing emissions control standards on vehicles and factories are both strategies.

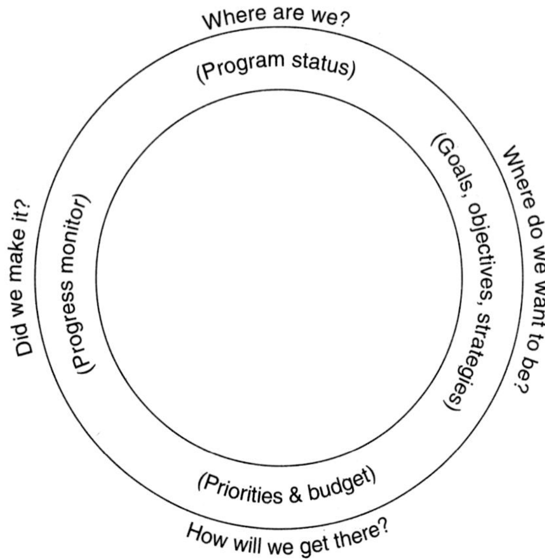

Figure 1.3
The elements of a planning system.

How do we get there? This part generally involves developing an operations plan which lists our objectives with strategies, time schedules, funding, and manpower.

Did we make it? To answer this we must see if our objectives were achieved during the time frame and with the budget and personnel designated (i.e., "Did we reduce the level of air pollution by the amount we prescribed?").

The planning cycle can best be visualized as a circular route. This continuous cycle must always be operational at many levels if we are going to sustain our environment (Figure 1.3).

Levels of Planning

Sound natural resource management is based on comprehensive planning that promotes the greatest overall benefits from our resources. This planning can occur at a number of different levels. Government officials deal with various issues and policies, making decisions based on political and economic balances that affect areas such as recreation, water quality and use, and zoning. At top-level national management, decisions are often made from a minimal amount of information because data must move from field-level personnel who collect the information, up to the regional and national levels. Often the complexity of natural systems makes adequate knowledge difficult to obtain and managers are frequently forced to make decisions based on quick surveys. When a plan based on short-term data is implemented, in-depth studies can later prove the plan less than optimal. Management must have time to acquire input from appropriate field biologists, forest rangers, and range managers to develop effective policies.

At intermediate and regional levels, decisions frequently involve energy and water development, the impact of alternative resource development on the economy and society, the rate of deforestation, and cattle grazing restrictions. These resource areas often have broad, general master plans. For example, New Jersey's pine land master plan directs the management of areas for wildlife, development, water use, recreational use, and timber production and provides for the availability of information to mid-level planners.

Ground-level management is in the field where plans must be executed. Local and economic factors, recreational needs, and detailed environmental impact studies must be considered, along with regulations regarding road building, local timber harvest, off-road vehicles, and park planning. When master plans are available, field-level people make decisions based on those guidelines.

A good example of the planning process is the Strategic Plan for Comprehensive Management of Wildlife in Wyoming. This plan is the "Where do we want to be?" phase of planning. It leads to operational planning. The plan, a five-year guideline for the Wyoming Game and Fish Department, states its objectives and addresses the problems and methods of managing fish and wildlife throughout the state. For example, strategies to deal with problems caused by the rapid influx of people include developing ways to provide recreational opportunities for more people while maintaining the state's diversity of wildlife; determining people's impact on wildlife communities; and developing and implementing standardized methods to monitor demand for wildlife-related recreation.

The second part of the plan discusses managing big game and nongame wildlife as well as sport fisheries. Each species is then examined on a district level and, when appropriate, as herd units within the districts. For example, the plan lists the status of antelope in each district, the number of antelope-hunter recreation days at the end of each year and the number of animals expected to be taken, and proposes management parameters until 1993. Population models are used to project the winter population, hunter success, and total recreation days per year. A program is then presented based on the available population, desired harvest, and habitat data in each district to achieve the program's goals using methods such as hunting quotas, length of the hunting season, land acquisition, and habitat management. In the past, most natural resource agencies concerned themselves with single species management such as tree species or a game animal. Now they are required by law to consider total resource management. This involves periodic inventories and assessments of natural resources including timber, land, and wildlife.

Tools for Planning

Field surveys, literature reviews, aerial photographs, and land charts are planners' tools to produce data banks, models, land classifications, and maps. In the process of planning, areas of limited expertise are often identified and plans are made to acquire more information.

Data Banks Computerized data storage and retrieval systems not only reduce the time and cost of handling data but also enable us to analyze large masses of data in different ways. For example, when a data retrieval system is used to determine which species of wildlife are found on a potential mine site, planners can be immediately alerted to evaluate the possible presence of endangered species. Computerized data systems help make essential field checks to find out if the species is present or if the habitat has changed.

A major problem with data systems is that information often comes from scientific literature and unpublished statistics that were not gathered for the specific purpose. For instance, wildlife data are entered into a bank based on potential or original vegetation. When this input is entered, it does not reflect the vegetation changes that result from farming, logging, or urbanization. Extensive field verification is required for the data systems to be useful. If data are gathered haphazardly, results can be misleading.

Models Model building is a tool frequently used by planners. In developing government policy and in making private decisions, a trial-and-error approach can be disastrous. To anticipate the ultimate result without tampering with the real system, we use something else that looks like or acts like the real thing, a **model.** Most of us are familiar with small models of airplanes, railroads, and autos. A photograph can be a model in communication—"one picture is worth a thousand words." Sketches are used by architects and blueprints by carpenters. If boards are cut as shown on the drawings, a home can be built that will look like that in the drawings. Small replicas of furniture can be placed on the drawings to test various window and door arrangements. This is much easier than building a house and later tearing out the walls and rebuilding if a window is in the wrong place for the furniture.

Other changes in living systems can be modeled by mathematical equations. By varying the input on one side of the equation, we can predict the outcome symbolized by the other side. Game biologists use such models to determine how fawn production might vary if doe hunting is allowed in the fall. To forecast downstream oxygen levels and species diversity, a biologist needs a more complicated model using temperature, flow of water, use of oxygen by wastes and organisms, rate of replenishing oxygen, and the availability of nutrients in rivers.

Model building is not only predictive; it can also be descriptive. The important thing to remember is that the less complicated the system and the more information available, the greater the likelihood of a successful model. Input from models can be incorporated into a computer data system and used to predict possible changes. Based on information in a data system, the effects of a transmission line corridor cutting through a habitat could be evaluated and an indication made of what changes in species composition might occur from reduced habitat size, increased water runoff, and soil compaction. If the system is set up correctly, it is also possible to quickly test and predict the effects of the alternative management strategy of keeping the transmission line in grass, shrubs, or short trees.

Land Classification Another tool used by resource managers is a land classification system. This system can be exemplified by classification types such as drainage systems, soils, wildlife habitats, and areas of potential vegetation. The U.S. Fish and Wildlife Service's classification of wetlands and deep-water habitats of the United States is a typical wildlife classification system that can be updated periodically with aerial photos and ground checks.

Geographic Information Systems (GIS) are now being developed by many resource agencies. This system consists of a series of distribution maps. The computerized maps may show topography, vegetation cover, distribution of different wildlife species, and other features. The maps can be overlapped in a computerized system to show correlations between different facets such as pheasant distribution and vegetation. Biologists can then predict the presence of the birds based on vegetation type. The GISs can each have its own set of unique maps that allows the user to ask many questions.

Planning Applied to Natural Resources

Resources that are consumed as they are used are *nonrenewable resources.* Coal, oil, and minerals are examples of nonrenewable resources because no additional amounts of these materials will be added to the earth's system. As we shall see in the section on minerals and in Chapter 4, the extraction and use of nonrenewable resources must be carefully planned so they are not wasted.

Some natural resources are renewed by growth or reproduction. These resources are referred to as *renewable.* As we will discuss in Chapter 10, trees can be removed from a forest and used for lumber and, with proper management techniques, the forest will grow back. Wildlife populations can be managed so that animals that are killed by hunters or die from other causes are replaced by natural reproduction (Chapter 11).

The unique nature of renewable resources requires a different kind of planning. As with minerals, there are questions of supply and demand and how to use the resources without degrading the environment. The growth and reproduction characteristics of renewable resources make it possible to manage them so that production can be regulated, however, and resource use can be sustained indefinitely.

This can be accomplished by managing on a *sustained yield basis,* whereby the rate at which the resource is removed is balanced by the rate at which it is replaced by new growth or reproduction. Under sustained yield management, a deer herd that is expected to increase by twenty animals per year could withstand the removal of twenty animals by hunters and other causes of death without showing a decline. The concept of sustained yield is also applied to the management of fisheries, rangelands, and forests. A great deal of information is needed to apply the concept effectively. In the case of forests, for example, it is difficult to apply sustained yield on an annual basis or for a small area. Instead, it is common to try to achieve an average sustained yield of forest products for a relatively large forest region over a span of several years.

While we sometimes assume that renewable resources can be maintained indefinitely and ignore concepts such as sustained yield, current research shows that we can push a renewable resource beyond its limit to renew itself. Actually, it is important to look at both renewable and nonrenewable resources under the concept of sustainable development introduced earlier. This leads to the idea of *sustainability.* Some fish populations are harvested intensively and decrease in numbers. Although considered a renewable resource, when harvesting stops the fish population fails to return to its former level. The reasons are not known but could be related to changes in the species' behavior so males can find females during reproduction. Chemical changes could occur in the habitat, or new species that change the interactions might be present.

Sustainability of natural resources usually involves interaction of renewable and nonrenewable resources. It generally means reproduction of species, passing genetic material from one generation to the next. Reproduction, of course, depends on some nonrenewable resources as well as on many other organisms.

THEMES

We emphasize three important strands or ideas in subsequent chapters.

1. Planning is illustrated as an important tool in environmental management. We point out how we can decide where we are,

ANTARCTICA—AN EXPERIMENT IN GLOBAL COOPERATION

Antarctica represents one of the last resource frontiers left on the planet Earth. Its great scientific potential and aesthetic value as a wilderness have fascinated humans for decades. More recently, its abundant biological resources and the potential for mineral exploitation in the Southern Sea have attracted the attention of the world community.

Antarctica is 98 percent covered by ice, and its low annual precipitation of less than 8 centimeters makes it a virtual desert. Its snow and ice hold 90 percent of the world's freshwater. The ice-covered land is inhospitable to life, but the edge of the ice and the surrounding offshore areas have a rich fauna—whales, penguins, seals, seabirds, fish, and an economically important shrimplike species known as krill. These offshore areas have recently become sites of intensive exploration for oil and gas deposits and other mineral resources.

Exploration of Antarctica began in the 1800s. Sealers and whalers were the first to exploit Antarctic resources. In the early 1900s several nations claimed territory in Antarctica. Scientific interest in the region grew, and research activities evolved into cooperative research agreements among twelve nations (the United States, the Soviet Union, the United Kingdom, Belgium, Japan, South Africa, New Zealand, France, Australia, Norway, Chile, and Argentina), who were joined by four others (Poland, the Federal Republic of Germany, Brazil, and India) in signing the Antarctic Treaty that took effect in 1961. Sixteen more nations subsequently joined in the treaty. The treaty established Antarctica as a demilitarized, nonnuclear area and guarantees freedom of scientific investigation. The treaty also provides for free exchange of information among all researchers, and the participating nations agree not to assert any territorial claims while the treaty is in effect. As more nations become interested in the Antarctic and its resources, the treaty and its provisions are likely to be severely tested.

All around Antarctica the coast is dotted with corrugated metal buildings, oil storage tanks, and garbage dumps. At least sixteen nations have established permanent bases on this, the only continent that belongs to the whole world. Most of the bases are set up to conduct scientific research. The area has, however, become a magnet for tourists who come to see ice and penguins. There are many proposals for mining minerals and drilling for oil. Regulations for disposal of garbage and sound environmental use of this fragile area are needed.

Recent attempts to negotiate a treaty called the Convention for the Regulation of Antarctic Mineral Resource Activities (CRAMRA) failed when France and Australia refused to ratify it. These countries proposed a ban on mining and a more comprehensive treaty to protect the Antarctic environment.

The Antarctic ecosystem is based on krill, a small (6 centimeters or less in length) shrimplike crustacean. The importance of krill once centered on its role as the principal food source for the Antarctic whale and seal populations which were harvested in commercial operations. With the decline of whale populations, interest has grown in exploiting krill as a potentially harvestable food for humans. It is estimated that harvestable production of krill could reach 150 million metric tons a year. Some believe krill might best be used to support a commercial fishery or to support populations of the other krill consumers such as seals, whales, and birds.

Krill clearly plays a key role in the Antarctic food web, but we must first understand its life history and reproductive potential before we can determine its potential as a harvestable resource. This knowledge will come from detailed ecological studies combined with ecosystem modeling; for example, krill distribution and abundance are difficult to analyze because krill are not evenly distributed and apparently migrate long distances. Furthermore, there are large year-to-year and seasonal variations in krill distribution. Assessment of reproductive potential is also complex. Research thus far shows that the krill's food supply (phytoplankton) is a key factor in regulating its population growth.

Besides ecological understanding, sustaining the Antarctic ecosystem will require the continued cooperation of all the nations that have an interest in the Antarctic resources. The Antarctic Treaty and other cooperative efforts already underway provide hope that international cooperation in resource development and ecosystem management is possible. It is an experiment in global cooperation that could lead to similar cooperative efforts in other regions of the earth where such efforts are badly needed.

where we want to be, how we will get there, and whether we made it.

2. In the boxed Legislation sections we describe some of the laws that apply to topics discussed in the text.
3. Specific examples of how your involvement as an individual can make a difference in the way our environment is managed are shown. These are only meant to be a few ideas. You and your fellow students can discuss these ideas and generate many other ideas to manage and live more effectively in our environment.

SUMMARY AND CONCLUSION

Environmental science is a multidisciplinary field that requires a broad understanding of life

LEGISLATION

National Wilderness Preservation System Act (1964). Sets aside land where motorized travel is prohibited. Individuals can nominate such areas to federal land management agencies or their congressional representatives.

National Wild and Scenic Rivers System Act (1968). This legislation allows protection of free-flowing rivers from dams or other obstructions and protects the area within a quarter of a mile of the river from unsuitable development. Areas can be nominated by letters to agencies managing the land or congressional representatives.

and how life responds to changes in the environment. Life evolved over millions of years, and the influence of humans on the living world is a relatively recent phenomenon. We now understand some of the complex interactions involved in natural systems. Concepts such as homeostasis, energy, limits, symbiosis, and models form the basis for understanding environmental systems and how to sustain them.

We have reached a point where environmental modification is monitored at a global level. It has become apparent that economic development and ecologically sound environmental management are closely linked. A World Conservation Strategy has been developed to guide environmental planning worldwide.

Systematic environmental management requires balancing preservation of the earth's resources with wise use. It is important that we carefully use nonrenewable resources, such as minerals and fossil fuels, so that they are not wasted. This involves understanding patterns of supply and demand and ecologically sound methods of extraction. Renewable resources, such as forests or wildlife, require a different kind of planning so that they will be continually replenished by growth and reproduction.

FURTHER READINGS

Crowe, D. M. 1984. *Comprehensive Planning for Wildlife Resources.* Cheyenne: Wyoming Game and Fish Department.

Leopold, A. 1949. *A Sand County Almanac.* New York: Oxford Press.

Nash, R. 1989. *The Rights of Nature.* Madison: University of Wisconsin Press.

Nicol, S. 1990. The Age-Old Problem of Krill Longevity. *BioScience* 40:833–36.

Oparin, A. I. 1962. *Life: Its Nature, Origin, and Development.* New York: Academic Press.

Orians, G. H. 1990. Ecological Concepts of Sustainability. *Environment* 32:10–39.

Talbot, L. M. 1984. The World Conservation Strategy. In *Sustaining Tomorrow*, ed. F. R. Thibodeau and H. H. Field. Hannover, NH: University Press of New England.

STUDY QUESTIONS

1. Why does the study of environmental science require a multidisciplinary approach? How does it differ from ecology?
2. How does the concept of feedback relate to homeostasis?
3. How does a closed system differ from an open system?
4. How did life apparently begin on earth? Do you think life exists on other planets?
5. How do you account for the fact that humans have been on the earth for a relatively short time, but have had a tremendous impact on the earth's ecosystems?
6. Distinguish between preservation and conservation. Give examples of resources that should be preserved rather than conserved.
7. Explain the difference between goals, objectives, and strategies in planning.
8. How do planning and the use of models relate to systematic environmental management?
9. How can models be used in planning?
10. Why is it necessary to manage renewable resources on a sustained yield basis?
11. Do you think it is practical to have a World Conservation Strategy? Explain your reasoning.

SUGGESTED ACTIONS

1. *Attend Public Meetings.* Most states and federal agencies have public meetings to present their budgets or parts of their management program. In addition, federal projects, listings of endangered species, and changes in federal legislation are printed in the federal register.
2. *Attend Planning Sessions.* If you feel strongly about an idea, get your friends to attend and comment on the subject.
3. *Write Letters.* Find out from the newspaper when legislation of interest to you is being proposed. If it relates to a particular agency, write the agency for information. Write your congressperson or senator to express your ideas on a subject. Likewise, letters to agency heads can help to get your ideas across to government agencies in the federal and state governments.

PART ONE

NATURAL SYSTEMS

2

ECOSYSTEMS

Together with all living things—plant or animal, seen or unseen—people are an important part of an ecosystem. All nonliving substances—air, water, soil, and minerals—comprise another important part of this ecosystem. The living and nonliving are intimately related to each other through energy flow and mineral cycles. No organism exists in a vacuum, but depends on other life and its physical surroundings.

People like to think of themselves as the highest form of life, yet they are not free from constraints placed on other life. While they have the ability to destroy a whole forest and thereby drastically alter the ecosystem, the oxygen they breathe, the atmosphere that supports them, and the animals and plants they eat make them intimately dependent on the natural system.

By studying the dynamics of this natural system, we lay the foundation for discussions of the interactions of the biological and physical systems, basic principles of environmental management, and the natural laws that ultimately limit the growth of the human population.

WHAT IS AN ECOSYSTEM?

An ecosystem is composed of all the living organisms in an area plus the surrounding physical environment with which they interact. Because no sharp boundaries exist between different ecosystems, studies must necessarily be limited both in area and content. After the purpose of a study is defined, arbitrary limits must be set. While it would be impossible for you to study the entire ocean ecosystem, you could investigate specific aspects affecting the decline of one fish population. This might include food supply, predators, and environmental pollutants.

The purpose of thinking in terms of an ecosystem is to link obligatory, interdependent, and causal relationships that form the whole—much as individual ingredients go together to make a cake. Shifts in the constituents of either change the end product.

Biosphere is the term used to denote the collective ecosystems of the world. Biosphere studies are difficult to undertake because of the magnitude of interactions involved, yet examination of carbon dioxide levels, radiation changes, and energy flow must be applied to the biosphere as a whole.

Ecologists refer to all living organisms in a given ecosystem as a **community** (Figure 2.1). Life within a forest is called a forest community. Sometimes we study subdivisions of a community, such as the plant, vertebrate, mammal, or rodent communities. Life above ground and in

Figure 2.1
Natural communities. (a) Coral reef off San Andreas Island in the Caribbean. (b) Eucalyptus grove near Cleveland National Forest, California. (c) Cobra lilies, which are able to digest insects. (d) Mangrove trees in a Costa Rican forest. (e) Swamp forest in Florida. (f) Mount Rainier alpine community.

the soil may be called terrestrial communities, while plants and animals in bodies of water may be called aquatic communities.

Within the community, organisms are grouped into populations. A **population** is a group of one species of organisms occupying a particular space at a particular time. Each population has characteristics, such as birth rate, death rate, age distribution, and genetic composition, which no individual in the population has by itself. Population characteristics are frequently expressed statistically. The ecologist uses populations and communities (Figure 2.2) to tell us more about the ecosystem.

HOW ENERGY MOVES THROUGH AN ECOSYSTEM

All living organisms require energy to live. The sun supplies the basic source of this energy. As animals eat food in the form of plants or other animals, energy passes from one organism to another in the ecosystem.

Photosynthesis

Sunlight is stored by green plants as chemically bound energy during the process of **photosynthesis.** In the overall process, carbon dioxide and water are used as raw materials to produce sugar and oxygen. This is a summary of the equation for photosynthesis:

$$\underset{6CO_2}{\text{carbon dioxide}} + \underset{6H_2O}{\text{water}} \xrightarrow{\text{sunlight}} \underset{C_6H_{12}O_6}{\text{sugar}} + \underset{6O_2}{\text{oxygen}}$$

Photosynthesis is a complex process involving many chemical reactions. The reactions take place in small green organelles **(chloroplasts)** containing chlorophyll, which gives plants their

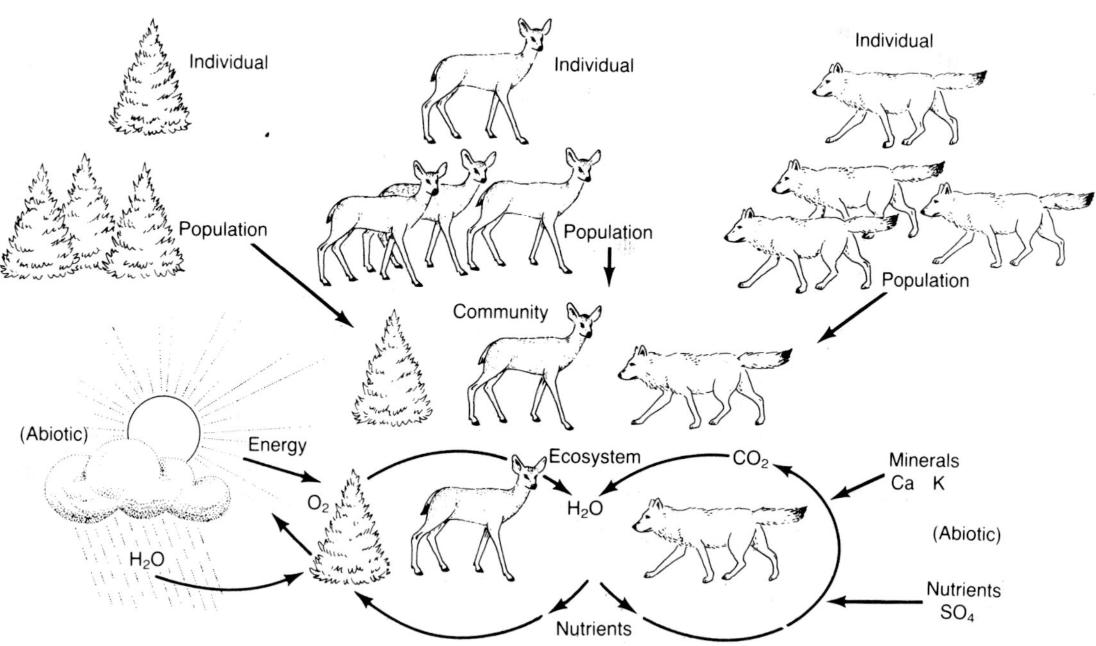

Figure 2.2
Components of an ecosystem. Groups of interacting plants and animals form populations, and two or more populations in the same place at the same time form communities. The community forms the living, or *biotic*, part of the ecosystem. Energy, minerals, nutrients, and water form the nonliving, or *abiotic* component.

characteristic green color (Figure 2.3). Photosynthesis occurs in two stages—a light-dependent stage **(light reaction)** and a light-independent stage **(dark reaction).**

In the first step, chloroplasts absorb light and transform it into chemical energy. The chemical energy is coupled with the bonds of chemical compounds in the chloroplasts. **Chemical bonds** are the forces holding chemical elements and compounds together.

During the dark reaction, compounds formed by bonds of chemical energy (originally solar energy) are used to break water and carbon dioxide into free elements. In a series of steps the carbon is combined with the hydrogen and oxygen of water to form sugar. Free oxygen is released.

Respiration is the process whereby sugar produced by green plants is broken down into energy that living organisms can use for growth, reproduction, and tissue repair. As discussed here, respiration is cellular and should not be confused with the breathing process of inhaling oxygen and exhaling carbon dioxide. Cellular respiration combines oxygen, with sugars to form carbon dioxide, water, and energy:

$$\underset{C_6H_{12}O_6}{\text{sugar}} + \underset{6O_2}{\text{oxygen}} \rightarrow$$

$$\underset{6CO_2}{\text{carbon dioxide}} + \underset{6H_2O}{\text{water}} + \text{energy}$$

Respiration (Figure 2.4) takes place largely in another specialized organelle of the cell, the **mitochondrion.** When the glucose is broken down, energy—packaged in high energy bonds—is released. Plants use some of the bound energy for survival and pass the rest on to animals that eat the plants. (We will discuss the movement of energy in the next section.)

Most organisms use the free oxygen gas (O_2) released in photosynthesis for body respiration. This is called **aerobic** (requires oxygen) respiration. People, fish, earthworms, and most bacteria are aerobic organisms.

Anaerobic organisms, which include some species of bacteria, can grow in the absence of free oxygen. Anaerobic respiration uses oxygen from the breakdown of compounds such as nitrate or sulfate. Those organisms that must grow in the complete absence of free oxygen are *obligate anaerobes;* those that can grow in the presence or absence of free oxygen are *facultative anaerobes.*

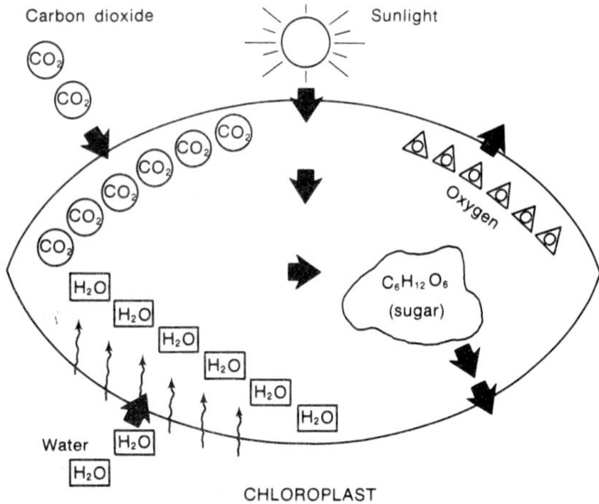

Figure 2.3
The chloroplast is the site of photosynthesis. The sun's energy is used to bind carbon with hydrogen and oxygen. Free oxygen is released.

Anaerobic organisms are essential to life. They help to break down food in digestive tracts of many organisms, including humans, and decompose organic matter in lake sediments, landfills, and sewage treatment plants. Swamp gas, tetanus, and gangrene are also caused by anaerobes.

Fermentation, a specialized form of anaerobic respiration in which oxygen is supplied by an organic compound, satisfies many human needs. Yeast fermentation produces carbon dioxide that makes bread rise. Alcohol production for beers, wines, and other liquors involves fermentation. Fermentation can be a preservative process when its products inhibit the growth of microorganisms that cause decay or spoilage of food. Silage is made by allowing green hay, grass, and/or cereal crops to undergo fermentation that preserves the food value of crops and provides feed for animals during the winter season.

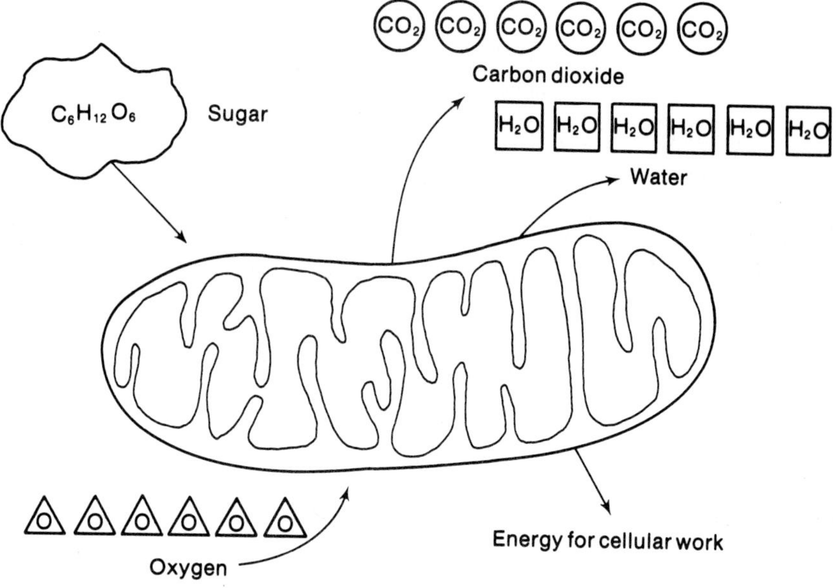

Figure 2.4
Cellular respiration in living organisms. Sugar and oxygen are taken into the mitochondria of living tissue. Complex chemical reactions release carbon dioxide, water, and energy.

Food Chains

Energy flow from green plants to consumer organisms, as each population is eating and being eaten, is called a **food chain.** Food chains are relatively uncomplicated, involving energy movement from one population to another. Complex or interlinked food chains, where one population feeds on a number of other populations, are called **food webs** (Figure 2.5). Food is the means by which energy moves from one organism to another.

Autotrophs, including green plants and a few species of bacteria, convert the sun's energy to chemically bound energy—food used for life—through photosynthesis. All other forms of life, called **heterotrophs,** depend on autotrophs either directly or indirectly for their life's energy. Green plants, then, are the **primary producers. Consumers** feed on producers, but not all consumers feed directly on plants. Animals that eat only plants are **herbivores;** others that feed only on animals are **carnivores,** or flesh eaters. **Omnivores** feed on both plants and animals.

To trace the sequence of energy flow in ecosystems, ecologists superimpose **trophic levels** on food chains or food webs (Figure 2.6). All green plants (producers) are members of the first trophic level; all herbivores constitute the second trophic level; animals that feed primarily on herbivores make up the third trophic level. The fourth and fifth levels are composed of animals that feed on the consumers of the trophic level just below them. Some animals, including humans, can occupy more than one trophic level. Very few populations occupy the fifth level or higher.

Natural systems have two types of food webs—grazing and detritus (Figure 2.7). The terrestrial **grazing food web** involves moving energy and minerals from green plants to herbivores to carnivores. The decomposition or **detritus food web** becomes operative when organisms die. Millions of decomposer organisms break down dead bodies, using energy and releasing nutrients from plant and animal matter back into mineral cycles. Organisms such as

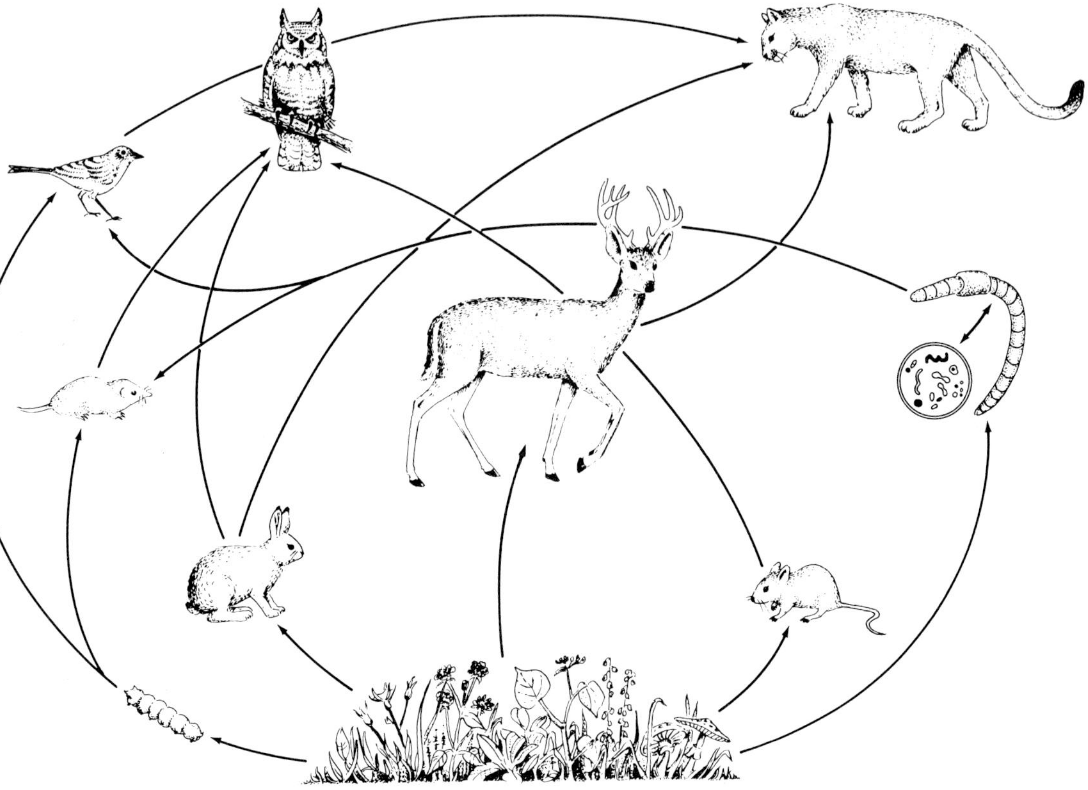

Figure 2.5
Terrestrial food web showing some of the many interconnecting links of energy flow in an ecosystem.

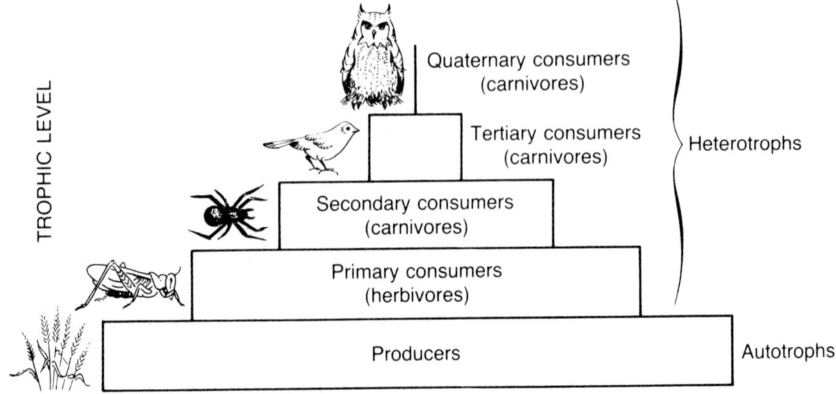

Figure 2.6
Trophic level organization. Energy flowing through different populations that feed on one another is called a food chain; food is the mechanism by which energy is moved.

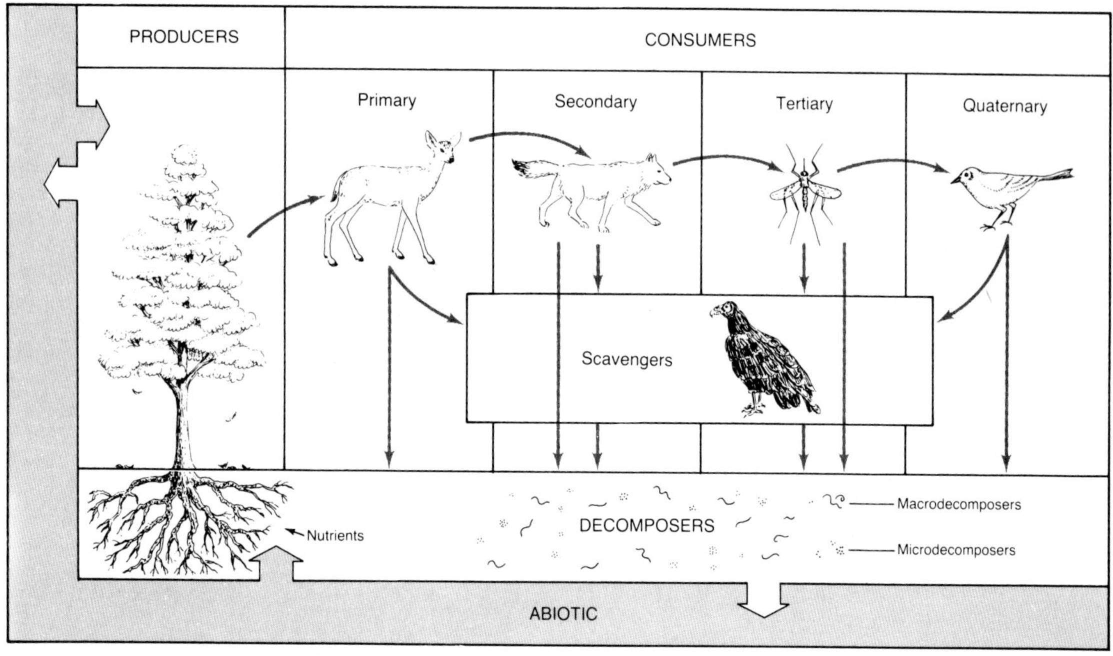

Figure 2.7
Interrelation of grazing and detritus food chains. Producers and consumers are part of the grazing food chain; scavengers and decomposers are part of the detritus food chain. Producers are green plants; primary consumers are herbivores; secondary and tertiary consumers are carnivores such as lions, humans, sharks, and some mosquitoes. Scavengers, such as vultures and crabs, eat the remains of other organisms. Macrodecomposers are primarily earthworms and insect larvae, and microdecomposers are bacteria and fungi. (Adapted from C. Benjamin Meleca, Phyllis E. Jackson, Roger K. Burnard, and David M. Dennis, *Bio-Learning Guide*, 2nd ed. Minneapolis: Burgess Publishing Company, 1975.)

earthworms and beetles, called *macrodecomposers*, begin the process by removing large pieces of the dead organism. *Microdecomposers* such as bacteria and fungi then finish the process.

Phytoplankton, minute floating plants, form the base of the grazing food web in aquatic systems. They are eaten by small floating animals called **zooplankton,** which in turn are food for small fish and filter feeders. Filter feeders obtain their food by straining plankton from the water. They are then eaten by other animals (Figure 2.8). Decomposers, including crabs, worms, and bacteria, tend to operate rapidly in the aquatic system by beginning to break down organic matter immediately after death or sometimes even before death.

Energy Conversion

The use of the sun's energy to form new *biomass* (weight of living organisms), called **productivity,** varies in different types of ecosystems. Typically, it is expressed as the amount of usable energy produced per unit of area per unit of time; examples are kilocalories per square meter per day (kcal/m^2/day) or grams of food per square meter per year (g/m^2/yr). The **gross primary productivity** is the rate at which green plants convert solar energy (by means of photo-

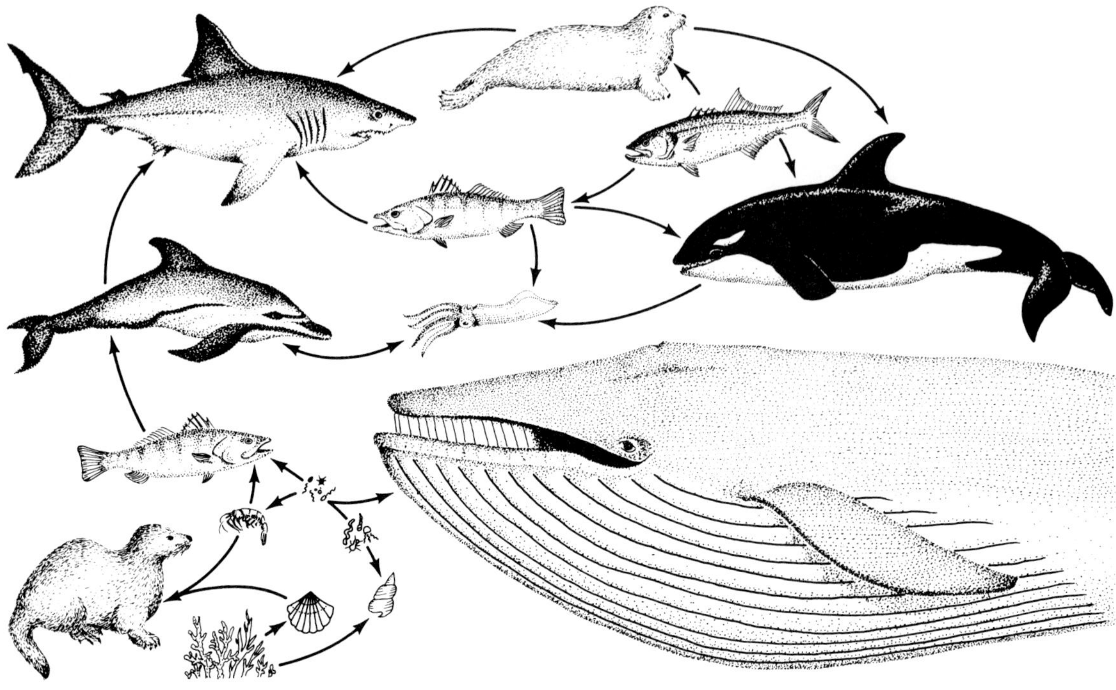

Figure 2.8
Marine food web showing some of the many interconnecting links of energy flow in the ocean. Where do humans fit in this web?

synthesis) to chemical energy usable by life. Plants use much of this converted energy to maintain respiration. The **net primary productivity,** or the energy available for consumption by the plant itself for growth and reproduction, equals the gross primary productivity minus the rate of plant respiration:

Solar energy $\xrightarrow{\text{photosynthesis}}$

Gross primary productivity ↗ Respiration
↘ Net primary productivity

Communities like estuaries, springs, marshes, and eutrophic lakes can have relatively high rates of net productivity, but they constitute a relatively small proportion of the earth compared with deserts, deep oceans, or other areas with low rates of productivity (Figure 2.9).

Both the grazing and detritus food webs are important in energy flow. While it takes energy to accumulate biomass, it also takes energy to break down biomass. Some decomposers in the detritus food web, such as algae or other plants, are able to convert energy absorbed from the sun. Thus, organic matter is an energy source for the algae at the same time the algae are converting the sun's energy during primary productivity. While it is convenient for us to separate energy processes in living systems, keep in mind that they are intimately linked in a homeostatic process (see Chapter 1) which, when disrupted, alters the entire system.

Net primary productivity is an important concept to apply to food production. To supply food for people worldwide, we must grow crops to yield the highest net productivity in each area. Energy used through the additions of water and

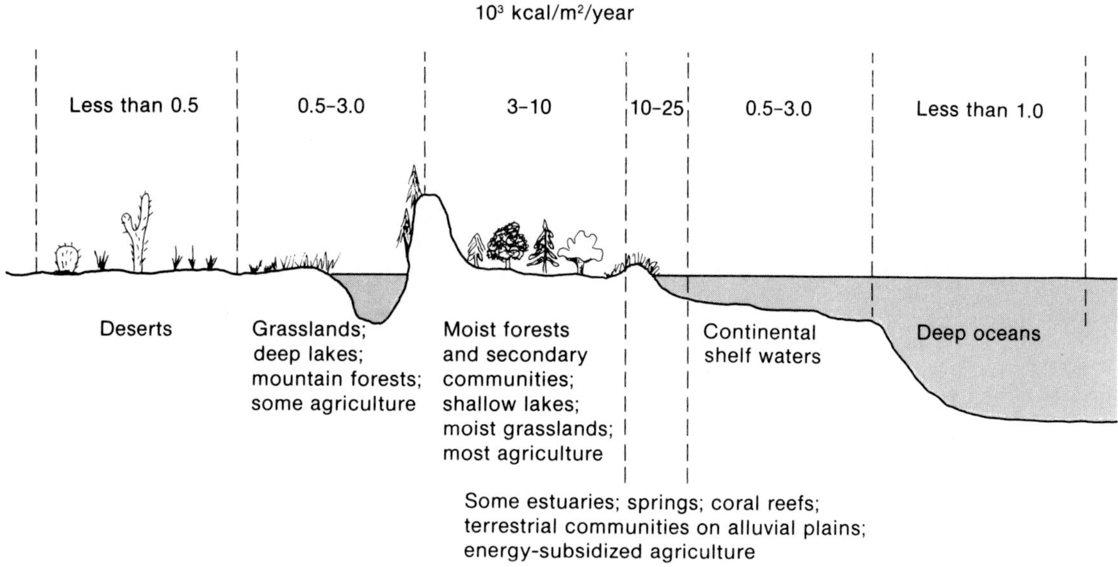

Figure 2.9
Primary productivity in terms of annual gross production in some world ecosystems. Notice that the highest productivity occurs in some estuaries, springs, coral reefs, and terrestrial communities, which constitute a very small part of the total world area. Much of the world, including the deep oceans and deserts, has very low productivity. (From *Fundamentals of Ecology*, 3rd ed., by Eugene P. Odum. Copyright © 1971 by W. B. Saunders Company. Copyright 1953 and 1959 by W. B. Saunders Company. Reprinted by permission of Holt, Rinehart and Winston, CBS College Publishing.)

fertilizer must be subtracted from the apparent increase in net primary productivity before we can know the true new productions. Although using temperate or tropical estuaries for aquaculture tends to provide high returns on sea crops, their restricted area and great susceptibility to pollution limit their potential.

The *efficiency* of green plants in converting solar energy varies with location. It is about 0.3 percent on land and 0.13 percent in the ocean. Although some spring-fed ponds have recorded efficiencies of up to 7 percent, most ecosystems remain below 1 percent. Even though ecological techniques for measuring efficiency vary considerably, we can see that a very small proportion of incoming solar energy is diverted into food chains and webs. Capturing energy by managing ecosystems is one method of increasing the world's food and energy supply.

Transfer of energy from one trophic level to the next is not 100 percent efficient because the animals in each trophic level require energy for survival and reproduction. Energy is also lost as organisms consume one another. Not all of the animals in each trophic level are eaten by others; some die and decay, transferring this energy to the detritus food chain. Producers use energy for respiration and lose energy as heat in the photosynthetic reaction. Energy uptake by herbivores represents the total amount of energy available not only to herbivores but to all animals.

To summarize, three things can happen to energy assimilated at each trophic level:

- It can be used for respiration of organisms on that trophic level and lost as heat.
- It can become part of the detritus food chain either when the organisms of that

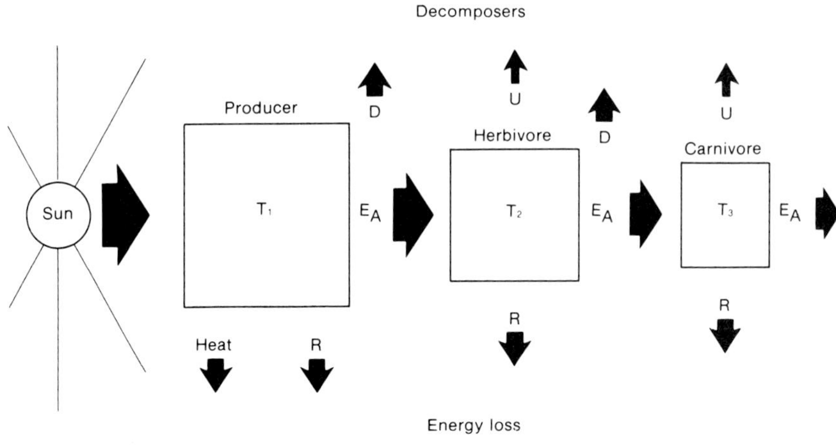

Figure 2.10
Input and loss of energy at each trophic level. The sun's energy provides fuel for operation of a food chain. During photosynthesis, this heat is lost. At each trophic level (T_1, T_2, and T_3), fuel is lost because of respiration (R). Some of the energy available to the next trophic level (E_A) is removed by decomposers (D), and some energy taken in by the next trophic level is not used (U).

level die and decay or as it passes through the bodies of other animals without being assimilated.

- It can be passed on to the next trophic level when animals are consumed and assimilated (Figure 2.10).

During transfer of energy from one trophic level to the next, 80 to 90 percent of the energy is lost through respiration or decay, leaving only 10 to 20 percent. Thus energy available at each trophic level is shown as decreasing pyramids (see Figure 2.6). Energy, however, is basic in our examination of trophic levels, for it provides the power necessary to sustain life.

Ecological Efficiency

The principle of **ecological efficiency,** or the ratio of the energy received divided by the energy available at each trophic level, is important to many managed ecosystems, such as agriculture. The lower the trophic level from which people derive food, the greater is the available energy. If people lived entirely on green plants, a much larger world population could be supported since energy would not be lost in converting plant products to meat products. Dietary deficiencies would probably be more prevalent, however. Only a very small biomass of top-level consumers such as lions, hawks, and sharks can be supported in comparison with the total biomass produced. The total efficiency of the energy flow in living systems therefore dictates the quantity and type of available biomass in a given area.

Smaller organisms use more energy per gram of body weight than larger organisms. This phenomenon is partially accounted for by the fact that larger plants and animals have less surface area per gram of weight, so less heat is lost.

MINERAL CYCLES

Each mineral or element within a mineral has a natural cycle that involves changing the mineral from one chemical state to another. Some are necessary constituents of living tissue, while others accumulate in tissue and disrupt physiological processes. Most studies of natural mineral cycles are conducted by ecologists, who observe relationships between the living and nonliving components of elements essential to life. Cycles of elements essential to life are called *biogeochem-*

ical cycles. Mineral cycles are a part of the dynamics of an ecosystem. While energy moves from the physical component, minerals circulate back and forth in different chemical states.

In its natural cycle, a mineral often goes through a **reservoir stage** in which a large amount of the mineral can be found. Reservoir stages can be gaseous, liquid, or sedimentary. Minerals such as oxygen, carbon dioxide, and nitrogen move freely between gaseous and liquid states. Many of the common soil chemicals, such as silica, represent a sedimentary reservoir.

Carbon Cycle

Carbon is the major mineral component in coal, oil, and natural gas. All living organisms contain carbon; it is the basis of life on earth. Carbon is found in the atmosphere as carbon dioxide (Figure 2.11). From this reservoir it is used by green plants in the process of photosynthesis and some of it is returned to the atmosphere as carbon dioxide from plant respiration. Other carbon is passed along the food chain or forms carbon dioxide via the decay process. The carbon cycle is not only part of the terrestrial web of life, but also a part of the aquatic web as carbon dioxide diffuses in and out of both fresh and salt water. In the ocean, photosynthesis is confined to the light zone, where a large proportion of carbon becomes bound in the shells and exoskeletons of ocean invertebrates as calcium carbonate ($CaCO_2$). When these organisms die, some of their body coverings are buried in the mud and sand and are thereby isolated from biological activity. Through geological time, some of these carbon deposits become part of coral reefs or limestone rock and some are slowly transformed into fossil fuels.

Carbon dioxide is currently building up in the atmosphere. Some scientists believe the buildup is a result of burning fossil fuels; others feel the loss of forests which use carbon dioxide is the cause (see Chapter 7).

Oxygen Cycle

Oxygen is necessary for life, yet as a free gas it is toxic to organisms such as anaerobic bacteria. It

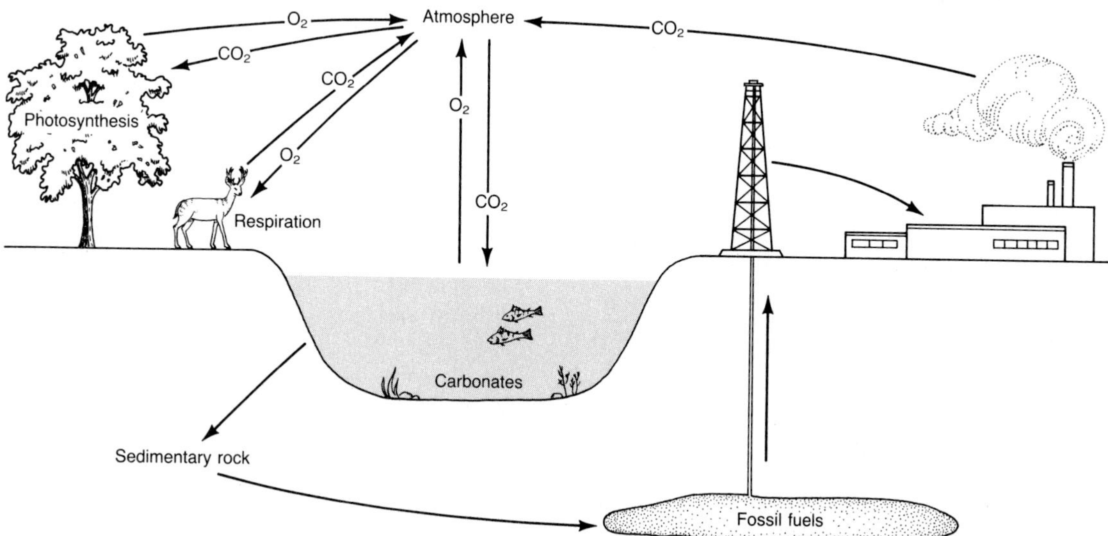

Figure 2.11
The carbon cycle. Carbon dioxide formed from fuel combustion and respiration is released into the atmosphere, where green plants use it in photosynthesis to produce oxygen and sugars. Carbon dioxide cycled into water combines with metals and hydrogen to form carbonates. The carbonates and decaying organic matter deposit carbon in the bottom sediment of waterways.

is freed from water in photosynthesis and reconstituted into water during plant and animal respiration (Figure 2.12). The oxygen cycle is quite complex because oxygen is highly reactive. We know oxygen is necessary for the combustion of organic matter and that it is a component of water. It also circulates freely as carbon dioxide and combines with some metals, such as iron, to form oxides.

Nitrogen Cycle

Even though nitrogen makes up approximately 78 percent of the atmosphere, it is not usable by most living organisms in its atmospheric form (N_2). Life processes generally require nitrate or some other nitrogen compound. A few microorganisms, including some bacteria, fungi, and the blue–green algae, can use atmospheric nitrogen to form substances usable by other organisms. This process, called **nitrogen fixation,** also occurs to a small degree as a result of lightning or other electrical discharge, but it is a biological nitrogen fixation by microorganisms that provides most of the usable nitrogen (Figure 2.13).

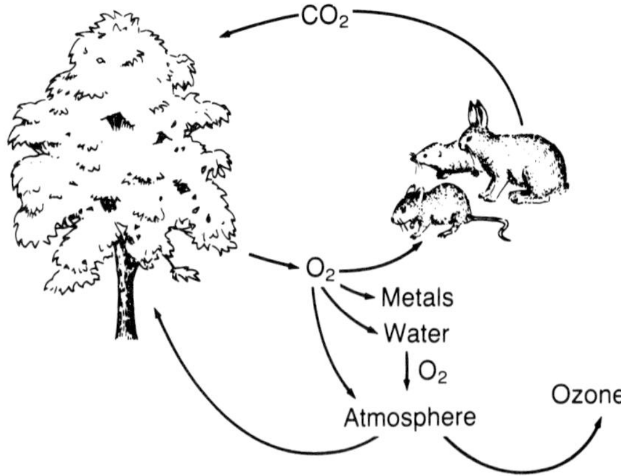

Figure 2.12
Oxygen is involved in many reactions. It is released as a gas by green plants and used in respiration by living organisms. It can combine with metals, form water with hydrogen, or remain as a free gas in the atmosphere.

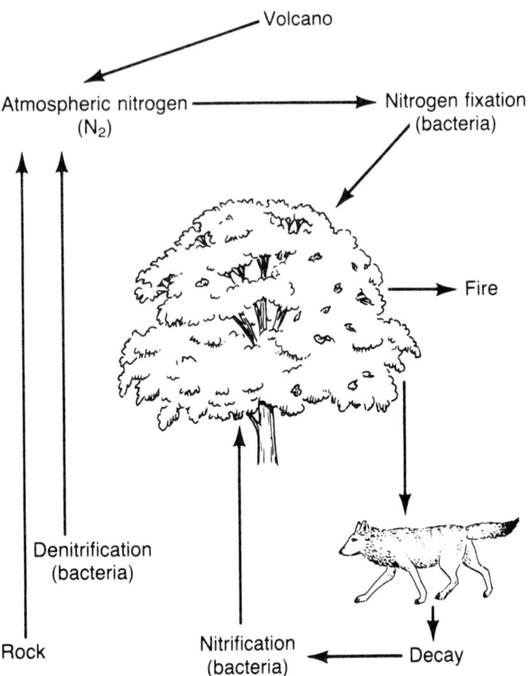

Figure 2.13
The nitrogen cycle. Atmospheric nitrogen (N_2) is fixed by some bacteria so that it is available to plants. Other bacteria in the soil convert nitrogen to nitrates usable by plants. A third group of bacteria (denitrifying) converts ammonia and nitrates to free nitrogen (N_2).

Some species of plants, primarily legumes (members of the pea family such as alfalfa, clover, and beans), live in a symbiotic relationship with nitrogen-fixing bacteria. The bacteria enter plant roots via the root hairs, where they are stimulated by secretions from the roots to grow and multiply and eventually form a swollen nodule. There the bacteria become immobile and carry on the process of nitrogen fixation. Soils low in nitrogen can have the nitrogen supply replenished by legume crops. Other microorganisms able to fix nitrogen are also found living free in the soil and water.

Nitrogen-fixing organisms have the ability to utilize free nitrogen to form ammonia (NH_3).

These organisms can then use some chemical state of ammonia to form other nitrogen compounds, or they can release the ammonia into the soil and water to be available for other organisms to use. Legumes benefit from association with nitrogen-fixing bacteria because they receive a supply of fixed nitrogen directly from the root nodules. Legumes can therefore grow in nitrogen-poor soils, which often occur in areas that have been disturbed and are eroding.

Some grass and tree species use ammonia, but most flowering plants utilize another nitrogen compound, nitrate, as their main source of nitrogen. Nitrates are produced in the soils by certain bacteria in a process called **nitrification.** Several groups of nitrifying bacteria are involved in this process, each forming a product that can be used by the next. When nitrates are released into the soil, they are quickly taken up by plants that store them.

Another group of bacteria and some species of fungi can convert ammonia and nitrates into free nitrogen and release it into the atmosphere. This is **denitrification,** a process particularly important in the decay of dead organisms. Denitrifying organisms are not restricted to breaking down nitrogen products of dead organisms, but utilize all sources of soil nitrogen. The nitrogen cycle therefore involves removal of nitrogen from the atmosphere, utilization of nitrogen by living organisms, and removal of nitrogen from the soil and decaying organic matter.

Living organisms use nitrogen mainly in the formation of proteins, the basic building blocks of all life (Chapter 12). Many of the cellular bodies (including chromosomes, the structures that transmit genetic information) also contain proteins. Fibers in connective tissue and muscles are important proteins that give strength and elasticity to these tissues.

Before nitrogen fertilizers were manufactured, the amount of nitrogen removed from the atmosphere by natural processes was closely balanced by denitrifying processes. Industrial intervention in the nitrogen cycle now appears to fix more nitrogen than can be denitrified. These additional amounts of nitrogen in the soil often disrupt natural balances, resulting in large runoffs of nitrogen compounds into rivers, streams, and lakes. The runoff problem is part of the sedimentary phase of the nitrogen cycle. Nitrogen not used by living organisms can be incorporated into sediments of the soil. When water percolates through the soil, these very soluble nitrogen compounds are leached, or washed out into the neighboring water supply.

Phosphorus Cycle

The phosphorus cycle (Figure 2.14) does not involve an atmospheric phase, since the primary reservoir of this element is rock or sediment. As rocks erode, phosphate (PO_4) is released into the environment. Plants then absorb phosphate through their roots. Animals ingest this essential element in food and excrete any excess in feces. In the coastal areas of western South America, birds take food containing large amounts of phosphate from the sea. Then they perch along the coast and deposit large amounts of *guano,* composed mostly of phosphate. In areas where rainfall is low, the phosphate remains on the surface and does not wash away. Guano deposits also accumulate on the floor of dry caves in New Mexico and Arizona where millions of bats roost (Figure 2.15). These deposits are major sources of the world's phosphate supply.

Phosphorus is essential for two important body functions: the transfer of energy and as a component of genetic material. When organisms die and decay, the phosphorus is degraded and released as inorganic phosphate through the action of microorganisms. In this phase it can immediately be taken up by plants and recycled through living organisms. The phosphate not taken up by plants goes into the sedimentary phase of the phosphorus cycle.

Unlike nitrogen, phosphate is not very soluble in water, but it does react chemically with a number of soil compounds. These reactions produce some compounds with elements so tightly bound that they cannot be utilized by plants. Some soil elements, including calcium, alumi-

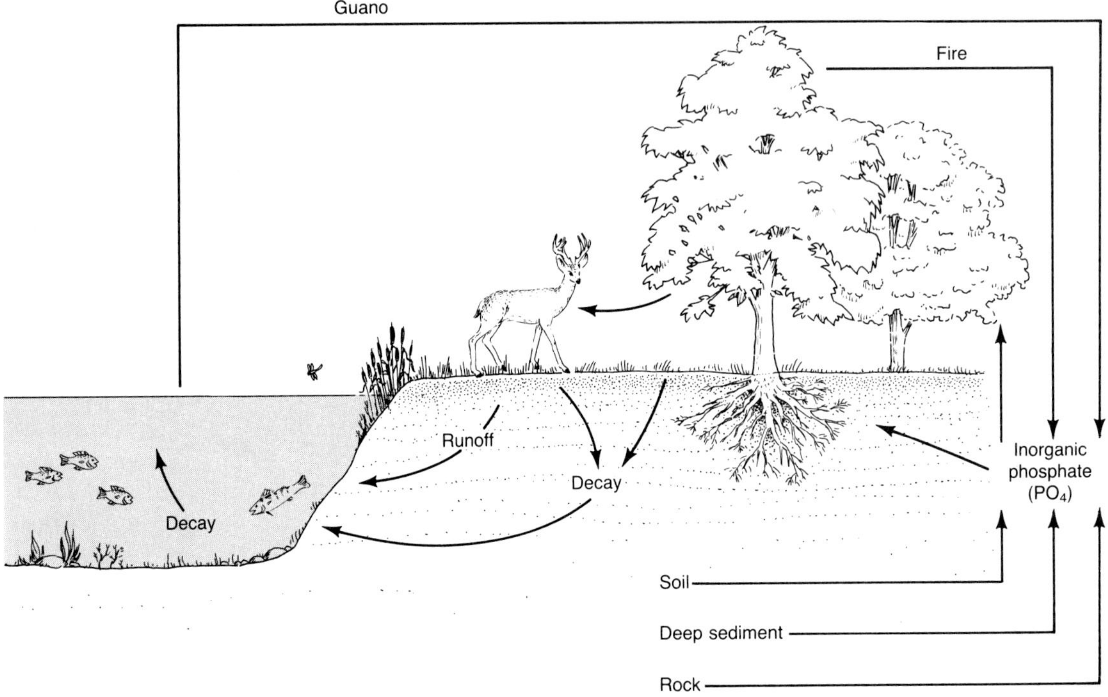

Figure 2.14
The phosphorus cycle. The primary reservoir of phosphorus is phosphate in soil and rock. Plants absorb phosphorus through their roots and pass it on to the animals. During the decay process, phosphorus is returned to the soil.

num, iron, and manganese, bind with phosphate to form insoluble compounds. This can produce phosphate in the soil that is unavailable to living organisms. The resulting sedimentary phase of the phosphate cycle can be very slow in comparison with the biological phase.

When phosphate enters streams and lakes in large amounts, it can act as a stimulant to algae growth. Phosphate detergents are responsible for the pollution of many waterways because they stimulate these algae to grow and eventually usurp the oxygen that other forms of water life require. When sewage containing phosphates is placed in lakes or streams, it is taken up by the algae and later deposited on the bottom as plants and animals die. This phosphate can be stored as part of the sedimentary phase or returned to the biological cycle. In the oceans, dead organisms can bind phosphate in the bottom sediment. Phosphate is then brought to the surface from phosphate-rich deep waters in upwellings along coastal waters. These high concentrations of nutrients enable many forms of marine life to grow around the upwellings.

Sulfur Cycle

The sulfur cycle involves both an atmospheric and a sedimentary phase, with the sedimentary phase being the longer of the two (Figure 2.16). In nature, sulfur exists in several states: elemental sulfur (S), sulfides (SO_2), sulfur monoxide (SO), and sulfate (SO_4). It is used in the manufacture of many products, including chemicals, matches, and fertilizers. Sulfur is also required for the formation of some proteins in living organisms. When organisms die, sulfur is released

Figure 2.15
Bat guano deposits, New Cave, Slaughter Canyon, Carlsbad Caverns National Park. (Courtesy National Park Service.)

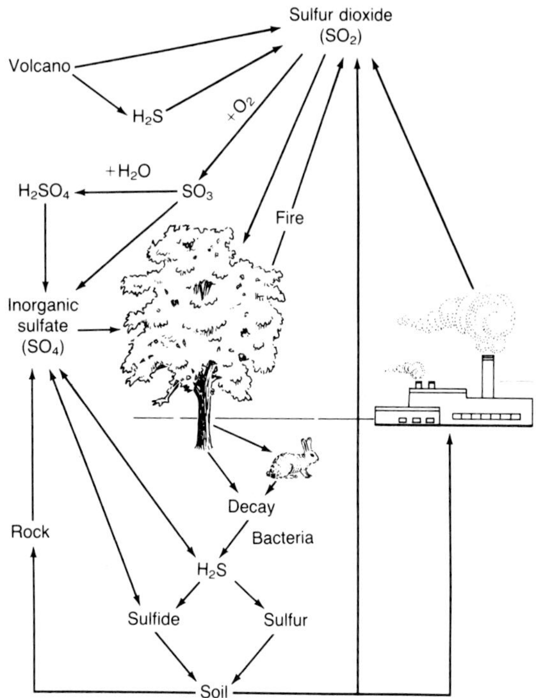

Figure 2.16
The sulfur cycle. Sulfides and sulfur in the soil provide sulfate for plant growth. Because many fossil fuels contain high levels of sulfur, atmospheric pollution with sulfur compounds is common. Atmospheric sulfur can be returned to the soil, where it becomes available to plants.

in several forms by bacteria. In the absence of oxygen, hydrogen sulfide (H_2S) is released, which gives the characteristic rotten egg odor in the decomposition process. When oxygen is present, the hydrogen sulfide forms sulfates that can be reused by plants. Some bacteria living in anaerobic conditions can use the hydrogen sulfide for energy. These bacteria live primarily in the ocean depths or in anaerobic portions of eutrophic lakes and form sulfate compounds from the hydrogen sulfide.

As a constituent of coal and petroleum products, sulfur causes atmospheric contamination when its products are burned. About one-third of the sulfur entering the world's atmosphere comes from industries in the form of sulfur dioxide (SO_2). Sulfur dioxide reacts with oxygen to form sulfur trioxide (SO_3), which combines with water to form sulfuric acid (H_2SO_4) mist. High concentrations of sulfuric acid in the air can be destructive to a wide variety of building materials. Marble, limestone, roofing slate, and mortar, for instance, become pitted and discolored. Sulfuric acid reacts with the carbonates (CO_3) in these building materials and leaches them out, causing structural weakness. In areas with high levels of sulfur dioxide, acid rains occur, which can mar buildings and automobiles and damage plants. Atmospheric levels of 0.1 to 0.5 parts per million are particularly harmful to the human respiratory tract, and plant leaves start turning brown at these levels.

Mercury Cycle

Between 1953 and 1960, industries around Minamata Bay, Japan, emptied mercury into the bay, causing 111 deaths or serious injuries to those who ate contaminated fish. In 1970 mercury was the focus of much public attention when toxicologists found abnormally high concentrations in freshwater fish from Lake Erie and saltwater tuna and swordfish. People became further alarmed when dreadful effects, including deaths, were reported from accidentally eating grain seed coated with mercury. These examples showed us how a disrupted mercury cycle could harm people.

Mercury is liquid metal at ordinary temperatures. When metallic mercury is washed into the water, it remains in the sediment under anaerobic conditions. If oxygen is available, mercury sulfide is formed, which is converted by microorganisms to a sulfate and then to methyl mercury compounds absorbable directly from the water. Methyl mercury is toxic to living organisms. Most organisms—fish, birds, mammals—that receive excessive doses of mercury show neurological symptoms. Methyl mercury compounds damage all tissues, including the brain. They also have long retention times—half the original amount remains in the human body seventy days after exposure. Prolonged exposure to mercury creates a variety of neurological symptoms, including temporary insanity. The phrase "mad hatter" came from the practice of chewing hat bands soaked in mercury to shape men's hats. As a result they became temporarily intoxicated. Mercury compounds from metal products such as mercuric sulfate and mercuric chloride accumulate in the human kidney, liver, and brain. There is also evidence that mercury causes genetic damage.

Mercury becomes an environmental contaminant because of many of its uses (Figure 2.17). As an algicide, it is used to clean vats in paper mills and then is washed into neighboring bodies of water. In Sweden, much coastal water is contaminated from the paper making industry, resulting in high levels of mercury in fish caught in that area. Mercury is also used in batteries, thermometers, and fungicides.

Lead Cycle

Lead occurs naturally primarily as lead sulfide and is mined in many parts of the world. It is a useful metal because it is very dense, has a low melting point, and is highly malleable. Lead is used for many metal objects and paint pigments.

This mineral is normally found only in the earth's crust, but the use of leaded gasoline brought a great increase in the atmospheric lead concentration. Ninety-eight percent of the lead in the atmosphere comes from combustion of gasoline products. It settles on the earth's surface and concentrates near the source of the pollution, so urban areas have lead levels up to 100 times higher than rural areas. As a result of distribution by atmospheric wind currents, higher than normal levels have been recorded in the Greenland ice cap beginning in 1950 (Figure 2.18).

Plants obtain lead from soil, water, and air, and animals obtain lead from food, water, and air. A small portion (5 to 10 percent) is absorbed in the digestive tract while 40 percent is absorbed in the lungs. Lead then enters the blood system, where it can be deposited in tissue to disrupt enzyme systems and neurological processes, or it can be excreted in the urine. In the blood, lead interferes with the synthesis of hemoglobin, causing anemia from impaired oxygen transport. Lead deposits also interfere with normal kidney function. Furthermore, behavioral changes occur from lead poisoning. Some people become listless, tired, and seem to lose some mental capacities. Others exhibit symptoms associated with drunkenness. Some historians suggest that the Roman Empire declined partly because lead poisoning affected the Romans' mental capacity. Traces of lead were found in the bones of these people, probably from drinking wine stored in lead-glazed pottery vessels.

Because of high levels of atmospheric lead and misuse of some products, lead poisoning occurs in humans more frequently than it should. Lead from paints often gets into food, and most lead poisoning in children comes from eating lead paint chips. For adults, it can be an occupational hazard. People working near lead smelters and police on traffic duty in tunnels often suffer distinct lead poisoning symptoms. Those who drink moonshine liquor subject themselves to lead poisoning because the solder on the cop-

BIOSPHERE 2

In December 1990, the air locks were sealed on a 3-acre enclosed ecological system in the foothills of the Catalina Mountains north of Tucson, Arizona. This project, known as Biosphere 2, had been in the planning stage since 1984. A part of the field of *biospherics*, involving an understanding of the earth's biosphere and its component parts, Biosphere 2 will be sealed for 2 years.

Biosphere 2 consists of a structure that covers 1.27 hectares (3.15 acres) or 12,727 square miles (137,000 square feet) in floor area and is 198,100 cubic meters (7 million cubic feet) in volume. Eight people will live in a five-story, white-domed building inside Biosphere 2. There are laboratories, computer and communications facilities, workshops, libraries, and recreation facilities. Access to radio, television, telephone, video, and computer telecommunications networks exists. The people will have individual apartments for privacy and personal hobbies such as painting; wilderness areas are available for recreational activities such as scuba diving in the ocean and camping in the desert or rain forest. A small biomedical facility—equipped with diagnostic equipment and telecommunications—links with outside medical services.

Within the Biosphere 2 structure are agricultural communities, a tropical rain forest, tropical savannah, fresh and saltwater marshes, a marine system, and a desert community. Food crops are grown on the broad terraces; domestic animal areas for chickens, pygmy goats and pigs, and aquaculture are behind the plant crop area. A total of 150 cultivars (varieties) of tropical and warm climate food crops will be grown throughout the year, with about 50 crops in cultivation at any one time. These include papaya, figs, bananas, guavas, strawberries, a wide range of vegetables, sweet potatoes, and even coffee trees. A year-round cropping schedule has been developed for timing of planting and harvests to ensure a steady supply and good nutrition throughout the seasons. No pesticides are used in the intensive agriculture area; the water recycles in approximately one week—insufficient time for any pesticide to degrade. Pesticide in Biosphere 2's water system would endanger the human and animal inhabitants, including beneficial insects such as bees. Instead, integrated pest management—the use of natural insect predators, plant intercropping and other techniques—is used to control plant pests without pesticides. Waste material is treated and composted to maintain soil fertility.

The tropical rain forest is patterned after the Amazonian rain forest, and includes about 300 plant species. Surrounding the perimeter of the rain forest is a belt of plants related to the gingers and bananas, with a lowland forest at the interior, and a cloud forest with myriad ferns, tree ferns, and bromeliads. A stream plunges down the waterfall on the mountainside to a pool, through a floodplain forest, and flows south into the tropical savannah.

At the top of the rock cliffs, the tropical savannah is a composite ecology of plants and animals from tropical savannahs of Africa, South America, and Australia. A large termite community assists nutrient recycling here, as in the savannahs of the Earth, by eating and breaking down dead plant material. Termites and the approximately 250 other insect species in Biosphere 2 perform such functions as pollination, decomposition and transport of nutrient material, as well as serving as food sources for wild animal species.

The stream then flows into a freshwater and saltwater marsh, a complex estuary ecosystem of plants, insects, small animals such as frogs, turtles, and crabs, modeled on the Florida Everglades. The water grades from fresh to salt water along the winding course of the stream as it approaches the saltwater ocean whose tides, vital to the maintenance of this rich and diverse ecosystem, rise and fall periodically to bring salt water into the marsh mangrove system.

Based on a Caribbean coral reef ecosystem, the marine system contains coral reef, lagoon, beach, and deep-water ecosystems. The ocean includes about 1000 species of plants and animals normally associated with a coral reef, including living corals, parrot fish, starfish, and anenomes. Wave action—required to circulate nutrients that might otherwise accumulate at the ocean floor and to bring oxygen to the coral reef ecology—is generated mechanically.

The desert is patterned on a coastal fog desert, such as the Vizcaino Desert in Baja California, and populated with plant and animal species adapted to low rainfall but high humidity conditions. This desert is a winter rainfall ecosystem which can remain dormant in the summer—when other communities are using carbon dioxide for their rapid growth period—and become active in the winter to balance the demands on atmospheric carbon dioxide for plant growth throughout the year.

Wild animal species, including hummingbirds and the nectar-eating bat for pollination, the monkeylike bushbabies or galagoes, lizards, skinks, tortoises, toads and frogs, can move freely among the communities.

Biosphere 2 should help scientists determine if they can create a functioning closed system. It can assist in developing purification techniques for water and air. If space stations are established, the results of Biosphere 2 could aid in their design.

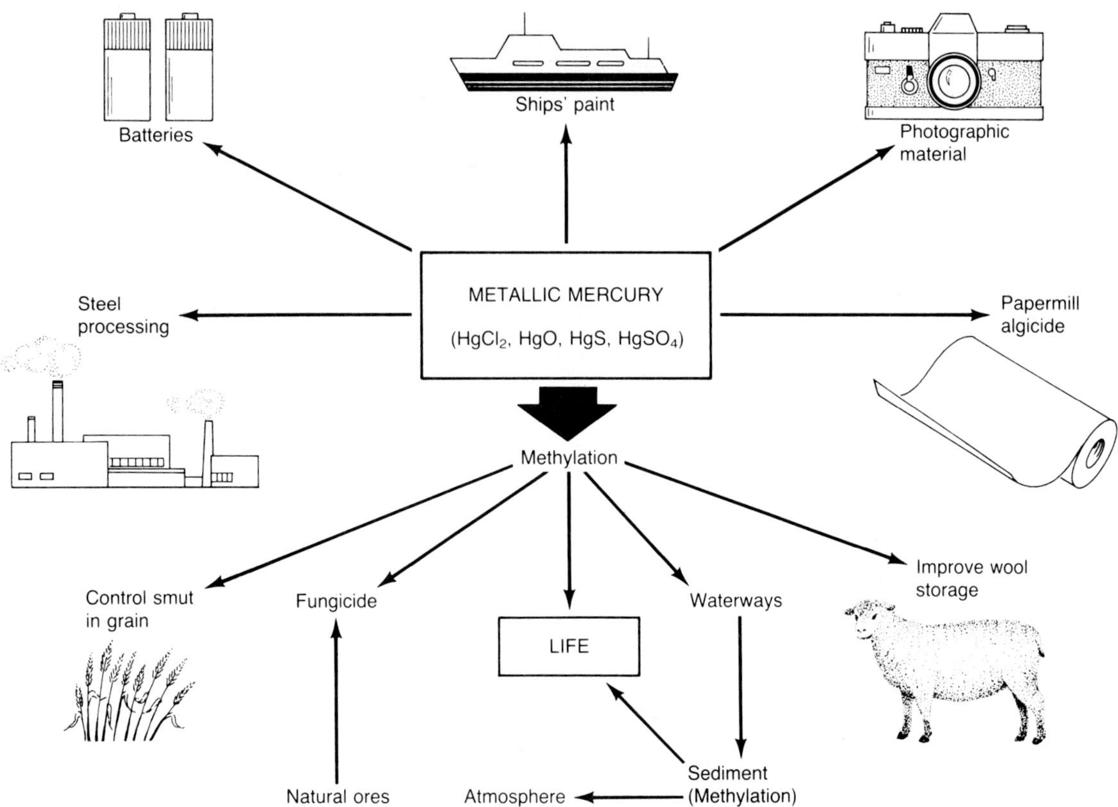

Figure 2.17
The mercury cycle. Various forms of metallic mercury are used in industrial processes. Methylation, or combination with carbon compounds, allows many other uses. Methylation can be caused by bacteria in the bottom sediment of lakes or in streams. Methyl (carbon-containing) mercury is readily absorbed by living tissue and produces harmful effects.

per tubing and automobile radiators used in stills contains lead. Some cooking utensils are still made with leaded solder, and pottery incorrectly glazed releases lead into the food or liquid it contains.

Cadmium Cycle

Cadmium is another metal that can be detrimental to living systems if the natural cycle is disrupted. It occurs as a constituent of zinc, copper, and lead ores rather than by itself. Environmental contamination is often the result of mining these ores.

Cadmium is used in alloys and paint pigments, plastic stabilizers, rubber products, and metal plating. The disposal of industrial waste products from their manufacture releases additional amounts into the environment. Once in the environment, animals take up cadmium through food and water. It accumulates in their liver and kidney tissues, where it can be retained for long periods. In human beings, cadmium is associated with hypertension, bone decalcification, arteriosclerosis, emphysema, and the reduced ability of the kidneys to reabsorb water. The effects of cadmium on life are not fully understood; thus, it is

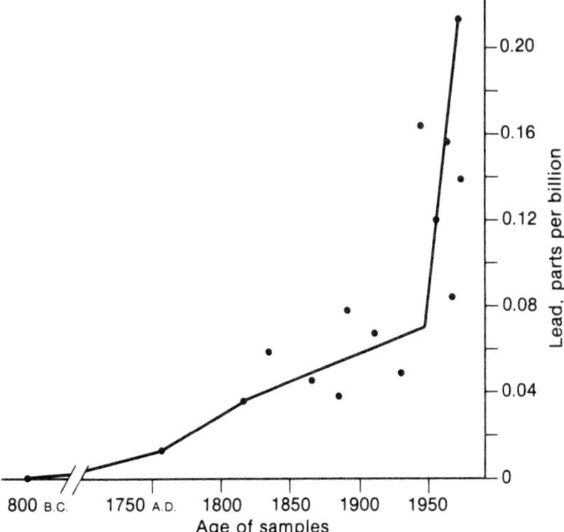

Figure 2.18
Increase of lead recorded from samples taken on the Greenland ice cap. Industrial nations add lead to the air in combustion processes, and the lead can move to the poles via wind currents. It settles out as air descends to the earth's surface. (From the National Academy of Sciences. *Lead: Airborne Lead in Perspective.* Washington, D.C., 1972.)

difficult to assess any long-term changes that might result from contamination.

Chromium Cycle

Chromium is fairly abundant on the earth's surface and occurs in low concentrations in sea water. In the natural cycle of chromium, it is passed from soils to plants to animals and back into the soils as organisms die and decay. Many soils contain low levels of chromium.

Biologists use chromium to study red blood cells. It is also used as a coating for metal car bumpers and other materials to protect them from corrosion. Combustion of most materials, including coal and wood, releases small amounts of chromium into the atmosphere. Recent shifts to other forms of synthetic materials in industry are reducing the amount of chromium in the atmosphere of urban communities, but the environment is still being contaminated at alarmingly high levels.

When ingested or inhaled, chromium enters the bloodstream and is carried to the liver and spleen. A normal intake of chromium at harmless levels is used with the hormone insulin to promote glucose utilization. In fact, lack of chromium causes an impairment in tolerance to glucose. Higher levels, however, are definitely harmful and are known to cause skin lesions, gastrointestinal ulcers, and lung cancer.

Trace Elements

Generally, the minerals present in large amounts in the environment are required by living organisms. Elements present in small amounts in the geochemical environment are called **trace elements.** Most of the biological consequences of continuous exposure to limited amounts of these trace elements are not well-known although it appears that disruptions of the natural cycles cause most problems.

Many trace elements are found in coal. Most occur at concentrations which approximate those of the earth's continental crust. Trace elements released by coal combustion that are of the greatest environmental concern in at least one chemical form are arsenic, beryllium, cadmium, copper, chrominum, fluorine, lead, mercury, nickel, selenium, silver, thallium, tin, and zinc. Others suspected of having detrimental effects are antimony, barium, bismuth, cobalt, gallium, and tellurium.

Fluorine is a particularly controversial element. A highly reactive gas, it interferes with some physiological processes in the body. When placed in drinking water at levels of 1 part per million, it helps prevent dental cavities; but at 2 parts per million and higher levels, it can cause brown mottling on teeth. At very high levels, such as 30 parts per million, more serious health problems occur.

Elements like cobalt, copper, chromium, fluorine, iodine, iron, molybdenum, vanadium, selenium, and zinc are essential for life and so have biological and nonbiological components in their cycles. Industrial use of these elements, as discussed in the chromium cycle, can produce harmful atmospheric levels. Altered metabolic activity occurs in living organisms that take up these elements and their compounds. A mineral like coal, which contains many elements, must be used carefully to avoid environmental contamination.

Data now available indicate that each species of organism responds in a unique way to concentrations of toxic materials. Age, sex, state of health, genetics, and diet as well as the amount and time of exposure to the element affect the degree of response.

ECOSYSTEM SUCCESSION

Examining an ecosystem over a period of years shows that it is in a constant state of change. Population interactions, species composition, organic structure, and energy flow do not remain constant. Plants and animals are gradually eliminated and replaced by other species.

Ecologists find that orderly, predictable changes occur in the populations of undisturbed ecosystems. This process occurs over a period of time and is called **succession.** The time sequence of succession is different for each ecosystem. As one community of animals and plants grows and utilizes a space, it gradually makes the area unsuitable, thereby allowing another community to become established.

Two forms of succession occur. *Primary succession* occurs in terrestrial areas where no soil exists. As plants become established, soil formation or deposition occurs, allowing other plants to grow. Rocks and cliffs are one example. Rocks can be colonized by plants as soil accumulates in crevices. As plants spread out, more soil accumulates, allowing additional plants to become established. When pulverized rock is deposited by winds and water in the form of sand dunes, plants can become established. Communities change with time as the soil becomes established and continues to accept other plants (Figure 2.19). *Secondary succession* occurs when a disturbance to a community has caused it to revert to an earlier stage in succession. An old abandoned field provides an excellent example. First, annual grasses colonize the area. After two or three years, perennial grasses appear; shrubs and a few trees follow ten to fifty years later. Finally, a forest develops (Figure 2.20).

In western Oregon, the Oregon white oak develop dense forests at the eastern base of the Coast Range. Eventually, the canopy becomes so dense that the oak seedlings underneath cannot grow for lack of sunlight. As a result, Douglas fir seeds blown or carried into the woods begin to germinate without competition from the oak seedlings. Thus, a Douglas fir forest slowly replaces the oak forest (Figure 2.21). Eventually the Douglas fir chemically alter the soil so that their own seedlings cannot germinate. Then, depending on slope, moisture, and altitude, other trees become established. Thus, succession occurs as species within the system change their own environment. Such changes generally take hundreds of years.

In any one physical environment, succession ends in a relatively stable **climax community.** Generally the populations have balanced birth and death rates, although fluctuations can occur. A climax community can live so that its habitat remains suitable for continued reproduction by constituent species. It maintains a type of equilibrium or steady state, is highly organized and stratified, and has a net productivity lower than the earlier developmental stages because the organisms use most of the energy it assimilates. In a climax community, most energy needs are supplied by the organisms within it and losses of energy and nutrients are minimal. During early developmental stages, such as grasses taking over a field, the respiration rate of a community is lower than its gross productivity. This makes new productivity high and leaves a lot of energy that can be removed from the community. In climax communities such as mature forests, the

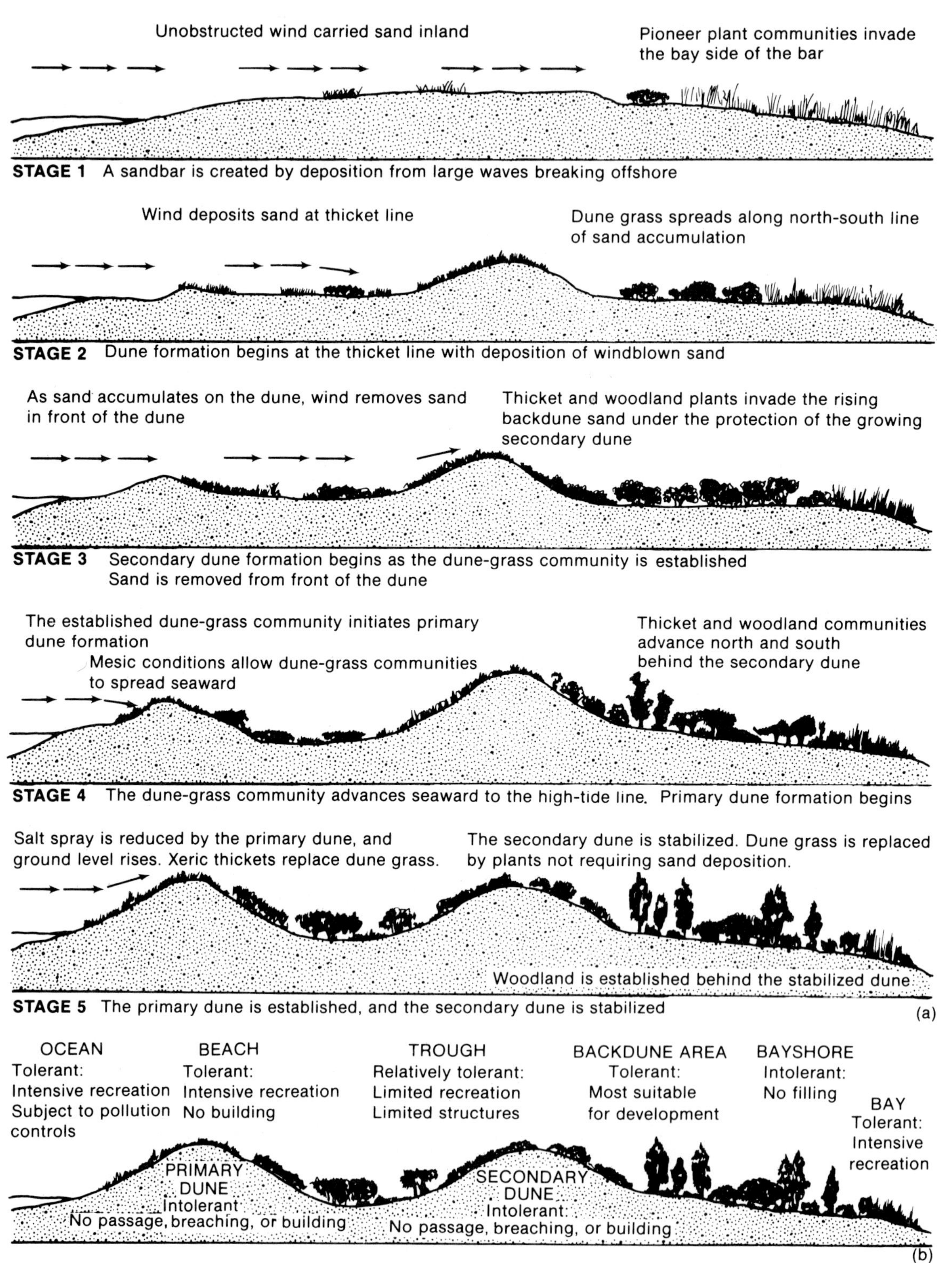

Figure 2.19
Primary succession establishes communities on sand dunes. (From *Design with Nature* by Ian L. McHarg; Copyright © 1969 by Ian L. McHarg. Reprinted by permission of Doubleday and Company, Inc., and from Ian L. McHarg, "Best Shore Protection: Nature's Own Dunes," *Civil Engineering* 42 [1972]: 66–67.)

Figure 2.20
Secondary succession from an old field through forest.

gross productivity supplies the needs of life with little energy left over. Minerals tend to cycle among components within the community, and species diversity is usually high. The principle of homeostasis with positive and negative feedback loops (Chapter 1) operates to maintain the conservation of resources and stability of the community.

This tells us something about the management of natural systems. If we must remove energy in the form of food or wood products, we create the least disturbance by using the developmental stages in a successional sequence. For example, a community logged early in the successional stage will reestablish itself faster than one logged in a late or climax stage.

Eutrophication

The life of any lake or reservoir is limited because ultimately natural erosion will fill the lake or reservoir with sediment, causing succession. Organic matter is washed from forests and fields along with the sediments into the lake.

No lake or reservoir remains in an equilibrium state. The physical and biological components of lakes (or their *ecosystem*) are constantly changing. Consequently, the life forms in these bodies of water also change as stages in succession occur.

In the early life of a lake, the concentration of nutrients is very low, causing the populations of living organisms to be low. Such lakes are called **oligotrophic.** Oligotrophic lakes are usually deep, have an adequate supply of oxygen to sup-

Figure 2.21
Douglas fir growing in white oak forest in western Oregon.

port some cold-water bottom fish such as lake trout, and are remote from human settlements. The low concentration of nutrients precludes extensive fish life in the lake, but the lack of algae may give the lake a blue cast that is aesthetically appealing.

An increase in concentrations of nitrogen, phosphorus, and other nutrients is a natural process that occurs in the waters of the world. This process of enrichment, called **eutrophication,** occurs over a long period of time (on the geological time scale) as nutrients are washed from the land into the water. The activities of people may also add to the process of eutrophication. The disposal of human waste material, washing of fertilizers from fields, and the disposal of organic industrial by-products speed the process, so that lakes age much faster now than they would naturally—perhaps during a person's lifetime. The impact of people on freshwater lakes is called *cultural eutrophication.*

Lakes that have high levels of nutrients and dense forms of vegetative growth are **eutrophic** lakes (Figure 2.22). Such lakes are generally shallower than oligotrophic lakes, but they have abundant fish populations and other life because of the high nutrient content. This process of change may mean that lakes with little or no fish of economic value will become very productive as they become eutrophic. In the 1950s, an oligotrophic lake (Lake Dalnee) in Alaska was enriched with a variety of different nutrients for several years. The result was an increase in phytoplankton production and juvenile sockeye salmon. In other cases, lakes producing fish palatable to people may become unfavorable for the growth of these fish as eutrophication occurs. Lake Erie is an example of a lake in which populations of valuable species declined as a result of the overall increase in productivity due to eutrophication. The collapse of the Cisco fishery in 1925 was in part attributed to the introduction of large amounts of wastes from Detroit and Toledo at the western end of the lake.

Methods for retarding or controlling eutrophication involve removing nutrients from

Figure 2.22
Lake with dense growth of water lilies.

sewage and regulating the discharge of untreated sewage; improving agricultural practices so that nutrients do not wash into lakes; removing nutrients from lakes by dredging sediments; and otherwise controlling the availability of nutrients in the lakes. Measures may also be undertaken to kill the algae in lakes. Combinations of control measures used in Lake Erie are reported to be improving conditions and to some extent reversing the eutrophication process. To retard the eutrophication of Lake Tahoe, which is still oligotrophic, waste flows to the lake are being restricted.

PHYSICAL FACTORS THAT LIMIT DISTRIBUTION

Life develops within the constraints of each set of physical conditions such as moisture, sunlight, and air movement. Each species changes the environment to some degree, frequently creating microhabitats that encourage new species to colonize the area. While succession can be followed in each ecosystem, it differs greatly with changes in the amounts of moisture, sunlight, and soil type, among other things. Physical factors in combination with the biology of the organisms interact to influence the type of community that can exist in each ecosystem. When one physical constituent such as a soil mineral necessary for the survival of a plant or animal population is not present in sufficient amounts to support that population, the population cannot exist there. This is referred to as the *law of the minimum.*

Water

Nothing can live without water, nor can it live without the proper amount and kind of water. The evolutionary invasion of land occurred when plants and animals developed an autonomous means of retaining water. Very refined adaptations of water retention are found in desert life. Animals often avoid the heat of day by being most active at night. Kangaroo rats and others conserve moisture by reusing their body water. Cacti retain moisture because of their tough outer coating. Most plants remain as seeds until enough rain falls for them to grow, flower, and form more seeds. Plants might not reproduce in a dry year but may do so several times during a wet year. Polar areas on the earth have their own type of desert. Although water exists in the form of ice, it is not available to support life.

Aquatic life needs water to regulate salt uptake. In freshwater lakes and rivers, some organisms prevent the loss of salt and other nutrients by means of special forms of kidney and salt absorption systems. Other animals have tough outer coatings that control water flow.

Saltwater organisms have to prevent water loss. You may have noticed some plants along the sea coast with very rubbery outer coatings. These coverings keep the salt spray from desiccating the plants. Kidneys help maintain the proper level of salt in animals. Sharks retain urea to keep a slightly higher concentration of salt than the ocean water. Some ocean birds have special salt glands in the surface of their foreheads that extract and excrete salt.

Because of the gradual change from fresh to brackish to salt water, estuary life displays distinct gradations of organisms existing in the various degrees of salinity. The different forms of estuaries depend on the type and degree of mixing of fresh and salt water. Coastal regions, where evaporation is high and water mixing is low, can have a higher salinity than the ocean. High salinity also occurs in some tidepools due to evaporation. Animals and plants adapt to these conditions by preventing salt from concentrating in their bodies.

Water movement also controls the distribution of aquatic life. Organisms in fast-flowing rivers and streams or tidal areas must be able to either move so they are not washed away or to attach themselves to a stable object. Sometimes this object turns out to be another organism. In the ocean, major upwellings and currents influence the distribution of life. Upwellings bring additional oxygen to deep water and additional nutrients to the surface. Ocean currents move marine organisms throughout the world.

Solar Radiation

Solar radiation varies considerably according to location, time of day, and time of year. Even in one area the solar radiation can differ because of trees and mountains. For example, solar radiation reaching the bottom of the forest is much less than that striking the tops of the canopies. Each habitat, then, has certain microclimatic conditions that different species favor. Thus, a particular community composed of a variety of microhabitats is to a large extent the result of solar radiation.

Maximum yearly temperature, minimum daily temperature, and mean annual temperature have major effects on the distribution of organisms like the cold-blooded animals. Reptiles and amphibians are found primarily in warmer parts of the world because their bodies must obtain heat from the surrounding environment. Amphibians require moist areas as well. Plants are also limited in distribution by temperature. (We will discuss the effects of temperature further in the section on population distribution.)

Day Length

The length of daylight, called the **photoperiod,** is measured by chemical means in plants and controls their flowering and sprouting. Many plants flower when daylight hours increase, signalling the arrival of warmer weather. Photoperiods vary from a few minutes in the tropics to hours in the northern temperate regions. When plants are moved a long distance either north or south, their photoperiod mechanisms frequently operate incorrectly, causing the plant to flower when weather conditions may not be opportune. Photoperiod differences also exist in the same species of plant growing at different altitudes on the same mountain slope.

Some animal activity is also triggered by daylight. Some birds, for example, have special photosensitive cells in their eyes. Migration of these species is triggered by day length as registered by the cells. This gives us a clue as to why the swallows return to Capistrano on the same day each year.

Because plants must produce enough biomass to sustain themselves throughout a year, the length of time when the light, temperature, and water are sufficient to allow the plant to grow is very important. A short growing season restricts the species of plants that can locate in an area. Growing seasons also influence animal distribution. In colder regions most animals are warm blooded; they maintain their body heat by expending energy. Some animals conserve heat by going into a state of dormancy or hibernation, allowing them to reduce their body heat requirements.

Air Currents

Air currents have a major effect on many forms of life. Updrafts from valleys provide major routes along mountain ridges for migratory birds, particularly raptors. Winds distribute plant seeds and some microorganisms, and local winds affect soil, moisture, and humidity. Winds are also responsible for many undesirable effects. Forest fires spread rapidly and remain uncontrollable because of wind. Areas of high wind depress plant growth. Plants above the timberline or on mountaintops are usually stunted and have more spreading characteristics (Figure 2.23). Winds of hurricane or tornado force create major changes in the communities they strike.

Soil Nutrients

Soil nutrients are important controlling factors on life. While most of the details on soil are presented in Chapter 12, it is important to recognize the best agricultural lands are areas with the highest levels of soil nutrients. Many of the nutrients are components of mineral cycles dis-

THE NEW JERSEY PINELANDS, AN EXAMPLE OF PLANNED ECOSYSTEM MANAGEMENT

The New Jersey pinelands or pine barrens comprise a mosaic of upland aquatic and wetland communities occupying approximately 400 hectares of sandy acid coastal plains soil in the southern portion of New Jersey. The popularity of pinelands for recreation and development causes many impacts on the communities. Plant nutrients move relatively rapidly through the sandy upland soils, and are thus in short supply. The area is dominated by pitch pine, with a variety of oak species and a few other pines. The upland forest composition relates closely to fire frequency; pitch pine is the forest's most tolerant fire species. We find some 39 species of mammals, 59 species of reptiles and amphibians, 91 species of fish (including estuarine species), 229 species of birds, and over 800 taxa of plants in the inland ecosystem, including several relatively rare species.

Because of the tremendous impact of people on this relatively fragile system, legislation was initiated in the late 1970s to manage this area in cooperation with the state and federal government. A comprehensive management plan was developed based on the types of flora, fauna, geology, soils, and hydrology, as well as certain cultural features. The area was designated to maintain essential characteristics of the pineland environment and to encourage appropriate patterns of development in and adjacent to the area. Each type of land in the pineland area has specific sets of rules governing its use. Both fire and nutrient deficiencies in the soil were considered in the development plan. A consideration of the plan was long-term preservation of an ecosystem with units sufficiently large to include a genetic diversity of population, allowing genetic flow between populations.

This is one of the few examples of an overall attempt to manage a major ecosystem in the United States. This kind of approach will undoubtedly become more common as we attempt to preserve natural systems against a tremendous human impact.

Figure 2.23
Stunted plant growth above timberline (U.S. Department of the Interior).

cussed earlier. Thus mineral (nutrients) must be present in a chemical state that makes them available to plants.

Pressure

Atmospheric pressure is based on the air pressure at sea level. All life is exposed to the same pressure as other life at the same altitude. The higher the altitude, the lower the pressure, making respiration for animals more and more difficult as the altitude increases. Exploration of higher altitudes such as Mounts McKinley and Everest is highly restricted without artificial means of forcing air into the lungs. Space exploration requires the use of a mini-environment, or full-pressure suit, to maintain normal pressure for astronauts. As we move down into the ocean the pressure increases. We know from hard experience some of the effects of too rapid a change in pressure. Eardrums can burst from descending too rapidly, and "bends" caused by nitrogen bubbles in the blood can occur from too quick an ascent from ocean depths.

Atmospheric pressure serves as a clue to impending weather. Rapid changes in pressure alert animals to coming storms. Hurricanes and tornadoes bring drastically lower atmospheric pressure, often causing destruction.

COMMUNITY ORGANIZATION

Stratification

Populations living and interacting in one ecosystem collectively show characteristics that can be analyzed on a community level. All communities have some form of structure that normally reflects the physical environment. In forest communities, this structure is easily recognizable as vegetation layering, or vertical **stratification** (Figure 2.24). Several layers can occur, with a canopy of different tree species dominating various layers and an herb and shrub layer found near the ground. The incoming sunlight is scattered at each layer, with less than 1 percent penetrating through the foliage and reaching the ground. The species at each level adapt to the amount of sunlight penetrating their area, creating microhabitats throughout the forest. Plants and animals also evolve unique adaptations to enable them to survive successfully on the material at each layer.

A forest community develops as the vegetation of each layer grows to the level required by its photosynthetic needs. Larger trees of the upper canopy receive more sunlight, but they must expend energy in the form of woody tissue growth to support themselves. Other trees cannot tolerate open sunlight and so grow well under the larger ones. Herbs on the ground require less energy for support and receive less energy from the sun.

In open ocean communities, all primary production occurs above a depth of 100 meters (328 feet) in the *euphotic zone.* Although many organisms live at greater depths, they are all dependent on food produced in the euphotic zone. At least one-third of all photosynthesis in the world takes place in the ocean.

Pattern

Pattern, another community characteristic, involves the spacing of plants and animals. Community pattern can be uniform, clumped, or random (Figure 2.25). Desert sage places a chemical

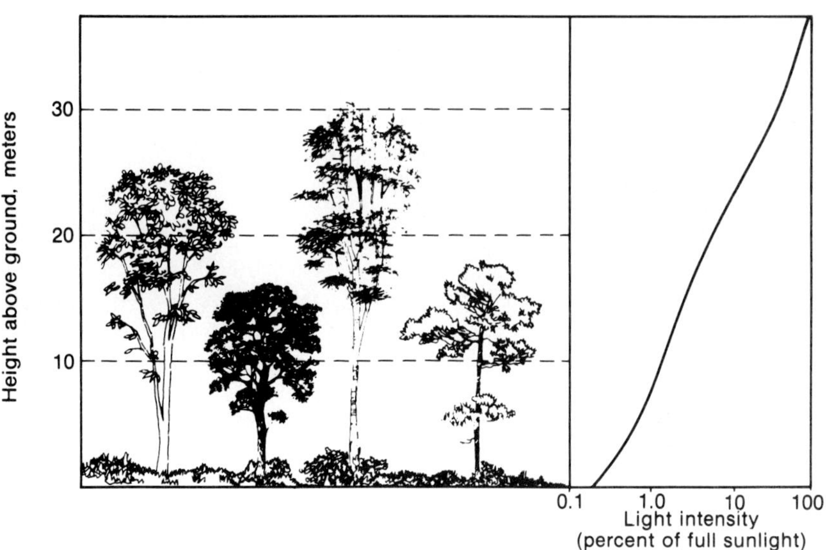

Figure 2.24
Stratification and light intensity in a forest. Different species of trees, shrubs, and herbs are adapted to different light intensities. (From R. H. Whittaker, *Communities and Ecosystems.* New York: Macmillian, 1970.)

toxic to other sage in the surrounding soil, creating uniform spacing. In dry regions, plants and animals are found in clumps around water holes. Randomness means that the location of each individual organism is independent of the location of other individuals in the community. A windstorm might randomly scatter seeds over a field, or an ocean current might randomly distribute the eggs from an oyster.

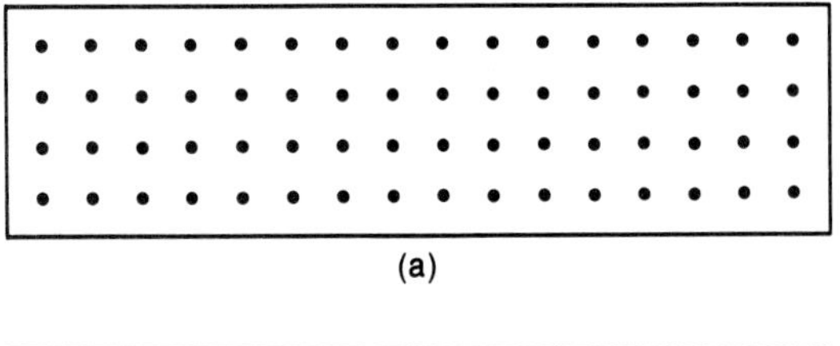
(a)

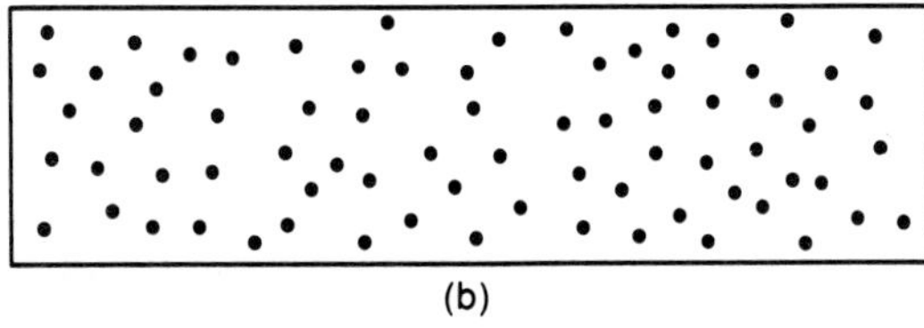
(b)

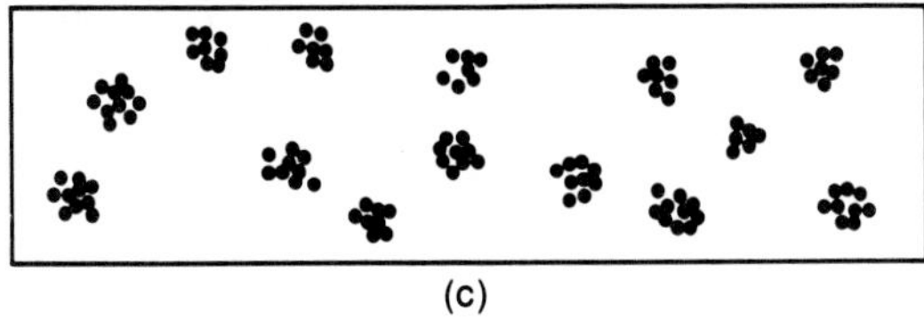
(c)

Figure 2.25
Patterns that can be established by natural plant or animal populations: (a) uniform; (b) random, (c) clumped.

Species Diversity

Most communities have large populations of a few common species and smaller populations of many less numerous, sometimes rare, species. We can establish an index to compare the number of species with the number of individuals in each species. This index, called **species diversity,** tends to be low in harsh environments (e.g., tundra), where the vast majority of individuals belong to a very few species, and high in milder environments (e.g., rain forests). In high diversity areas, individuals are more evenly distributed among a large number of species.

Artificial ecosystems usually have a very low species diversity compared with the surrounding natural communities. Thus, species diversity can often be used as an indicator of human influence on natural communities as, for example, tropical forests. Under natural conditions, decaying plant fibers and wastes from animals are recycled in rivers without buildup. If wastes keep increasing, species diversity decreases. This indicates that the natural system eventually is no longer effective in recycling.

Ecotones

Where a field, grassland, or desert meets a forest, a marked distinction can be seen (Figure 2.26). When we come to a new community in the woods, we find the two forest communities integrated. These boundaries between communities, whether sharply defined or integrated, are called **ecotones.** As conditions in the ecotone often favor life in both communities, more species can live there than in either of the bordering communities. The overall tendency for an increased variety and density at the junction of two communities is known as the *edge effect.* When alterations in a community create additional edges, new species often move into the area. This happens when roads or transmission lines through forest land open up edges, attracting different species of birds, insects, mammals, and plants. Managers of natural systems must understand the effects of ecotones and edges on populations and communities.

Figure 2.26
Ecotone between forest and desert in Oregon.

Stress

Changes that occur in the ecosystems as the result of human intervention are now being evaluated. Terrestrial and aquatic ecosystems are exhibiting signs of stress, including increases in community respiration and nutrient turnover, decreases in life span of organisms, shorter food chains, and decreases in species diversity.

Studies in marine ecosystems show that human impact is changing food webs and species diversity, although the full impact cannot be completely understood because of the size of the ocean and the mechanisms of energy transfer, beginning with small zooplankton that reproduce very rapidly.

Any disturbance to ecosystems and communities causes changes in the energy flow and organization in the community. Populations are likely to experience stress. This may be of particular concern if the disturbance occurs when young are being raised or when the environment is causing natural stress (e.g., a cold winter).

Planning, therefore, becomes an important tool. We need to know when our actions may cause ecosystem disturbances. What is present in the system? How do we accomplish the change with the least disruption? Finally, after the impact has occurred, did we succeed in our plan to minimize disturbance?

Because of the lack of planning in the past, restoration ecology has become an important new discipline. Restoration ecologists are trying to restore ecosystems that have often been damaged by unwise or poorly planned action. People entering this field need an extensive understanding of how the ecosystem process operates.

SUMMARY AND CONCLUSION

Energy in the form of food is the basis for life on the earth. The sun's energy, converted by green

LEGISLATION

National Environmental Policy Act (NEPA). One of the most broad federal laws protecting the environment is NEPA, passed in 1969. This law promotes efforts to prevent or eliminate damage to the environment. The act requires that all federal agencies prepare detailed environmental impact statements for proposed major federal actions. The proponents of NEPA hoped that avoidable adverse environmental effects could be foreseen and avoided. Environmental impact statements must identify possible adverse impacts on the environment and propose methods to reduce or avoid the adverse impact. Draft environmental impact statements are made available for public comment.

plants to energy usable by other organisms, begins the flow of energy from one group of organisms to another. This flow of energy is called a food chain or food web, and each level in a food chain is called a trophic level. The net primary production, or the ability to convert the sun's energy to living biomass, is high in such areas as estuaries, springs, and marshes. As energy moves from one level to the next, some is dissipated in the form of heat and some is used to maintain the biomass of each trophic level. This limits the number of possible links in most food chains to three, four, or five. When organisms die, their bodies are broken down by decomposers and certain materials and energy are released. Decomposers include organisms that use free oxygen and those that use oxygen combined with another element or oxidizing power from other compounds.

Ecosystems are natural levels of organization consisting of living and nonliving components. As a result of the actions of the living components, ecosystems are constantly changing. This group of living organisms, called a community, is formed by the combined characteristics of the populations living in one physical area at the same time. Populations within a community interact, establishing a natural balance of energy flow, mineral cycles, productivity, and population numbers.

The nonliving component and living organisms in a particular area function together as an ecosystem. Components of the ecosystem are unified through mineral cycles and energy flow. Human interference with the dynamic interrelations in ecosystems can lead to disrupted mineral cycles and broken links in the energy chain. As people use more nonrenewable resources such as minerals, they reduce the total amount available. People often change habitats, enabling organisms of different competitive abilities to become established, displacing organisms present.

Maintenance of natural communities must be based on the principles of ecosystem dynamics, community formation, and population interaction. Our knowledge of energy flow and ecosystem dynamics will help us control many environmental pollution problems discussed later in the book. Management based on the principles developed in this chapter provides the most effective means for people to feed their own and to improve their living conditions.

FURTHER READINGS

Begon, M., J. L. Harper, and C. T. Thompson. 1990. *Ecology: Individuals, Populations, and Communities.* Cambridge, MA: Blackwell Scientific Publications.

Brewer, R. 1988. *The Science of Ecology.* New York: Saunders.

Good, R. E., and N. F. Good. 1984. The Pine Lands Natural Reserve: An Ecosystem Approach to Management. *BioScience* 34:169–73.

Jordan, W. R., M. E. Gilpin, and J. D. Aber (eds). 1987. *Restoration Ecology: A Synthetic Approach to Ecological Research.* New York: Cambridge University Press.

Odum, E. P. 1985. Trends Experienced in Stressed Ecosystems. *BioScience* 35:419–22.

Ray, G. C. 1985. Man and the Sea—The Ecological Challenge. *American Zoologist* 26:451–68.

Tangley, L. 1985. A New Plan to Conserve the Earth's Biota. *BioScience* 35:334–36.

STUDY QUESTIONS

1. What is the difference between an ecosystem and a community?
2. Humans can create large cities. Why can't they create a large efficient ecosystem?
3. What is ecological efficiency? Why does it concern us?
4. How do mineral cycles such as the carbon, nitrogen, or oxygen cycle, relate to the ecosystem?
5. Do all mineral cycles have a living component? Explain.
6. What do we mean by pathways in the ecosystem?
7. What does species diversity tell us about a community?
8. When would it be desirable to increase ecotones? Decrease them?
9. What is the importance of rainfall in delineating biotic regions of the world?
10. With all the water in the Arctic, why don't more animals live there?

SUGGESTED ACTIONS

1. *Recycling.* Recycling has become a major theme among environmentally aware people. You can encourage your local garbage pickup company to recycle cans, glass, and paper. Fast food outlets can recycle paper and styrofoam. Local recycling centers can be set up with volunteers. This is likely to make the operation profitable.
2. *Food Selection.* People can reduce their dependency on food high on the food chain. Many plant products provide excellent and nutritional food items. If Americans reduce their meat intake by 10 percent, the savings in grains and soybeans could feed about 60 million people a year.
3. *Plant a Tree.* Because green plants are essential in the oxygen cycle, their replacement is essential. Deforestation in many areas is proceeding at a runaway rate. If everyone planted a tree each year, we could reduce this loss of vegetation, provide our neighborhood with added beauty, and reduce noise. Plant trees in your yard and encourage the development of parks and greenbelts in your state.

3

POPULATIONS

Populations, as defined in Chapter 2, are the basic units that most biologists study to evaluate life. Populations grow, move, and become extinct. They can be the total component of a species, or a species may be composed of a few to a large number of populations. It is the population that people like to manipulate to produce more and better food, to control food, to control pests, and to maintain for our aesthetic appeal. How do these populations interact? How do they change? Can they be manipulated? These are some of the subjects of this chapter as we examine population biology. Population biology can be divided into population ecology, genetics, and evolution, all components of our study.

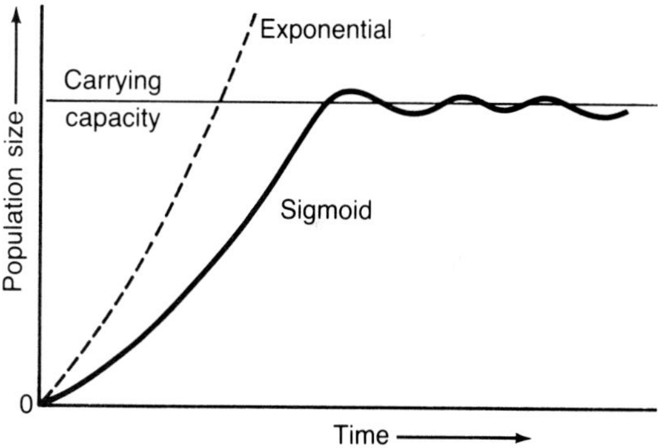

Figure 3.1
Exponential and sigmoid growth patterns of populations.

POPULATION DYNAMICS

Population Growth

How rapidly does a population grow? All populations grow in an environment until some force, either internal or external to the population, slows or stops growth. An internal force would be stress from many encounters with other members of the population. External forces can be *physical,* including a reduced essential resource such as water, air, food, or shelter, or *biological,* as is the case when populations compete for similar resources or one population preys on another.

When a population is introduced to a new environment where survival conditions are good, growth is generally slow at first and then becomes rapid as the population gains a foothold. This is called **exponential growth** (Figure 3.1). Exponential growth cannot continue indefinitely. Generally, some combination of environmental forces causes a population either to decrease rapidly when it reaches a certain size or to achieve an approximate state of equilibrium with the environment. An example of the former is the lemmings. This small mammal of the arctic region shows rapid population growth until stress, the lack of food and cover, or adverse weather cause it to decline. The numbers begin to increase again after a year or two of decline.

The ability of a population to grow in an unrestricted environment where no forces are acting to slow the growth rate is called the **biotic potential.** It has been calculated that a single female housefly laying 120 eggs at a time, of which 60 develop into females, would produce close to 6 trillion flies after seven generations, and flies might have seven or more generations per year! On the other hand, Charles Darwin calculated that a single pair of elephants, with a gestation period of 600 to 630 days, would have 19 million descendants after 750 years. The biotic potential is not recorded for many populations. The maximum growth rate of a population does not continue for long because of carrying capacity limits.

Most animal populations exhibit a growth pattern wherein the numbers fluctuate about some mean size. This mean size, called the **carrying capacity,** is usually the number of individuals the habitat can support at a given time. The growth form exhibited by populations increasing rapidly and slowing as the size approaches the carrying capacity is called **sigmoid growth** (see Figure 3.1). Actually the population may increase slightly beyond the carrying capacity and then fluctuate around that number. This growth is difficult to measure in the field because the carrying capacity in an ecosystem can change.

In each habitat the organisms become specialized to use the resources most effectively, and the resources in turn limit the size of each population. In effect, the "balance of nature" is established. Only those individuals most effective at using the resources can become established and reproduce. When the habitat is disturbed through natural or human actions, the constraints of the habitat are altered and new populations can be established. But other constraints quickly become effective, and a new balance of life is established.

When the volcano on the island of Krakatoa in the East Indies erupted in 1883, almost all life on the island was destroyed. By 1886 a few plants, including some blue–green algae and mosses, became established. Flowering plants and grass followed shortly. By 1906 a small woodland had appeared, and in 1908, 202 animal species were recorded. This number grew to 770 species in 1924 and 1100 species in 1933. The arrival of each new species altered the homeostatic mechanisms. Succession was occurring (Chapter 2). New symbiotic relationships were established and began to grow. The interactions of the natural system and the biological component combined to constantly influence the growth rate of every population present.

Population Interactions

Many forms of interactions occur among individuals and populations living in the same habitat. They can result in positive forces that aid the growth and survival of one or both of the populations, or negative forces that result in a decrease or extinction of one or both. A population can grow in an environment with all the necessary materials for survival until space, food, or other resources become scarce. As members of a population seek to use resources in short supply, they compete with each other or with other populations.

Competition Competition between members of the same species, called **intraspecific competition,** occurs as each member tries to get the best of everything: the best food supply, the best mate, the best nesting materials, and so on. For example, during the breeding season most male birds defend a territory against other males of the same species. Territories of red-winged blackbirds are easy to spot in a marsh because the males perch on top of the grasses and fly at intruders.

TAMPERING WITH A FOOD WEB IN AN AQUATIC ECOSYSTEM

Flathead River–Lake ecosystem is a diverse ecosystem in north central Montana. This system, like most in North America, has been subject to the introduction of non-native fish and fish food organisms. More than 25 percent of the sport fishes caught are non-native.

Between 1968 and 1975, opossum shrimp (1 to 2 centimeter long) were introduced into the Flathead ecosystem. The shrimp ate many of the small invertebrates that were the foundation of the fish food web in Flathead Lake, causing major reductions in the invertebrates. In 1986 shrimp densities, which had been on the increase, began to decline. At the same time, mineral nutrients were changing because of the presence of the shrimp. As a result, populations of algae began to increase.

Kokanee (landlocked sockeye salmon) were introduced into Flathead Lake in 1976 and replaced native cutthroat as the major sport fish. Kokanee did not take to eating shrimp; thus, as other invertebrates declined, kokanee also declined because they were unable to compete with the shrimp for the same food source.

The kokanee populations provided an abundant source of food for bald eagles, as well as several tern and duck species, coyotes, river otters, and grizzly bears as they spawned and died in the fall. The fish normally live three to four years before spawning in the lake tributaries. The fish eggs were also an important food source for birds and mammals. Since the collapse, the population of these birds and mammals along the lake tributaries has declined. Eagles probably do not have an alternative food source, thus they may be stressed during migration and will likely perish. Bears and coyotes will seek other food sources, some of which will involve competition with people.

Meanwhile, Glacier National Park, where some of the tributaries are located, has noticed a drop in fall visitation. Many people who were previously attracted to the large congregation no longer come.

Spencer, C. N., B. P. McClelland, and J. A. Stanford. 1991. Shrimp Stocking, Salmon Collapse, and Eagle Displacement. *BioScience* 41:14–21.

Interspecific competition involves two or more populations vying for limited resources. The outcome of such rivalry can be the extinction of one population, coexistence, or the triumph of one population over another because of environmental conditions such as temperature or humidity. One study involving two species of *Paramecium* showed that each species grew well separately, but when raised in the same test tube, one species always became extinct.

Competition is difficult to document because populations tend to evolve methods of coexistence in natural communities. Robert MacArthur's 1958 study of five species of warblers in New England shows that these birds avoid competition by subdividing the living area and foraging for food in different parts of a tree (Figure 3.2). While some overlap does occur, both contact and competition for food among species are reduced).

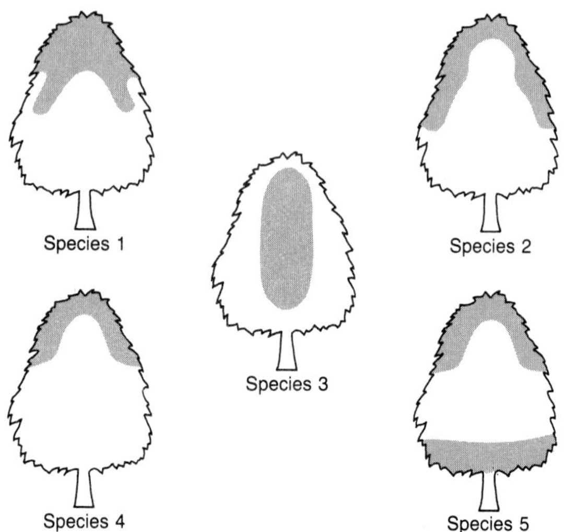

Figure 3.2
Parts of deciduous trees where warbler species spend more than 50 percent of their time foraging. (From R. H. MacArthur, "Population Ecology of Some Warblers of Northeastern Coniferous Forests," *Ecology* 39 [1958]: 599–619.

When people move into new areas, they frequently impose spatial restrictions on animals. Figure 3.3 shows that bighorn sheep of the western United States are severely limited in range (less than 1⁄20 of the area occupied before the arrival of white settlers) because of competition with people and their grazing stock). The settlement of isolated areas is one factor instrumental in the decline of the grizzly bear population in North America. Grizzly bear require large expanses of undisturbed wilderness. Roadways, fencerows, hunting pressures, and other products of human civilization cause them to seek other living areas. Both the grizzly bear and the bighorn sheep populations are declining to such low levels in the continental United States that it is difficult for mates to find each other.

Symbiosis Dissimilar organisms or species living together is called **symbiosis.** Some biologists stress the mutual dependency between two species which benefits both species in their definition of symbiosis. Tree roots and fungi have this form of relationship. Specific species of fungi grow around the roots of the tree, particularly the very small rootlets. The fungi provide protection and some nutrients for the tree; the tree, in turn, provides nutrients for the fungi. Neither the fungi nor the tree can grow properly without the other. When soils are devoid of fungi spores, trees are unable to grow.

We shall use the broader concept of species or organisms living together in a compatible relationship as our definition of symbiosis. However, it is also appropriate to talk about a symbiotic relationship between an organism and its environment. Homeostatic mechanisms maintain this symbiosis. When natural catastrophic events such as earthquakes, floods, or volcanic eruptions occur, the balance of nature is disrupted. Feedback loops are altered, resulting in a breakdown of many symbiotic relationships. Generally the disrupted system is influenced by the forces of the surrounding system and the homeostatic mechanisms are reestablished, resulting in a return to previously existing symbiotic patterns.

Figure 3.3
Distribution of bighorn sheep in the western United States (a) prior to white settlement and (b) in 1960. (From H. K. Buechner, "The Bighorn Sheep in the United States: Its Past, Present and Future," *Wildlife Monographs* 4 [1960]: 3.)

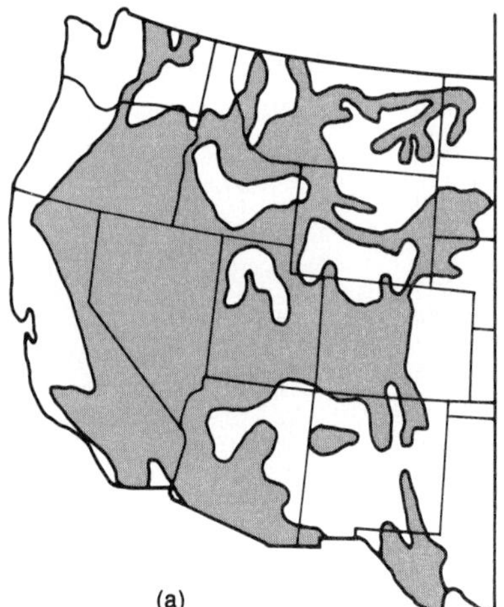

Figure 3.4
Feeding elk on the National Elk Refuge, Jackson, Wyoming, in the winter.

Predation Predation is one means of maintaining the balance of nature in an ecosystem. The word **predation** is used to describe the eating of one individual by another: carnivores eat herbivores, large carnivores eat smaller ones, cannibals eat their own kind. Such interactions occur between different trophic levels and within the same trophic level. Energy flow in an ecosystem is very dependent on **predator–prey interactions.**

In a natural community, predator–prey relationships evolve over a long period of time. Changes in the ability of a predator to capture a prey often result in prey populations developing new methods of escape. When people hunt or fish, they act as predators; but they often use advanced techniques which are developed so rapidly that the prey population cannot respond. Modern fishing fleets, with their new methods of locating and capturing schools of fish, can reduce the population to levels at which it cannot reproduce at replacement rates.

Such changes in predator and prey populations illustrate an important concept in natural ecosystem balance—natural control. Most populations evolve toward a state of self-regulation, in which most avoid extreme fluctuations in size through elaborate positive or negative feedback mechanisms (see Chapter 1). The growth of a prey population is positive feedback to a predator population. More prey means more food for the predators, so more predators can survive. The growth of a predator population is a negative feedback mechanism to a prey population. More predators means more prey are eaten, leaving fewer to reproduce.

Human interference with natural systems often creates disruptive behavior in animals. Elk in northwestern Wyoming migrate each year from their concentrated winter range near Jackson to summer feeding areas in Yellowstone and the Grand Teton National Parks as well as national forest land. Natural predators such as wolves have been removed from most of the area. Hunting occurs in the fall for part of the elk herd but is not allowed on elk in Yellowstone National Park. These animals can be hunted, however, during the few days they move beyond the park boundaries to the National Elk Refuge in Jackson. In addition, outfitters seeking a quality hunt for their clients place salt blocks at the edge of Yellowstone to attract elk to surrounding forest lands during the hunting season.

The National Elk Refuge was created because the city of Jackson expanded onto the prime elk winter grounds. An extensive elk winter feeding program is undertaken each year on the refuge, to maintain the elk. The program was designed for approximately 7500 elk. In recent years the number of animals wintering on the refuge has been well in excess of 10,000 animals. This expands feeding beyond proposed agency budgets (Figure 3.4).

The removal of predators and reduction of hunting in National Parks have acted in combination with a decreased size of winter range to affect the elk population. Normal predator–prey cycles no longer exist, thus human management techniques (winter feeding) must be instituted to maintain the elk. In all likelihood the total population would be smaller if predators existed. The proposed reintroduction of wolves in Yellowstone will have some impact on this elk population. However, the wolves will not likely attack the elk on the refuge.

Another example of people's interference with the natural system is the introduction of the sea lamprey into the Great Lakes. The lampreys, which were common in the St. Lawrence River, moved into the Great Lakes when the Welland Canal was built, in 1829, to bypass Niagara Falls. In the lakes they virtually exterminated the lake trout—a major fisheries resource. Efforts to eliminate the lampreys have been successful in many areas; thus, the lake trout population has been increasing. In addition, salmon from the Pacific Northwest have been stocked in the Great Lakes and a very successful fishery has been established.

Hunting (Chapter 10) is also a form of predation. Controlled hunts can serve to replace other mortality factors and remove excessive animals from a population. Some populations, particularly big game species, are maintained for hunters as most of their natural predators have been removed. If hunting were to stop, die-offs would occur because of lack of food and shelter. The population would probably reach a point of stability at numbers lower than current levels.

Some people today are questioning hunting as a form of wildlife management. They feel that it is not ethically right for people to kill or harm wild animals. Some states like California have nonhunting coalitions larger than the number of hunters.

Control of excess animals by shooting has also been the subject of legal action. Bison removal in the Grand Teton and Yellowstone National Parks has been the subject of much debate. Park officials feel the bison population has exceeded the land's carrying capacity. Cattlemen claim bison pass disease, particularly brucellosis, to cattle. Antihunting groups, on the other hand, feel that bison should be a self-regulating population in the area.

Parasitism A special case of predation is **parasitism.** The parasite (a predator) is smaller than its host (the prey) and does not destroy the host. In parasitism, the tissue or food supply of the host is used by a parasite for survival. Such interactions are part of a balanced ecosystem. From an evolutionary viewpoint, any parasite that destroys its prey also destroys itself.

A number of diseases caused by parasites, including malaria, schistosomiasis, and dysentery, are responsible for a tremendous amount of human suffering in the world today. In their manipulation of the environment, people sometimes alter the natural balance of these organisms. Cattle, for example, are a reservoir for the parasite causing sleeping sickness that is transmitted by tsetse flies to people. Suffering can be reduced by managing environmental factors upon which parasites are dependent.

Other Interactions Several other forms of population interaction affect the growth and survival of populations. When two populations interact so that both benefit, the interaction is called **mutualism.** A good example of this is the relationship between nitrogen-fixing bacteria and their leguminous hosts, which we discussed in Chapter 2.

Commensalism is a relationship that benefits one population and has no effect on the other. For example, cattle egrets and cowbirds follow large grazing animals around. As the herbivores pull up grass and cause disturbances with their hooves, many insects are dislodged. Thus, an easy meal is provided for the birds, but there is no benefit for the herbivores.

Population Cycles

The cyclic nature of population growth often involves predator–prey interaction. The lemmings of the arctic region are prey for the snowy owl. The lemming populations follow a three- to five-year cycle, with the owls lagging a year behind. When the lemming population increases in size to the extent that it destroys its own food and shelter, the population declines rapidly during a cold winter. The snowy owl population, which increases in response to a large lemming population, is then left without a source of food, causing many owls to starve or fly south in search of food.

Predator–prey interactions are but one component of the study of population cycles; regulation of population density is another interesting aspect. Some ecologists believe that populations regulate themselves by *density-dependent factors.* In effect, they say that more deaths occur at higher population densities because predation, disease, and food shortage can act more severely when numbers are high. Food shortage appears to be the chief natural factor limiting any animal, such as the snowy owl. Another group of ecologists believes that the effect of any mortality factor depends on its severity and the population's susceptilibity to it. This group accepts the idea that populations are regulated by *density-independent factors*, so regardless of the population size— 100 or 10,000 —a given perturbation will result in killing the same percentage of the population.

Activity Patterns

The **activity patterns** of animals play a dynamic part in any community. Each species is active during a specific part of a 24-hour period *(temporal activity)*. Fish are feeding just after sunrise and just before sunset. Around midday, they are more likely to be found exploring the lake bottom. In the spring just after sunrise, the woods are alive with songbirds. Owls increase their activity at sunset, primarily because they feed on small mammals that are active then. As can be seen by these examples, community studies cannot be confined to one time period or they are likely to produce biased information.

Some animals use parts of their habitat on a weekly, monthly, or seasonal basis. At times large mammals move repeatedly through the same vast area **(home range),** always staying within its confines. For example, a tiger's home range in Asia is so large that it takes several days for one animal to go from boundary to boundary. Yet songbirds might nest and defend a specific site in a forest for several weeks, then move in a nomadic fashion throughout the forest during other periods.

Migration, movement from one community to another on a seasonal basis, is another pattern exhibited by some animals. The arctic tern is a champion "globe trotter" and long distance migrator. It breeds near the North Pole and winters in the Antarctic, 7700 kilometers (11,000 miles) away. North American elk migrate each spring from low mountain meadows to high alpine habitat; while in winter antelope move from open grassland to the foot of mountains because the snow is not as deep. Along the west coast, whales present a spectacular show as they travel from one ocean community to another.

POPULATION DISTRIBUTION

A population must have all of its essential needs met in order to survive in any location. The **habitat** is the area where all the needs of a population are satisfied and where we would find a member of that population. Although the habitat is the physical environment of the species, the species might not be present in all parts of its habitat at all times. Buffalo habitat, for example, extends throughout most of the mountain states, central plains, and eastward to New York, but buffalo are no longer found in the greater part of this area.

To determine the habitat we must consider the physiological ability of the population to withstand physical factors of the environment. Essentially, the required heat, mositure, light, and nutrients are available in the habitat; but populations in different habitats are affected by different physical factors or combinations of factors, as described earlier. Figure 3.5 shows how

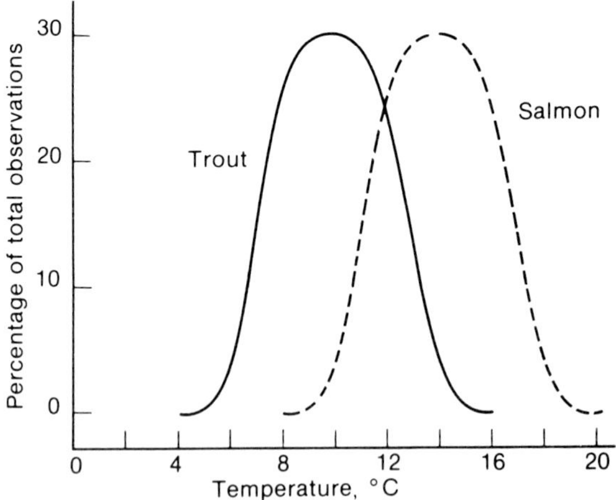

Figure 3.5
Different temperature ranges for speckled trout and Atlantic salmon. (From K. C. Fisher and P. F. Elson, "The Selected Temperature of Atlantic Salmon and Speckled Trout and the Effect of Temperature on the Response of an Electric Stimulus," *Physiological Zoology 23* [1950]: 27–34. Copyright 1950 by the University of Chicago Press.)

two populations of fish respond to temperature. Most brook trout are found in waters around 11°C (52°F). However, the trout can live in waters from 2° to 18°C (35°–64°F), indicating that this range is their **zone of tolerance.** Waters below 6°C (42°F) and above 14°C (57°F) are **zones of stress,** where the trout can survive for a time but apparently cannot reproduce. Atlantic salmon, on the other hand, live in waters from 6° to 20°C (43°–68°F) but are most abundant at 14°C (57°F).

The absence of a population in a suitable habitat could mean that it has failed to reach the area. Sometimes people help transport species to new areas, as in the case of the European starling. Originally found in Eurasia from the Mediterranean north to Norway and east to Siberia, the European starling is now well established throughout the United States. This happened because eighty birds were released in New York City's Central Park in 1890 and eighty more in 1891. People are also responsible for introducing to new areas a number of disease-causing pest organisms, such as black rats and chestnut blight. In Chapter 14 we discuss attempts to control some of these pest species by biological means.

While the habitat is the physical area where all the needs of a population are satisfied, a population's **niche** is its functional role in the habitat—what it does in a particular area. Is the population a producer, herbivore, or carnivore? What link does it occupy in the food chain? How does it fit into the mineral cycle? Some species occupy broad ecological niches by feeding on many kinds of food. Others have a very specialized niche and feed on only one or a few food types.

Because species arrive at their respective niches through long periods of evolution, no two species in the same community have the same niche. Furthermore, quite unrelated species can occupy similar niches. For example, kangaroos, bison, and cows are all grass eaters. They occupy similar, but not identical, niches in different ecosystems. When a population is forced to leave a particular habitat, its niche is taken over by another. The loss of large grazing animals might result in more insects that feed on herbaceous material. Also when a population enters a habitat, it can take part of the niches of others. The cattle egret most likely encroaches on the niches of some small mammals, insects, decomposers, and others. The habitat is compared to a person's address; the niche to a person's profession or occupation.

Environmental alterations can result from physical changes, such as a variation in the amount of light, moisture, or temperature, or from the appearance of a new population that is more efficient in using the habitat. When the environment changes so that it is no longer favorable for a particular population, the population may move to a suitable environment, adapt to the new conditions, or become extinct. The option of moving to a new habitat is not always open, as dispersal avenues are frequently limited. However, the extinction of species has occurred millions of times during evolutionary history. In fact, less than 0.01 percent of all known species are alive today. Consider the Age of Reptiles, when those cold-blooded animals occupied many of the earth's niches.

When communities are altered, habitats necessary for the survival of some populations are destroyed. For example, many wild animals in the Everglades are in danger of extinction because of water control projects undertaken in the Lake Okeechobee area. Water from the latter normally flows southward into a vast expanse of marshlike area sometimes called the River of Grass. The annual periods of high water during the winter months provide abundant food, enabling many birds to breed. As the water flow subsides, other species utilize the habitat. When the water does not flow, holes dug by alligators hold pools of water as the allocation of water becomes a concern for wildlife (Figure 3.6).

In other cases, people reduce natural populations to such low numbers that males are unable to find mates. This situation is now occurring

Figure 3.6
Alligator hole in Everglades National Park. This source of water sustains many wildlife species during dry seasons.

for some of the large animals, such as mountain lions and whales.

To maintain natural communities, it is critical for us to combine comprehension of the biology of natural populations with effective planning of changes in natural systems. We now understand that species in natural communities evolve over millions of years in response to environmental conditions and that community succession occurs over hundreds of years. Rapid alteration of these conditions reduces the time available for the populations to respond, causing many to become extinct.

Population Genetics

Genetics is the study of the biological transmission of characteristics from one generation to the next. Not all the characteristics a particular individual exhibits, however, result from transmission of biological information from the parents. Some characteristics are the result of the individual's interaction with the environment. Because characteristics are generally influenced by both genetic inheritance and the environment, it is appropriate to discuss genetics in an environmental science book.

Genetic Makeup

An individual's genetic makeup is the **genotype,** which includes all the genetic information passed down from both parents. Individual appearance is referred to as **phenotype.** The genetic makeup or genotype sets the limit to which an individual can respond. This limit can be thought of as a scale. Within that scale, different responses or appearances, the phenotype, are possible depending on different environmental characteristics. Thus, in humans, a difference in diet can result in different phenotypical expressions. Skin color is also partially determined by the environment. Exposure to the sun generally darkens human skin, but a limit based on the individual's genotype determines the degree to which the skin can darken. Pigment beneath the skin may be quite limited in some fair-skinned individuals and they may not darken at all, but may burn quickly. On the other hand, eye color is influenced entirely by genetics.

Chemistry of Inheritance

How is genetic material passed from parent to offspring in each generation? Recent studies clearly show how information moves from generation to generation via chemical structures in the body. This revolutionary approach in biology is one of the major discoveries of the twentieth century and, as we see throughout this book, will be the springboard for major changes in our ability to develop new sources of food, to respond to diseases, and to control the environment.

An individual's chemical units of inheritance are **genes.** All of the individual's genes constitute its genotype. Genes combine in most plant and animal cells on long rodlike structures called **chromosomes,** long molecules of deoxyribonucleic acid (DNA) associated with a number of proteins. The DNA and proteins are organized in a specific way, consisting of a double helix which coils around continuously and is connected by other chemical elements (Figure 3.7).

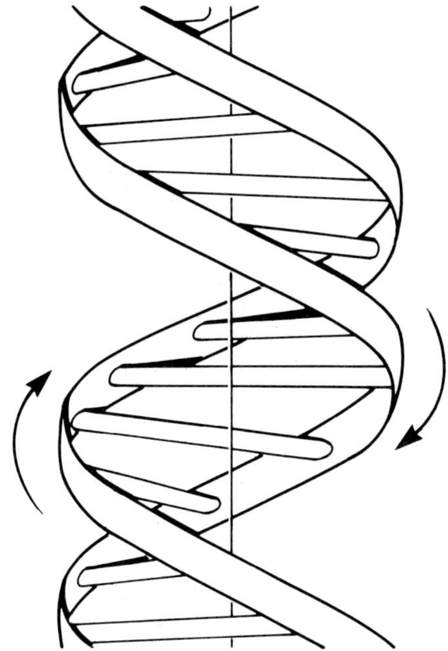

Figure 3.7
Double helix DNA.

At least one complete set of chromosomes is found in all cells of most plants and animals. The number of chromosomes varies from one species to another. Some plants have only 4 chromosomes, and there are animals with as many as 500. The number of chromosomes is usually, but not always, consistent within a species and is not correlated with either the size of the individual, the size of cell, or the evolutionary advancement of the species. Chromosomes generally appear in **homologous pairs** that align together during cell division. Each chromosome within the pair has identical loci with genes that control the same characteristics. The chromosomes are duplicated as the cell divides and an identical set goes to each new cell. Cell division resulting in the reproductive cells—egg and sperm—varies from cell division in nonreproductive cells. The reproductive cell contains only one member from each homologous pair of chromosomes in a body cell. During sexual reproduction, the reproductive cells unite to form the basic cell of a new individual, a **zygote** (Figure 3.8). A zygote therefore contains one set of chromosomes from each parent. These chromosomes then unite or pair up in the zygote, and cell division begins as the new individual develops with a complete set of chromosomes in each of its cells. This resulting process means that not all a parent's individual characteristics are passed on to its offspring—half are. Different forms of reproduction occur in some organisms.

Specific sites on chromosomes are occupied by genes. A site is generally called a **locus** (plural *loci*). Geneticists have been able to determine the location of some individual genes on chromosomes and have, in some organisms, actually mapped gene loci. In the tomato plant, for example, we know the location of 247 genes. Some genes are known to control specific characteristics or expressions of an organism. As geneticists continue to investigate the location of different genes and their controlling factors in humans and animals and plants, they develop some ability to control the destiny of populations.

Gene Pool

Members of a species are often distributed over a wide geographic area, but because they can

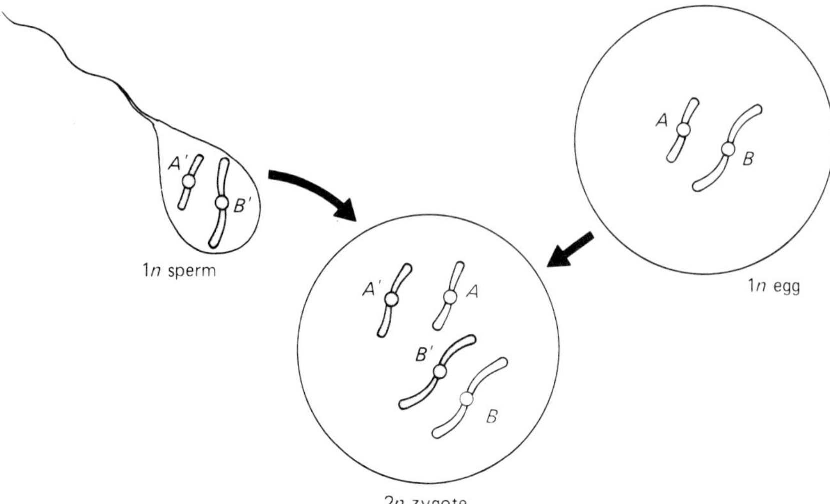

Figure 3.8
Formation of a zygote by the union of egg and sperm.

interbreed, they are considered members of the same species. As a result the species can be composed of many local populations. An individual's genetic makeup is often different from another's in the population. A local population can have a different combination of genetic material from another population, but both populations are the same species. Within a population, the total genetic information carried by all interbreeding members is the **gene pool.** The gene pool is a population, not an individual, characteristic. Where a population's genetic makeup is diverse, an individual may have only a small representation of the total gene pool.

An individual gene can have a number of chemical forms that cause different expressions of the characteristics the gene influences. Each state of a gene is an **allele.** There may be one, two, or a number of alleles for each gene in the population's gene pool. Obviously, if there is only one allele, the expression controlled by that gene should be the same in each individual in the population unless some factor from other genes or the environment overrides the expression of the characteristic.

Population genetics focuses on the total group or population rather than on an individual. We will examine different genes present (allelic) and genotypic frequency from one generation to the next in a population rather in individual's. The idea is to determine how gene frequency affects the way a population lives and evolves. Rapidly growing populations are occupying many new areas or concentrating heavily in a particular habitat. It is possible that these populations have a different gene frequency from populations that decline because of major changes in the habitat. Likewise, the environment may change in such a way that a population's genetic makeup allows only certain members to survive. When this occurs, gene frequency can change considerably.

As an example, we will use MN blood groups in humans. Most people are familiar with the ABO blood types. In the case of the MN blood type, a locus on a human chromosome can have one of two gene alleles, M or N. Each controls the production of an antigen on the surface of the red blood cell which can be tested and indicated by chemical means. Since chromosomes are found in pairs, each individual will have two alleles. An individual's blood type can thus be MM, NN, or MN. If parents are both MM or NN, all their children will be MM or NN also. Can you calculate the probability of each child's

Table 3.1
Frequency of M and N Alleles in Various Populations

Population	Genotype Frequency (%)			Allele Frequency	
	MM	MN	NN	M	N
U.S. whites	29.16	49.38	21.26	0.540	0.460
U.S. blacks	28.42	49.64	21.94	0.532	0.468
U.S. Indians	60.00	35.12	4.88	0.776	0.224
Eskimos (Greenland)	83.48	15.64	0.88	0.913	0.087
Ainus (Japan)	17.86	50.20	31.94	0.430	0.570
Aborigines (Australia)	3.00	29.60	67.40	0.178	0.822

From Klug and Cummings, *Concepts of Genetics*, 2nd ed. 1986. Merrill, p. 600.

blood type if both parents are MN? Genotype frequency in different individuals varies with populations, both in a country and in the world (Table 3.1). This example shows how genetic makeup controls a particular trait, in this case, the MN blood type.

Dominant/Recessive Genes

Even though two alleles of a gene might be present, the phenotypical expression of each may not be the same. In some cases one allele masks another; the gene that masks the other is **dominant.** As an example, each parent in the human population passes on to a child one gene for eye color. The child has two sets of genes. If one gene is the allele for brown eyes (B) and the other is the allele for blue eyes (b), the child's genotype is Bb. The child's eye color will be brown, because the brown allele is dominant over the blue allele. Thus, a child with a genotype BB will have brown eyes, with Bb will have brown eyes, and one with bb will have blue eyes. This shows us that brown-eyed individuals can produce blue-eyed offspring; however, when both parents are blue-eyed, all their offspring will most likely be blue-eyed. Individuals in whom alleles of the same gene are identical are **homozygous** for that allele (Figure 3.9); those in whom the alleles differ are **heterozygous** for the gene. Thus in the case of eye color, BB and bb are homozygous and Bb is heterozygous; in the case of blood type, MM and NN are homozygous and MN is heterozygous. We can see that population characteristics may exist, but may be masked by a dominance characteristic. Further complicating factors result in masking traits that may exist when the **multiple gene** effect is considered. Many of an individual's characteristics including eye color are based on a number of genes, and it is possible for one gene to mask the effect of another.

Sex-Linked Characteristics

Some genetic material passed from parents to offspring is associated with one sex or the other. In some animals, one set of chromosomes is re-

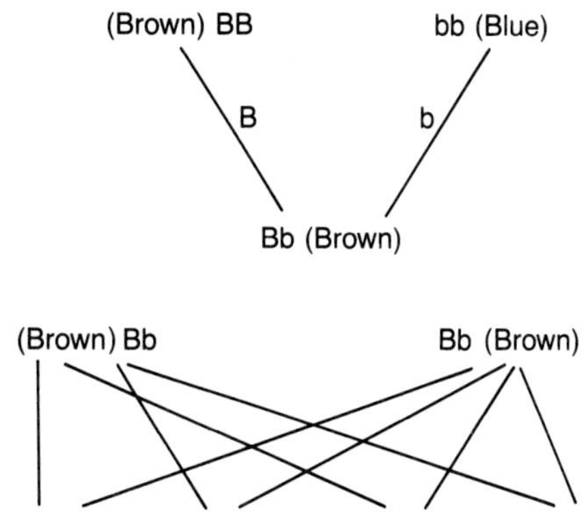

Figure 3.9
Genetic makeup of blue-eyed and brown-eyed people.

ferred to as the **sex chromosomes.** Women have two female chromosomes in this pair, and males have one male and one female chromosome. The passing of sex chromosomes to offspring is similar to the division of genetic material. Each male has one male chromosome designated Y and one female chromosome designated X, and each female has two X chromosomes. When the egg and sperm cells form, chromosomes divide, and each egg cell contains one X chromosome from the sex chromosome pair, while each sperm contains either an X or Y chromosome. When the egg and sperm combine, a male or female is formed (Figure 3.10). Some genes are associated with these sex chromosomes, and the characteristics these genes control are **sex-linked.** These characteristics cannot be determined based on normal distribution. One sex-linked trait in the human population is a form of color-blindness. The gene for this condition is carried on the X chromosome, and the allele for color-blindness is recessive to an allele for normal vision. If a male receives an X chromosome from his mother that has an allele for color-blindness, he will be color-blind because the Y chromosome does not carry an allele for normal vision. Because a female has two X chromosomes, she may be heterozygous for the allele—she would have normal vision but be a carrier of the allele for color-blindness. A female with two X chromosomes that contain the allele for color-blindness would be color-blind (Figure 3.11). Other characteristics influenced by sex-linked genes include pattern baldness (Figure 3.12), hemophilia (a condition in which normal blood clotting does not occur), and a type of hair tufts on the ears.

Figure 3.10
Sex chromosomes in the fruit fly.

The Hardy–Weinberg Law

Population biologists use the **Hardy–Weinberg Law** to determine a population's genetic state. To apply this simple mathematical law, we must evaluate the total population. The Hardy–Weinberg Law assumes that

- The population is infinitely large or large enough that there will not be unusual sampling errors
- Mating within the population occurs randomly
- There is no selective advantage for any particular genotype (all genotypes are produced by random mating and are equally viable)
- There is an absence of other factors that might alter the genotype (such as changes that occur in the genes or the exodus of individuals in the population with all of certain genetic characteristics).

Let us apply the Hardy–Weinberg Law to the case of blue-eyed individuals. Let p represent the frequency of the dominant allele B and q the frequency of the recessive allele b in the population. When we add the frequency of all the Bs and all the bs, we get 1; in other words, $p + q = 1$. We can express the results of the population randomly mating as number of individuals with BB $= p^2$; number of individuals with Bb $= 2pq$; and number of individuals with bb $= q^2$. Adding together, $p^2 + 2pq + q^2 = 1$. This expression can be used to predict the genotype frequency in the next generation. Consider that the individuals in the first generation contribute to a gene pool that will form the zygotes of the next generation. Remember that the frequencies of B and b are p and q, respectively.

The probability of one of the reproductive cells that carries B uniting with another carrying the same allele to form a BB offspring is the

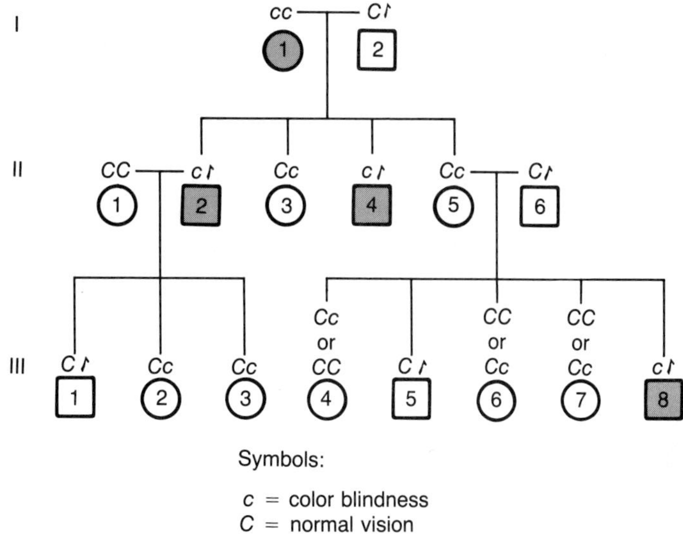

Figure 3.11
Transmission of color-blindness on sex-linked chromosomes.

product of the separate frequencies, or $p \times p = p^2$. Therefore, the probability of a sperm's uniting with an egg carrying b is $p \times q = pq$. Since heterozygotes may also be formed by an egg carrying B and a sperm carrying b with the probability $q \times p$, the total frequency of the heterozygotes will be $pq + qp = 2pq$. Calculation of the frequency of the genotype bb is $q \times q = q^2$. If 70 percent of the alleles in a population are B and 30 percent are b, then $p = 0.7$ and $q = 0.3$.

The genotype frequency in the second generation thus depends only on the allele frequency of the parents and not on the genotype frequency in the population. The genotype frequency established in the offspring following the single generation of random mating are the expected Hardy–Weinberg equilibrium frequencies, which will remain the same throughout future generations as long as the four assumptions of the Hardy–Weinberg equilibrium hold true. We can assume in population genetics that if there are no changes, the population will continue to remain the same because the gene frequency will remain the same. We know, however, that this is not true. Mating may not be random, and a number of things can occur to alter the total genetic makeup or the genetic equilibrium for a particular gene in the population. This is the basis of changes that occur in the population and therefore an important concept in our discussion of population genetics.

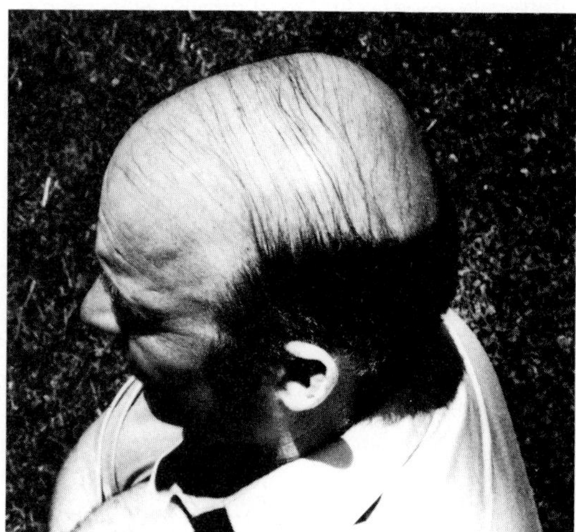

Figure 3.12
Pattern baldness is transmitted on sex-linked chromosomes.

Mutations

Because of the number of possible genetic combinations, the members of a population at any given time represent only a small fraction of all possible genotypes. Changes can occur as the gene pool is reshuffled each generation to produce new combinations in the genotype of the offspring. No new alleles are produced in this process, only *recombinations.* The production of a new allele can occur, however, and is referred to as a **mutation.** Mutations actually increase the genetic variation in the population. Mutations occur as the result of a change in one of the chemicals at the gene or large changes in the chromosome. Changes at the individual gene loci are called **point mutations.** Spontaneous mutations can occur.

Some genes and chromosomes seem to be more susceptible to mutations than others. Massive changes in the environment, on the other hand, have accelerated the mutation rate in some species. Specific chemicals found in air, water, and soil also cause some mutations. The use of radioactive chemicals has increased the mutation rate in some organisms. In a number of cases, mutations have proven deleterious or harmful to the organism. In Chapter 15 we will examine the impact of radiation on humans that resulted from atomic bomb testing. Many people believe that mutations resulting from radiation produce cancer or other diseases in humans.

Migration

Different mutation rates in different subpopulations or populations of the species can create allelic frequencies in those populations. As a result, environmental pressures may allow different characteristics to appear in different populations. Migration of genetic material in the population occurs as individuals move between the two populations and intermix. Migration can thus be regarded as the flow of genes between two populations that were once geographically isolated (Figure 3.13).

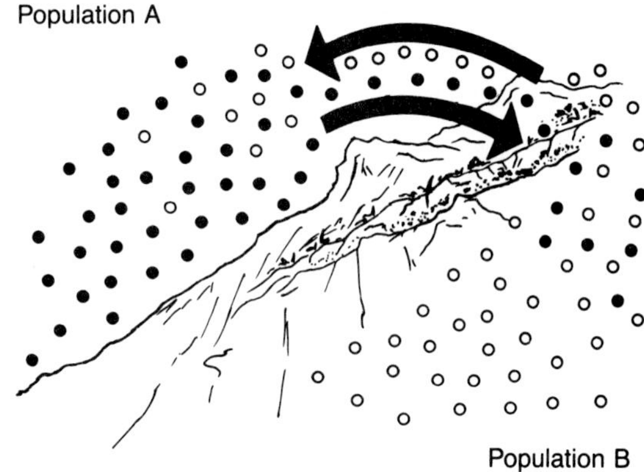

Figure 3.13
When geographic barriers break down, gene flow between two formerly isolated populations may occur.

Inbreeding

Isolation of populations can create another phenomenon that would cause a variation in the Hardy–Weinberg frequency, referred to as inbreeding. When a small group of individuals isolate themselves and mate with individuals locally rather than on a random basis, serious inbreeding occurs. In other words, individuals in small populations may have homozygous characteristics for certain forms. As these individuals continue to mate, any form of heterozygousness may disappear. The potential for genetic variability therefore decreases. Inbreeding results from isolation caused by geographic barriers, religious barriers, and behavioral barriers. Inbreeding is of great concern when we are dealing with endangered species.

EVOLUTION

Selection

Mutations and migration introduce new alleles into a population. Natural selection, on the other hand, is the force that shifts gene frequency within a population and is therefore the driving factor in evolution.

CROWN OF THORNS: BALANCE IN A NATURAL SYSTEM

In the 1960s and early 1970s, a coral-eating starfish species called the Crown of Thorns invaded areas of the Pacific and turned patches of tropical reef into water moonscapes. Major efforts were undertaken to eradicate this starfish, considered to be the ocean's number one enemy.

Recent research indicates that starfish are a cyclic invader of the ocean ecosystem and serve as a sort of homeostatic mechanism. Marine biologists contend that the starfish predator is necessary to maintain the ecosystem's high species diversity. The demand for space to live and grow results in a bigger, stronger species of coral, thereby monopolizing a great deal of the area. Consequently, small or rare species cannot get a start; or if they do begin to breed, they get pushed out by other coral. When the starfish population increases, its members clear off some of the established coral thus allowing the number of rare species to multiply.

Scientists think that the normal starfish population density on a balanced reef is about six to twelve per square mile. The population begins to increase, however, every eight to ten years causing an infestation by thousands of starfish that eat their way across an average coral reef at the rate of approximately one-half of a mile per month. As a result, the entire area turns into a pile of rubble. Fish and other reef dwellers quickly leave the area, creating concern among people who depend on the reef for food and income. Nevertheless, coral reef communities need to be renewed periodically to sustain healthy surroundings. Natural systems maintain this balance through the interplay of starfish predation on coral, allowing coral species to compete and thereby produce a balanced environment.

When a particular genotype/phenotype confers an advantage to an organism in competition with others that have a different combination, *selection* occurs. This means that a larger number of offspring of the organism with the advantage will survive. The relative strength of the particular selection varies with the degree of the advantage. The probability that a particular phenotype will survive and leave offspring is a measure of its **fitness,** which refers to the total reproductive potential or efficiency. Fitness is usually expressed in relative terms by comparing a particular genotype/phenotype combination to one regarded as optimal. Fitness is a relative concept because as environmental changes occur, so do the advantages conferred by a particular phenotype. Any gain in fitness by one unit of selection is generally balanced by losses in fitness of others.

Adaptations

Some members of a population develop characteristics, or **adaptations,** that make them better suited to their environment. Human ancestors (small, rodentlike creatures known as prosimians) adapted to the extensive forests which stretched over the earth about 70 million years ago and evolved basic structures that set the limits to human performance today. Sense organs (including binocular vision), flexible, grasping organs at the end of limbs, changes in the reproductive system, and an enlarged brain for motor coordination evolved during that period. After a time, these early ancestors began to wander out into the savannalike grassland areas bordering forests on the African continent. Anthropologists speculate that long periods of drought reduced the types of forests over millions of years, giving advantages to those individuals who could adapt to the more open savanna.

Adaptations that made early humans better able to live on the savanna included upright posture, rapid movement on two legs, use of hands for grasping material, loss of hair through increased activity, and a dark skin pigment to prevent harmful effects from the sun. Sexual promiscuity was probably common in the tree dwellers, with bands of individuals moving from place to place. Life on the savanna changed this pattern. Pair bond formation, where a single male established a relationship with a single female, was an advantage in that the female raised the young and the male provided food and protection.

As hunting became a way of life, offspring required more training and therefore a longer period of dependency on adults. A larger cranial cavity developed, but was limited by the size of the mother's birth canal. Thus, more development and growth occurred after birth. During this period of change, the brain size increased greatly. Individuals who were intellectually more adept, able to recall information, and learn by observation had a selective advantage. Furthermore, offspring who received survival information from their parents and other adults had a greater chance of survival.

A more recent example of human adaptation is found in certain areas of central Africa. There the inhabitants have been exposed to years of malaria. During the reproductive cycle of the malaria parasite in the victim's bloodstream, red blood cells are destroyed. However, there are persons who have some red cells deformed in the shape of a sickle (Figure 3.14). Apparently, those individuals are more resistant to the malaria parasite. Over the years they have become dominant in the area, so this trait was passed on to their descendants. While this special trait served the people well who were exposed to malaria, the same sickle cell does not function well in carrying oxygen throughout the body. Thus, under stress situations which require more oxygen, those with this adaptation are vulnerable and in danger of dying from the sickle cell trait as well as from anemia and complications.

Each adaptation making individuals in a population better suited to their environment allows them to leave more offspring and so contribute more genetic material to future generations. Random genetic forms appear regularly, but the

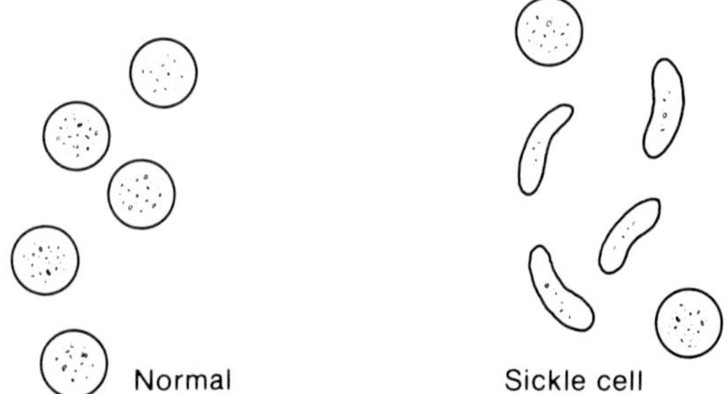

Figure 3.14
Schematic diagram of red blood cells affected by sickle-cell anemia.

survival of the new individual depends on how well it can adapt to its environment. The dominance of a new genetic form as a result of environmental change is called **natural selection.** A change in the genetic makeup of a population over time through natural selection results in **evolution.**

Evolution commonly occurs over hundreds or millions of years, but microevolution can be a very rapid process. In a laboratory situation, if most of a population of bacteria is destroyed by irradiation, the survivors reproduce and form a new colony within a few days. The housefly population reacts to the insecticide DDT in such a manner. Those flies unable to survive in the presence of DDT die. The few flies unaffected by DDT reproduce rapidly, and in a period of a very few years, the housefly population is composed primarily of the immune individuals.

Speciation

In an environment that is changing, new genotype/phenotype frequencies may become dominant. Changes such as habitat alterations, earthquakes, and volcanic eruptions isolate some individuals in population from others. Currently, parks, reserves, and refuge populations are becoming increasingly isolated from other populations by habitat modification. Population isolation can occur both geographically and behaviorally.

A population's gene pool can then change through mutation and selection. As this change takes place over a long period of time, speciation can occur—that is, new species are formed from the original population. The members of the original species can interbreed within their own groups, but if the established barrier is broken down, they can no longer breed with former members of their population. Each of the new species has characteristics that make them unique and able to live in their respective environments. As a result, the world has birds that have conquered the air, whales that use the ocean, worms that burrow, insects that eat wood, and so on. Each isolated land mass shows some of the uniqueness of evolution—Australia with its marsupials, for example, was apparently isolated millions of years ago from Asia. Speciation has resulted in a diverse array of genetic material, so that the living components of ecosystems can better interact and withstand stress factors.

Actually, isolation rarely produces new species; it may make a population susceptible to predators, diseases, or forms of competition that impede population growth.

Hybrids

Behavioral differences in individuals can also create isolation. The courtship pattern of some waterfowl species isolates them from other species. When the courtship pattern is broken down, these populations can interbreed. When two populations that were considered different species interbreed, the resulting offspring is **hybrid.** In waterfowl species, mallards can form hybrids in the wild with eight species of ducks. Many of the ducks in parks began as hybrids between mallards and other species. Hybrids can sometimes breed back with members of either of the original populations, or they sometimes cannot breed at all. If hybrids can breed back to members of the original population, the flow of genetic material can occur between two populations.

If hybrids can breed between each other, they may or may not be able to form a new species. In an effort to save the gene pool of the endangered dusky seaside sparrow which is now only males, sperm are being collected and artificially bred with eggs of closely related sparrows. Translocation of eggs or hatchlings of sea turtles has resulted in hybridization of different populations. Many plants used in agriculture are hybrids that combine desirable food characteristics (Figure 3.15).

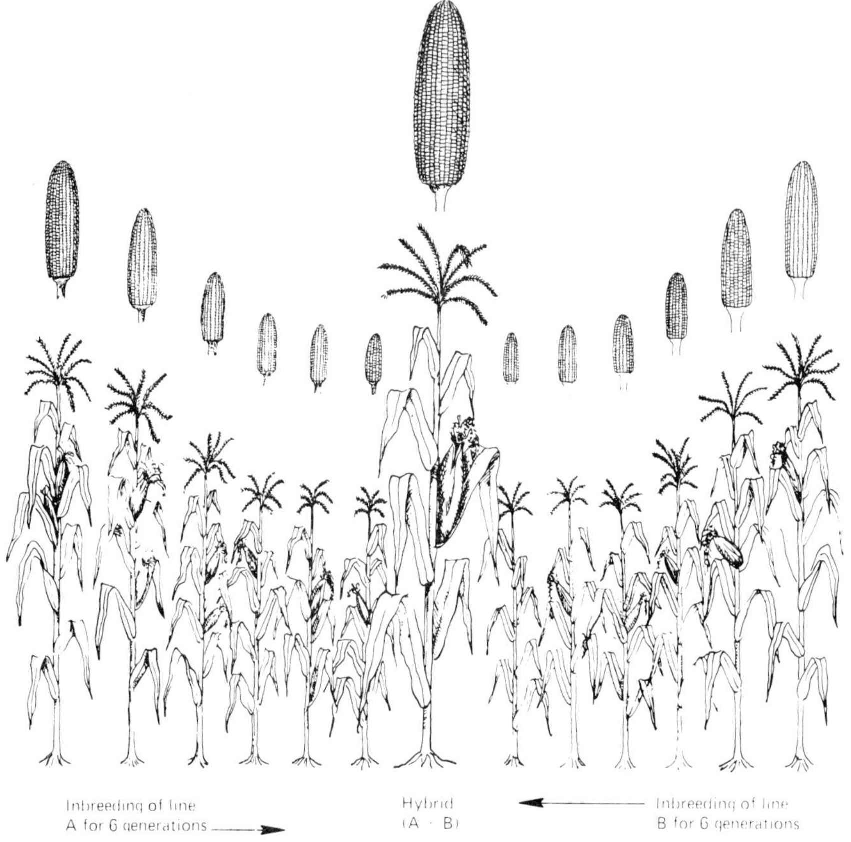

Figure 3.15
Inbred and hybrid maize, *Zea mays* (diagrammatic). The production of a hybrid from two inbred lines is illustrated. (After H. G. Baker, 1970, *Plants and civilization*, 2nd ed. Belmont, Calif.: Wadsworth Publishing Co.)

GENETIC ENGINEERING

Genetic Counseling

Manipulation of genetic material has been undertaken for many years with food crops (Chapter 12). In the last twenty years many changes have been made to improve human health. Advances in procedures for detecting genetic diseases have allowed genetic counseling. Through genetic counseling, prospective parents can be informed about the probability of defects in children they may have. Other parents may want to know their chances of having more children with genetic diseases if they have one. In very high risk cases, parents may choose not to have children or they may adopt.

Tay-Sachs disease is caused by recessive genes. These genes prevent the production of an enzyme that processes some faulty substance. As a result, fatty material accumulates and chokes off passage of nerve impulses. An individual who has both a dominant and a recessive gene for production of the enzyme is normal. If two individuals who are heterozygous for this gene marry, they could have a child (one in four chances) with Tay-Sachs disease. Thus, genetic counselors might inform couples whose families have a traceable history of the disease what their chances would be of having children with the disease.

Today we know that some factors affect a developing embryo, causing it to be defective.

These factors are called *teratogenic agents*. Agents operate in different ways during fetal development.

German measles (rubella) is one agent. When it occurs during pregnancy, particularly during the first trimester, it causes defective senses or organ formation such as poor hearing, poor sight, or defective hearts.

Drugs such as thalidomide can cause abnormal extremities. Some narcotics, alcohol, and smoking are thought to influence developing embryos. Ionizing radiation and some chemicals that are part of air pollution can cause defects.

We know many other agents that can cause abnormal development. Mothers can now be advised of these factors, thereby reducing the probability of abnormal children. Genetic counseling can be very valuable to prospective parents. They can help people have a healthy pregnancy.

Today some genetic diseases can be detected by *amniocentesis*, the process of removing some amniotic fluid from around the fetus by means of a slender needle inserted through the mother's abdomen. This fluid is evaluated to determine if abnormal chemicals are present that might be indicators of an abnormal child.

Some people feel that the human population can be improved by genetic counseling and engineering. These people are part of an *eugenics* movement. Extreme members of these groups feel that forced sterilization should be performed on people with abnormal traits or poor intelligence.

Genetic Alteration

When actual alteration of genetic material at the molecular level occurs, we have *genetic engineering*. Genetic engineering began when people perfected methods to grow tissues outside the body cells. Cells can grow under artificial conditions when the proper nutrients and temperature are provided in a sterile environment.

Today we are moving into an era when some of our dreaded diseases and birth defects may be controlled by genetic engineering. Hemophilia is a disease that affects about one male in 10,000. It is caused by sex-linked genes in which the male receives a recessive gene from his mother and is linked to the male sex chromosome. This gene fails to produce a protein that causes blood clotting. Now researchers are finding that they can insert the human gene that produces the protein (called factor VIII) into animal tissue, which then produces the protein in large quantities. The protein can be purified and used in humans that have undetected hemophilia.

Many other research efforts are underway in genetic engineering related to human health, cattle production, and plant propagation. When we learn more about the genetic system we are likely to find more traits that can be manipulated by genetic engineering.

Many of the new developments in genetic engineering present legal implications. As you read and listen to the news, think about how these changes can be used to alter the evolution of people and our environment.

EXTINCTION

A population that no longer exists is **extinct.** Many animals have become extinct over the last hundred years, including the dodo birds, passenger pigeons, and the Carolina parakeet.

Extinction can occur in the species or in a population. Often, environment changes in such a way that a particular population's genetic pool can no longer adapt and therefore produce offspring. This form of extinction has probably occurred many times in history. While climate, habitat change, and other factors can cause local extinctions, a chronic presence of low-level pesticides or major environmental changes can cause declines throughout a species' range. Many biologists feel that dinosaurs became extinct because of changes in climatic conditions that prevented them from reproducing. Generally, however, people have probably been the major factor of extinction in the last 200 to 300 years. By introducing different forms of predators that can either destroy or effectively compete with a population, people have eliminated a number of species. Habitat destruction has probably been the major cause of extinction to populations.

One example of human hunting that has caused the extinction of a species is the case of the passenger pigeon. These birds typically foraged in large flocks throughout the midwest, and the large flocks facilitated reproduction. Humans exterminated the birds faster than they reproduced, so the populations declined and the large social flocks could not form. As a result, the population of passenger pigeons rapidly became extinct.

Human impact on species is often so sudden that natural systems are unable to buffer themselves. Overharvesting and thus reducing the gene pool prevents species from adapting to changing environmental conditions. Habitat destruction means there are fewer areas for species to live. At the same time, human interference inhibits speciation. The loss of the species or a large portion of a species gene pool produces a more homogeneous environment, with the result that the evolutionary process is slowed. Populations that have limited genetic makeup are often unable to adapt and thus unable to survive as environmental conditions change.

Some people point out that extinction is not really a problem today because large groups of animals became extinct in the past. Using the dinosaurs (Figure 3.16), they point out that life changed and flourished following that loss some 65 million years ago. This is partly true; however, the dinosaurs' genetic materials were limited. Other organisms survived. The current loss of such a large amount of genetic diversity could have more profound affects on human survival. Biologists believe the current extinction rate is much higher than in the past.

Conservation efforts throughout the world are now directed at preventing extinction of many species, and habitat management seems to play a key role. Habitats have changed so that they no longer support the species, or the species can no longer survive in relation to changing conditions, such as the influx of human pop-

Figure 3.16
Dinosaurs became extinct as the environment changed. (Peabody Museum of Natural History, Yale University.)

LEGISLATION

Animal Welfare Act (1985). This act provides guidelines for proper and humane methods of handling and disposing of animals used in laboratory and field research. The Animal and Plant Inspection Service (APHIS) of the U.S. Department of Agriculture is responsible for enforcing the act through the Regulatory Enforcement and Animal Care organization (REAC). This organization requires that all facilities using animals or people handling wild animals have their procedures approved in advance. Universities and other research facilities have usually set up committees to cover review procedures. REAC inspectors can check facilities and activities at any time.

California Ballot Initiative to Close Cougar Hunting (1990). In California, any citizen can campaign to get an initiative on the ballot. A group of people in the state succeeded in getting such a initiative on the ballot in 1990 that won and successfully closed all hunting of mountain lions in the state. This initiative also directed use of funds by the California Game and Fish Department, severely restricting some of their programs.

ulations. Zoos and other reserves are often used to preserve species. Nearly one-third of all bird species and one-sixth of all animal species have been bred in zoos in recent years. Zoos often maintain species under artificial conditions, but many feel that maintaining the gene pool of individuals in these isolated conditions may enable us to develop breeding programs and eventually release individuals back into the wild.

The Hawaiian Islands present examples of extinction of many native species. In this case the introduction of predators has taken a heavy toll on many native island species. These predators have become established in such a way that they are often difficult to remove. Only through intensive management practices can native species be maintained in these areas where introduction of exotics have become dominant (Figure 3.17).

Efforts are now under way to develop conservation programs that will preserve entire ecosystems and thus prevent wholesale species destruction.

Figure 3.17
Visitors to Hawaii can view exotics that have displaced native species.

SUMMARY AND CONCLUSION

Populations are composed of individuals of a species that occupy a particular place at a particular time. Population demographic characteristics differ from those of their individual members. Population characteristics include birth rate, death rate, age distribution, and genetic composition or gene pools. Habitat carrying capacity has a restrictive effect on population growth. Populations can interact through competition, predation, parasitism, and in other ways. Areas that satisfy a population's needs are called habitats. The population's role in that habitat is its *niche*. Lack of habitat limits a population's distribution.

The study of biological transference of characteristics from one generation to the next is called *genetics*. The genetics of a population shows us how populations can change or evolve in a changing environment.

Individuals pass genetic material from one generation to the next in the form of complex chemicals that make up genes and chromosomes, the units of inheritance. An individual's genetic makeup is its genotype; the expression of the genotype in the environment is the phenotype. Individual genetic characteristics are also controlled by dominance, sex-linkage, and multiple genes. Changes in the genetic material or mutations form the basis for selection when a portion of the population becomes isolated in a different habitat.

Natural selection occurs as some offspring of a population are more fit to survive and reproduce in the environment. When conditions change so a population cannot survive, extinction can occur. People have been the cause of extinction of a number of populations as they create large-scale habitat changes to which populations cannot adapt.

FURTHER READINGS

Avers, C. J. 1989. *Process and Pattern in Evolution.* New York: Oxford University Press.

Brown, M. H. 1991. *The Search for Eve.* New York: HarperCollins.

Emlen, J. M. 1984. *Population Biology.* New York: Macmillan Publishing Co.

Goodman, B. 1990. The Genetic Anatomy of Us. *Bioscience* 40:484–89.

Pianka, E. R. 1983. *Evolutionary Ecology.* New York: Harper & Row.

Strickberger, M. W. 1990. *Evolution.* Boston: Jones and Bartlett.

Tamarin, R. 1991. *Principles of Genetics.* Dubuque, IA: Wm. Brown.

STUDY QUESTIONS

1. Explain the conditions in which people are competitors, predators, or parasites to wild plants and animals.
2. Why do most populations that colonize new areas exhibit exponential growth and most populations that live in relatively stable communities exhibit sigmoid growth?
3. What is the difference between population genetics and population ecology?
4. Distinguish between an individual's genotype and a population's gene pool.
5. What is the difference between a dominant gene and a sex-linked gene?
6. Why doesn't the Hardy–Weinberg Law apply to all populations?
7. What is the difference between a gene mutation and a chromosome mutation?
8. Explain how natural selection works.
9. Can hybridization lead to speciation? How?
10. Distinguish between species extinction and population extinction.
11. What type of animal rights legislation could you support? Why?

SUGGESTED ACTIONS

1. *Trace the Genetic History of Your Family.* Determine how eye color has been inherited. Calculate what the eye color of your offspring might be. Consider deleterious genes that might be in your family. When might they show up?
2. *Genetic Engineering.* Read about and discuss genetic engineering as it applies to people. How should legislation be enacted to control genetic engineering involving humans? Collect a file of information from magazines and newspapers. Find local experts on genetic engineering. Pull together a fact sheet to help people understand what genetic engineering means and what it can accomplish.
3. *Planting for the Environment.* Collect a list of garden plants from your local nurseries that have been produced by hybridization. List such plants under different conditions such as "requires little water," "attracts wildlife," "low maintenance," and so on. Prepare such a list for distribution to people who wish to landscape their yards.

4

GEOPHYSICAL SYSTEMS

In the 1980s, a volcano eruption killed thousands in Columbia; earthquakes killed more than a hundred people in the San Francisco Bay area and thousands in Mexico City; hundreds of people lost their lives because of flooding in China; a hurricane destroyed beaches and homes in the United States. What can we do about disasters like these? Will southern California really physically split away from the rest of the United States? Why has the ground subsided in the vicinity of Galveston, Texas, and Wilkes-Barre, Pennsylvania? How do geological processes affect life? These are a sample of the issues involving geology, ecology, human habitation, and the quality of life that we will examine in this chapter.

THE HISTORY AND STRUCTURE OF THE EARTH

Unraveling the earth's mysteries is one of the most exciting scientific studies of this era. At the same time, understanding geophysical processes has a practical significance that goes beyond scientific curiosity.

Natural geophysical processes result in the building of mountains, the eruption of volcanoes, and associated earthquakes. Other physical and biological processes interacting with geophysical processes act to wear away mountains and form the soil in which plants grow. The results of all these interactions determine whether a place is suitable or hazardous for life.

Geophysical processes should not be considered in isolation. Weathering of rocks, for example, must be examined in the context of atmospheric, hydrologic, and ecologic systems. People also interact with the geophysical environment and through imprudent or thoughtless actions have at times precipitated geological events with wide-scale, sometimes disastrous results, such as landslides or subsidence of the land.

Because of the scale, magnitude, and power involved, the possibility of ever being able to control geological systems completely is only a distant vision. Knowledge and understanding of these processes, however, can be applied so that cities and homes are not built in hazardous locations, construction is suited to the geology, habitats for people and animals are preserved, and the resources of the earth are used to enhance the quality of human life. Since knowledge of geophysical systems forms the basis for planned land use and intelligent environmental management, selected aspects of geology which

Table 4.1
Geologic time

Time (millions of years ago)	Era	Period	Life		Physical Events
	Cenozoic	Tertiary	Modern life	Development of humans Large carnivores Monkeys Horses Development of mammals	Glacial advances and retreats Rise of Alps and Himalayas Rocky Mountain uplift
100 200	Mesozoic	Cretaceous Jurassic Triassic	Middle life	Extinction of dinosaurs Flowering plants Climax of dinosaurs Dinosaurs	Mountains form in western North America Present-day Atlantic Ocean forms Appalachian Mountains uplift
300 400 500	Paleozoic	Permian Carboniferous Devonian Silurian Ordovician Cambrian	Ancient life	Conifers Reptiles Coal-forming forests widespread Amphibians Earliest forests Land plants and animals Primitive fish Vertebrates Marine invertebrates	Appalachian and Ural Mountains begin Seas drain from North America Extensive seas and coal swamps cover North America Mountains and volcanoes in eastern North America Mountain building in Europe Extensive seas cover continents
600		Precambrian		Primitive marine life	Earth's crust forms Free oxygen accumulates Total age about 5 billion years

affect habitats or illustrate the effects of people on geological processes are presented as basic influences on all life.

Geologic Time

Natural processes that result in changes in the earth's climate or vegetation occur over thousands or millions of years. These changes are recorded in the sediments and rocks of the earth. To unravel this history, we must establish dates of events that have occurred over millions of years—what we call *geologic time* (Table 4.1 and Figure 1.2).

Developments in atomic physics in this century have been a boon to persons trying to establish reference points in geologic time. The technique used is based on the fact that some elements are not stable, that is, they readily disintegrate and in the process emit particles and energy. The process is called **radioactive decay** and such elements are said to be radioactive (see Chapter 14). In the process of disintegration, the unstable element will take on the characteristics of a different element, which in turn may also decay. This process continues until a stable element is formed. (The decay of uranium to lead is shown in Figure 13.17.)

The rate of decay is different for each element, but is constant for each radioactive element. The rate is measured by the length of time it takes one-half of the atoms present to disintegrate—the **half-life.** If an element has a half-life of 100 years, one-half remains after the first 100 years of decay. One-half of that, or one-fourth of the original material, will remain after 200 years. One-eighth remains after 300 years,

Table 4.2
Principal radioactive transformations used for measuring the ages of rocks[a]

Primary Natural Radioactive Element	Half-life (years)	End Decay Product Element
Uranium-238	4.51×10^9	Lead-206
Uranium-235	7.0×10^8	Lead-207
Thorium-232	1.42×10^{10}	Lead-208
Rubidium-87	5.0×10^{10}	Strontium-87
Potassium-40	1.27×10^9	Argon-40

[a]*See Figure 15.13 for decay of uranium-238.*

and so on. Some radioactive elements have a half-life of seconds; others, thousands of years.

When a rock solidifies from a molten state, radioactive elements may be *locked in*. If the rock remains intact, the decay products from a period of years, including the final stable element, are retained in the rock. If you knew the amount of a particular radioactive material in the rock at the beginning and the amount now remaining, it would be possible to estimate its age from the established rate of decay. In cases where we do not know the amount present in the beginning, the initial amount is estimated from the ratio between the amounts of unstable radioactive element remaining and the stable end-product element. These computations indicate the age of the rock since it solidified.

Several radioactive chains are used to date rocks, including some which start with uranium and end with lead (Table 4.2). Some rocks have been found that have been dated at 3.5 billion years old. The age of the earth is estimated at between 4 billion and 5 billion years. A similar process using radioactive carbon is used to date plants and other organic material younger than 30,000 years.

Another method that can be used in conjunction with radioactive dating is based on the magnetic orientation of beds of rock. Molten rock becomes magnetized and is oriented to the north and south poles as it solidifies. For some unknown reason, the polarity—the location of the north and south poles—of the earth has periodically reversed through time and these magnetic reversals are reflected in the magnetic orientation of the rock. This phenomenon can be used to determine which rocks on both the ocean floor and land were formed at the same time.

The relative position of a rock layer among other layers also helps in dating. Normally the upper layers would be the youngest (that is, the most recently deposited); however, patterns of deposition are often complex, and local geological formations have to be examined carefully to determine whether the deposits have been disturbed or if layers have been removed by erosion. Plant and animal remains may also be incorporated into sediment layers as fossils (Figure 4.1) and can provide additional clues about an area's age and history. Layers of volcanic ash occurring in sediment layers can also be used to determine when the sediments were deposited (this is called **tephrochronology**). More recent events, those occurring in the past few hundreds or thousands of years, can be dated by employing biological methods. These methods include using the size and growth rates of lichens as an indicator of substrate age (called **lichenometry**) and correlating events with the annual rings of trees (called **dendrochronology**).

The Structure of the Earth

The earth went through a molten period in its formation about 5 billion years ago. During the molten period, the heavier materials settled toward the center of the earth to form the core and the lighter materials *floated* on them to form the crust. Figure 4.2 shows a cross section of the earth's **core, mantle,** and **crust.** The core is in two layers: a solid inner core and a liquid outer core. The crust (lithosphere) is solidified material, but lighter than the core.

The rock **(basalt)** forming the ocean floors and found underneath the continents is heavier than the granite rocks that form the continents.

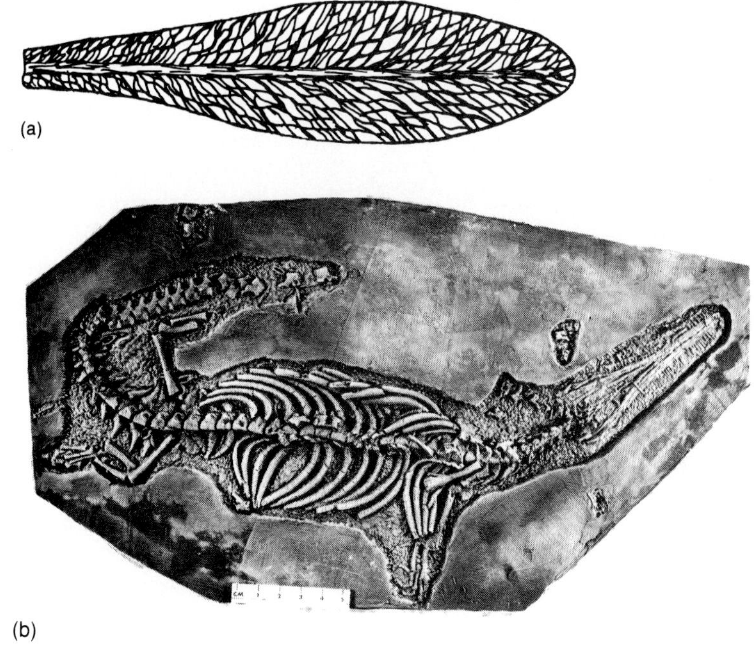

Figure 4.1
Establishing the age of sedimentary rock layers also fixes the age of fossils in the layers. The occurrence of like fossils in layers of undetermined age helps in estimating the age of those layers. Like fossils indicate when continents joined: divergent fossils show when continents separated. These fossils gave evidence that Africa, South America, Australia, and Antarctica were joined 250 million years ago. (a) *Glossopteris flora;* (b) *Mesosaurus* (Courtesy of Takeo Susuki, UCLA.)

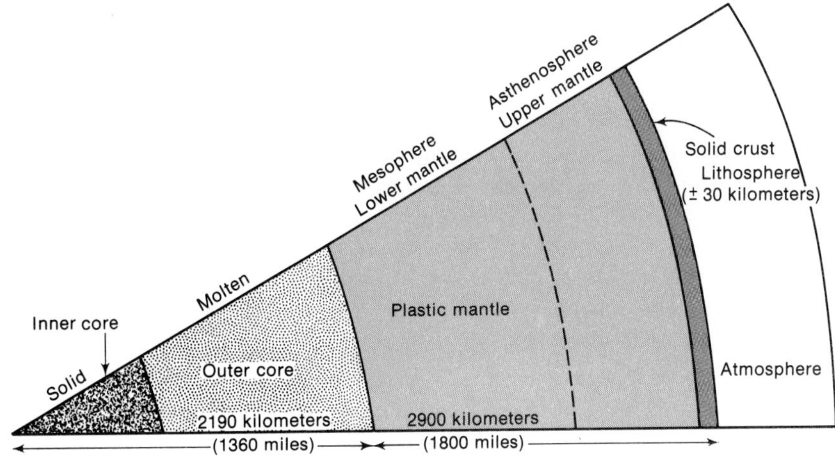

Figure 4.2
The earth's interior: a cross-sectional segment.

This difference in density is an important factor in the *floating* of continents on large segments of crust, called plates, which move and carry the continents along. Like a ship that rides high or low in the water according to how heavily it is loaded, continents protrude downward underneath mountains in proportion to the height of the mountain. Thus, the rock layer is relatively thin in lowland areas and thicker where there are mountains. As mountains wear away, the depressed material rises. Similarly, land depressed by the weight of ice rebounds when ice melts. The principle involved is called **isostasy.**

To help reconstruct the sequence of past geological events, we have fixed the approximate times for the formation of rocks and the deposit of sediments (Table 4.1). Theories about the spreading of the seafloor and the movement of continents have evolved from these data. An understanding of past events and when they occurred is also valuable in locating deposits of ores, oil, and gas. When coupled with fossil records, geophysical events which led to the expansion or decline of ecosystems may be understood. In the remainder of this chapter we discuss aspects of geophysical systems that influence life.

CONTINENTS ON THE MOVE

We have long accepted the fact that the earth is moving through space, but we have only recently recognized that the continents on which we live are actually moving about on the surface of the globe. This recognition came about as a result of observations of scientists over many years. As scientists traced the appearance, expansion and development, and extinction of individual species of plants and animals and ecosystems, they were puzzled by the similarity of certain species in areas of the world that are now remote from one another. For example, South America, Africa, and Australia show similar fossil species to a particular point in time and then begin to show divergence. The divergence in species began in the late Mesozoic times (see Table 4.1) after the continents began separating. As the land connection at the Isthmus of Panama rose above water, animals from North America began spreading throughout South America (and vice versa), introducing species into new areas. There is also evidence of significant changes in climate; there appears to have been tropical growth in areas now frigid and apparent dramatic changes in rainfall patterns. Many of these puzzling observations can now be explained by **plate tectonics** (sometimes called continental drift), a theory of geophysical processes that describes how continents move.

Many people have looked at maps of the world and noticed the similarity of the coastlines of western Africa and Eastern South America. They wondered, as Sir Francis Bacon did in 1620, if they were once joined like pieces of a jigsaw puzzle (Figure 4.3). In 1915, Alfred Wegener proposed the theory of the movement of continents, but it was not until the 1960s that Harry H. Hess of Princeton University and Robert S. Dietz of the U.S. Coast and Geodetic Survey independently developed an elaborate theory to explain the formation of new ocean floor, the movement of continents, and the building of mountains. It is now accepted that at one time the continents were joined together and at different times they have been separated and brought together in various configurations.

Seafloor Creation and Spreading

The generally accepted theory of plate tectonics states that forces in the earth's upper mantle **(asthenosphere)** cause molten material to rise, elevating the earth's crust and forming a ridge. At the ridge the crust splits, creating a rift in the ocean floor. The rift widens, and a rift valley is formed. The molten material spreads out and creates a new ocean floor. As the new rock (basalt) cools, it contracts, becomes more dense, and settles. Consequently, the seafloor slopes down away from the ridges and rift. The separation or tearing apart of the earth's crust at the

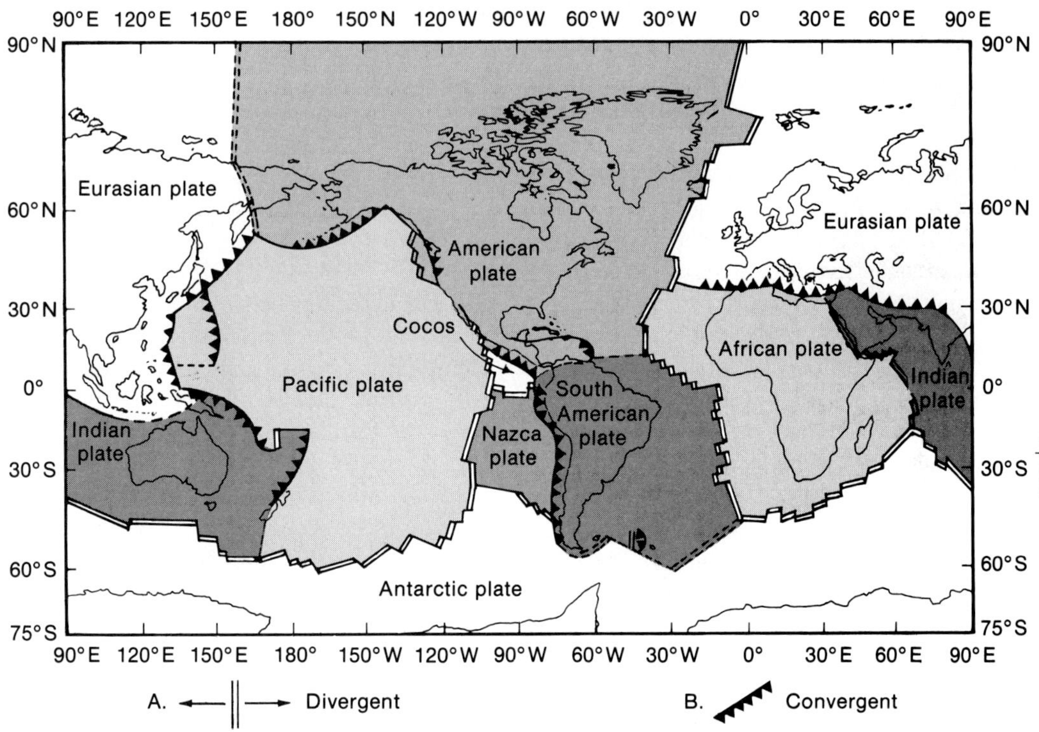

Figure 4.3
Major tectonic plates. (After F. J. Sawkins et al., *The Evolving Earth* New York: Macmillan, 1974.)

rift creates huge sections or plates of crust **(tectonic plates)** that move apart as new seafloor is formed.

At the edge of a plate opposite the rift–ridge axis, one crustal plate pushes against another. The resulting pressure causes a buckling similar to that caused by pushing a sheet of paper against an unmoving wall. As more new seafloor pushes out from the rift, one of the plates slips under the opposing plate into the mantle, where it melts. The area where the plate sinks, called a **subduction zone,** is characterized by deep trenches in the ocean floor.

Continents composed of granite or rocks of granitic origin are lighter than the basalt of which the crustal plates are made. Thus, the continental landmasses tend to float on and move with the crustal plates. The continents are not consumed at the subduction zones, but float and override the subduction area. Some 30 million years ago, the western portion of the North American continental crust overran a subduction zone and rift–ridge near the eastern edge of the Pacific.

A rift could occur under a continent, tearing the continent asunder as the plates carrying the continents separated and moved apart. The Atlantic Ocean was formed by such a rift. This action would explain the similarity between the coastlines of Europe and Africa and those of North and South America.

Life at the Rift

The development of submersible vehicles capable of operating at depths of more than 2.5 kilometers (1.6 miles) has enabled scientists to perform close-range visual investigation of ocean rift activity. The first observations occurred in

February 1977, along the Galapagos rift, or spreading center, off Ecuador. These were followed in January 1979 by dives at the East Pacific Rise located 3000 kilometers (1860 miles) northwest of the Galapagos site. The Galapagos rift is the boundary between the Pacific plate and a fragment of the North American plate—the Rivera plate. This rift area is spreading at 6 centimeters (2.4 inches) per year. Although that rate is three times the spreading rate in the Atlantic Ocean, another portion of the East Pacific Rise near Easter Island is spreading at the rate of 18 centimeters (7.1 inches) per year.

Along the axis of the East Pacific Rise rift is a young volcanic region about 1 kilometer (0.62 mile) wide, consisting of basaltic lava flows that have virtually no sediment cover. New crust begins to move horizontally away from the rift center and is stretched and cracked, forming major faults. Slippage along these faults causes frequent earthquakes.

In the rift valley water seeps into the cracked and fissured rocks, is heated as it nears reservoirs of molten lava, and rises in vents. Dissolved minerals precipitate to form chimneys called "black smokers."

Life becomes scarce at these depths (2650 meters or 8692 feet), but within the area of active vents an oasis of life abounds. Unique life forms exist without dependence on photosynthetic energy. Giant tube worms up to 3 meters (9.8 feet) long absorb food through blood-filled red tentacles. Other life includes clams 30 centimeters (12 inches) long and "dandelions" suspended by filaments.

Bacteria metabolize hydrogen sulfide (H_2S), carbon dioxide (CO_2), and oxygen (O_2) in the vents. The H_2S is oxidized and the energy released is used to convert CO_2 to organic matter. Bacteria grow in mats and clumps in porous rocks until sloughed off in flowing water. Suspended food concentration around vents is 300 to 500 times that of surface waters so animals in the area are attracted by nutrients rather than warmth.

Life in these areas is precarious, depending on the status of the vents. In the vicinity of inactive vents along the rift, remains of clam shells are evidence of life when the vent was active. However, since this is a new area of investigation, we do not know the extent or productivity of life along rifts, nor the relationships, if any, with life at the ocean surface.

Continental Movements

By using the magnetic orientation of undisturbed molten rock masses, tectonic data such as old mountain belts, sedimentary deposits, fossils, and climate history, scientists have speculated about the formation of continents and how they have moved about. Evidence exists of movement for 2500 million years. About 330 million years ago, the existing continents converged to form a single landmass—**Pangea.** After Pangea broke into two supercontinents some 200 million years ago, periodic divisions and collisions have occurred causing the opening and closing of ocean basins. Pangea was surrounded by the universal ocean **Panthalessa**—the ancestor of the present Pacific. A sea, the Tethys, intruded between Eurasia and Africa (Figure 4.4).

While the Pacific is an ancestral ocean (the remainder of Panthalessa), the Atlantic and Indian oceans were formed by the development of the rifts that split the continents and moved them apart. Where a rift splits continents apart, the rift and ridges will be situated in the middle of the newly formed ocean. However, rifts also occur in the ocean floors and these may be at any location. In the Pacific there is a rift, the East Pacific Rise, which is located near the east margin of the ocean.

There are now ten major plates and a few subplates in the world. A plate that moves over 3 centimeters (1.2 inches) per year is considered to move fast. Some 80 million years ago the plate carrying India broke away from Africa and moved at a speed of 16 centimeters (6.2 inches) per year to its present location.

The Pacific plate is presently being consumed at a rate of 8 centimeters (3.1 inches) per year—a

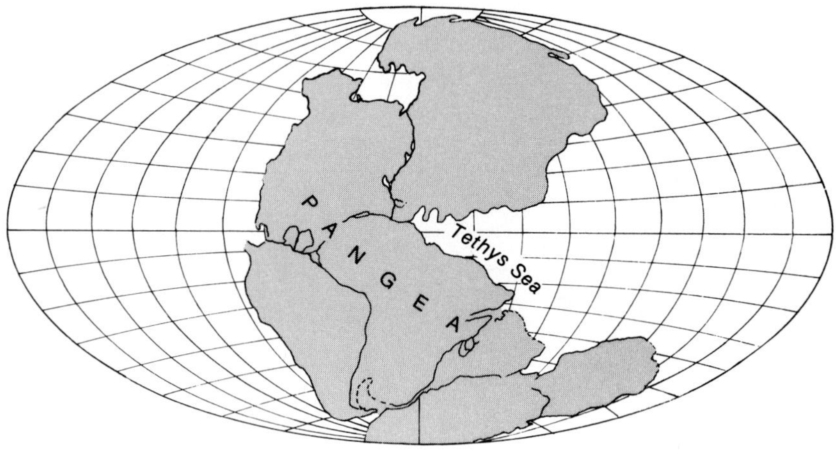

Figure 4.4
All continents were once joined in one land mass, Pangea, and surrounded by the universal ocean Panthalessa. The Tethys Sea intruded between Eurasia and Africa. The Pacific Ocean is a remnant of Panthalessa, and the Mediterranean is a remnant of the Tethys Sea. (After R. S. Dietz and J. C. Holden in *Journal of Geophysical Research* 75:4943. Copyright by American Geophysical Union.)

rate which will require replacement of its 15,000-kilometer width in 100 million years. One of the slowest current rates of seafloor spreading is in the Atlantic Ocean, where plates are moving away from the ridge between 1 to 8 centimeters (0.4 to 3 inches) per year. The floor, therefore, is spreading at twice that rate. The continental movements have constricted the Pacific, while the Atlantic is expanding.

Other tectonic plate movements affected the Mediterranean and the Red seas. At one time the Tethys Sea was squeezed between Europe and Africa, and the Mediterranean Sea is all that remains. Now an embryonic ocean is being formed where the Red Sea is located as a rift moves the Arabian plate away from the African plate.

Moving and colliding continents are responsible for the major landforms of the United States. Some 600 million years ago, North America and Africa split apart to form the ancestral Atlantic Ocean. About 250 million years later these continents moved back together. Sediments that accumulated off the eastern edge of North America were squeezed, crumpled, uplifted, and thrust westward over the prior continental shelf to form the Appalachian Mountains. The continents then separated again about 180 million years ago to form the present North Atlantic Ocean. The Rocky Mountains in the western United States (Cordilleran region) were uplifted when the Atlantic seafloor widened and caused the North American continent to be pushed slowly westward.

The major continental movement directly involving portions of the United States is occurring along the San Andreas fault in California, as shown in Figure 4.5. The San Andreas fault is a transform fault, that is, a fault produced because the Pacific plate is moving to the northwest with respect to the rest of the North American continent at a rate of about 5 to 8 centimeters (2 to 3 inches) per year. The San Andreas fault system is in line with the separation of Baja California from the rest of Mexico and extends past San Francisco as the eastern boundary of the Pacific plate. One reason that there is so much earthquake activity in the vicinity of Los Angeles is that the edge of the plate has to *bend* around the roots of the Sierra Mountains. Geophysical events associated with plate tectonics, such as earthquakes and volcanoes, are often threatening to human populations. In October of 1989 an earthquake struck the Bay Area of northern

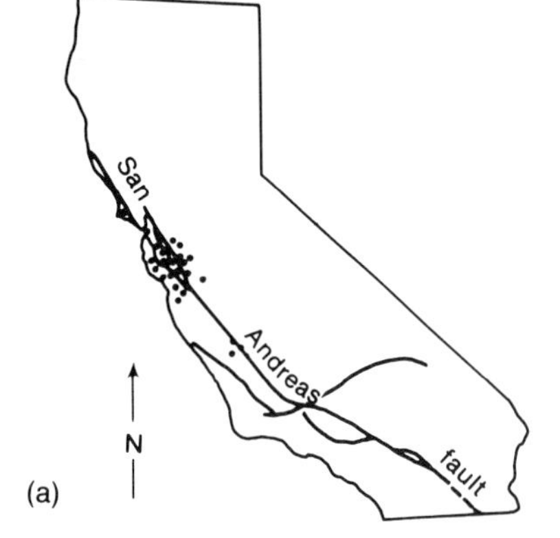

FIGURE 4.5
(a) Fault map of California showing the seismograph network. (b) An offset riverbed near the San Andreas fault. (From *Earthquakes*, U.S. Geological Survey.)

California, killing more than 100 people, leveling buildings and disrupting transportation systems. Authorities pointed out that, despite the devastation it caused, this earthquake was of only modest severity and provided an opportunity to assess the area's preparedness for the larger, more catastrophic earthquakes that could strike the area in the future.

VOLCANOES AND EARTHQUAKES

Improved knowledge of the more threatening aspects of volcanoes and earthquakes can be useful in avoiding catastrophes. In time, slippage along transform faults might be managed so that severe earthquakes resulting from sudden slippage could be avoided. Greater accuracy in predicting impending volcanic activity and plate movements will allow for early warning. This information could also be utilized in land use planning to avoid natural hazards, especially in locating sensitive facilities such as dams and atomic energy plants. Building codes could be specially adapted for construction in earthquake-prone areas (Figures 4.5, 4.6). Millions of dollars in property damage could be averted and thousands of lives saved by knowledge we now have (Figure 4.7).

Kinds of Volcanoes

Volcanoes are especially dramatic natural hazards, to which human responses are determined by several factors, including the size, intensity, and frequency of eruptions. We will compare two very different types of volcanoes, volcanoes found near subduction zones and volcanoes associated with oceanic hotspots.

A hotspot is a pool of molten rock in the upper mantle. As a plate moves across it, the molten material may erupt through the plate, producing a volcano. Yellowstone Park in the northwestern United States is a **caldera,** a basin left by an explosive eruption that blew away the top of a volcano. A hotspot under the continent formed this volcano and left a trail of extinct volcanoes across Utah. The hotspot is still there and could cause future eruptions. In fact, recent observations show evidence of ground swelling and uplifting.

Hotspots that occur under the ocean floor produce volcanic islands. As the seafloor plate moves, old volcanoes lose contact with the hot-

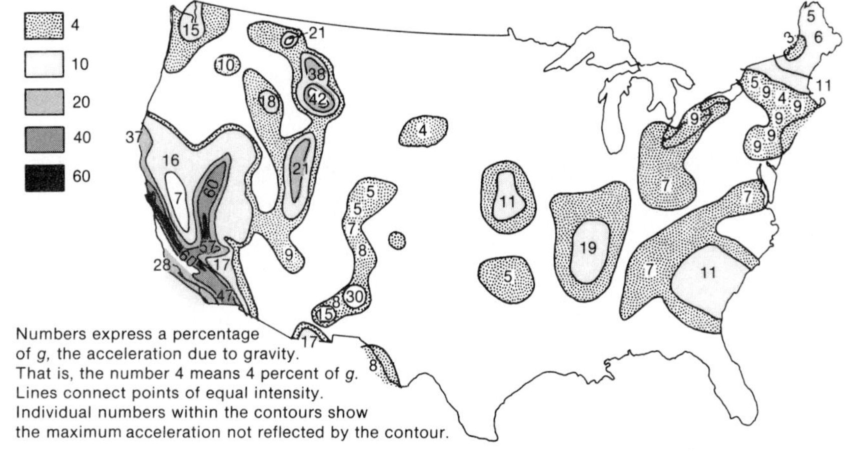

Figure 4.6
Earthquake risk in the United States. The numbers show the relative intensity of anticipated earthquakes in terms of horizontal acceleration, which is expressed as a percentage of the acceleration due to gravity (g). There is a 90 percent probability that these accelerations will not be exceeded in 50 years. (From S. T. Algermissen and David M. Perkins, "A Probabilistic Estimate of Maximum Acceleration in Rock in the Contiguous United States," U.S. Geological Survey Open File Report 76–146, 1976.)

Figure 4.7
Collapse of V.A. Hospital, San Fernando Valley, in 1971. Pre-1933 standards were used for construction. (Courtesy of the Los Angeles City Department of Building and Safety.)

PREDICTING EARTHQUAKES: WHEN, WHERE, AND HOW BIG?

If we could predict when and where earthquakes will occur, many lives could be saved. However, earthquake prediction is a very tricky business. For example, in 1975 Chinese scientists successfully predicted the time, place, and magnitude of an earthquake in Haicheng, China. However, one year later another earthquake in Tangshan, China, was not predicted and hundreds of thousands of people were killed. A noteworthy unsuccessful prediction was made in 1981 by a U.S. Geological Survey scientist. He predicted that three great earthquakes would occur in Peru. The country suffered economic losses as many people left the country to avoid the predicted disaster. However, the earthquakes never occurred and many people are now skeptical about earthquake predictions made by scientists. This is unfortunate because these Peruvian earthquakes could still occur in the near future and people may be less likely to respond to warnings.

Many methods for earthquake prediction have been explored. Researchers have even tried to determine if there is a correlation between predictions made by psychics and the occurrence of earthquakes, although this is not generally considered a promising avenue for research. Chinese scientists use unusual animal behavior to assist in earthquake prediction. Animals may be able to sense underground changes associated with earthquakes and become restless or exhibit bizarre behavior. It is possible that they are responding to some precursor of an earthquake such as subtle vibrations, electrical currents, small changes in the earth's magnetic field, the odor of gases forced out of fractured rocks, or sounds made by rocks under pressure.

Other possible indicators of earthquakes are under study. Earthquake prediction research in the United States has emphasized *dilatancy theory.* This theory is based on observations that rocks under stress swell or dilate just before they rupture. This dilation is associated with changes in physical characteristics that can be measured by geologists. These include changes in the level and chemical composition of groundwater and increased electrical currents generated by rocks under pressure. Earthquakes are also associated with abrupt changes in land elevation or tilt.

In another approach geologists monitor the number and strength of small earthquakes along a fault system. Since these small quakes could relieve the stresses associated with rocks sliding past each other, the cessation of this activity could mean that stress is building up and a large earthquake is likely to occur.

Sometimes earthquake activity is at an unusually low level in a localized area, called a *seismic gap*, while it remains at a normal level in the surrounding region. Seismic gaps are carefully watched because they are considered likely locations for earthquakes to originate.

To be successful, an earthquake prediction should provide information about the location, the time, and the magnitude of an impending earthquake. If only the location and magnitude are given, it is called an earthquake forecast. Scientists are working hard to identify accurate precursors to earthquakes so that human populations can avoid their disastrous effects.

For additional information see Tyckoson, David A. 1986. *Earthquake Prediction.* (Oryx Science Bibliographies, Vol. 5). Phoenix: Oryx Press.

spot and become extinct, but this allows for the formation of new volcanoes. The trail of extinct volcanoes is exemplified by the chain of islands and **seamounts** (a submerged, extinct volcano) in the Pacific that includes Midway and Hawaii (Figure 4.8).

The Hawaiian Islands are the most recent chain of volcanic islands produced as a result of the Pacific plate moving over a hotspot. If the Pacific plate continues its present movement, the active volcanoes on the island of Hawaii should become extinct; but new islands could appear to extend the chain.

Volcanoes form at subduction zones when the descending plate melts and some of the molten material finds its way through fissures back to the surface. It is common to have an arc of volcanic islands—*a rim of fire*—just beyond the subduction zone formed by these deep-rooted volcanoes, such as the islands of Japan. Various

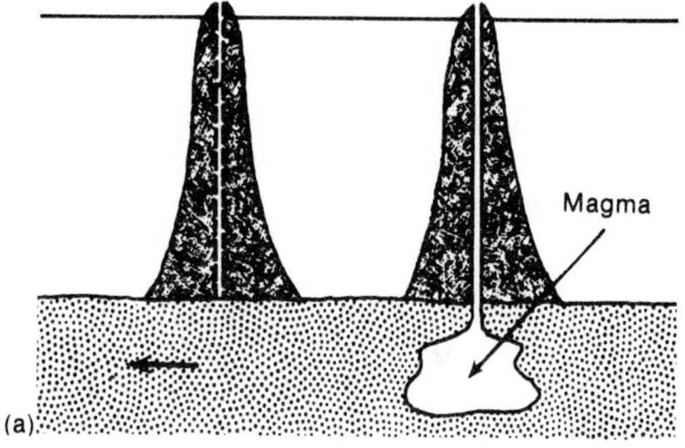

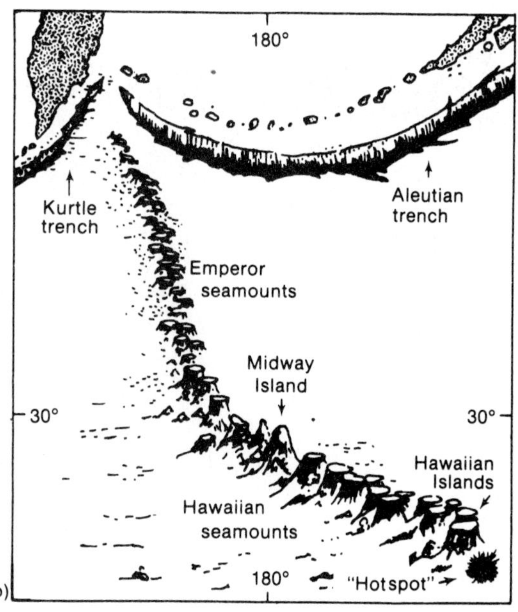

Figure 4.8
Islands and seamounts produced by a hotspot. (a) As the plate passes over a hotspot, a volcano develops that may reach the surface. As the plate continues to move, the old volcano is cut off from its source of molten rock and a new volcano is formed. (b) A chain of islands and seamounts in the Pacific Ocean formed as the Pacific plate moved over a hotspot.

types of volcanic origins and activity are depicted in Figure 4.9.

The volcanic mountains of the Cascade Range are associated with a subduction zone. This range, located along the western edge of the North American continent, began to form 7 million years ago and is caused by a subduction zone related to the small Juan de Fuca plate just off the coast (Figure 4.10). This plate is descending under the continent at a rate of 7.6 centimeters (3 inches) per year.

Volcanoes at Seafloor Hotspots

Volcanoes associated with seafloor hotspots have basaltic lava that is fluid and flows readily. Kilauea Volcano on the Big Island of Hawaii erupts rather frequently (every several years). Its lava flow eruptions are rather mild compared to the more explosive eruptions of other volcanoes. Even though the eruptions of Kilauea occur in areas of human habitation, lava flows are slow enough to allow people to move out of the way, and loss of life is not common. Damage to buildings, agricultural land, and other property from a single eruption, however, easily reaches millions of dollars. Since 1983 when it began erupting near the community of Kalapana, Kilauea Volcano has consumed most of the homes in the community. Residents generally had time to pack their belongings so that human lives were not threatened, although economic losses have been high. During the spring of 1990, a lava flow 13 kilometers (8 miles) long and 550 meters (600 yards) wide was advancing on two historic churches and a general store, traveling at rates up to 30 centimeters (1 foot) per minute.

Volcanoes at Subduction Zones

In contrast to Kilauea, volcanoes associated with subduction zones have more viscous lava and tend to erupt more violently, but less frequently.

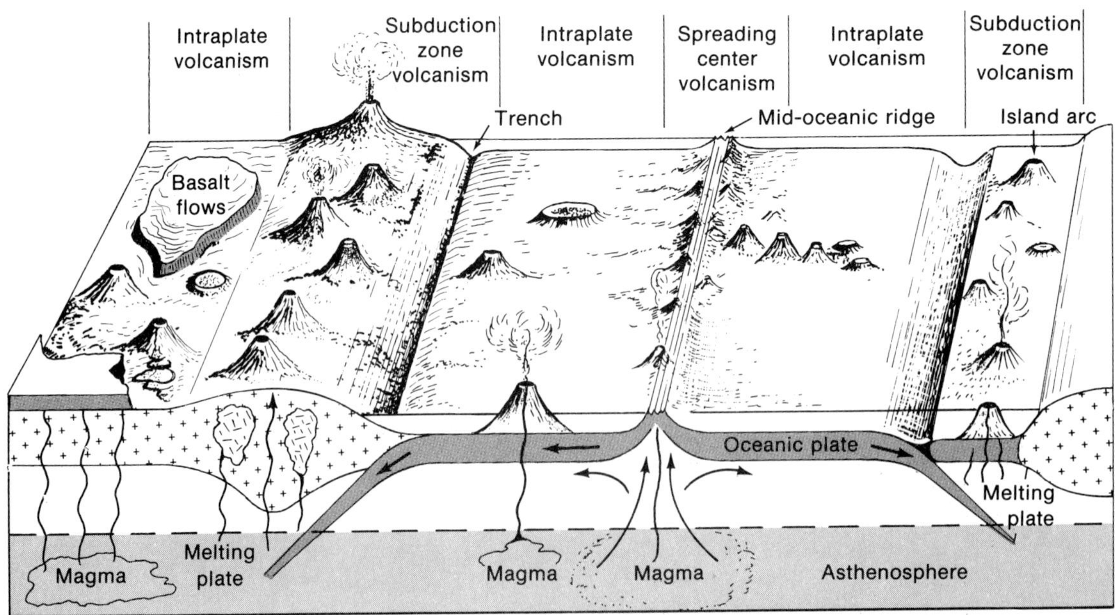

Figure 4.9
Composite illustration of various types of volcanic activity. Volcanic islands are sometimes formed at midocean rift-ridge. Subduction and melting of crustal plates cause volcanoes offshore (to the right above) and onshore (to the left above). Hotspots produce volcanoes on land (to the left) and in the oceans (center).

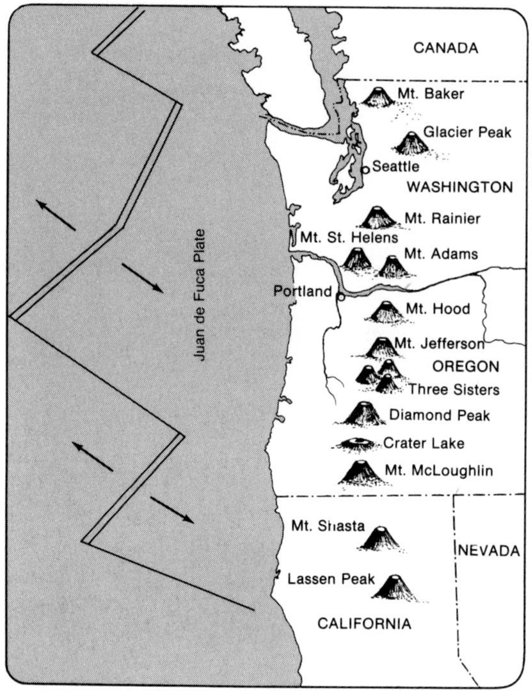

Figure 4.10
Cascade Range volcanoes associated with override of an oceanic plate. The subduction zone for this oceanic plate lies under the continent. Volcanoes are fed with molten rock (lava) when the plate melts. (After Tabor and Crowder, U.S. Geological Survey Professional Paper No. 604, 1969.)

Many volcanoes in the Cascade range continue to possess the potential to erupt. The entire area has experienced widespread volcanic activity for 50 million years. Mazama erupted 6700 years ago and left a caldera—Crater Lake. Sixty cubic kilometers (15 cubic miles) of its 3660-meter (12,000-foot) peak disappeared in a cloud of volcanic dust. Mount Rainier, 4393 meters (14,410 feet) high, has 26 glaciers and a history of massive mud flows. Steam vents along the rim of its crater have initiated predictions of an eruption before the end of the century.

The Eruption of Mount St. Helens

During the last 4500 years, Mount St. Helens has been more active and explosive than any other volcano in the coterminous states. In the spring of 1980 it ended a 120-year dormant period as molten rock began to rise under the volcano and frequent earthquakes signaled the events that started on March 20. The hot rocks caused ice on the summit to begin melting. As water seeped downward and came into contact with the underlying hot rocks, the water flashed to steam, causing **phreatic explosions** and forming a crater. In April the northeast flank of the volcano began to swell, creating a 100-meter (328-foot) uplift by April 23. A subsequent earthquake and debris avalanche sent 50,000 cubic meters (65,500 cubic yards) of rock and ice down the north flank. Finally, on May 18 another earthquake triggered a landslide in the area of the steepened bulge (Figure 4.11). Relieved of this pressure, the superheated groundwater flashed to steam by a violent explosion that released volcanic gases. The summit was lowered from 2950 meters (9680 feet) to about 2549 meters (8364 feet).

Landslides and mudflows caused mixtures of rock, debris, ice, snow, and water to fill valleys and lakes and clog streams. Log jams and floods destroyed bridges. Sediment reduced the navigation channel of the Columbia and Cowlitz Rivers from 12 to less than 4 meters (46 to less than 15 feet). As blasts produced **pyroclastic flows** (dense turbulent mixtures of hot rock fragments, steam, and gases that move rapidly over the ground surface) at about 500°C (932°F), an area reaching 10 kilometers (6.2 miles) was denuded and covered with ash. The force of this blast leveled trees up to 15 kilometers (9.3 miles) away (Figure 4.12).

An ash-laden plume ascended about 23 kilometers (14.3 miles) entering 11 kilometers (6.8 miles) into the stratosphere. Two and seven-tenths cubic kilometers (1 cubic mile) of rock were diffused in the eruption. About one-tenth of this rock settled on three states to the east, and a similar amount lightly dusted the central United States. Only a small amount stayed in the atmosphere, thereby producing no detectable influence on climate. (Eruptions of other volca-

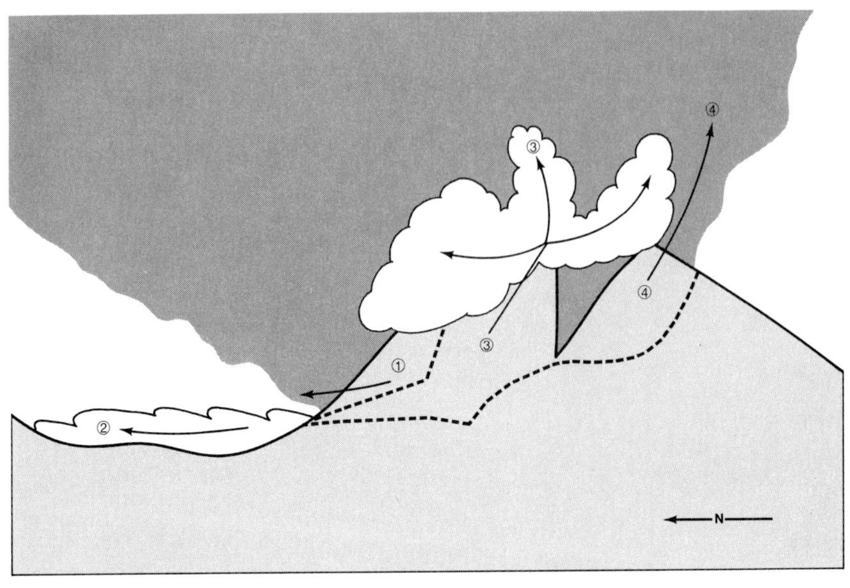

Figure 4.11
Diagram of Mount St. Helens' May 18 explosion. The eruption could have been caused by a landslide (1): superheated groundwater that flashed into steam because of reduced pressure (2): the steam explosion removing underlying rock, resulting in summit collapse (3): and an ash-laden plume enlarging the crater as gases lifted the plume from the breach (4). (Numbers indicate the areas involved: dashed rules outline the crater's rim after the eruption.) (From Charles L. Rosenfeld, "Observations on the Mount St. Helens Eruption," *American Scientist* 68[1980]: 498.)

noes have affected climate. See Table 4.3.) The catastrophic eruption of Mount St. Helens affected wildlife in an area of 414 square kilometers (160 square miles). Mammals and birds above ground had no protection and suffered massive losses; but subterranean animals, such as gophers, survived the direct blast. Twenty-six lakes were severely damaged and 11 million fish killed. In eastern Washington the 1980 crop losses from dust fall were estimated at $100 million. Sixty-one people were killed or missing.

Other Explosive Eruptions

The loss of human life in the Mount St. Helens eruption is relatively small compared to losses incurred by other explosive eruptions. In 1902, a pyroclastic flow from the eruption of Mt. Pelée destroyed the city of St. Pierre on the island of Martinique and killed its 30,000 residents. The volcanic island of Krakatoa, near Java, was devastated in 1883 during 2 days of explosive eruptions. A huge tidal wave created by the blast engulfed small islands and coastal villages, killing 36,000 people. More recently, the 1985 eruption of Nevado del Ruiz in Colombia killed more than 20,000 people.

Posteruption Ecosystem Recovery

The eruption of Mount St. Helens provided scientists with an exciting opportunity to study the ecological effects of volcanoes and the process of ecosystem recovery. At first the devastation was feared to be so severe that recovery would take a long time, but it has been surprisingly rapid.

In 1982 the area was protected for scientific research by designating it as the Mount St. Helens National Volcanic Monument. In the more than 10 years since the major eruption scientists have been fascinated by the recovery process. Birds and large mammals, including the Roosevelt elk, returned to the blast area within a

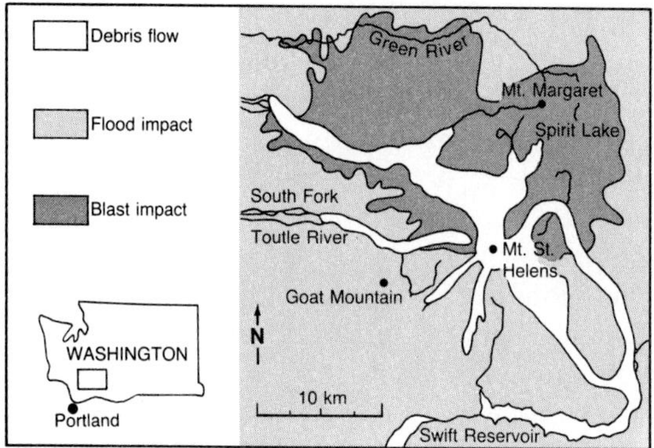

Figure 4.12
Map of the impact of Mount St. Helens' May 18 eruption. Although ridges provided some protection from the blast effects on the north side of the volcano, debris flows and flooding increased the range of damage beyond the area in the immediate vicinity of the blast. (From Charles L. Rosenfeld, "Observations on the Mount St. Helens Eruption," *American Scientist* 68[1980]: 498.)

few weeks. Smaller and less mobile animals such as reptiles, amphibians, and small mammals are also migrating back into the area, but it is a slower process. Fewer than half of the small mammal species living in the area at the time of the blast have returned. A constant rain of spores, seeds, insects, and the eggs and cysts of other small animals has provided a source of new life for the area. During the early phases, recovery was dominated by plants that had survived below—and occasionally above—ground. Three years after the eruption, 90 percent of preeruption plant species had returned to the site, although forest recovery will take much longer (100–200 years, or more).

Table 4.3
Discharge of ash and pumice from Mount Saint Helens (1980) compared with other major eruptions of volcanoes

Year	Volcano	Amount of Ash and Pumice in Cubic Kilometers
1980	Mt. St. Helens	1
1912	Mt. Katmai	12
1883	Krakatoa	18
1815	Tambora	80
1500	Mt. St. Helens	1
79	Vesuvius	3
1900 B.C.	Mt. St. Helens	4
4600 B.C.	Mt. Mazama	42

Research at Mount St. Helens is teaching scientists interesting and important lessons about ecosystem recovery. It turns out to be a complex process with a diversity of patterns, differing significantly from the descriptions of secondary succession in most textbooks. Local conditions vary widely from site to site; in some areas recovery is quite rapid and in others quite slow.

Unexpected linkages among organisms have also been found to be quite important. For example, pocket gophers, which survived the blast because they were underground, are commonly considered to retard plant succession. At Mount St. Helens they accelerate ecosystem recovery by mixing nutrient-rich soil with the relatively sterile volcanic ash. Usually destructive erosion associated with human and other animal use of hillsides is often beneficial because it releases buried seeds and provides a fertile seedbed for their germination. Soil recovery in some areas is occurring in a very interesting manner. The surface of volcanic deposits has hardened into a layer referred to as desert pavement. This layer protects the developing soils below it, and as it weathers it provides this soil with nutrients. Lakes and streams are expected to recover quickly, but sediment loads will retard recovery of fish populations. Fish surviving in streams and lakes isolated from the effects of the eruption have been sources of recolonization of affected aquatic ecosystems. The return of fish to lakes where there had been no survivors has generated some controversy. Most scientists want to observe the natural recovery of these lakes, while state wildlife officials want to establish a sport fishery by stocking the lakes. At the present time only lakes outside the borders of the national monument are being stocked.

Social Implications

Human responses to volcanoes, and to other natural hazards such as earthquakes, floods, and severe storms, depend on a complex mixture of factors. An eruption of Kilauea in Hawaii is relatively common and people can usually avoid the flowing lava, whereas the response to a more explosive eruption, such as occurred at Mount St. Helens, will be quite different. Social scientists are interested in understanding how people respond to natural hazards and how they evaluate the risk associated with exposure to potential hazards. A rather elaborate hazard management system has evolved in relation to volcanic activity on the Hawaiian Islands. It includes hazard mapping, warning systems, disaster relief, and land use controls. No management system was in place for Mount St. Helens because eruptions were so rare. Hazard maps of all the Cascade volcanoes, including Mount St. Helens, were available, but local officials had not developed a specific hazard response system.

GEOLOGICAL PROCESSES AND HUMAN ACTIVITIES

Other social implications of plate tectonics have nothing to do with volcanoes. How can we use what we have learned about the spreading of the seafloor and the building of mountains to improve our life? One fact is readily apparent: We live in a dynamic world where conditions are forever changing.

Understanding the dynamics of geophysical systems can aid in the search for minerals and petroleum deposits. By recognizing the conditions that led to a particular deposit on one continent, prospectors can look for a similar deposit or vein on another continent that may have been connected to the first. Furthermore, with the aid of satellite pictures, prospectors can predict where to look on the other continent.

The deposit of sediments heavily laden with organic matter in shallow marine areas adjacent to continents favors the formation of oil. Therefore, in the early formation of the Atlantic Ocean, the process of seafloor spreading may have formed shallow deposits favorable to the development of petroleum. Prospecting for these oil deposits is now underway off the Atlantic seaboard of the United States. Similar conditions preceding the raising of the Andes Mountains may also have been favorable for the formation of oil deposits.

The nonrecyclable solid wastes of urban societies could be dumped or placed in the subduction zones to be carried into the mantle. While toxic materials not readily degraded might be disposed of in this manner, it is not necessarily a practical solution for the disposal of nuclear wastes for several reasons. The nuclear material could be scraped off as the seafloor crustal plate descended into the mantle at the subduction zone, to be uplifted later rather than descending into the mantle with the plate. Or, the nuclear waste might find its way back to the surface through a volcano originating in the subduction zone.

Over a longer period of time, plate tectonics explains the raising of mountains and the movement of continents. These events produce dramatic changes in weather and climate, the effects of which ultimately influence the vitality of ecosystems and the evolution of various species. Imagine the change in vegetation that will occur as Los Angeles moves past San Francisco in the eons ahead!

Minerals and Ores

All substances are composed of chemical **elements.** Ninety-two elements are known to occur in nature, and ten of them constitute 99 percent of the earth's crust (Table 4.4). The basic units of elements are **atoms,** and two or more atoms may be joined together to form **molecules.**

Molecules may be made by joining two or more atoms of the same element. For example, an oxygen molecule is composed of two atoms of

Table 4.4
Average composition of the Earth's crust

Chemical Element	Chemical Symbol	Average Percentage by Weight in Crust
Oxygen	O	46.6
Silicon	Si	27.7
Aluminum	Al	8.1
Iron	Fe	5.0
Calcium	Ca	3.6
Sodium	Na	2.8
Potassium	K	2.6
Magnesium	Mg	2.1
Titanium	Ti	0.4
Hydrogen	H	0.1
Total		99.0

the same element—oxygen. Three atoms of oxygen produce a molecule of ozone. Many molecules, however, are made of different elements. A molecule of table salt, for example, has one atom of sodium and one of chlorine. We call large quantities of like molecules *chemical compounds.*

For convenience, a type of international shorthand notation using letters to represent the chemical element (such as O for oxygen, Na for sodium, and Cl for chlorine) is used to describe the composition of compounds. The symbolic letters are not necessarily related to the English word for the element, but frequently are derived from some other name. To indicate the number of atoms of an element in a molecule or a chemical compound, that number is placed as a subscript immediately after the letters. Therefore, oxygen is O_2, ozone is O_3, and table salt is NaCl.

Water is the closest thing in nature to the mythical *universal solvent.* Whether or not a rock dissolves readily in water determines how rapidly it weathers. When substances are dissolved, some of the atoms, or groups of atoms, dissociate in the liquid into what are called **ions.** The ions have positive or negative electrical charges. Thus, salt (NaCl), when dissolved, produces a sodium ion (written Na^+) and a chlorine ion (Cl^-).

Water (H_2O) molecules dissociate freely, ionizing into two ions: one is an atom of hydrogen (H^+); the other is composed of one atom of oxygen and one atom of hydrogen, called hydroxyl ion (OH^-). This reversible reaction is

$$H_2O \quad H^+ + OH^-$$

If an excess of hydrogen ions is present, the liquid is **acidic;** and if hydroxyl ions are in excess, it is **basic** or **caustic.** An index scale based on the relative number of hydrogen ions present has been developed to represent this relationship. Called a **pH** scale, it uses numbers from 0 to 14. A pH of 7 represents a neutral condition, where the hydrogen and hydroxyl ions are balanced. The pH decreases as the hydrogen ion concentration increases. However, the scale (a **logarithmic scale**) is such that a pH of 5 means that the concentration of hydrogen ions is ten times greater (more acid) than at a pH of 6. Similarly, a pH of 4 is ten times more acid than a pH of 5 and one hundred times more acid than a pH of 6.

These chemical interactions have applications in many environmental and life systems and are discussed in several places in this text. Limestone readily dissolves in water to form solution channels and caverns. Gold, on the other hand, ionizes very little and can be recovered (usually in very small amounts) from sediments in streams.

Rocks are mixtures of substances called **minerals.** About 90 percent of all minerals occur as ionic-type compounds and therefore do not consist of molecules in the usual sense. The mineral names are different from the names of chemical compounds, as in the case of calcite and calcium carbonate. Limestone rock is composed of one atom of calcium (Ca), one atom of carbon (C), and three atoms of oxygen (O_3), plus impurities. The chemical compound of that composition is known as calcium carbonate ($CaCO_3$) by chemists, but mineralogists have given the name calcite to rocks of that composition. As a result of leaching, deposition, chemical action, reaction to temperature, and other prevailing factors,

PERMANENT DISPOSAL OF NUCLEAR WASTE

The government of the United States is in the process of selecting stable geologic formations that can serve as permanent disposal sites for high-level nuclear wastes. These sites must be able to keep the wastes isolated for the 10,000 years it takes for the waste to lose its toxicity. The most promising kinds of formations are salt domes, basalt (solidified lava), and tuff (compacted volcanic ash). The Nuclear Waste Policy Act of 1982 specified that several potential sites be evaluated. Some of the sites were found to be unacceptable because they were geologically unstable or because there was a possibility that groundwater could be contaminated. There were also problems with the NIMBY ("Not In My Backyard") syndrome, the name given to public opposition to living near repositories for nuclear and other toxic wastes. For example, when citizens learned that salt domes in Texas were being investigated as potential nuclear waste sites, some of them formed an organization called Serious Texans Against Nuclear Dangers (STAND) to oppose the study.

In 1987 Congress voted to limit the evaluation to one site, Yucca Mountain in the Nevada desert. This site was chosen because of its remoteness, the dry climate, and its thick deposits of tuff. About 83 percent of the nuclear waste slated for Yucca Mountain will be spent fuel rods from the more than 100 U.S. nuclear power plants. At the present time these 3.7-meter (12-foot) metal tubes are stored in water adjacent to the power plants. The other 17 percent of the waste has been produced by military reactors.

Plans call for containers of nuclear waste to be hauled by truck down a gently sloping tunnel to a network of storage tunnels more than 300 meters below the surface. However, before this can happen, there will be extensive research to determine if earthquake or volcanic activity is likely to occur in the area, and to make certain that water cannot reach the stored wastes. It will also be necessary to convince the general public and state officials that the storage site will not constitute a public health hazard.

Some people have heard of radiation leaks at temporary storage facilities and question the certainty of waste containment over thousands of years. Even though officials point out that groundwater contamination is very unlikely, some people fear that aquifers could be contaminated if rainfall patterns change in the future. People who favor the project, however, view its construction as a possible means to economic stability. Because of limited employment opportunities in the area, a project that would alleviate economic concerns would be beneficial. Furthermore, the federal government will provide large financial incentives to Nevada if Yucca Mountain becomes the nation's permanent waste depository.

certain minerals occur in great abundance in deposits or veins, sometimes known as **ores.** The term *ore* is used to indicate that it is economically profitable to extract the mineral.

Mineral Resources **Mineral resources** are potentially usable concentrations of elements in a particular location in or on the earth. The U.S. Geological Survey (USGS) includes in its definition of mineral resources elements such as cadmium or tin and chemical compounds such as salt, rock, marble, or coal that can be extracted profitably. A fundamental concept in the evaluation of mineral resources is the distinction between resources and reserves. **Reserves** are identified deposits of mineral-bearing rock from which minerals can be extracted. **Resources** include reserves plus all other mineral deposits either not yet discovered or unobtainable for technological or economic reasons. Mineral resources and reserves are evaluated as to grade and extent and divided into the categories shown in Figure 4.13. Extensive investigations are going on to locate and map the reserves of most minerals.

Economic Importance of Minerals Minerals are important in the U.S. economy. The United States imports some as manufactured products (Figure 4.14). Others are found as natural resources within the United States.

The current use of reserves is leading to rapid depletion of many minerals. Industrial nations with only 30 percent of the world's population use over 70 percent of the minerals mined worldwide. The United States, with only 6 percent of the world's population, uses about 30 percent of the minerals produced annually in the world. As lesser developed nations grow, they compete for many of the materials developed

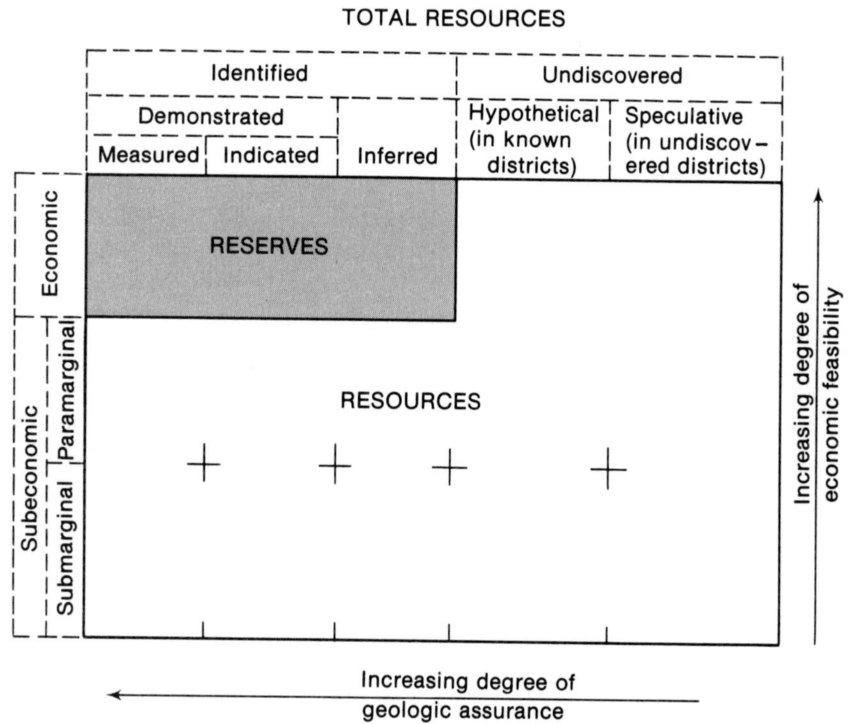

Figure 4.13
Classification of mineral resources. Identified resources consist of economically recoverable reserves and subeconomic resources. Undiscovered resources are divided into hypothetical and speculative resources. (From D. A. Brobst, and W. P. Pratt, eds. *United States Mineral Resources,* U.S. Geological Survey Professional Paper No. 820. Washington, D.C., 1974.)

Figure 4.14
The role of minerals in the U.S. economy. (U.S. Bureau of Mines.)

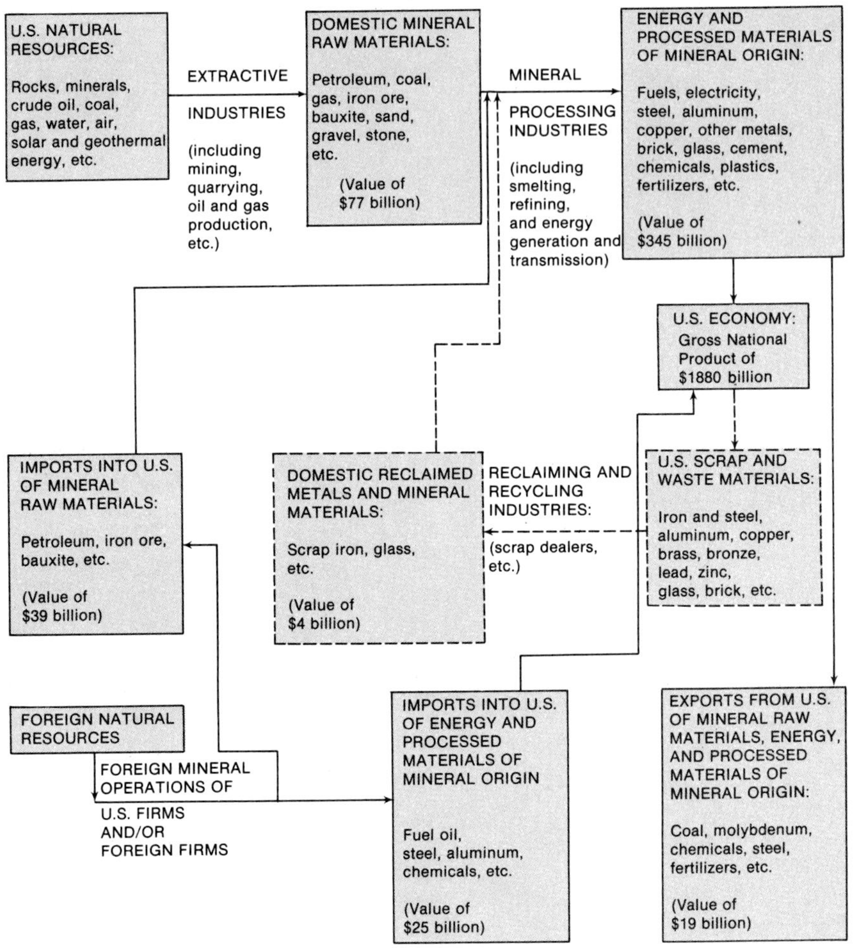

nations can presently obtain economically, causing supplies to diminish more rapidly and prices to increase.

The high level of technological development in industrial nations demands a far greater supply of minerals than is required by the rest of the world. Furthermore, during the past 20 years, the United States' mineral use pattern has been characterized by a shift away from self-sufficiency. Concern has been expressed about the United States' dependence on imports of some critical minerals (Figure 4.15). For example, critical minerals such as cobalt and chromium, used for such things as specialty steel, heat-resistant alloys, and aircraft and automobile parts, are imported from relatively insecure sources that may be subject to supply disruption. This pattern has been true for some other parts of the world. Europe, which was self-sufficient in fuels forty years ago, is now dependent on the Near East and Africa.

Shifts in mineral use occur daily. As the supply of wood decreased, people shifted to other fuels such as coal, oil, and gas. The tendency in the United States is to use mineral resources until they are in short supply; the price then increases and we seek alternatives. Presently we are seeking alternative sources of energy and

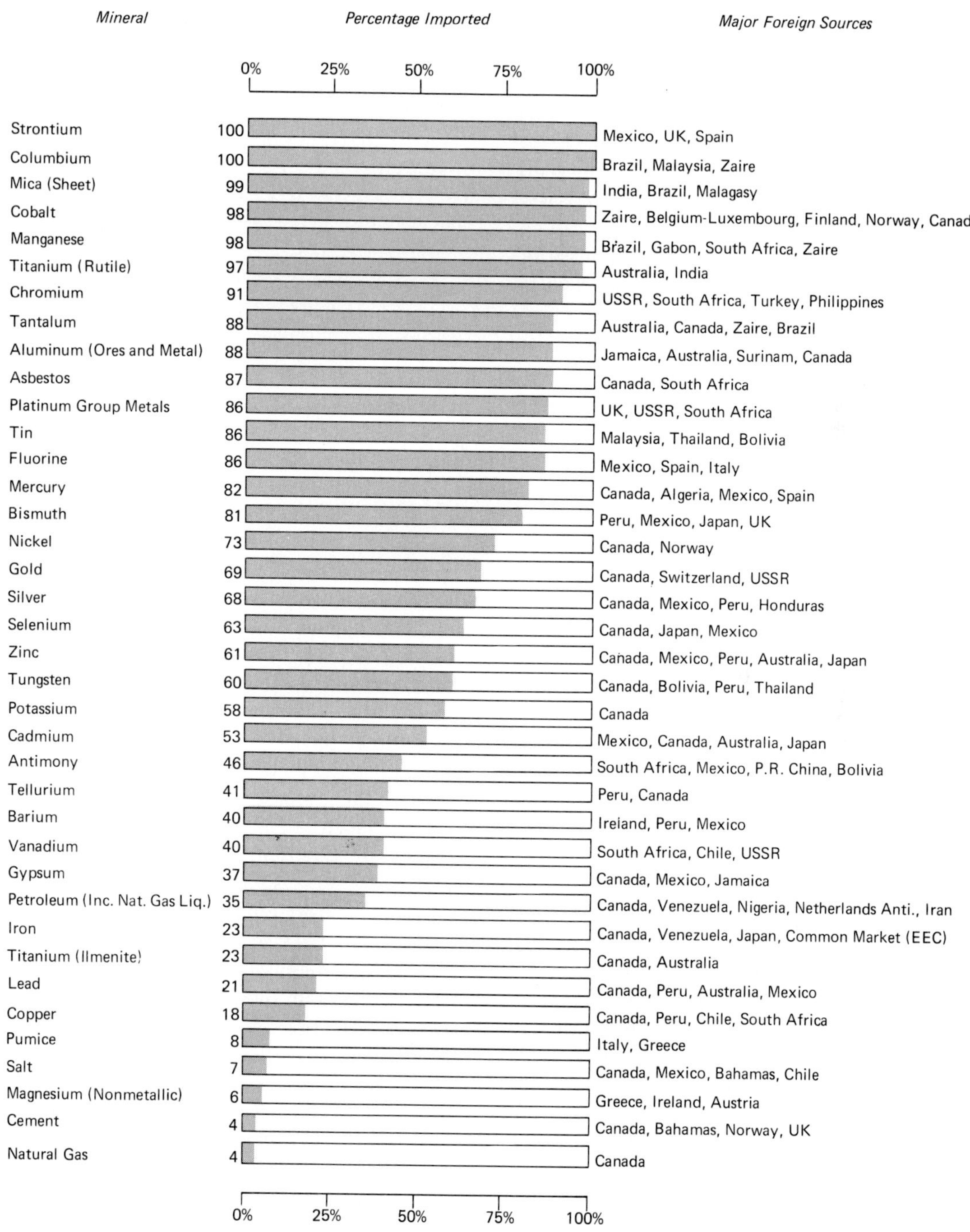

Figure 4.15
Percentage of selected minerals imported by the United States, with major foreign sources noted.

must shortly search for zinc, copper, and lead substitutes.

Many industries can switch to a number of different minerals as raw materials. For example, builders can use many forms of insulating materials or substitute plastic for copper piping. The food industry can preserve and package food in a variety of ways. It can be frozen, smoked, salted, or dried and packaged in aluminum, steel, plastic, glass, or paper containers. Major industrial shifts, such as changing a mineral used in making automobiles, can have a big impact on our economy.

In recent years the development and production of *new materials* has experienced very rapid growth, and there is intense international competition for leadership in this relatively new field. Many products made from traditional mineral sources have been outperformed and replaced by products made from new materials. New materials include metal alloys, ceramics, and polymers (plastics) that are widely used in construction and in the transportation and electronics industries.

Social or political changes can also have an impact on mineral use. The recent concern about the environment which began in the late 1960s is an example of such a social change. The energy industry is now under pressure to preserve the natural system and at the same time extract energy-producing minerals from the earth.

Supply and Demand Once used, no additional amounts of minerals can be created or added to the earth's system, making them *nonrenewable.* If we estimate the production rate of each mineral reserve, we can show the year in which peak production will occur. Further estimates can be made of the probable time of the practical depletion of that mineral resource. Data compiled on the years of peak production and depletion for selected mineral resources are listed in Table 4.5. Shifts can occur in these data as new reserves are discovered or price increases prompt consumers to shift to alternative materials.

Table 4.5
Year of peak production for selected minerals of the world

Mineral	Probable Year of Peak Production	Probable Year of Practical Depletion
Aluminum	2060	2215
Asbestos	2015	2105
Chromium	2150	2325
Coal	2150	2405
Gold	1980	2075
Lead	2030	2165
Crude oil	2005	2075
Tin	2020	2100
Zinc	2065	2250

From V. E. McKelvey, "Mineral Potential of the United States," in *Mineral Position of the United States*, 1975–2000, ed. E. N. Cameron. Madison: University of Wisconsin Press, 1973. Reprinted by permission.

Today we use minerals at a faster rate than the population growth rate. Since this use rate cannot continue indefinitely beyond the year of peak production, alternatives to many common minerals must be found by the year 2050. While reserve estimates of some minerals seem high, current use rates and projected demands are also high.

Individuals and companies have become more aware of the limited resources of many minerals. Minerals such as iron, lead, aluminum, and antimony are now derived in part from recycled minerals. The U.S. Bureau of Mines is investigating the conversion of materials such as urban refuse, industrial wastes, and stockyard wastes to recyclable minerals. For example, animal manure and organic materials are sometimes converted to oil; precious and common metals are recovered from demolished cars, aircraft, and industrial scrap; and used tires are heated to yield oil and gas. Some recycling projects are abandoned because of high cost, but research could eventually add measurably to our resources.

Although recycling must be considered in the cost of the material, it solves several problems inherent in our industrial society. It reduces waste material, prevents environmental contamination, and extends the lifetime of mineral resources. Recycling minerals and manufacturing products with longer lives add new dimensions to the future of mineral resources. Because of recycling, we can no longer predict usage from statistics of previous years. Shifts to different minerals can be expected as resources increase and decrease, as the economy causes changes in resource utilization, as other nations change their demands on mineral resources, and as our technological ability allows us to recycle and use alternative mineral resources. We can partially understand the future of our mineral reserves by looking at the available supply. Any future calculations of mineral demands must take into account such factors as clean air, restoration of mined land, and maintenance of wilderness areas. Furthermore, we must evaluate potential changes in our technology in light of shifts in demands. Can changes in technology and mineral use come fast enough to provide the materials necessary for our changing world and at the same time be compatible with our environmental concerns?

Rock Formation

To understand geologic processes, we must be familiar with three rock types. Rocks formed by the cooling of molten material are **igneous rocks.** The upwelling of molten material at ocean rifts forms rocks called *basalt.* The melting of continental debris carried into subduction zones with the plates and of continental rocks downfolded where plates bump into each other creates *granite* rocks. Molten granite can issue from volcanoes or, if it does not reach the surface, intrude as a layer between other beds of rock.

As mountains wear away, rock fragments are transported and deposited in various locations in layers that are usually flat or gently sloping. This loose material is subsequently consolidated by pressure or cemented together in a process called *lithification* to form **sedimentary rocks.** Often, large-scale earth upheavals accompanying volcanic action, the collision of continents, and earthquakes tilt and fold these layers.

When rocks are subjected to extremely high pressures and to temperatures short of melting, their physical and chemical characters are changed and they become **metamorphic rocks.** *Limestone* is a sedimentary rock, and marble is the corresponding metamorphic rock derived from limestone (Figure 4.16).

Weathering

Rocks begin disintegrating upon exposure to wind, sun, and rain. How fast the weathering proceeds depends on the type of rock, its mineral composition, and the chemical and physical processes to which the rock is subjected.

In general, igneous rocks are the most resistant and sedimentary rocks the least resistant to physical weathering. Temperature changes make rocks expand and contract, causing fractures. Water enters the cracks and freezes, with the expansion of the freezing water breaking up the rock further. Roots entering cracks can also break rocks apart as they grow and enlarge. When pieces of rock are moved by slides, wind action, rivers, and glaciers, the rubbing action of

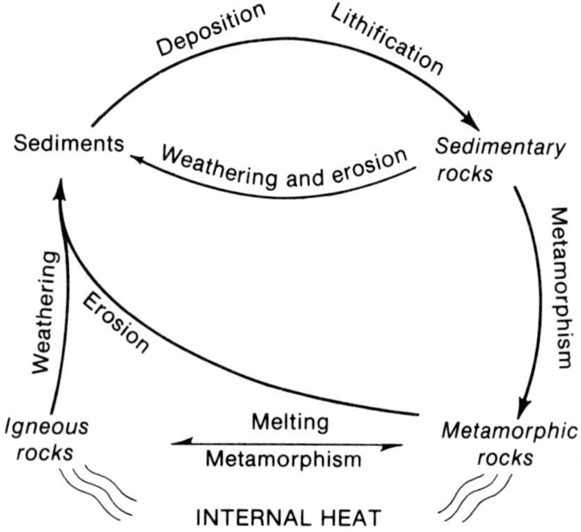

Figure 4.16
The rock cycle. (From Edward A. Keller, *Environmental Geology,* 4th ed. Columbus, OH: Merrill, 1985.)

one rock against another **(abrasion)** results in even smaller fragments. As we mentioned earlier, some rocks, such as limestone and dolomite, are readily dissolved by water and weak acids.

Soils

Disintegration and weathering of rocks reduces particle size and releases inorganic nutrients in rocks, resulting in deposits of unconsolidated material from which soil is developed. The uppermost layer of soil—topsoil—is composed of small rock fragments, sand and silt, the remains of plants and trees, and live soil organisms. Clay is a product of the disintegration of granite rock and is mixed with the sand and silt. Rain percolating through the topsoil removes (or *leaches*) the soluble materials such as calcium carbonate and iron oxides and deposits them in the subsoil below. Beneath the topsoil and subsoil is a transition zone to the unaltered base rocks or sediments which contains a mixture of base rock fragments and subsoil (Figure 4.17).

The topsoil is the *living* portion of the soil that is important to the growth of trees and other vegetation. The decaying remains of trees, plants, and other organic material is called **humus.** In addition to supplying organic nutrients to the soil, humus retains moisture and provides spaces for oxygen to reach the roots of growing plants. It takes ages to develop the topsoil, but only one generation to destroy it by human action. Naturally destructive processes are accelerated by removing trees, careless plowing, and failing to rotate crops and replenish the natural humus.

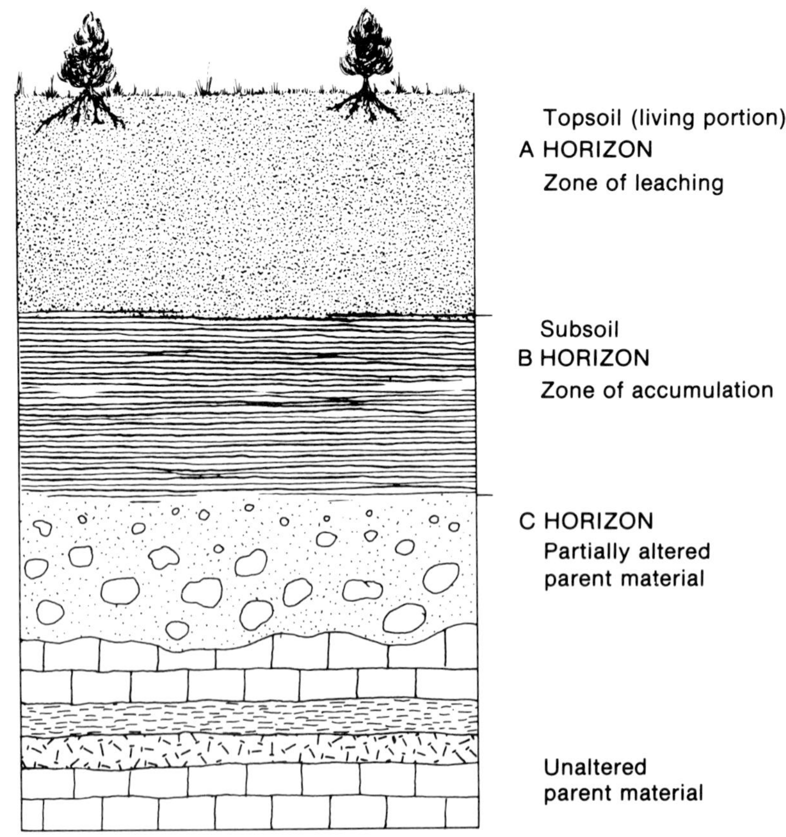

Figure 4.17
Diagram of soil horizons. (From Edward A. Keller, *Environmental Geology*, 4th ed. Columbus, OH: Merrill, 1985.)

Erosion

The general processes whereby crustal material is worn away and removed by weathering, solution, abrasion, and transportation is called **erosion.** In the process of erosion, the force of gravity acts on loosened rock and soil to restore upraised land to a base level. Water aids the process by transporting and depositing material downhill. The process of topsoil erosion begins with the splatter of the first raindrop. (In areas unprotected by vegetation and especially in arid areas, wind may cause erosion.) Rain which does not seep into the ground forms tiny rivulets that carry material downhill to be washed away by rivers. Ground cover, topography, and other factors influence the rate of erosion. In this chapter we will discuss mainly landslides and mudflows.

There is a natural angle of repose for rock and soil materials. As long as the slope of a hillside is less than that angle, the slope will remain fairly stable. The range in the angle of repose varies from flat layers of hard rock, that may be cut vertically without the resulting wall failing, to dry sand piles that will form a slope of about 45 degrees. When the ground is saturated with water, the angle of repose is usually much less than if the ground were dry. Earthquakes increase the pressure on the ground, causing an effect similar to saturation of the ground with water—liquefaction.

Where some event alters the natural slope by sharply steepening or cutting into the slope or removing material at the base of the slope (the *toe* of the hill), a landslide may occur. In a landslide, a section of ground slumps forward away from, and down the hill (Figure 4.18). At the toe, some material may even be scooped out from below the original ground level.

Landslides occur naturally where rivers eat away at the toe of a hill. The bulldozing of ter-

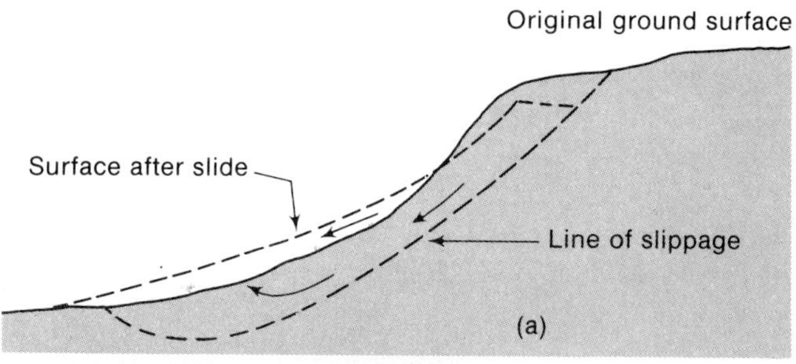

Figure 4.18
(a) Slope failure causes landslides. (b) The 1964 earthquake at Anchorage, Alaska, produced this slump. (Courtesy of W. R. Hansen, U.S. Geological Survey.)

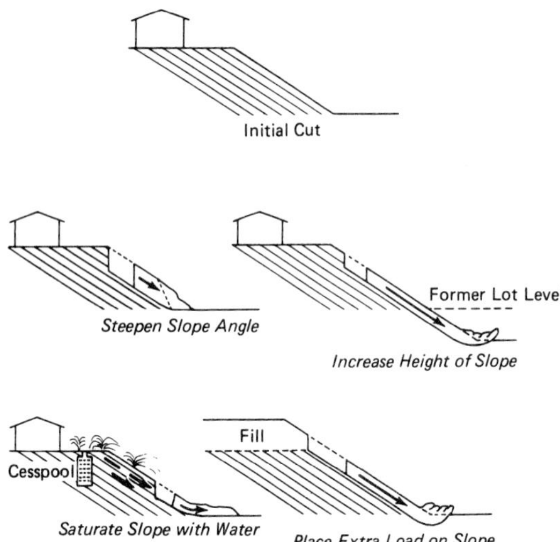

Figure 4.19
Diagram showing ways a slope can be made unstable by human activities. (Reprinted, by permission, from F. B. Leighton, "Landslides and Urban Development," in *Engineering Geology in Southern California.* Whittier, CA: Association of Engineering Geologists, 1966.)

races in hillsides for home construction and making deep cuts for highways and similar activities cause many slopes to fail (Figure 4.19).

Not only have slides been caused on the upper side of a terrace or highway by undercutting the slope, but slides on the lower side of the cut have been caused by piling up earth at an angle exceeding the natural angle of repose. In the Los Angeles area, where many landslides have occurred, geological surveys and reports are now required before such construction can begin. Leaking swimming pools and seepage from many septic tanks can provide lubrication to start a landslide. Rock slides can be caused by water seeping in between layers of rock resting on a slope.

A tragedy on October 9, 1963, killed 2200 people below the Vaiont Reservoir in Italy when a surge of water topped the dam and descended as a wall of water into the valley below. The cause was a huge slump of earth into the reservoir above the dam. The slope failure began when the filling of the reservoir caused the level of the water in the adjacent ground to rise. Rain saturated the ground even more, and the ground on one slope began to creep— 1 centimeter per week at first, then 1 centimeter per day in September, until the rate reached 10 centimeters per day early in October. A last-minute effort was made on October 8 to lower the water level in the reservoir, but it was too late. The landslide occurred the next day.

Other types of massive earth movements include debris slides, mudflows, and earth flows. Mudflows originate in small, steep canyons. Debris at the bottom of the canyon mixes with flood waters and moves down the canyon as a sloppy mass of mud and debris. Where mudflows occur periodically, the material accumulates at the mouth of canyons in cone-shaped deposits called *alluvial fans.* These are not safe places to live.

On May 31, 1970, an earthquake in Peru, which was disastrous in itself, resulted in even greater disaster when a slab of ice and rock one-half mile wide broke loose from the western face of Huascarán's northern peak. The avalanche took on the characteristics of a mudflow in its 18,000-foot descent to the Rio Santa. A wall of debris ten stories high rode over a 300- to 600-foot barrier ridge and destroyed the city of Yungay and 15,000 of its 19,000 inhabitants. A total of 70,000 were killed, 50,000 injured, and close to 200,000 buildings destroyed in the affected areas.

Earth flows result in areas where freezing pushes material up and away from the hillside. Upon thawing, the ground falls back into place slightly downhill. Repeated cycles result in the surface earth creeping downhill almost imperceptibly. Creep is especially troublesome in efforts to stabilize fresh highway cuts, where roots of grass do not penetrate to a sufficient depth to hold the earth in place. Employing deeper rooted vegetation that can reach below the depth of frost upheaval or artificial means may be helpful in stabilizing the banks.

Subsidence

Withdrawal of water from the ground faster than it is replenished by rain lowers the water table. Removal of the water allows the ground to be compacted by the weight of the material above, resulting in a settling or **subsidence** of the ground surface. The rate of subsidence is about 30 centimeters for every 6 to 10 meters (1 foot for every 20 to 30 feet) that the water table is lowered. Irrigation of loosely compacted soil will cause it to settle.

In California's San Joaquin Valley, groundwater has been withdrawn for irrigation and to drain land for cultivation since the 1850s (Figure 4.20a). In the last 40 years, groundwater levels have fallen 15 to 120 meters (50 to 400 feet) and 1400 square miles of ground have subsided, some as much as 6 meters (20 feet). Currently, the ground surface in that area is sinking 30 centimeters (1 foot) per year. Mexico, Tokyo, Las Vegas, Houston, London, and Venice are among other cities that have experienced subsidence as a result of water withdrawal.

In order to perform construction work below the normal groundwater table, an area may be temporarily dewatered. A subsidence of 13.5 centimeters (0.45 foot) was caused by dewatering for construction on Terminal Island in Los Angeles.

A similar subsidence results from pumping oil and removing natural gas from the ground. In the 1920s, the Galveston (Texas) Bay area subsided 1 meter (3 feet) or more. In the 1940s and 1950s, the land in a 22-square-mile area in Long Beach, California, subsided in a bowl-like depression up to 8 meters (Figure 4.20b). The ground was sinking at the rate of 60 centimeters (2 feet) per year when a remedy was introduced. The effect of pumping out oil was counteracted by injecting sea water as a replacement. An increase in oil yield was an additional benefit from this injection.

Subsidence resulting from pumping water and petroleum has serious implications for construction, especially in the rupture of pipelines, irrigation canals, sewers, and gas lines and the settlement of office buildings and houses. The Long Beach damage amounted to about $100 million. Areas near the ocean have also been flooded because of subsidence.

While excessive withdrawal of groundwater causes subsidence, the addition of large amounts of water, such as irrigation, results in the hydrocompaction of certain soils. In arid climates these soils are light, loose deposits comprised of about 12 percent clay. The addition of water destroys the bonding of dry clay films that hold grains together and provide supporting strength. Areas of known hydrocompaction include California, Wyoming, Montana, Washington, Colorado, Utah, and the Missouri River Basin.

Drainage of wetlands also causes subsidence. Usually these areas contain organic matter that oxidizes when water is removed. As a result, the clay dries and shrinks. Drainage of parts of the Florida Everglades resulted in 2.5 meters (8.2 feet) subsidence over a 30-year period.

Subsurface mining of ores, coal, sulfur, and salt may remove support for the overlying material and result in collapse of the surface. In 1963 an auto disappeared in a ground collapse in Wilkes Barre, Pennsylvania. For years many areas in Pennsylvania have experienced subsidence after coal mines were abandoned because support pillars were removed or destroyed. Subsidence also occurs when coal mine fires burn up the coal.

Expansive Soils

Many homes and small structures are built on foundations resting on soils containing clay. Clay expands (swells) when it is wet and shrinks when it is dry. The amount of swelling depends on the quantity and type of clay present. Montmorillonite is the most expansive type of clay. At the ground surface, the clay alternately becomes wet and dry, causing structures to be heaved up and settle back. Walls crack. Floors become uneven and doors and windows stick. In the United States damage from expansive soils amounts to 2.3 billion dollars per year. Areas most affected by these soils are shown in Figure 4.21.

(a)

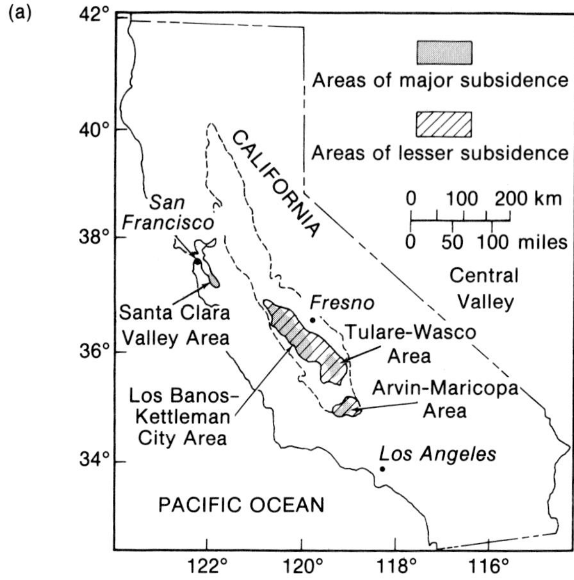

(b)

Figure 4.20
Land subsidence caused by withdrawal of water and oil. (a) Land subsidence caused by water withdrawal. (After W. B. Bull, Geological Society of America Bulletin 84, 1973. Reprinted by permission.) (b) Land subsidence caused by oil extraction in Long Beach, California. (Courtesy of the Port of Long Beach.)

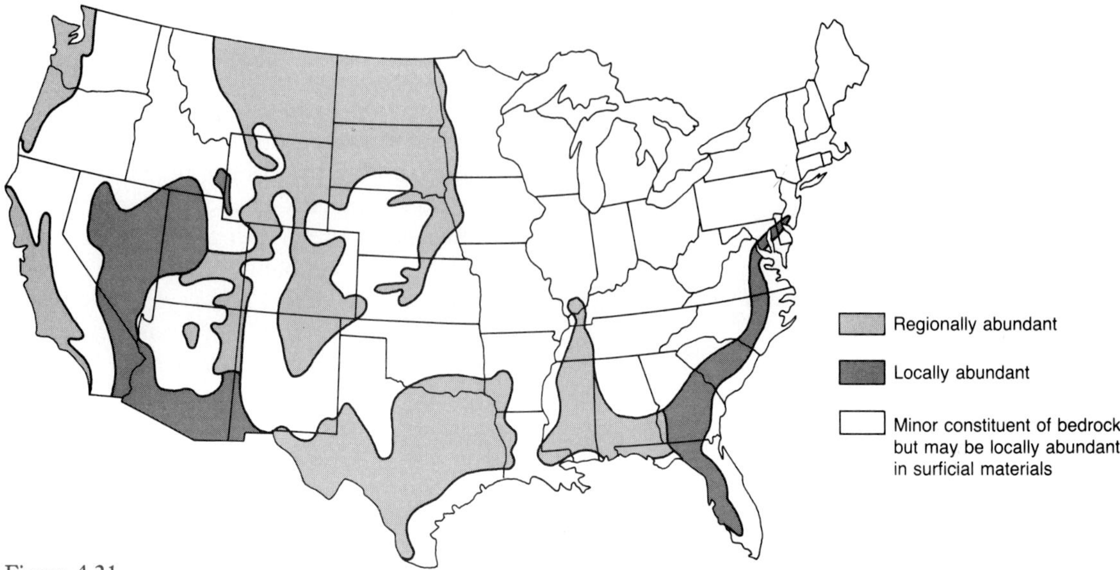

Figure 4.21
Deposits of swelling clay found in the United States. (Adapted from "Nature to Be Commanded," *Geological Professional Paper 950.* Dallas: Association of Engineering Geologists, 1978.)

SEASHORES

Continental Shelf

Along the seashore of most continents, there is a gently sloping portion of land extending out beneath the sea—the *continental shelf* (Figure 4.22). The width of the shelf varies in different parts of the world from practically zero on the western coast of South America to some 1200 to 1300 kilometers off the Arctic coasts of Europe and Russia. Off the coast of Florida, it is about 240 kilometers wide. On the average, the shelf is about 65 kilometers wide.

The shelf extends to an abrupt break where the slope becomes rather steep—the *continental slope.* (This break is the true edge of the continent.) At the seaward edge of the shelf, the depth of water is about 90 to 180 meters. The shelf is cut by submarine canyons, some of which appear to be extensions of great rivers, such as the Hudson or the Congo. These rivers cut the deep V-shaped valleys when the sea was lowered during the Ice Age. Other submarine canyons do not have such a simple explanation, and their formation is a matter of speculation.

Sediments washed from the land and carried by the rivers are deposited on the shelf. Organic material in these deposits may later be converted to oil and gas. So, shallow seas and continental shelves favor such deposits and oil production, as we can see in the Gulf of Mexico off the shores of Texas and Louisiana. Sand and gravel can also be mined from continental shelves.

Sea life abounds in the shallow waters above the continental shelf, nourished by the mineral and organic material washed from the land. Phytoplankton, such as algae, grow here because of the availability of nutrients, whereas the deeper oceans are a *biological desert* because of lack of nutrients. Estuaries—the flooded mouths of rivers—serve as breeding grounds for many species of fish and, along with marshlands flooded by tides, are very valuable ecological systems.

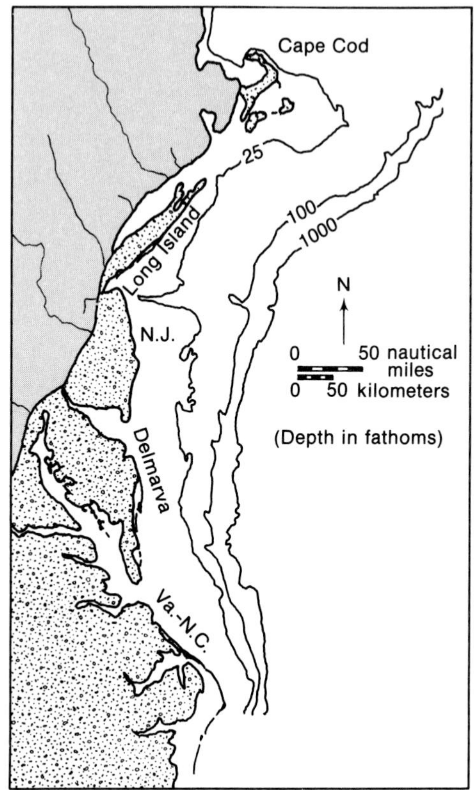

Figure 4.22
The Atlantic coastal plain and continental shelf at the mid-Atlantic bight of the eastern coast of North America. The continental shelf is the submerged area of the coastal plain. The shoreline varies from occasional headland or high land impinging on the sea and undergoing erosion to lagoon-barrier coastline with large areas of marsh surrounding the lagoons and estuaries. (From John C. Kraft, "Geological Processes in the Coastal Zone," in *Proceedings* of the Environmental Management of the Coastal Zone Conference, Drexel University, Philadelphia, 1976.)

Beaches

Beaches represent a dynamic interplay between the land, vegetation, and the sea. Wave action tends to build and erode beaches. Gentle waves tend to push sediments deposited on the continental shelf up onto the land. Stronger waves, such as those in a storm, tend to carry the sediments back toward the ocean. The sediment carried back by wave action may be redeposited to form a submerged offshore bar. The presence of the bar causes waves to break and deposit their sand, thus rebuilding the beach.

If the offshore bar continues to be built up, the bar will eventually extend above the surface, creating a bay between the bar and the land (Figure 4.23). The wind blows some of the sand toward the bay side of the bar. As vegetation takes root, the wind deposits more sand to form a dune. The dune continues to increase in height and stability and provides protection for more vegetation to grow on the bay side. As the dune grass grows, it advances toward the sea side, and the wind scoops up more sand to form a second dune in front of the original dune.

In addition to the action of waves, currents running parallel to beaches, called **littoral currents,** also erode beaches. The effect of these currents is to move the sand in the direction of the current. Where the littoral current encounters a bar or island, sand is removed from the head end of the island and deposited at the trailing end. This transport, called **littoral drift,** results in long arms *(spits)* extending in the direction of the current (Figure 4.24).

Because of their economic investment in beachfront development, hotels, and cottages, people would like to stabilize the beaches; however, efforts to counteract wave action and littoral drift have not been very successful. Bulkheads erected against waves generally are not effective. **Groins,** artificial barriers extending from the land into the sea, have produced unexpected results. The groin causes the littoral current to deposit sand on the upper approach, but on the downcurrent side of the groin the beach will be eroded. Even with groins, the beach is likely to be eroded during storms.

A *headland* is a piece of land that rises above the sea offshore. The beach between the headland and shore will be protected, although the headland itself is subject to erosion. The area behind the headland will be filled in. Old-timers

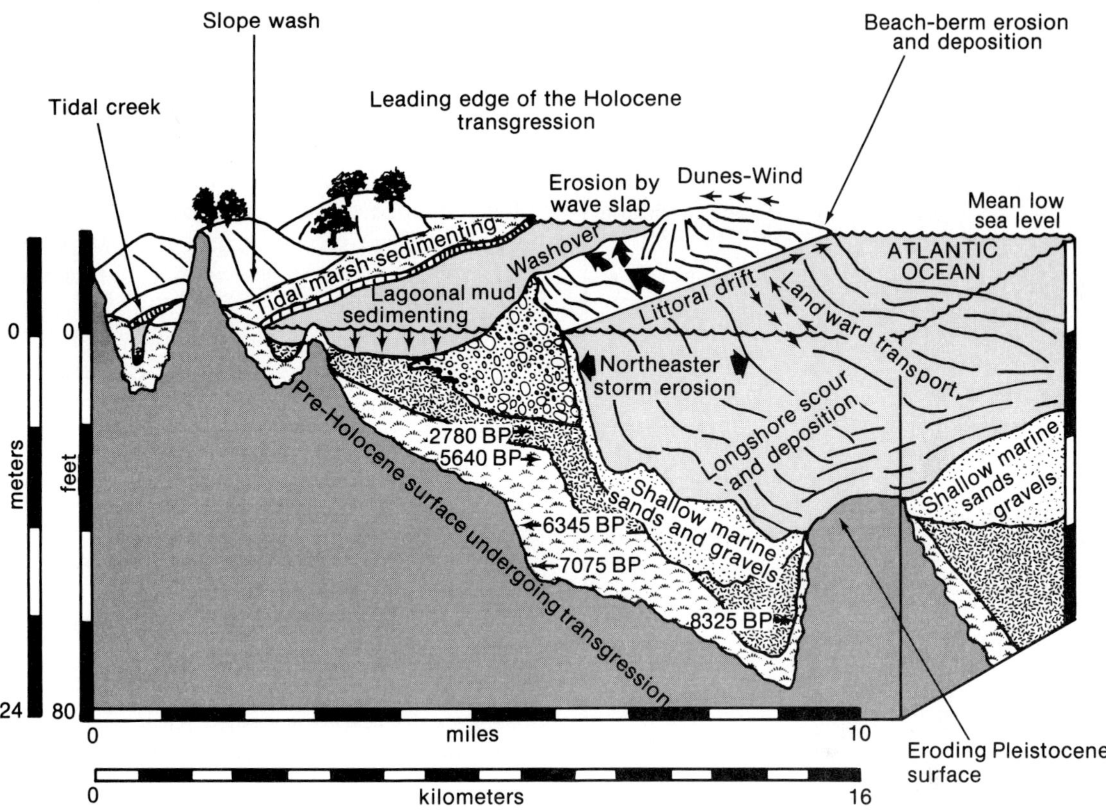

Figure 4.23

Schematic diagram of Delaware's coast showing offshore barrier and lagoon. The coastal barrier overlies the lagoonal and marsh muds dating from several millenia ago to 7500 years before present (BP). Besides indicating times of deposition of materials, the diagram shows methods of erosion and transportation of sediments that alter the structure of the coast as it migrates landward. (From John C. Kraft, "Geological Processes in the Coastal Zone," in *Proceedings* of the Environmental Management of the Coastal Zone Conference, Drexel University, Philadelphia, 1976.)

living around seashores noticed that shipwrecks tended to cause the area between the wreck and the land to fill in. This principle was used to construct some artifical headlands offshore at Singapore in order to fill in the land behind the headlands.

Barrier reefs are deposits of sandy material stabilized by plants and animals (including coral, the calcium-rich secretions of some marine organisms), and cemented or anchored to a surface. Physical and biochemical processes convert these deposits, including the calcium in remains of the dead organisms, to a type of limestone. Enormous barrier reefs have been formed off Australia and around Pacific islands. The Florida Keys are essentially barrier reef formations. The barrier reef is a living ecosystem—one of the oldest types of ecosystems in the world. The organisms involved in reef building thrive in areas where there is little seasonal change in climate. Siltation, pollution, and predators (such as starfish) can kill the reef organisms. At the present time the reefs of the Florida Keys are being killed, perhaps by siltation from dredging or from the pollution from the Palm Beach–Miami area.

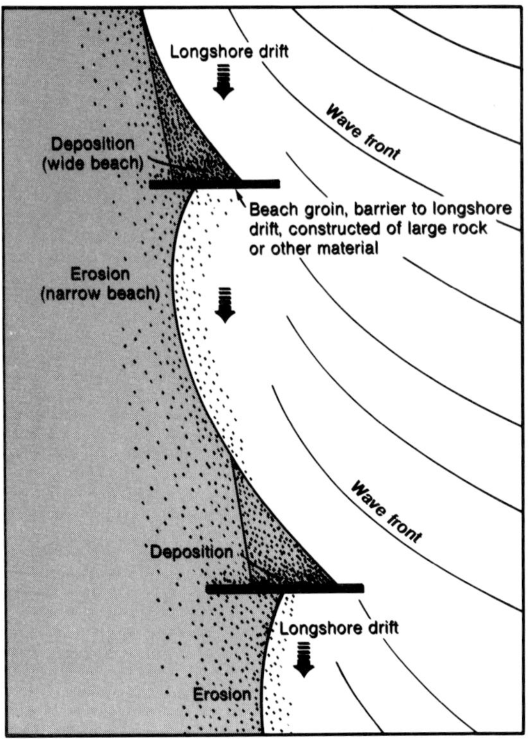

Figure 4.24
Effects of littoral currents. Currents parallel to the shore erode the head end of the island and deposit sand at the trailing end to form a spit. Where groins are constructed, sand is removed from behind the groin and deposited in front of the next one, producing a scalloped effect.

Barrier Islands

Along the Atlantic and Gulf coasts of the United States, the shore zone, (i.e., the interface between land and sea portions of coastal plains) consists of a series of barrier islands. Some stretches along the Atlantic coast have only mainland beaches open to the sea. Off South Carolina and Georgia, however, *sea islands*, mainland segments surrounded by water when melting glaciers caused the sea level to rise, have appeared.

The long chains of barrier islands were built when the continental shelf was flooded. They vary in size from 3 to 30 kilometers (1.86 to 18.6 miles) off shore, 2 to 5 kilometers (1.24 to 3.1 miles) wide, 10 to 100 kilometers (6.2 to 62 miles) long, and have low elevation, usually 3 to 6 meters (9.8 to 19.7 feet) above sea level, with some dunes up to 30 meters (98.4 feet). Although vegetation on the islands is mostly grasses or shrubs, maritime forests of pine and oak survive in sheltered areas. Lagoons or bays on the sound side are shallow and may have large tidal mud flats and marshes that teem with life, particularly nesting birds such as herons and gulls.

Under normal conditions, beaches are built seaward between storms. However, when storms erode beaches, storm and wave surge can drive water and beach sand across the islands and cut new inlets. These inlets tend to fill up if there is no stream behind them to keep them open.

Because of dam construction and erosion control measures, sediment that is transported by rivers has decreased, causing beach sand to move landward at about 1.5 meters (4.9 feet) per year. Land development and the slow rise in sea level have also contributed to beach erosion. In the past 10 years the rise in sea level has been more than 30 centimeters (11.7 inches) and the rate is increasing.

Area used for urban development on barrier islands has increased from 5.5 percent in 1945 to 14 percent in 1975, and it is continuing to climb. Much of this growth can be attributed to federal subsidy programs such as sewers, water supply, roads, bridges, disaster relief, shoreline stabilization, and flood insurance. However, since efforts to stabilize beach movement have generally been unsuccessful, the development creates a potential for loss of life and property from storms and wave surge (Figure 4.25). For example, 60,000 people now reside in the Florida Keys where a 1935 hurricane killed more than 400 persons.

To prevent these problems from occurring, some states have adopted coastal zone management acts. California's act provides public access to the shore, ensures that coastal developments serve the public as a whole, protects and restores marshes, preserves coastal farmlands, and requires stringent environmental safeguards, particularly for energy facilities.

(a)

(b)

Figure 4.25
Storm damage from hurricanes. (a) An oceanside resort hotel. (Courtesy of Lacey Photography.) (b) The same hotel after storm surge of a hurricane has overswept the area. (By Robert Madden © 1980 National Geographic Society.)

LEGISLATION

Resource Conservation and Recovery Act (1976). Calls on the Environmental Protection Agency to identify and regulate hazardous wastes and provide incentives for states to improve solid waste management programs.

Surface Mining Control and Reclamation Act (1977). Regulates the surface mining of minerals and requires reclamation of the land after mining has been completed.

General Mining Law (1872). Establishes procedures for filing claims for minerals on public lands. The law is very controversial because it lacks environmental protection clauses and no royalties are collected on the minerals. In addition, claimholders can hold a claim indefinitely without paying rent. Calls for reforming it have been heard for years.

Coastal Zone Management Act (1972). A law designed to protect coastlines by calling on states to develop protection plans and providing funds for purchasing land and carrying out their plans.

SUMMARY AND CONCLUSION

Understanding the physical systems of the earth is essential to intelligent land use planning. In the chapters to follow, we will discuss the interactions of the physical systems with the atmospheric and hydrological systems that provide an environment for life. All are dynamic systems, influencing as well as being influenced by human activities.

Our description of the earth's physical systems began with the powerful, slow, but relentless forces deep within the earth. The earth's crust is torn apart into plates that are moved about, renewed by molten rock rising to fill the rift, and ultimately consumed where one plate collides with another and is diverted underneath to sink back into the mantle. Continents are torn asunder, volcanoes born, and mountains raised in the process. Other forces—wind and rain, freezing and thawing—erode the mountains and develop the soil in which plants grow to nourish life.

While humans cannot alter the ultimate course of these events, these phenomena are considered in the location of human settlements in order to avoid or minimize the effects of such natural hazards as volcanoes, earthquakes, landslides, and subsidence. Furthermore, some human activities accelerate natural processes that are detrimental to natural ecosystems and the quality of human life. Among these human activities are those that disturb the ground, accelerating erosion or inducing landslides; the extraction of water, oil, and minerals in ways that precipitate subsidence of the ground surface; and the development of coastal areas in ways that interfere with ecosystems of estuaries and marshlands.

The U.S. Geological Survey has produced geologic maps showing the characteristics of exposed rocks and loose surface materials, topographic maps indicating ground elevation and contours and drainage, and maps illustrating land use and land cover. Images and pictures from spacecraft and ordinary aircraft reveal alignment of structural geological features, such as faults and fracture zones. This information can be used for rational land use planning in order to plan for future extraction of minerals and building materials such as stone, sand and gravel, and clay; to avoid land development in geologically hazardous areas; and to protect environmentally sensitive areas.

FURTHER READINGS

Bureau of Mines. 1983. *The Domestic Supply of Critical Minerals.* Washington, DC: U.S. Department of Interior.

Bureau of Mines. 1990. *Mineral Facts and Problems.* Washington, DC: U.S. Department of Interior.

Dayton, Leigh. 1990. Something Stirs on Mount St. Helens. *New Scientist* 126 (1717): 53–56.

Dolan, R., B. Hayden, and H. Luis. 1980. Barrier Islands. *American Scientist* 1: 16–25.

Franklin, J. R., J. A. McMahon, F. Swanson, and J. R. Sedell. 1985. Ecosystem Responses to the Eruption of Mount St. Helens. *National Geographic Research* 1: 198–216.

Laycock, G. 1991. 'Good Times' Are Killing the Keys. *Audubon* 93: 38–51.

Lutgens, F. K., and E. J. Tarbuck. 1990. *The Earth. An Introduction to Physical Geology,* 3rd edition. Columbus., OH: Merrill Publishing Co.

Pollock, C. 1987. Mining Urban Wastes: The Potential for Recycling. Worldwatch Paper 76. Washington, DC: Worldwatch Institute.

Vink, G. E., W. J. Morgan, and P. R. Vogt. 1985. The Earth's Hot Spots. *Scientific American* 252: 50–57.

Whittow, John. 1979. *Disasters. The Anatomy of Environmental Hazards.* Athens: University of Georgia Press.

STUDY QUESTIONS

1. What methods can be used to determine the age of rocks? Why is it important to know their age?
2. In Chapter 2 we learned that life is based on photosynthetic energy. How can life exist in a rift zone where there are "black smokers," but no direct sunlight?
3. What evidence is there that seafloors are spreading and continents are moving?
4. What are the implications for people of the movement of continents?
5. How would you explain the differences between a volcano that forms at a subduction zone and a volcano that forms over an oceanic hotspot?
6. Some geophysical events such as volcanoes or earthquakes can be hazardous to human populations. What kind of hazard management system would you develop to deal with an earthquake, a volcano, or some other geophysical hazard?
7. Why do you think the Mount St. Helens area is recovering more rapidly than expected?
8. Can you name some items that have become important resources in the last hundred years? What materials were, but are no larger, important resources?

9. How do the concepts of speculative reserves, supply and demand, and recycling relate to management of nonrenewable resources?
10. How do people alter the processes of erosion?
11. How can planners use information about geophysical systems?
12. How do human activities cause ground subsidence?
13. Should government funds be used to subsidize development of offshore islands in the Gulf of Mexico and along the Atlantic Coast?

SUGGESTED ACTIONS

1. Buy products that can be recycled and recycle them.
2. Find out if your college or university reuses and recycles materials, and purchase recycled materials when they are available. If they don't, find out why. In particular, encourage dormitories to set up recycling programs.
3. Look for articles about environmental issues in newspapers and news magazines and either clip them out, photocopy them, or make notes about their content. When you have accumulated articles over a 2 to 3 week period, look for patterns in the kinds of issues they describe, what people think should be done about the issues and other pertinent factors.
4. Where do geologic hazards cause the most problems? Keep track of newspaper accounts of earthquakes, volcanos, landslides, and other geologic hazards. Mark the location of the events on a world map. What patterns do you observe? How do you account for these patterns?

5

THE HYDROLOGICAL SYSTEM

Living organisms consist of more than 70 percent water. Water is important to life because it serves as a medium for moving nutrients into plants, regulating heat, and supporting aquatic systems.

Although there is no shortage of water on a global scale, water is not always in the right place at the right time and of sufficient purity to meet our needs.

Civilizations usually flourish where there is water. Without water, mineral-rich topsoil dries up and blows away. Transportation, recreation, and electric power facilities develop around water. But people may not want to live near water or the water may be so polluted that it is not readily usable. Los Angeles requires an immense amount of water from the Colorado River. Because it is still not enough for all its residents, a tremendous project brings water from northern California to southern California. There is ample water in the Hudson River, but because it is polluted at its lower end, New York City brings its water supply from the Catskill Mountains and the Delaware River Basin.

Water courses may be described as the arteries of civilization, yet unmanaged water can bring death and destruction. Floods bring havoc; the dripping of water wears away mountains; and polluted water may contaminate ecosystems and transmit disease organisms and toxic chemicals to humans.

WATER IN CIRCULATION

The Hydrologic Cycle

Water in circulation on earth and in the atmosphere constitutes the **hydrologic cycle.** Some new water, called **juvenile water,** is added to the system when water in rocks is released as volcanoes spew out molten rock. Water molecules are dissociated in the upper atmosphere, allowing some hydrogen ions to escape to outer space. Nevertheless, the total amount of water on earth remains about the same. Ninety-seven percent of all water is liquid salt water in the seas and oceans. Two percent is fresh water in polar ice and glaciers, while only 1 percent is flowing in streams, underground, or exists as water vapor (Figure 5.1).

Water evaporates from the seas, rivers, lakes, trees, and land surfaces, adding moisture to the atmosphere. When the moisture in the air cools, it condenses to form water droplets or ice crystals. As these become larger, they fall toward the earth as precipitation. Some of them evaporate while falling or from intercepting tree leaves, but some reach the ground. A portion of the water that reaches the ground evaporates.

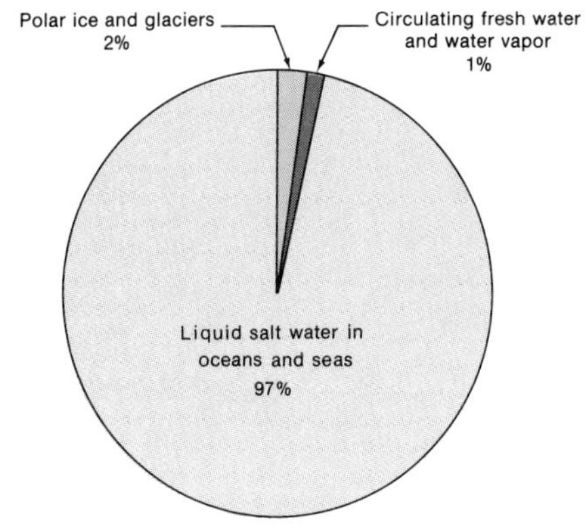

Figure 5.1
World's water supply. Life processes on land depend on the 1 percent of fresh water and water vapor in circulation.

As water accumulates on the ground, some of it runs off to form rivers and lakes, with most eventually flowing back to the sea. Evaporation from all of these surfaces completes the **cycle** (Figure 5.2). Some water filters into the ground **(infiltration)** to add to the moisture available for plants and to replenish the groundwater. Capillary action in the soil then brings water to the surface to evaporate. Groundwater may ultimately seep into rivers, lakes, and seas. Water in the atmosphere is recycled every 2 weeks. All water in the atmosphere and on the earth takes about 4000 years to recycle.

Evaporation of water from the leaves of plants is called **transpiration.** Water enters roots of plants, rises up the plant body, and evaporates from leaf surfaces. Transpiration is closely tied to mineral cycles because many mineral nutrients enter living systems via plant roots (see Chapter 2). Plants adapted to desert conditions have a reduced leaf surface and a thick outer layer of cells

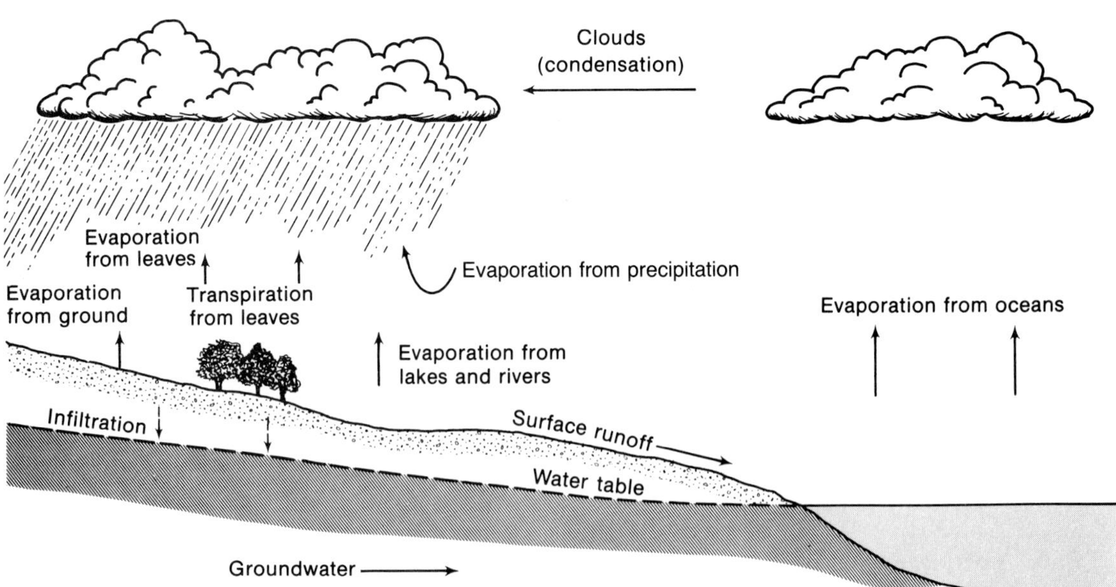

Figure 5.2
The hydrologic cycle. Some of the water falling on land seeps into the ground to replenish groundwater, and some runs off to rivers, lakes, and oceans. Evaporation from the surfaces of leaves, ground, and water returns water vapor to the atmosphere, where it condenses to fall as rain or snow.

to minimize water loss. Loss of water from a vegetated land surface is often called *evapotranspiration* because of the difficulty in separating evaporation from transpiration for measurement.

Groundwater

Precipitation seeps downward through the ground until it reaches a layer of rock or other material that is impervious, that is, it will not let water pass. Clay, when wet, tends to swell and become an impervious barrier. As rain water percolates downward, it accumulates above the impervious layer. The level to which the water rises is the **groundwater table** (Figure 5.3). Water rises in a well to the level of the water table. Where the impervious layer intercepts the ground surface, as on the side of a hill, a spring will form.

Groundwater flows down the slope of underground impervious layers, but much more slowly than water flows on the surface—sometimes only a few meters per year. The zone that is saturated with water is known as an *aquifer*. Where the groundwater is confined between two impervious layers, one above and one below, the flow is similar to that in a pipeline. The weight of water at a higher elevation puts pressure on water at a lower level. If a well is drilled into such a formation, the pressure of the water backed up to higher elevations kilometers away will force the water to the ground surface forming an **artesian well,** as in Figure 5.3.

When water is pumped from a well in an area where there is no artesian pressure, the level of the water falls, creating a **cone of depression** of water level around the well. This, in turn, causes the surrounding groundwater to flow toward the well to restore the level. In many areas, especially arid ones, water is sometimes removed faster than it is replenished by seepage of rain falling on the recharge area of the aquifer. This is called *mining* of water. If a well is located near the seashore and the rate of pumping is greater than the replacement of the groundwater by rain, salt water from the ocean may flow in the ground toward the well. Where this has occurred in coastal areas, the aquifer has been made unfit for use. This **saltwater** intrusion may be stopped by pumping wastewater back into the ground between well and seashore.

Where the surface soil has a high clay content, the soil swells shortly after rain begins, and most of the rain runs off rather than seeps into the ground. In urban areas, where buildings and paving cover much of the ground surface, very little rain seeps into the ground, and the recharge of water into the aquifer is shut off. Urbanization thereby accelerates depletion of the groundwater.

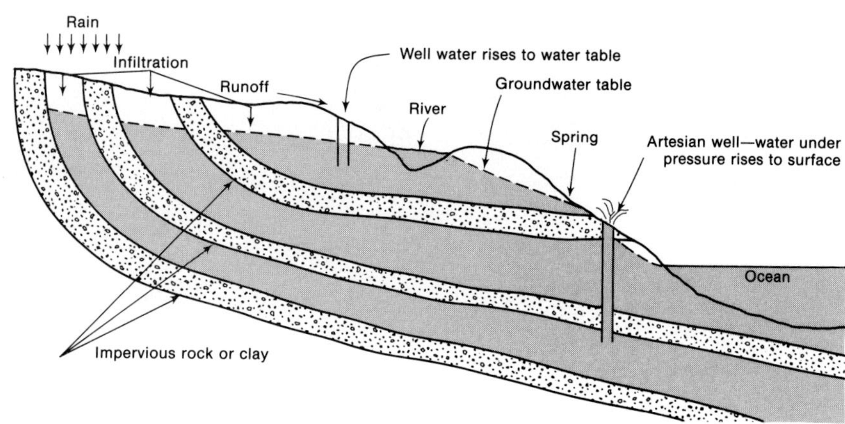

Figure 5.3
The groundwater table is the top level at which the ground is saturated with water. Where water is confined between two impervious layers, water pressure causes water to rise to the surface in an artesian well.

Limestone and Sinkholes

Limestone and dolomite are carbonate rocks and are readily dissolved by water. In carbonate rock, solution channels or caves develop in places where water seeps along joints and faults. After a while, enough material is removed so that the overlying ground surface is no longer supported and the surface collapses or sinks, forming a **sinkhole** (Figure 5.4). The Karst region of Yugoslavia abounds in this type of formation; thus, the term *Karst topography* is applied to places that have similar characteristics. Such regions exist in the United States in Tennessee, Kentucky, southern Indiana, Alabama, northern Florida, Texas, and elsewhere. These areas present hazards for construction because of the possibility of subsidence. Beautiful natural wonders, such as the Mammoth Cave in Kentucky and the Carlsbad Caverns, form in carbonate areas, but the term *Karst topography* is not usually applied to them unless surface depressions at sinkholes are also present.

Whereas topsoil filters bacteria out of water as it seeps through, water flowing directly into a sinkhole receives no filtering. As in a sewer, the underground flow in the solution channel can carry pollution many miles. Untreated water from springs and wells in carbonate areas has been responsible for several outbreaks of typhoid fever. For example, in the 1940s in Tazewell, Tennessee, a typhoid fever outbreak due to contaminated water from a spring resulted in more than one hundred cases of illness, resulting in ten deaths.

The collapse of sinkholes can swallow houses, highways, and cars and ruin cropland. Rubble and concrete are frequently dumped into sinkholes to try to stabilize them, sometimes to no avail. When the ground is saturated with water, it gives some support; but the lowering of the groundwater table can precipitate collapse of the surface in an area prone to sinkholes. In 1981 an extended drought in Florida led to a rash of sinkholes—engulfing buildings, streets, and cars on automobile sales lots. Therefore, geological surveys are advised before planning developments in carbonate areas.

Figure 5.4
Home destroyed by collapse of large sinkhole. (U.S. Department of Housing and Urban Development.)

Streams and Glaciers

Water falling on high places flows by gravity downhill to depressed areas and the oceans. Initially, the rain runs off as a sheet of water flowing across the surface; however, some soils and rocks are eroded faster than others, altering the basic surface topography. Irregularities in the surface tend to concentrate the flow of water in channels. Water, carrying pieces of rock, sand, and debris to grind away additional material in the channel areas, cuts grooves or gullies in the ground surface. Where the surface is composed of uniform material, the drainage pattern that develops is like that of tree branches and is called a dendritic pattern (Figure 5.5). Where bands of rocks at the surface that are resistant to erosion alternate with others less resistant, a drainage pattern similar to a trellis develops. From a conical rise, such as a volcano, a radial pattern forms.

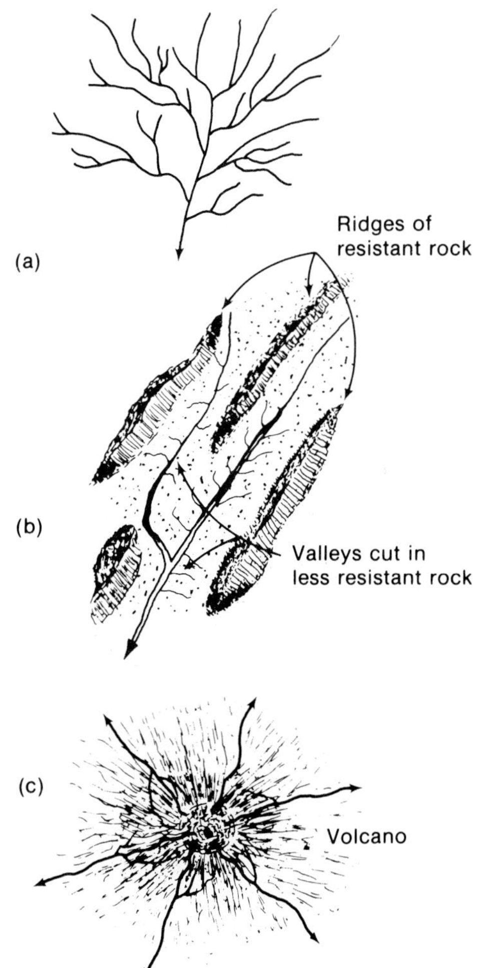

Figure 5.5
Drainage patterns: (a) dendritic; (b) trellis; (c) radial.

In the initial stages of a stream's development, the young stream moves swiftly down steep slopes, cutting the underlying rock rapidly and forming a narrow V-shaped valley (Figure 5.6a). Where resistant rocks are encountered, a waterfall is formed. Rapids develop as the resistant rock is broken off the edge of the waterfall. As the stream matures, the downward cutting slows and erosion takes place laterally, undermining the valley walls (Figure 5.6b).

This broadens the valley and the stream begins to meander back and forth across it (Figure 5.6c). An old stream is characterized by a broad valley and meandering river channel.

A river is dynamic, constantly changing course within the adjacent low-lying area of land called the **floodplain.** It is common for meander patterns to change during high water, and the river may even cut a new channel across a meander, isolating a section of the meander as an oxbow pond. The satellite image in Figure 5.7 shows the scars of old river channels of the Mississippi River in Tennessee. Riverine ecosystems are ecologically interesting because of their constantly changing cycles of erosion and deposition. Floodplain ecological succession occurs along the banks and forms oxbow ponds and backwater areas.

A river and its floodplain constitute a single ecological unit—"the floodplain belongs to the river." A flooding river removes sediment and nutrients from one area and deposits them in another. Because of their nutrient-rich deposits, floodplains are often ideal lands for agriculture. Floodplains are also popular sites for subdivi-

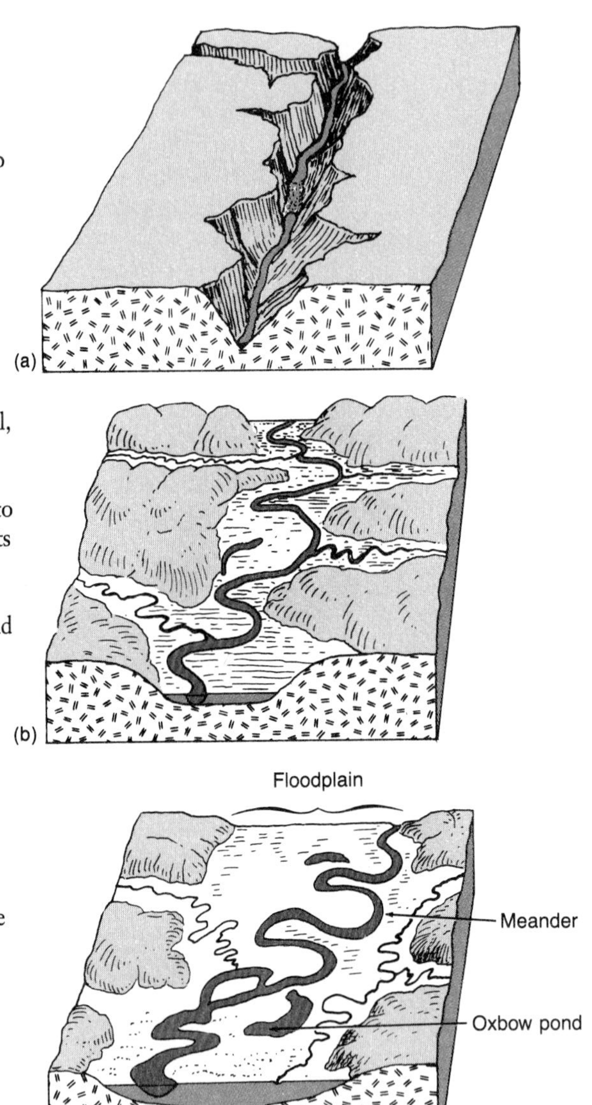

Figure 5.6
River valley development. (a) Early stages of a young river show banks with a steep slope and narrow, V-shaped valley. The river cuts downward rapidly at this stage. (From Edward J. Tarbuck and Frederick K. Lutgens, *Earth Science*, 4th ed. Columbus, OH: Merrill, 1985.) (b) A mature river extends its valley horizontally, cutting into adjacent slopes. (c) In its old age, the river has a broad valley. The gradient of the river and valley has lowered to such an extent that the river can easily flow in any part of the valley. Consequently, after floods new channels develop and the river meanders across the valley. (Courtesy of Ward's Natural Science Establishment, Inc., Rochester, N.Y.)

sions and urban development, although these uses often require flood control or confinement of the river to an artificial channel in a process called *channelization.*

If some event occurs which causes the ground surface to rise slowly after a stream pattern has been established, the river will tend to cut through the rising rocks, following the existing channel. In this way, such features as the Royal Gorge on the Arkansas River are created.

Fast-flowing streams carry soil and other material suspended in the water. When a river floods its banks, the flow of water is slowed and suspended material settles out. Flooding also enriches the soil by depositing organic material. In fact, although it was anticipated that flooding

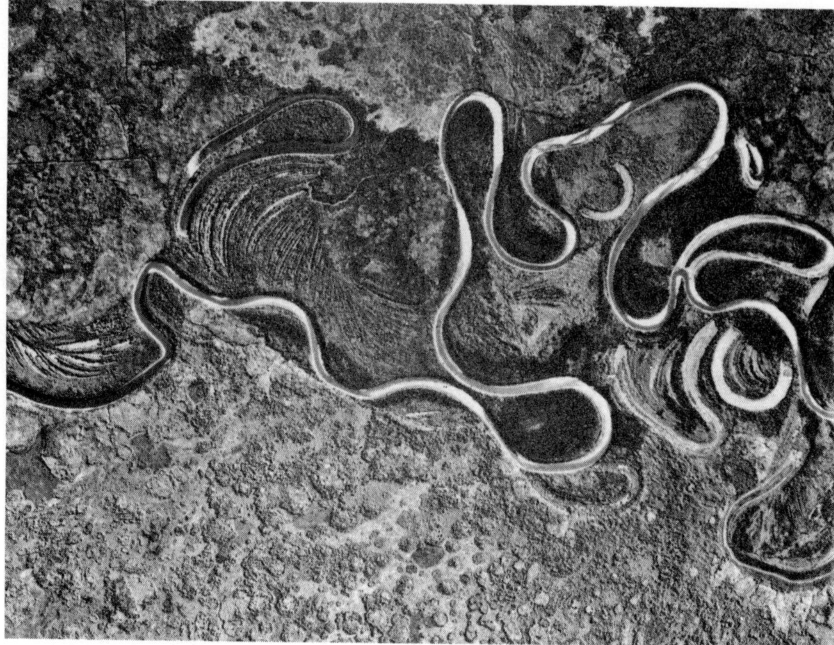

Figure 5.7
A satellite image of the Mississippi River as it meanders past Memphis, Tennessee, shows the existing channel along with the scars of old channels.

would be controlled as a result of building the Aswan Dam on the Nile River in Egypt, an unexpected result was the loss of the periodic enrichment of land that had been flooded previously. Consequently, chemical fertilizers had to be used to replace nutrients deposited by flood waters.

Next to the river, the sediments deposited from flooding form a bank or natural **levee** (Figure 5.8). Sometimes an exceptional flood will cut a path through a levee, making a new channel and abandoning the old channel. An old stream with broad valleys meanders in this fashion. The Snake River in Jackson Hole, Wyoming, for example, is higher than the surrounding plain.

When the snowfall during the year does not all melt in the summer, a snowpack accumulates. Over a period of years, the snowpack will build up, with increasing pressure exerted on the bottom layers. Ultimately, the pressure is so great that the packed snow and ice, now a **glacier,** begins to move ever so slowly, like a stream. Glaciers moving down mountain slopes carry along rocks and boulders which grind away the rock beneath, carving out a broad U-shaped valley (Figure 5.9). These so-called alpine glaciers represent a relatively small volume of the world's glaciers. More massive accumulations

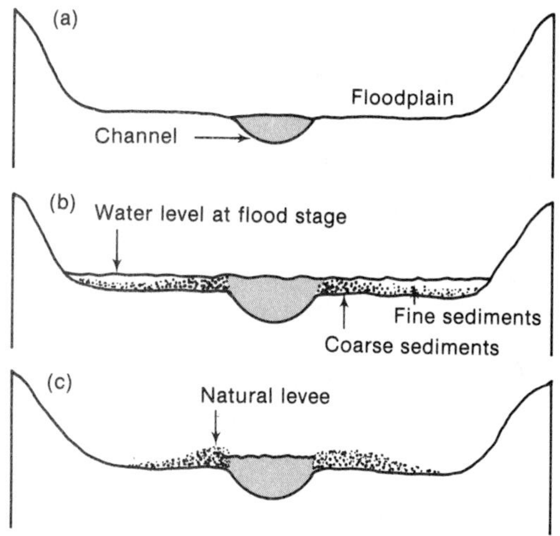

Figure 5.8
Formation of a natural levee.

Figure 5.9
U-shaped valley carved by a glacier in Yosemite National Park (P. W. Purdom).

form continental sheets of ice, as in Greenland and Antarctica (Figure 5.10), which pluck large blocks of fractured bedrock and scratch and gouge them into the surface. The finely ground material produced by glaciers is deposited where the glacier melts or is carried in the melted water.

Glaciers seem to occur with a cyclical regularity (Table 5.1), although the processes responsible for them are not completely understood.

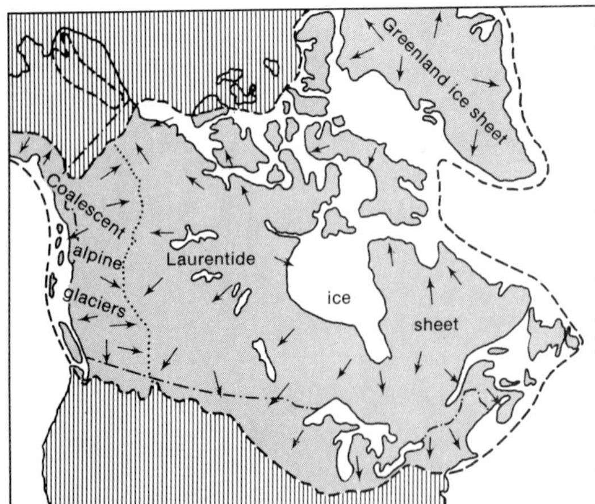

Figure 5.10
Maximum extent of Pleistocene glaciers in North America.

Apparently, glacial periods are not directly related to any long-term cooling off of the earth. Twenty percent of the earth's land area was covered with glaciers during the Wisconsin age; now 10 percent is covered—96 percent of that is in the Antarctic, and most of the remainder in Greenland.

During the last glacial epoch, the sea level was about 137 meters lower than now. Glaciers have been erratically retreating and advancing for several thousand years. Periods of melting cause the sea level to rise and flood the edges of the continents **(continental shelves)** and the mouths of the rivers **(estuaries).** The sea level

Table 5.1
Glacial Record in North America

Geologic Age (Pleistocene epoch)	Time (years ago)
Recent	10,000
Glacial—Wisconsin	100,000
Interglacial—Sagamon	225,000
Glacial—Illinoian	325,000
Interglacial—Yarmouth	600,000
Glacial—Kansan	700,000
Interglacial—Aftonian	900,000
Glacial—Nebraskan	1,000,000+

rose at the rate of 8.75 centimeters per century prior to the 1920s, but there has been a sudden increase to 60 centimeters per century. A significant rise in sea level will occur if current trends in global warming continue (see Chapter 7). If all glacial ice melted, the sea level would rise 60 meters and flood seaports and coastal areas around the world (Figure 5.11).

As glaciers melt, a great weight is removed from the land and the land rises like a ship that has been unloaded (due to isostasy, mentioned earlier). In Finland, the land is rising from 0.3 to 0.9 meter per century. The bottom of the Hudson Bay in Canada, now 150 meters below the sea level, is predicted to rise above the surface of the ocean as it fully recovers from the depression caused by glaciers in the past.

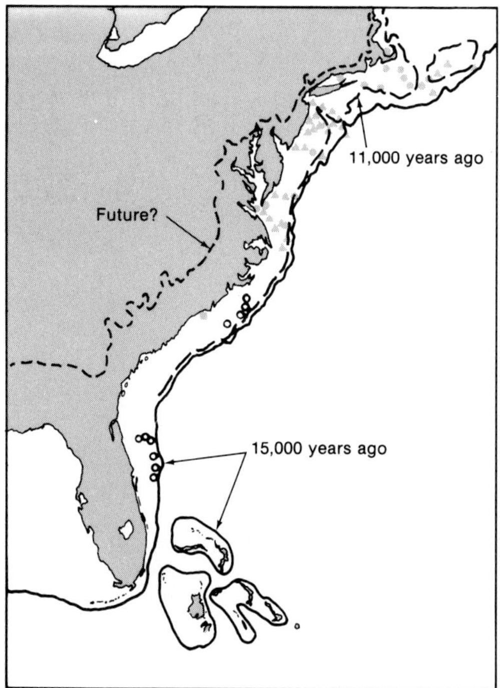

Figure 5.11
Projected rise in sea level if present-day glaciers melt. (From K. O. Emery, "The Continental Shelves." Copyright © 1969 by Scientific American, Inc. All rights reserved.)

LAKES

Stratification

During normal flow, irregularities in a river's banks and bottom cause turbulence, mixing the water from top to bottom. This continual mixing brings water into contact with air to absorb oxygen and allows gases from decomposition of organic material in sediments, such as carbon dioxide, hydrogen sulfide, and methane, to escape into the air.

Where the flow slackens, as in a lake, turbulence is lacking and there is no mixing. The same thing occurs in a reservoir. Consequently, the water in a lake forms layers *(stratifies)* because as surface water cools, it becomes denser and sinks. Water has a tendency to form latticelike crystals with lots of open space. In warm water, these bonds are broken as fast as they form; however, when water is cooled to 3.98°C, the bonds are stronger and the cold water becomes less dense. Ice is even less dense and floats on top. Thus, lakes do not freeze all the way to the bottom and fish life can survive.

In temperate zones during the summer, water circulates only in the upper warm portion—the **epilimnion.** Water in the lower cold layer, the **hypolimnion,** is separated from the atmosphere and may stagnate if all its oxygen is depleted. Between these two layers is a transition zone called the **thermocline.**

In winter the temperature at the surface may range from below freezing where there is ice to about 4°C. Because the maximum density of water occurs at about 4°C, water at that temperature sinks to the bottom. Thus, the temperature tends to be uniform below the surface at about 4°C.

In the spring, the ice melts and water at the surface warms to 4°C, whereupon it sinks to the bottom. This results in a circulation of the water, or spring turnover, which brings the stagnant water to the surface with all the decomposition gases it accumulated over the winter. Some ice-covered northern lakes experience winter kill when heavy snow cover prevents light from penetrating to the water below. Algae and other vegetation cannot photosynthesize and oxygen is depleted, causing fish and other organisms in the lake to die. In fall the surface water cools to 4°C and sinks, causing fall turnover. Tropical and subtropical lakes, where the temperature never falls to 4°C, have only one turnover a year, in the winter season when the surface water is cooled.

Like rivers and streams, lakes and reservoirs are dynamic and constantly changing. As mentioned in Chapter 2, natural erosion tends to fill in lakes and reservoirs with sediment and organic matter; the organic matter contributes to the process of eutrophication.

INFLUENCES ON THE WATER CYCLE

Precipitation

Many factors influence the water cycle. Of these, weather and land surface are the most important. Rainfall is influenced by wind action and mountains. While short storms may be quite intense, storms of lesser intensity but longer duration may actually produce a larger volume of precipitation. In general, the larger the area over which rain falls, the lower will be the average rate over the entire area.

Precipitation varies considerably over the United States, with two-thirds falling in the east and one-third falling in the west (Figure 5.12). Seventeen western states, however, consume 85 percent of the water in the United States, mainly for irrigation. Worldwide patterns of precipitation are shown in Figure 5.13. Notice the arid areas at about 30° north and south latitude.

Human activities also alter the natural water cycle. People withdraw water from rivers and the ground for cleaning, cooking, manufacturing, cooling, irrigation, and other uses. After many of these processes, the water is returned to the river or ground for further use downstream. In its use, the water may be polluted by sewage, industrial wastes, and chemicals. As water flows downstream it may be partially cleansed by the actions of mi-

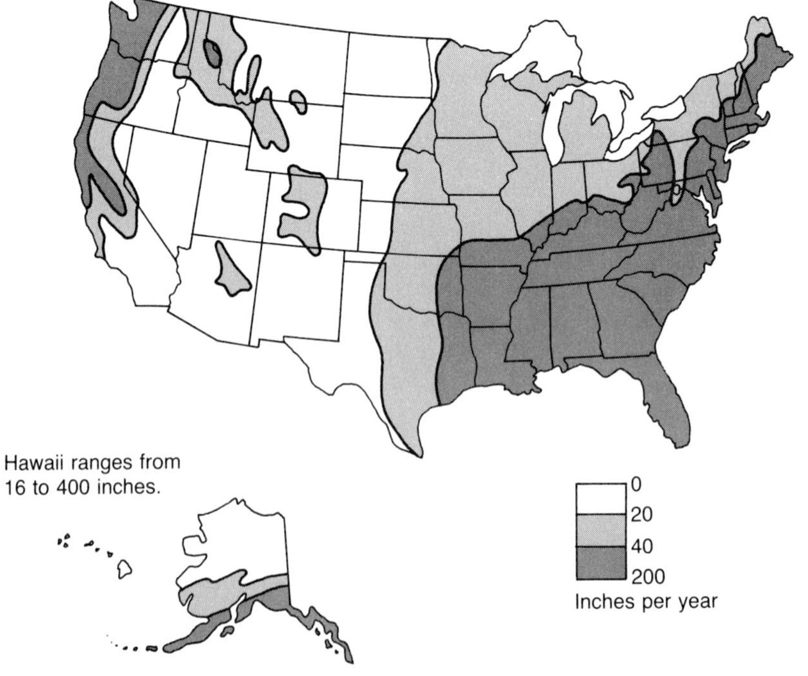

Figure 5.12
Average annual precipitation in the United States. (From Water Resources Council, *The Nation's Water Resources.* Washington, D.C., 1978.)

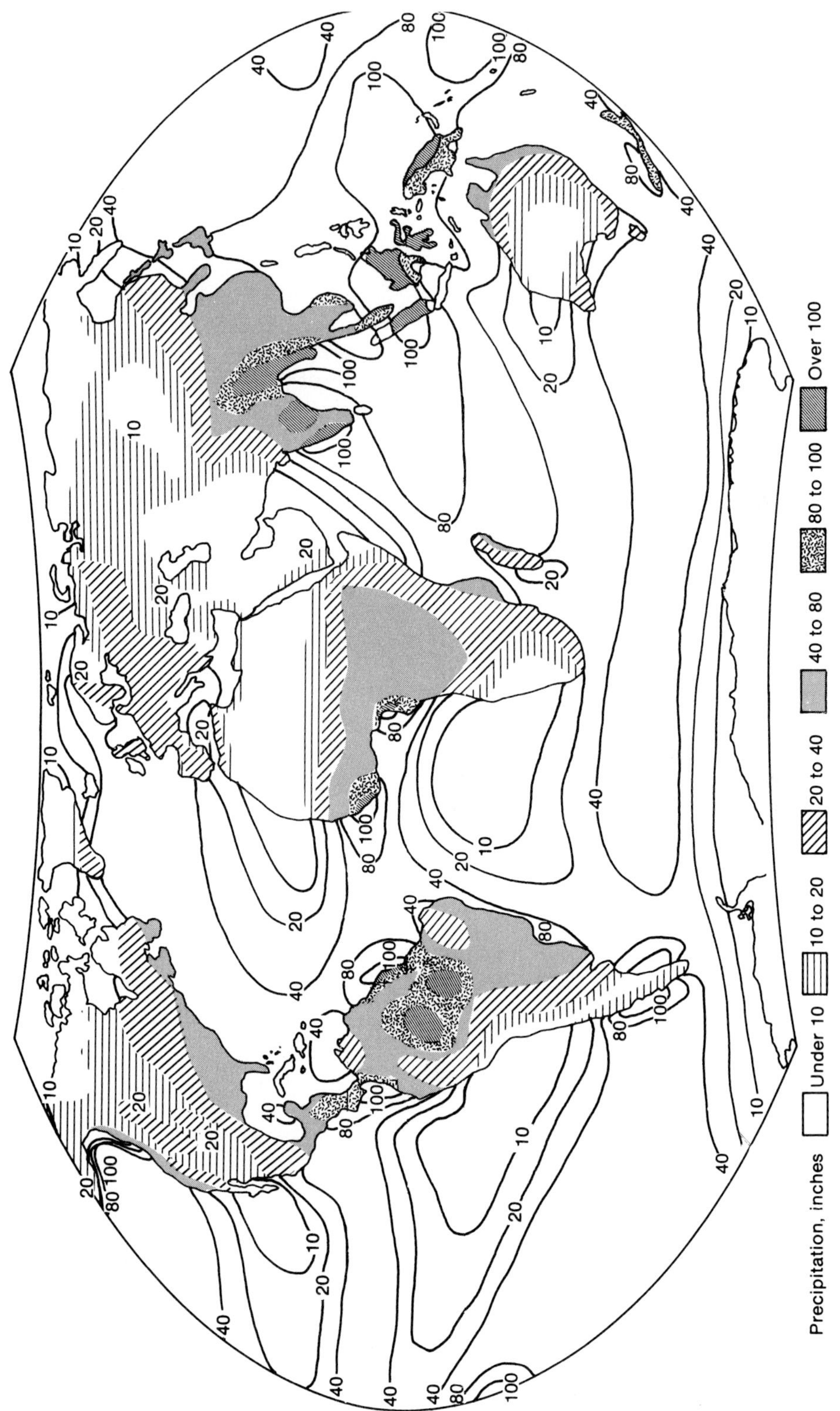

Figure 5.13
General pattern of annual world precipitation. Note the arid areas at about 30° north and south latitudes. (From U.S. Department of Commerce, *Climates of the World*. Washington, D.C., 1969.)

croorganisms and settling of solids. Dams, which store water for release in drought periods, also provide an extensive surface for evaporation. Human activities related to agriculture, forestry, and urban development influence the amount of water that seeps into the ground or runs off into rivers.

Effects of Soil

In addition to the intensity and duration of rainfall, three characteristics of the soil affect the rate of seepage into the ground, or infiltration: composition, compaction, and amount of moisture. Most ground contains some clay, which swells when wet and thus reduces the rate of infiltration. As infiltration decreases, water runoff increases as rainfall continues. Rain also causes soil compaction. The effect of a 5-centimeter (2-inch) rain on an area of land is equivalent to the compression that results from dropping 64 kilograms (140 pounds) a distance of 30 centimeters (1 foot). Compaction also reduces the rate of infiltration. The length of time since the previous rain determines the amount of moisture in the soil. High moisture content can reduce the rate of infiltration.

The composition of the soil also affects its structure, or its ability to stay in a ball after it is compressed. When pressure is removed, a structureless soil crumbles. Looser—that is, less structured—soil has a greater rate of infiltration. Tillage breaks up the soil structure and permits more infiltration, but long rains will compact uncovered soil and reduce infiltration.

Soils are classified as **sand, clay,** or **loam.** Sandy soils contain 80 percent or more of sand particles, which are 0.05 to 2 millimeters in diameter. It is structureless and will not remain intact after squeezing in the hand. Clay is composed of particles less than 0.002 millimeter in diameter. Clay soils are more than 30 percent clay and are sticky and plastic when the moisture content is high. Soils heavy in clay will crack and shrink when dry and swell when wet. Initially during a rain, clay soils may hold a lot of water in the fissures and cracks, but very soon these are filled. The clay soil will swell and there will be little, if any, infiltration.

Loam is composed of less than 20 percent clay and a considerable amount of silt (particles 0.002 to 0.05 millimeter in diameter) and fine sand. Because it contains a variety of particle sizes and some organic matter, this type of soil is *friable.* That is, it will tend to form a ball when compressed, but it can be broken up easily or crumbled.

Table 5.2 shows the effect of soil composition on the rate of infiltration for bare soils. The higher rates of infiltration are associated with sandy soil and friable loams containing little clay. The soils containing dense clay have 10 to 100 times less infiltration capacity.

The soil surface also affects infiltration rates. In very short rains, little indentations in the surface retain water and allow it to soak into the

Table 5.2
Infiltration Rates for Selected Combinations of Soil and Ground Cover

Soil Categories	Infiltration Rates in cm/hr (1 cm/hr = 0.4 in/hr)
Soils with high rates (sandy soils, friable loams)	
Bare soil	1.3–2.5
Forest and grass	3.8–19.1
Close-growing crops	3.2–7.6
Row crops	1.7–3.8
Soils with intermediate rates	
Bare soil	0.3–1.3
Forest and grass	0.8–9.5
Close-growing crops	0.6–3.8
Row crops	0.3–1.9
Soils with low rates (dense clays, clay loams)	
Bare soil	0.03–0.3
Forest and grass	0.08–1.9
Close-growing crops	0.06–0.6
Row crops	0.02–0.4

ground. The slope of the surface of the ground determines how fast water runs off and how much time there is for the rain to filter into the soil. If the ground is covered with paving or other impervious materials and structures, there is no infiltration.

Effects of Vegetation

Vegetation covering the ground helps to retain water and increase the infiltration rate. Rain falling on bare soil breaks up loose, coarse particles on the surface, leaving a layer of fine material that blocks openings in the soil and retards infiltration. A cover of vegetation protects soil from the pelting of the rain. Ground cover may increase infiltration rates up to 3 to 7 times (Table 5.2). Row crops provide some soil protection as they mature, but not as much as close-growing crops. Humus collected on the forest floor protects soil in forests. The trees alone contribute little to the increased infiltration.

Conservation practices also influence the infiltration rates. If a forest is *clear cut*, removing all of the forest canopy, the humus might wash away. On the other hand, clear cuts can be planned so that water runoff increases without excessive erosion. Because snow often becomes trapped in the branches of coniferous trees and evaporates before it reaches the ground, tree removal allows snow to accumulate on the ground and reduces evapotranspiration. The logging operation must be done carefully to minimize damage to the soil surface. Clear cutting is an important management practice in arid regions of the Rocky Mountains where people depend on snow melt for much of their water supply.

Overgrazing of grasslands will cause grass to die so it can no longer retain the soil. After the ground is plowed for planting and before the plants grow, the infiltration rate is like that of bare soil. Contour plowing helps retain the water and increases infiltration. Rotation of crops and fertilization make the plants grow more rapidly, restoring cover quickly.

RUNOFF

The flow of water from springs and seepage from groundwater persist over long periods of drought. Water from these sources provides the base flow of a river that persists in dry weather. In some urban and agricultural areas, water is pumped from wells over long periods of time at a rate exceeding the rate of recharge from rainfall, causing the groundwater table to fall. The normal flow pattern is then reversed; that is, water in creeks and streams seeps into the ground. In dry weather, ponds and creeks in areas of excessive pumping dry up unless they receive municipal and industrial waste water discharges.

When there is rain, surface runoff augments the base flow of rivers. Rain falling near a river runs across the ground and directly into the river. More remote rain flows into gulleys and creeks that lead to the river.

If the storm is of short duration, the rain falling near the river will reach the river and flow downstream before water from more remote areas reaches the same point. If the rain lasts long enough, water from a remote area will reach that point on the river while local water is still running in, adding to the flow of water in the river. When the river channel cannot contain the runoff, flooding occurs. The more widespread and the longer the duration of the rain, the more likely there will be a flood (Figure 5.14).

In mountainous areas, the runoff fills valley river channels very quickly, causing flash floods that sometimes drown hikers, campers, and motorists trapped in the valley. Desert areas adjacent to mountains are especially deceptive because walls of water resulting from rain in the mountains can come rushing from normally dry gulches.

Somewhere an upward slope of the ground breaks and slopes downward, forming a **topographic divide.** A drop of rain falling on one side of the divide flows on the surface to one river, and a drop on the other side flows to another river. Geological features underlying the

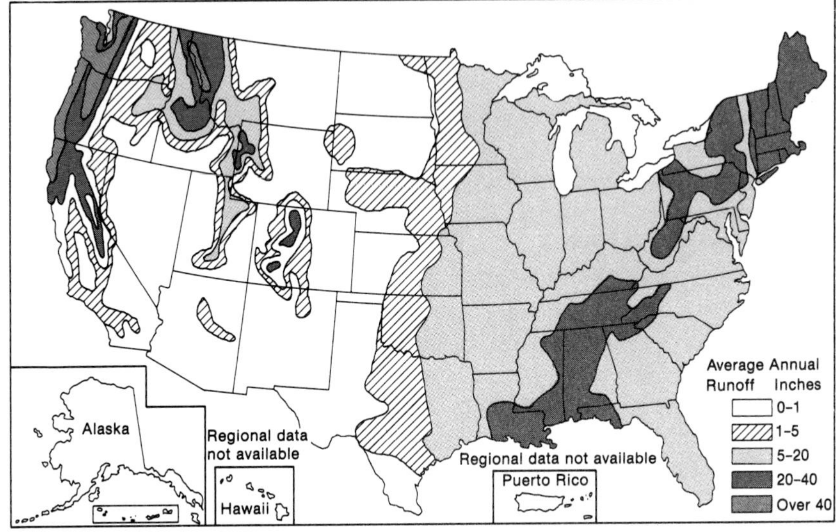

Figure 5.14
Average annual runoff in the United States. (From Water Resources Council, *The Nation's Water Resources.* Washington, D.C., 1968.)

surface sometimes cause groundwater flow opposite to the surface flow, creating a groundwater divide that does not exactly coincide with the topographic divide (Figure 5.15). The area that feeds runoff to a river or a lake—the watershed—is determined by tracing these divides, usually the topographic divide. Thus, some rain falling in the Allegheny Mountains (in the eastern United States) ends up in the Mississippi River and the Gulf of Mexico, while water falling on the other side of the divide flows to the Atlantic Ocean in the Potomac, Susquehanna, or other river. Watersheds are used as a basis for planning water resource uses and water pollution control activities. For convenience, extensive watersheds are broken into segments and several small watersheds are grouped together (Figure 5.16).

The rise and fall of the rate of flow in a river at a certain location is plotted in a **hydrograph.** For small watersheds with steep slopes and rapid runoff, the hydrograph rises and falls sharply. For extensive river systems such as the Ohio, Missouri, and Mississippi, rise and fall is less steep and the top of the hydrograph is broadly rounded.

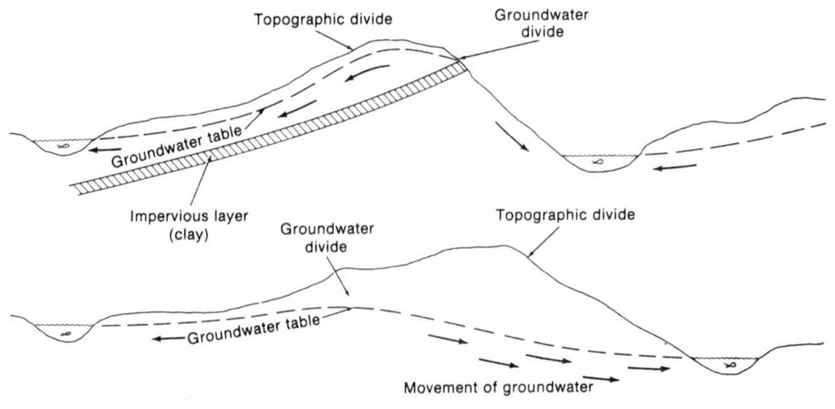

Figure 5.15
Diagram of a topographic divide. Surface water flows in opposite directions at highest elevation. The flow of groundwater depends on the geology below the surface. (From Edward A. Keller, *Environmental Geology,* 4th ed. Columbus, OH: Merrill, 1985.)

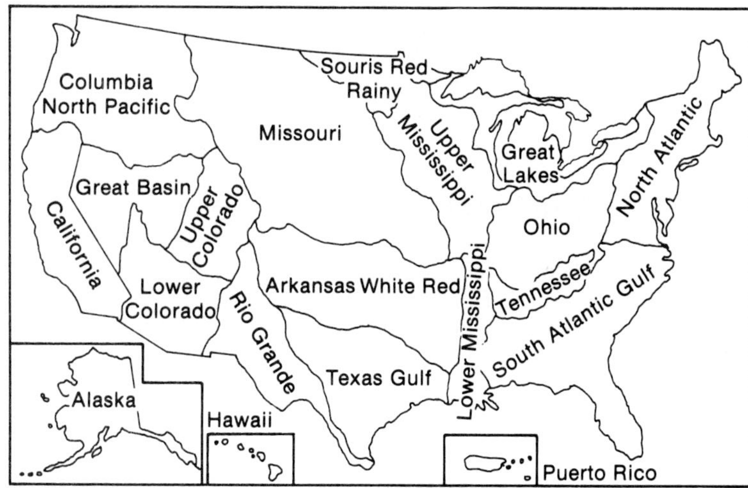

Figure 5.16
Water resource regions of the United States. (From Water Resources Council, *The Nation's Water Resources.* Washington, D.C., 1968.)

Hydrographs may be used to predict floods. Past flood hydrographs for various rainstorms on a particular watershed are plotted to see how the runoff was distributed over time. This information is used to construct a *unit hydrograph*, that is, one that shows the flow for a given depth of rain. In the United States 2.54 centimeters (1 inch) of rain is used for this calculation. The Weather Bureau measures the rain and notifies the Army Corps of Engineers. Using the unit hydrograph and the rainfall data, the Corps calculates and plots a hydrograph which shows the expected flood flow peak. For large watersheds, several days may elapse before the peak is reached and people living downstream may be warned of the impending flood. Heavy rainstorms on small watersheds may cause severe local flooding before a warning is broadcast.

Over time the characteristics of a watershed could change as a result of clearing forests, planting trees, urban sprawl, and agricultural practices. The infiltration and runoff patterns will therefore change, and the unit hydrograph used in predicting flood flows must be adjusted.

Examining past statistics on the height of flood peaks reveals the frequency with which certain floods can be expected to occur. One flood peak may occur once in 50 years, another every 100 years, and still another every 200 years. It is possible for a 100-year flood to occur one year and the 200-year flood the next. Since we are considering possibilities, a 200-year flood could occur at any time in the 200-year period, but on the average would only occur at that frequency. Often planners of flood control measures use the 100-year flood as a guide.

Flood Protection

Rivers flood periodically and "recapture" their floodplains. In some areas, such as the Mississippi Valley, floods leave deposits of fertile soil that are beneficial to agriculture. At other locations, floods erode the land and destroy growing crops. Downtown Pittsburgh and Chattanooga used to be ravaged by floods before flood control dams were built upstream. Buildings on the floodplain obstruct the flow of water, adding to the level of water behind the obstruction. The flood may wash away the buildings, or if they remain intact, flood the interior and damage goods, furnishings, and the building itself. Harrisburg and Wilkes-Barre, Pennsylvania, were severely damaged by floods in recent years.

Flood protection activities are aimed at reducing the peak height of flood flows in order to keep the river within its banks. Failing this, other measures may minimize damage from floods. Flood protection measures are sometimes classified as either structural (dams, levees, or canals) or nonstructural (restricting use of floodplains or managing watershed to reduce runoff).

Reducing Peak Flows Watershed management, such as planting forests and using soil conservation practices, can reduce flooding by increasing infiltration and reducing runoff. Flood peaks are also reduced by diverting some runoff to groundwater recharge areas where it seeps into the ground. These can be simple open land areas surrounded by embankments to hold the water. The soil composition of a recharge area must permit infiltration. Small recharge basins can be built adjacent to parking areas or even underground if necessary.

Holding back some of the runoff in temporary storage and releasing the stored water after the flood peak passes will reduce flooding. A simple approach is to construct a basin similar to a recharge basin. The storage basin may be used as a park or a playfield in dry weather and flooded temporarily, as necessary, following a storm. Other examples are the design of temporary rooftop storage for some new buildings in Denver to avoid overloading the existing storm sewers, and a storm water basin for temporary storage deep underground considered by the Chicago Sanitary District. In the latter, the falling water would generate electricity, partly offsetting that used in pumping the water out after the storm.

Structural approaches to flood control involve improving channels to allow the water to move on rapidly and the building of levees or dikes to protect low-lying areas from flooding. Low dams, built on the upper reaches of a watershed, hold water back and then release it during periods of low flow. A system of such dams was built by the Tennessee Valley Authority (TVA) on the Tennessee River watershed. However, one disadvantage is that these dams cannot intercept water flowing directly into the river downstream. Systems of high dams, such as those on the Colorado River, contribute to flood control in addition to generating electricity and storing water for irrigation, municipal needs, and other uses.

High Dams To adequately control floods, it might be necessary to build large dams on the main stem of a river, but this action might also permanently flood some land and reserve other land for flooding when needed. Large dams can serve multiple purposes—flood control, generation of electricity, public water supply, irrigation, augmentation of low flow in the river, and others. Water sports, boating and fishing, and other recreational activities may be developed in conjunction with other uses of the reservoir. Boats may pass around some dams through a system of locks, which allows a boat to enter a channel that is blocked off by the locks while the water level is raised or lowered as necessary for the boat to proceed to the next level.

Some persons are concerned about the ecological effects of building large dams. Permanent flooding behind the dam destroys the existing ecosystem, prime agricultural land is permanently flooded, and silt accumulation behind the dam will ultimately fill the reservoir, making it useless. A new reservoir may provide ideal breeding grounds for disease-carrying mosquitoes; however, fluctuation in the water level may expose mosquito larvae to fish predators.

The decomposition of organic sediments accumulating in the depths behind the dam depletes the amount of oxygen dissolved in the water. Consequently, water discharged downstream may not have enough oxygen to sustain game fish, such as trout. This condition may be corrected by mechanically reoxygenating the water. Construction of the Aswan Dam severely altered the ecosystems at the mouth of the Nile, reducing commercial fishing opportunities.

The U.S. Supreme Court in 1978 halted the construction of the Tellico Dam in Tennessee because it would have destroyed the habitat of the snail darter, a species of fish protected under the 1973 Endangered Species Act. Considerable debate ensued, and the cabinet-level Endangered Species Committee was asked to review the case. The committee agreed with the Court's decision, concluding that dam construction should be blocked on economic grounds, in addition to its threat to the snail darter. Subsequently, an amendment to an energy and water bill passed by Congress exempted the Tellico Dam from the Endangered Species Act and allowed its completion. The snail darter still survives, however, because the fish were transplanted to another river, and, in 1980, additional small populations were located in other rivers of the region.

Dams also prevent the migration of fish, some of which use the headwaters for spawning. *Fish ladders* permit some of the migrating fish to reach the spawning areas. A fish ladder contains small pockets of water in a steplike sequence. Water from above the dam flows from one pocket to the next to the river below. Fish are used to negotiating rapids and similar small obstructions. They follow the onrushing flow of water and will jump from one pocket to the next one upstream if necessary.

Other objections to dams include the creation of mud flats in the upper reaches of a river when the level of water in the reservoir is low. The scenic beauty of a natural river may be destroyed. In 1976, Congress added the New River in North Carolina to the National Wild and Scenic Rivers System after a 15-year battle to stop plans to build a power dam. The proposed Tocks Island Dam on the Delaware River at the Water Gap was controversial for years and was finally rejected in 1975 because it would alter the river corridor's natural beauty. Not only scenic beauty but historic value as well may be impaired. When the Aswan Dam was built, some archeological treasures were moved, but others were flooded.

Because both social benefits and costs are associated with large dams, their construction has become a controversial political issue. The value of flood control, irrigation, power, recreation, and other benefits has to be weighed in each case against permanently flooding prime agricultural land, changing ecosystems, altering natural vistas, breeding mosquitos, and other costs.

Land Use Control Another approach to flood protection is to restrict land use on the floodplain so that flooding will cause little or no damage. Agricultural and recreational uses might be permitted as they would not obstruct the flow of flood water. The construction of dwellings and other buildings that could be damaged by floods and would impede the flow of flood water may be prohibited.

Structural modification of buildings already built on the floodplain can be considered. Doors, windows, and other openings have to be made watertight and walls made strong enough to withstand the water pressure. Glass openings must be protected from floating logs and debris, and drains have to be equipped to prevent backflow of water. Floors may have to be strengthened to prevent buckling from increased groundwater pressure.

Flood Management Systems In Chapter 4 we discussed how humans have developed management systems to deal with natural hazards such as earthquakes, landslides, and volcanoes. Flooding is also a natural hazard. Flash floods, which occur suddenly with little warning, are in many respects similar to an explosive volcanic eruption like that at Mount St. Helens in 1980. Frequently recurring floods that develop more slowly and predictably are comparable to the flowing lava eruptions of Kilauea on the island of Hawaii. Each type of hazard requires a different management system to avoid or minimize its effects.

Flood hazard management systems involve a combination of the flood protection activities we have discussed, including the following:

- Watershed management, construction of dams, and other measures to reduce the level of peak flows.
- Levees, dikes, and diversion canals to keep water within a confined channel.

- Land-use controls and flood-proofing of buildings to minimize property damage.
- Flood prediction to provide early warning of impending flood.
- Emergency protection measures such as evacuation plans or sandbagging.
- Disaster relief.
- Insurance programs and other assistance to victims.

WATER SUPPLY

The fresh water in rivers and lakes constitutes only about 0.01 percent of the total world water supply, with groundwater up to a depth of 4000 meters (13,000 feet) adding another 0.61 percent. While the quantity of water in rivers and lakes at any moment is a small amount of the total, it is continually renewed by the hydrologic cycle of evaporation and precipitation.

The actual flow in a river depends on the runoff from precipitation and seepage from groundwater into rivers where there is a high groundwater table. The average annual precipitation around the world is 76 centimeters (30 inches), but this amount is not evenly distributed. In the United States, about 70 percent of the precipitation is lost to evaporation and transpiration and about 30 percent feeds the streams. Some of the stream flow is withdrawn and lost through consumptive uses, but about two-thirds of the flow reaches the ocean. About two-thirds of the precipitation in the United States falls in the east and one-third in the west (see Figure 5.12). Although the water supply is renewed and consequently inexhaustible in a sense, the supply of water for human use in any one location may be limited because of low rainfall in that area, prior consumptive use of available water, an excess of population with respect to the quantity of water resources, or the pollution of the available water by wastes from people and their activities.

It is possible to increase the supply of fresh water by *desalination* of sea water, although the process is very expensive. Even so, oil-rich desert countries such as Kuwait and Saudi Arabia have come to rely on desalination plants for a major portion of their water supply. Saudi Arabia has more than twenty of these plants, some of them using salt water from the Persian Gulf. The world became very aware of this during the early stages of the 1991 Middle East War when the Iraqis released oil into the gulf. The resulting oil slick had devastating ecological effects and threatened to contaminate Saudi Arabia's water supply. Desalination could become a very important future source of fresh water in other countries as competition for water increases and as we continue to pollute our other sources at current rates.

Water Rights

Laws regulating water use differ in the eastern and western parts of the United States. In the East, the concept of **riparian rights** prevails. The riparian system of water law developed from English Common Law where water was relatively abundant. Under this system the right to use water is based on the ownership of land bordering a stream or other body of water. Under this concept, the right of the person downstream to use the water in a river is protected. The person upstream cannot divert, diminish, or pollute the entire flow but must leave enough for the downstream user. New York City obtained the right years ago to use some water from the Delaware River; however, the Delaware River Basin Commission ordered New York City to limit its use of the water from that source during droughts to prevent salt water from coming up the Delaware to the Philadelphia water works intake.

The riparian system would not work in the more arid western states because water is often needed in areas that are distant from water sources. Consequently, the principle of *prior appropriation* ("first come, first served") is the basis of most western water law. One's water right status is determined by the date the water was first diverted from a stream and used in some beneficial way, such as for irrigation or a town's water supply. In a dry year, those who hold the oldest (senior) water rights will have water while those with junior water rights may not.

In the early days of western settlement, range wars were fought over water rights. More recently, western states have established commissions to define and regulate water use in the public interest. When there is a dispute, certain water uses, such as that for human consumption, receive preference over other uses like irrigation, even if the water right for irrigation is a senior right. Disputes over water can still become intense. In recent years Indian tribes have claimed water given to them long ago by treaty, and agencies like the U.S. Forest Service are claiming the right to manage water on federal land. They argue that laws establishing national forests, national parks, and other federal lands also gave the right to claim the water needed to manage them properly. These new claims, along with continuing disputes among states and private interests, make water resource management complex and interesting.

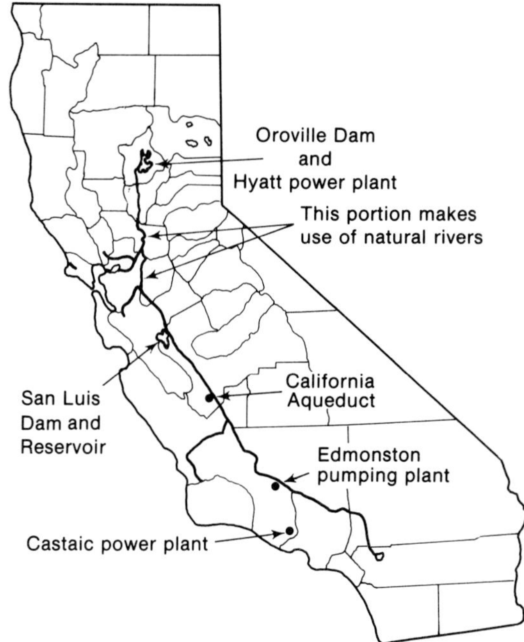

Figure 5.17
California Water Project. Stretching some 1126 kilometers (700 miles) and costing $2.3 billion, this project brings water from the northern to the southern part of the state.

Transbasin Diversion of Water

Although the prevalent approaches to water resource management involve using a watershed or a river basin, there are examples of diversion from one watershed to another or transbasin diversion. Cases involving Los Angeles and the Colorado River and New York City and the Delaware River have already been mentioned. Another example is the gigantic Feather River project in California, which conducts water from the northern to the southern part of the state (Figure 5.17). There is even a proposal to bring water from melting glaciers in Alaska and Canada to supply the western and Great Plains states. This grandiose scheme, proposed by the North American Water and Power Alliance (NAWAPA), would bring water to parts of Canada, thirty-three states in the United States, and regions of Mexico. It is so complex, however, that the possibility of its approval seems remote. Towing gigantic icebergs, or ice islands that break off the shelf ice, from Antarctica to the western part of South America, the Middle East, and southern California has been suggested. In the future, water use planning could encompass vast regional areas, transcending the normal river basin.

China and the Soviet Union are planning large-scale river diversion projects to transfer water from one basin to another over great distances. The Chinese plan to divert water from the Chang Jiang (Yangtze) River to the food-producing region of the North Plain, a distance of 1150 kilometers (714 miles). The Soviets intend to divert Siberian rivers 2500 kilometers (1553 miles) south to their grain-growing regions. However, the Soviets have removed these diversions from their most recent 5-year plan and the future of the project is in doubt. Because of their size and complexity, there has been much speculation as to the possible environmental effects of these large-scale water transfers.

Water Uses and Allocations

In the United States 1.4 million million liters (370,000 million gallons) per day of water are used for various purposes: 102,000 ML (27,000 MG) for public supplies; 492,000 ML (130,000 MG) for irrigation; and 567,000 ML (150,000 MG) for self-supplied industry.* Public supplies serve a number of purposes, including both domestic and industrial needs. The amount of water for most domestic purposes (bathing, cooking, waste disposal) is about 190 liters (50 gallons) per capita per day. The quantity of water required for commercial and industrial purposes will vary according to the types of industrial processes in the community. The average public water supply use in the United States is 628 liters (166 gallons) per capita per day. The principal requirement for a public water supply is that it be safe to drink from both a bacteriological and chemical standpoint. Other requirements are dictated by aesthetics or economics. Public water supply use may pollute the water, but the wastewater can be treated to remove pollutants before return to the ground or river for further use.

In an area where water is scarce, demand for water among public, agricultural, industrial, and recreational uses may exceed available supplies. Governmental institutions have been created to allocate surface and groundwater among competing uses. Some operate on a river basin basis, as does the Delaware River Basin Commission, which was created by a compact between states and approved by Congress. Others are run by the state, as is the Oklahoma Water Resources Board.

*ML is the abbreviation for million liters; MG for million gallons.

Irrigation and Its Effects

Water is necessary on the farm for drinking and domestic purposes and for animals, but irrigation, or *wet farming*, represents by far the greatest use of water in agriculture. To produce 20 tons (fresh weight) of crops, 20,000 tons of water must pass through the plant roots. Wherever the annual rainfall is less than 50 centimeters (20 inches), there is a risk to *dry farming*. By the use of wet farming methods it is not only possible to farm in arid areas, but it is possible in other areas to apply water to crops at crucial times in their growth (supplemental irrigation) when there is no natural rain.

There are two significant problems with irrigation. One is that a substantial amount of the water applied is evaporated from the land and transpired from plant leaves. This is a consumptive use, and the water is lost to the watershed; from 50 to 70 percent of the irrigation water is lost. Furthermore, evaporation of irrigation water from land leaves dissolved mineral salts behind. Over time, the salts accumulate in the soil, (salinization), rendering it useless for growing crops. The salts may be removed by flushing the land with large amounts of water, or by installing underground drains, but these are expensive and can raise the mineral content of the water in a receiving river to the point that is not usable for irrigation downstream.

In the western United States, the water from the Colorado River has been used extensively for irrigation. The salts in the return flow to the river have polluted the river to such an extent that relations with its downstream neighbor, Mexico, have been strained. Treaties between the two countries have attempted to deal with this problem. As a result of an agreement negotiated by the two countries in 1973, the United States implemented a plan to reduce the salinity of the Colorado River before it flows into Mexico. The plan involves salinity control projects associated with irrigated regions of the Colorado River Basin (see Figure 5.16) and the construction of the Yuma Desalting Plant in Arizona.

EASTERN UNITED STATES: PARTIAL RECYCLING OF WASTEWATER

Because of severe droughts that are expected before the end of the century, plans to deal with water shortages in the Washington, D.C., metropolitan area are being reviewed. Preventive actions under consideration to meet water deficits include conservation of water use, more efficient use through interconnections among water utilities, and an increase in the total water supply.

The Potomac River, located above the salt line, has never been used as a source of water supply because it is heavily polluted. An experimental water treatment plant has been built to determine if safe drinking water can be produced from a 50/50 mixture of polluted water from the Potomac River estuary and a secondary sewage treatment plant effluent.

The experimental treatment plant combines standard water treatment processes with additional optional systems such as predisinfection with ozone or chlorine, sludge recycling to a rapid mix tank, upflow and downflow carbon columns, demineralization using ion exchanges, electrodialysis or reverse osmosis, and final disinfection with ozone, chlorine, or ultraviolet radiation.

The microbiological quality of the effluent in total coliform meets drinking water standards, but is inferior to the conventional local water plant. All water quality goals for metal levels are being met, synthetic organic chemicals are comparable, and the total trihalomethanes and organic halogens are below those of the local water treatment plant. Limited toxicological tests with cells and mammals continue to be tested; however, preliminary results of finished water quality indicate that all EPA primary and secondary drinking water standards are being met by the demonstration plant.

Scheduled to be completed in late 1991 or early 1992, this desalination plant will remove enough salt so that Colorado River water can be used to irrigate crops in Mexico.

One-half of the land area in the contiguous United States has less than 50 centimeters (20 inches) of precipitation per year and is considered arid. Construction of dams to retain flood waters and divert water from rivers has permitted extensive irrigation in the arid areas.

Some irrigation problems can be solved with better management. In some cases, flood irrigation can be replaced with more efficient (although more expensive) methods such as sprinklers or drip irrigation systems. Even flood irrigation can be made more efficient by leveling the land to be irrigated or by providing systems of furrows, ditches, pipes, or siphons to achieve better distribution of water. Irrigation systems can also be planned so that water returning from irrigated land to a stream (return flow) can provide groundwater recharge or wetland habitat for wildlife.

Increased and poorly managed irrigation has accelerated desertification, the development of unproductive, desertlike conditions. Thirty-seven percent of arid lands in North America (2.85 million square kilometers; 1.1 million square miles) has undergone severe desertification (Figure 5.18), including 27,200 square kilometers (10,500 square miles) that is designated very severe. Most of this desertification has occurred in that portion of arid land located in the United States.

The process of desertification is associated with lowering of water tables, shortage of surface water, salinization of existing water supplies, and wind and water erosion. Contributing factors are overgrazing, mining groundwater for irrigation, poorly managed irrigation, and irrigation of poorly drained soils. Agricultural practices, industrial development, and urban growth may also accelerate desertification. Even the fertile San Joaquin Valley in California is experiencing desertification that results in a yearly 10 percent loss of crop yield valued at $31.2 million.

Desertification is an even more serious problem worldwide. Soil scientist Harold Dregne estimates that about 80 percent of the world's rangelands and 30 to 60 percent of other dryland areas are at least moderately desertified.

Lowered water tables resulting from excessive water withdrawals has also been recognized as a problem of global significance. Besides the examples we have mentioned, it has been identified as a serious problem on nearly every continent.

Water Conservation

Historically, water resource management has emphasized development of dams and reservoirs to increase water supply. It has become apparent, however, that we must also manage water demand, rather than just increase supply to satisfy the demand. Water demand can be managed by using water more efficiently, recycling and reusing water, and by water conservation.

Community water conservation programs require creativity. They usually involve a combination of water-saving technologies (for example, water-saving toilets, showerheads, and other appliances), economic incentives (water-rate

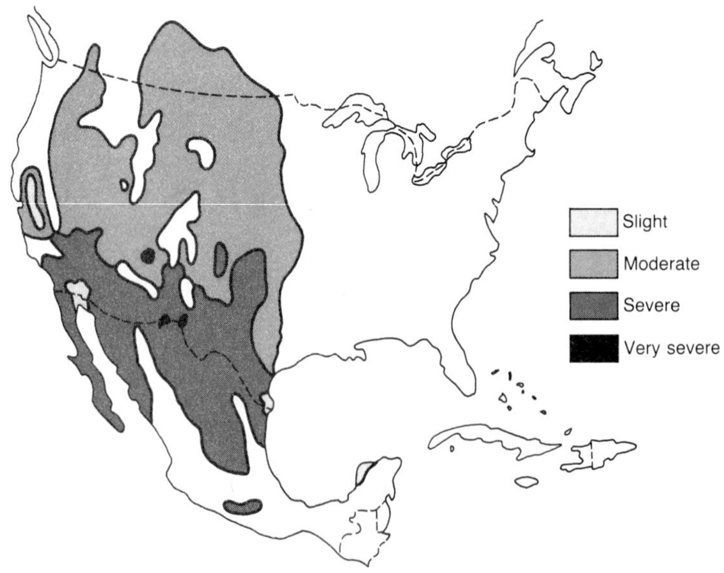

Figure 5.18
Status of desertification in North America. (From Harold Dregne, "Desertification of Arid Lands," *Economic Geography* 53[1977]:325. Copyright © by Clark University. Printed by permission.)

LEGISLATION

Reclamation Act (1902). Established a mechanism for developing irrigation projects, especially in the arid west. The Bureau of Reclamation was created as the agency responsible for carrying out the reclamation programs.

Flood Control Acts (1917, 1936, 1944). Gave the U.S. Army Corps of Engineers responsibility for all federal flood control programs and the authority to build reservoir projects for flood control, navigation, and recreation. The 1944 law authorized a large number of federal water development projects.

Wild and Scenic Rivers Act of 1968. This law protects free-flowing rivers by establishing a national system of wild and scenic rivers. Congress may designate pristine rivers as wild rivers, while less pristine rivers may be protected as scenic or recreational rivers.

structures), regulations, and consumer education. Cleverly employed, these measures are mutually reinforcing and can lead to substantial water savings.

In a 1983 study of water supply and conservation alternatives for southern California, the Environmental Defense Fund found that economic incentives to encourage conservation and more efficient irrigation could save 370 million cubic meters per year. A counterproposal to provide 327 million cubic meters annually by building a reservoir and diverting water from the north would cost almost twice as much.

Water Use and Purity

Certain industrial uses of water require a higher degree of chemical purity than does a public water supply. Water used for drinking, cooking, and cleansing, of course, must be free of disease-producing organisms. However, to be acceptable to most people it must also be free of color and turbidity (suspended material) and objectionable tastes and odors. Dissolved chemicals that can be neither tasted nor seen can have serious health effects.

An urban community uses an amount of water for cooling approximately equivalent to that used for domestic purposes. Ninety percent of self-supplied industrial use of water is for cooling purposes—a consumptive use. The generation of electricity by steam uses 2 liters (0.5 gallon) of water for cooling for every kilowatt-hour produced. Before cooling towers were used to dissipate the heat into the atmosphere as steam, heated water was often dumped into waterways causing thermal pollution.

Both community and industrial use contaminate water, and treatment is required before returning it to the river. Municipal wastewater may contain disease-producing organisms, and municipal and industrial wastes deplete oxygen dissolved in river water. In Chapter 6, we will discuss in greater detail the special concerns about chemicals in industrial wastes.

SUMMARY AND CONCLUSION

Since water is vital to all life processes, a dependable supply of water is essential for agriculture and permanent human settlements. The hydrologic cycle can renew the water supply, but human pollution can render the available water unfit for aquatic life and human uses.

We cannot consider the hydrologic cycle as separate from the atmospheric and physical systems of the environment. Soil and forest management have implications for the hydrologic cycle, but they are also related to erosion, ecological systems, and food production.

Flooding and flood control are often related to land use policy. Floods may be prevented or reduced by watershed management practices that reduce runoff. Dams provide storage for flood water and other related benefits, but the benefits are offset by ecological costs. Levees and channel improvements are other structural approaches to flood control. Flood protection programs try to utilize these various approaches while restricting the use of land on the floodplain.

As an important part of the hydrological system, lakes can have a major impact on ecological systems. Water in lakes will stratify, which influences the quality of water at various levels.

Water supply used to be a limiting factor in determining the number of people an area could sustain. By moving large quantities of water from one area to another, however, large populations can be maintained comfortably in arid areas.

Irrigation makes it possible to farm in arid areas, but on a long-term basis the evaporation of irrigation water leaves behind salts that could ultimately ruin the soil for farming. When one considers the consumptive use of water for irrigation along with the salts problem, the need for improved irrigation practices is obvious.

Reducing demand for water by means of water conservation programs is becoming more and more essential and in some cases is cost effective.

Pollution is considered in Chapter 6, although we have alluded to its impact with respect to eutrophication and the safety of water supplies.

In this chapter, we reinforced the idea that the physical systems of the earth cannot be considered apart from the chemical and ecological systems and human activities. At the same time, however, it is apparent that thoughtful action can result in sustainment of the natural system—by preventing floods and offsetting droughts. As other subjects are explored, we will consider more opportunities for using wisdom in environmental management.

FURTHER READINGS

Dregne, H. E. 1985. Aridity and Land Degradation. *Environment* 27: 16–20, 28–33.

Mather, J. R. 1984. *Water Resources: Distribution, Use and Management.* New York: John Wiley and Sons.

McPhee, John. 1989. *Control of Nature.* New York: The Noonday Press.

Micklin, P. P. 1985. The Vast Diversion of Soviet Rivers. *Environment 27:* 12–20, 4–45.

Postel, Sandra. 1990. Saving Water for Agriculture. Chapter 3. In: *State of the World, 1990.* New York: W. W. Norton.

Postel, Sandra. 1985. *Conserving Water: The Untapped Alternative*, Worldwatch Paper 67. Washington DC: Worldwatch Institute.

Postel, Sandra. 1985. Managing Freshwater Supplies. Chapter 3. In: *State of the World, 1985.* New York: W. W. Norton.

Reisner, Marc. 1986. *Cadillac Desert. The American West and Its Disappearing Water.* New York: Penguin Books.

STUDY QUESTIONS

1. Is it possible to intervene in the hydrologic cycle in a way that would benefit people?
2. Why is water from limestone solution channels not considered safe?
3. Why does two-thirds of the rainfall in the United States occur east of the Mississippi River?
4. Explain the impact of irrigating deserts on the hydrologic cycle.
5. Explain the major causes of flooding, particularly recent flooding. What are potential means of long-term flood control?
6. What is meant by the statement, "the floodplain belongs to the river"?
7. Why do northern lakes stratify?
8. What factors affect infiltration rate?
9. What is an example of a nonstructural method of flood protection?
10. What are the essential elements of a flood management system?
11. How does the riparian system of water law differ from the prior appropriation system?
12. What are the possible elements of a community water conservation program?
13. Should water be moved from one watershed to another or should people move to the water?
14. Should farmers in the Southeast be taxed to build dams for irrigation in the West?
15. Should farming and forestry practices be regulated by government?
16. Is it better to build dams for flood control or to prohibit building on the floodplain?

SUGGESTED ACTIONS

1. Locate the river basin you live in on a map. Contact your state engineer or local water authority to find out the most important water management problems in this basin. Discuss these problems and what should be done about them with your family and friends.
2. Keep track of your water use for a week. Use this information to develop and carry out a plan to use less water.
3. Instead of using sprinklers that spray water into the air where it can be blown away or evaporated, use soakers or drip irrigation systems that deliver water directly to plant roots.
4. Use water saving appliances (e.g., washing machines and dishwashers) and plumbing fixtures that use less water (e.g., water-saving toilets and low-flow faucets and showerheads).
5. Inspect the plumbing in your home. Fix any leaky pipes and faucets.

PART

TWO

•

POLLUTION

6

WATER QUALITY AND POLLUTION

Early human settlements tended to be near sources of abundant, pure water. As the size of settlements increased, human wastes polluted the rivers, causing ravaging outbreaks of typhoid and cholera. Aqueducts in ancient Greece and Rome, which were engineering marvels for their day, conducted pure water supplies from remote, uninhabited mountains to cities. Beginning in 312 B.C.*, the Romans built eleven aqueducts to supply approximately 1 million citizens with 40 million gallons per day—comparable to today's strictly domestic use.*

As the people of the world multiply, finding remote water sources in unpolluted areas becomes increasingly difficult. Society is faced with the alternatives of curbing pollution, mounting costs for treatment to remove pollutants, or suffering other social costs such as disease and environmental damage.

In recent years, ill effects of public, agricultural, and industrial pollution (excessive nutrients, metals, synthetic organics, and hot water) on ecosystems have been recognized. This chapter explores pollution sources and their effects, reviews standards for water quality, and describes measures to combat pollution.

WATER POLLUTION: ITS SOURCES AND EFFECTS

Human Health

Biological Agents Human diseases carried by water are caused by certain microorganisms—**bacteria, viruses, protozoa,** and **helminths**—that can be found in the feces or urine of a person ill with the disease. Sometimes a person who is exposed to a disease but shows no sign of illness discharges these organisms, transmitting the disease to other persons. Such persons are called **carriers.** Intestinal and urinous discharges from both ill persons and carriers, if not properly disposed of, can contaminate water subsequently used for drinking, cooking, washing dishes, or bathing. A person consuming food or water containing these organisms will contract the disease unless he or she is immunized naturally or by vaccination. For diseases caused by organisms such as protozoa and helminths, there are no immunizations. Swimming in water polluted by sewage is another way of contracting intestinal diseases. Freezing water does not kill these disease organisms.

Organisms found in intestinal discharges cause such diseases as **typhoid, cholera, salmo-**

nellosis, and **shigellosis.** Some dysenteries, such as **amoebic dysentery,** are caused by protozoa—microscopic organisms that move by protoplasmic flows or by using cilia or flagella. **Infectious hepatitis,** caused by a virus, can be transmitted by polluted water or by eating raw shellfish grown in polluted water. **Undulant fever** and **tularemia** are examples of other diseases, usually of animal origin, that could conceivably be transmitted by water but seldom are.

There are some parasitic wormlike infections that can result from drinking or swimming or wading in contaminated water. One of these is **schistosomiasis,** which usually occurs in tropical and subtropical areas where snail-infested fresh water polluted by human waste is used for irrigation. The **schistosome,** which lives part of its life cycle in humans and part in freshwater snails, enters through the skin as workers wade in the water. This has been a special problem associated with the irrigation systems of the Aswan Dam in the Nile Valley. A related schistosome affects waterfowl in the Great Lakes region. Its larvae are adapted to penetrate the relatively thin skin of birds, but cannot bore through the thicker skin of humans. They make a valiant attempt, however, and cause an irritation known as **swimmer's itch.** In the United States, a protozoa called *Giardia lamblia* has become a prime cause of waterborne disease. This organism prevents the lining of the small intestine from absorbing nutrients from food and can cause diarrhea, weakness, weight loss, abdominal cramps, nausea, and fever. It has spread to almost every state and has been responsible for waterborne epidemics from public water supplies in New York, Colorado, Washington, and New Hampshire. *Giardiasis* often called "beaver fever" because it can be carried by animals including beavers, is also a problem in recreation areas. Authorities have been trying to educate people not to drink what looks like clear mountain water while hiking. Mixing water with 8 milligram/liter of chlorine for 30 minutes or boiling water for 1 minute is required to kill *Giardia.*

Chemical Pollutants The concern of those people responsible for the protection of human health has turned increasingly to problems associated with chemical pollution. Industrial wastes can contain a variety of toxic inorganic, metallic, and organic compounds such as **arsenic, mercury, chromium, zinc, cyanide, chloroforms,** and **pesticides** of various kinds. In the Clean Water Act of 1977, Congress identified certain toxic pollutants to be controlled by July 1, 1984, by applying the best available technology (Table 6.1).

An outbreak of a severe central nervous system disease that occurred among people living in Minamata, Japan, in 1953 led to exploration of how metallic mercury and metallic salts are converted to toxic forms in nature. Previously, the presence of metallic mercury in waste discharges was considered harmless. Now we know that organisms in sediments can convert metallic mercury and mercury salts into an organic form—**methyl mercury**—that can then be concentrated in the food chain (see Chapter 2). This discovery raises a new specter concerning the fate of metals and metallic compounds in the water environment (Figure 6.1).

Excessive use of nitrogen-rich fertilizer results in **nitrate** residues which dissolve and seep into the groundwater. Seepage from septic tanks can add to the concentration. Organisms found in soil (and a baby's intestines) can convert nitrates to **nitrites.** Water polluted by nitrates or nitrites can be hazardous to health. If baby food is prepared with such water, the hemoglobin in an infant's blood combines more readily with nitrite than oxygen, causing **methemoglobinemia** *(blue babies).* Most surface waters contain less than 15 milligrams/liter of nitrates, but children have been affected at 18 to 257 milligrams/liter (usually over 100 milligrams/liter).

The worldwide use of pesticides has resulted in widespread distribution of synthetic organic compounds in the water environment. The discharge of *kepone,* a pesticide that causes nervous system disorders in humans, to the James River in Virginia caused fishing to be prohibited there. Some studies suggest that persons drinking wa-

Table 6.1
129 Priority Toxic Pollutants Regrouped into 10 Types of Pollutants

Pollutant	Characteristics	Sources	Remarks
Pesticides Generally chlorinated hydrocarbons	Readily assimilated by aquatic animals, fat soluble, concentrated through the food chain (biomagnified), persistent in soil and sediments	Direct application to farm- and forest-lands, runoff from lawns and gardens, urban runoff, discharge in industrial wastewater	Several chlorinated hydrocarbon pesticides already restricted by EPA; aldrin, dieldrin, DDT, DDD, endrin, heptachlor, lindane, and chlordane
Polychlorinated biphenyls (PCBs) Used in electrical capacitors and transformers, paints, plastics, insecticides, other industrial products	Readily assimilated by aquatic animals, fat soluble, subject to biomagnification, persistent, chemically similar to the chlorinated hydrocarbons	Municipal and industrial waste discharge disposed of in dumps and landfills	TSCA ban on production after 6/1/79 but will persist in sediments; restrictions on many freshwater fisheries as a result of PCB pollution (e.g., lower Hudson, upper Housatonic, parts of Lake Michigan)
Metals Antimony, arsenic, beryllium, cadmium, copper, lead,mercury, nickel, selenium, silver, thallium, and zinc	Not biodegradable, persistent in sediments, toxic in solution, subject to biomagnification	Industrial discharges, mining activity, urban runoff, erosion of metal-rich soil, certain agricultural uses (e.g., mercury as a fungicide)	
Other inorganics Asbestos and cyanide	**Asbestos** May cause cancer when inhaled, aquatic toxicity not well understood **Cyanide** Variably persistent, inhibits oxygen metabolism	**Asbestos** Manufacture and use as a retardant, roofing material, brake lining, etc.; runoff from mining **Cyanide** Wide variety of industrial uses	
Halogenated aliphatics Used in fire extinguishers, refrigerants, propellants, pesticides, solvents for oils, greases, and in dry cleaning	Largest single class of "priority toxics," can cause damage to central nervous system and liver, not very persistent	Produced by chlorination of water, vaporization during use	Large-volume industrial chemicals, widely dispersed, but less threat to the environment than persistent chemicals
Ethers Used mainly as solvents for polymer plastics	Potent carcinogen, aquatic toxicity and fate not well understood	Escape during production and use	Though some are volatile, ethers have been identified in some natural waters

Table 6.1 *(continued)*

Pollutant	Characteristics	Sources	Remarks
Phthalate ethers Used chiefly in production of polyvinyl chloride and thermoplastics as plasticizers	Common aquatic pollutant, moderately toxic but teratogenic and mutagenic properties in low concentrations; aquatic invertebrates are particularly sensitive to toxic effects; persistent; and can be biomagnified	Waste disposal vaporization during use (in non-plastics)	
Monocyclic aromatics (excluding phenols, cresols, and phthalates) Used in the manufacture of other chemicals, explosives, dyes and pigments, and in solvents, fungicides, and herbicides	Central nervous system depressant; can damage liver and kidneys	Enter environment during production and byproduct production states by direct volatization, wastewater	
Phenols Large volume industrial compounds used chiefly as chemical intermediates in the production of synthetic polymers, dyestuffs, pigments, pesticides, and herbicides	Toxicity increases with degree of chlorination of the phenotic molecule; very low concentrations can taint fish flesh and impart objectionable odor and taste to drinking water; difficult to remove from water by conventional treatment; carcinogenic in mice	Occur naturally in fossil fuels, wastewater from coking ovens, oil refineries, tar distillation plants, herbicide manufacturing, and plastic manufacturing; can all contain phenolic compounds	
Polycyclic aromatic hydrocarbons Used as dyestuffs, chemical intermediates, pesticides, herbicides, motor fuels, and oils	Carcinogenic in animals and indirectly linked to cancer in humans; most work done on air pollution; more is needed on the aquatic toxicity of these compounds; not persistent and are biodegradable though bioaccumulation can occur	Fossil fuels (use, spills, and production) incomplete combustion of hydrocarbons	
Nitrosamines Used in the production of organic chemicals and rubber; patents exist on processes using these compounds	Tests on laboratory animals have shown the nitrosamines to be some of the most potent carcinogens	Production and use can occur spontaneously in food cooking operations	

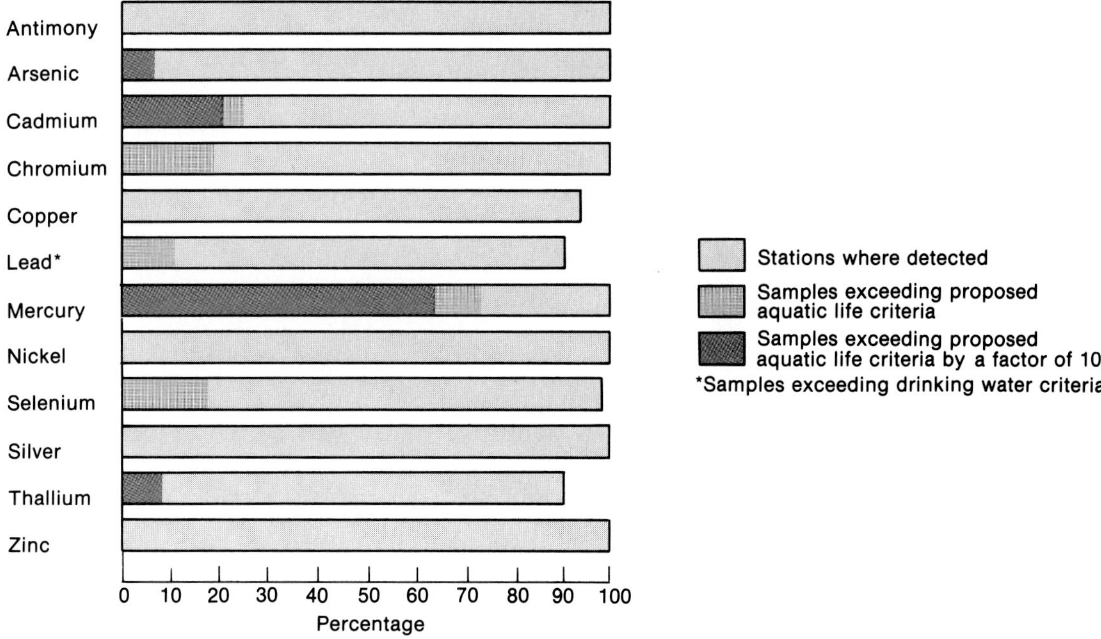

Figure 6.1
Inorganic toxic pollutants detected in a survey of river water quality in selected U.S. industrial areas (U.S. Environmental Protection Agency, Office of Water and Waste Management).

ter from rivers polluted with organic compounds develop more kidney and bladder cancer than persons who obtain water from unpolluted wells.

Four major types of organic chemicals are suspected as possible health hazards: (1) natural decomposition products, (2) substances resulting from interaction between chlorine and decomposition products, (3) organic chemicals from **point sources** (discharges from one pipe; in this case, mainly industrial), and (4) organic chemicals from **nonpoint sources** (the result of surface wash from rainfall running across land to enter along the banks of creeks and rivers; in this case, mainly agricultural) (Figure 6.2). Substances produced by normal organic decomposition are not known to be harmful in themselves. However, reactions between chlorine used in water and wastewater treatment and the normal decomposition organic chemicals produce synthetic organics such as **carbon tetrachloride, chloroform,** and **hexachlorothane.** There are a host of industrial and agricultural synthetic organic chemicals, including **carbon tetrachloride, vinyl chloride, chlorobenzenes,** various pesticides, **benzene,** and PCBs (polychlorinated biphenyls). PCBs are persistent chemicals with a structure similar to some pesticides (see Chapter 14). They contaminate ecosystems and pose a health threat to humans. Some organic toxic pollutants are examined in Figure 6.3.

The EPA has targeted more than 100 toxic pollutants for priority action. (These have been grouped into the ten categories shown in Table 6.1.)

While the presence of some pollutants is hazardous to human health, a deficiency of certain substances is also harmful. If the concentration of **fluorides** in drinking water is less than 0.5 milligram/liter, the incidence of tooth decay is likely to be high. Water containing about 1 milligram/liter of fluoride will help prevent tooth decay in young children and could possibly ben-

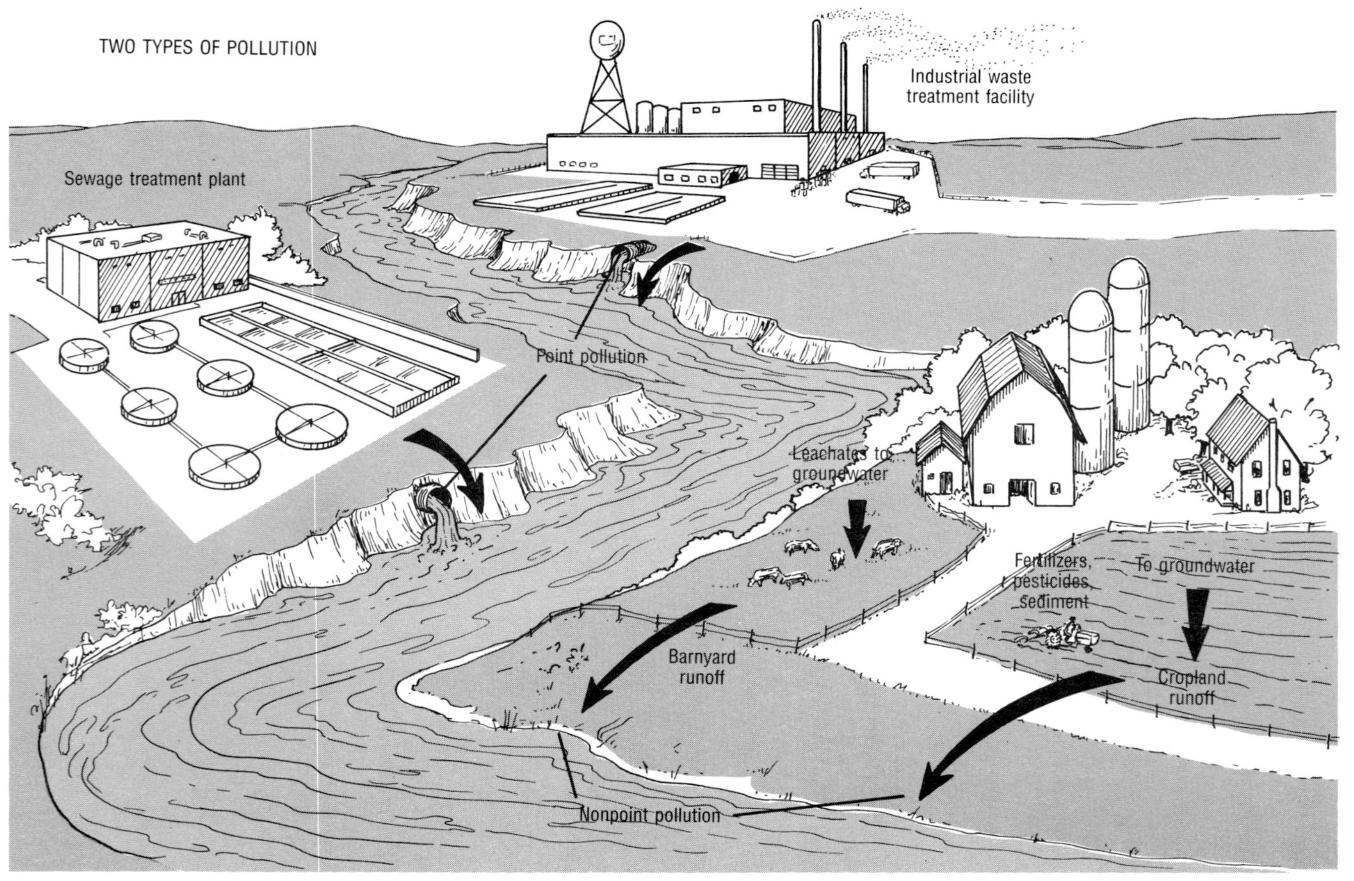

Figure 6.2

Examples of point and nonpoint pollution. (From O. S. Owen and D. D. Chiras, 1990. *Natural Resource Conservation. An Ecological Approach*. New York: Macmillan Publishing Company, p. 152, Fig. 8-3.)

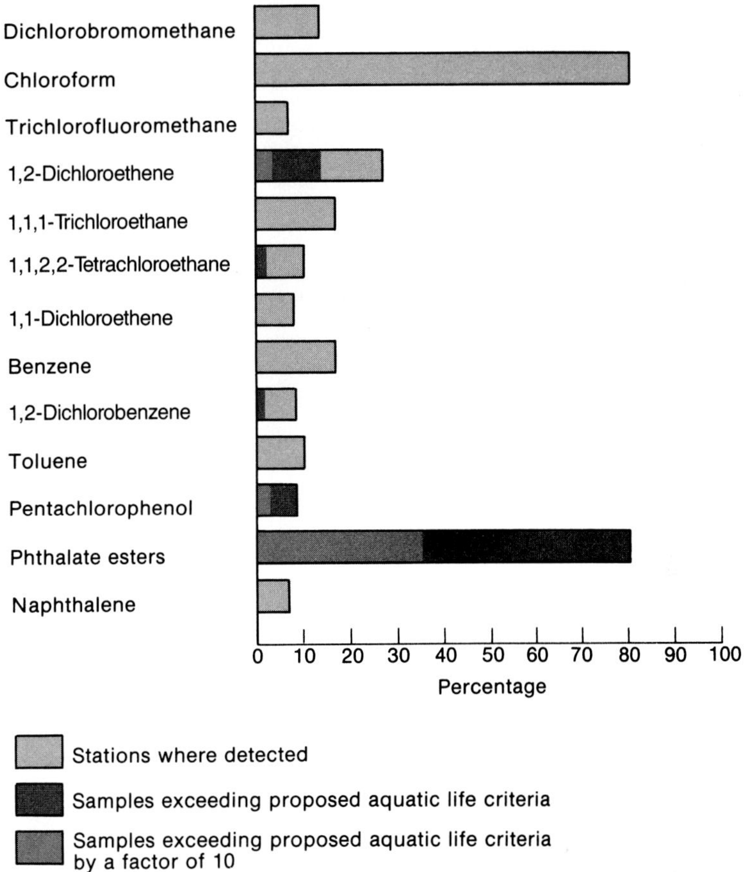

Figure 6.3
Organic toxic pollutants detected in a survey of river water quality in selected U.S. industrial areas (U.S. Environmental Protection Agency, Office of Water and Waste Management).

efit older people by retarding deafness and brittle bones. On the other hand, too much fluoride causes mottling of teeth. Most natural waters contain some fluoride; however, it is frequently less than 1 milligram/liter. Many communities adjust fluoride to the optimum amount. While there have been allegations of ill effects from fluoridation, scientific investigations have failed to confirm such effects at levels used in drinking water.

One incident of accidental fluoride intoxication took place on November 11, 1979, at the Annapolis, Maryland, water treatment plant. An open valve to a storage tank allowed 1000 gallons of 22 percent hydrofluosilic acid to overfill the feeder container and drain into a system that returned wash water to the raw water inlet. Excess fluoride passed through the plant and the water distribution system. Two days later eight patients undergoing renal dialysis became ill; one died. Subsequent sampling revealed "softened" water used for dialysis on November 13 contained 50 milligrams/liter fluoride.

Chemical constituents in water may have other effects on people's health. The Council on Environmental Quality, using maps and statistics, has confirmed reports that death rates from acute coronary heart disease and stroke are higher among users of soft water. The cause of the phenomenon has not been established. Most water-softening treatment involves exchanging sodium ions for calcium and magnesium ions. The influence of drinking water on heart disease may be small, however, compared with other factors such as nutrition and socioeconomic status.

Aquatic Life

Oxygen-Consuming Wastes The organic part of domestic wastes serves as food for bacteria and sewage fungi. Under normal conditions, aerobic bacteria obtain oxygen from that dissolved in water, called **dissolved oxygen (DO).** The amount of DO in water is determined by a combination of factors including temperature and water turbulence. Cold, flowing water generally has a higher DO level than still, warm water. As the oxygen is depleted by bacteria and other aquatic organisms, more oxygen from the air above is dissolved in the water. If the river is turbulent, water at lower depths is constantly being brought to the surface, where the oxygen is replenished. Organic wastes may be dispersed and decomposed in a flowing river as long as the cleansing capacity of the river is not exceeded (Figure 6.4). A large, well-mixed lake can have the same cleansing effect. This is often called the "dilution solution to pollution." However, if too much waste is discharged into the river or lake at one point, the need for oxygen exceeds the rate of replenishment and the entire oxygen supply may be consumed (Figure 6.5). The amount of oxygen consumed in decomposing organic wastes is called **biochemical oxygen demand (BOD).**

When a river becomes devoid of oxygen, the bacteria that can live without free oxygen increase. Because these anaerobes obtain their oxygen from sulfates and nitrates, they release hydrogen sulfide gas and ammonia—causes of the odors accompanying anaerobic decomposition.

Fish, however, require free dissolved oxygen. The larger game fish require about 4 milligrams or more per liter of water, while scavenger fish may be able to survive at levels of 2 to 4 milligrams/liter. The amount of oxygen fish need varies with temperature. The lower the temperature, the less oxygen is required. Thus, carp might survive at less than 2 milligrams/liter if the temperature is near 0°C (32°F). As the temperature rises, the life processes speed up and

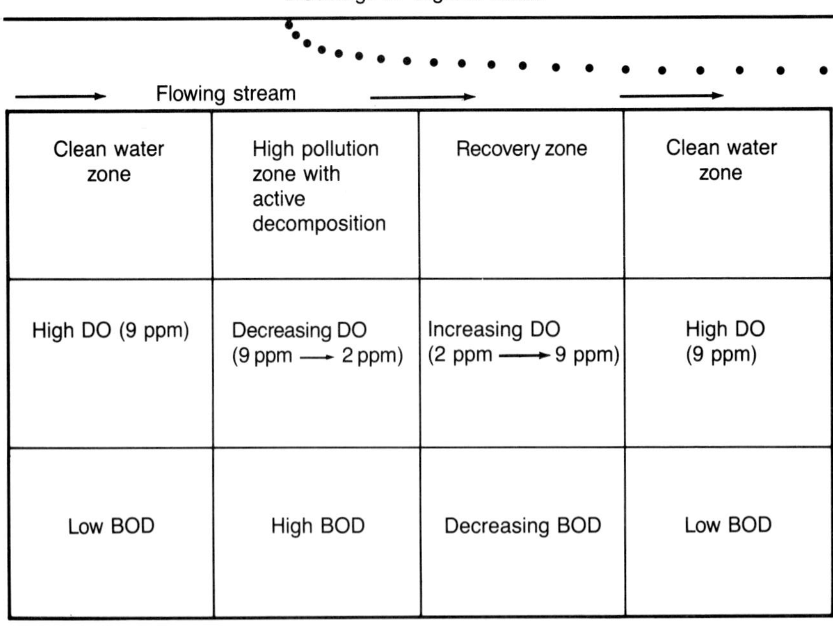

Figure 6.4
Effects of discharge of organic waste with a high level of BOD on the oxygen levels in a flowing stream.

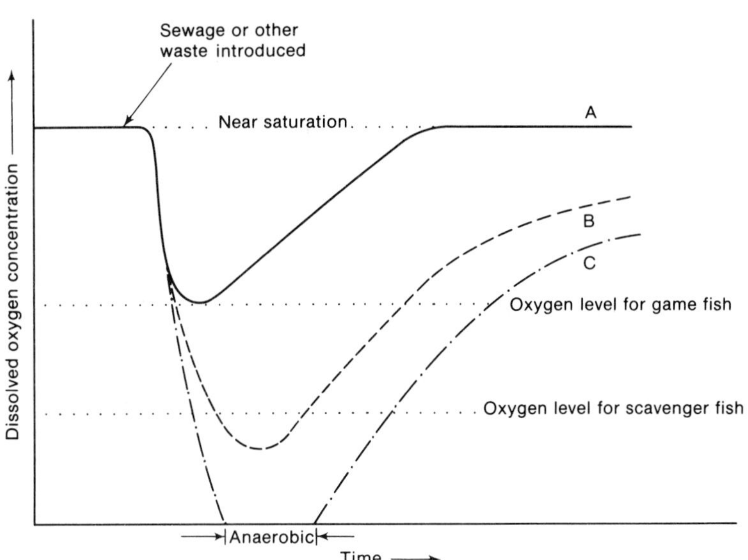

Figure 6.5
Dissolved oxygen sag curve. *Curve A:* When oxygen-consuming waste is introduced, the oxygen is depleted. The river will recover naturally if no more waste is added. In this instance, oxygen is sufficient for fish to live. *Curve B:* Oxygen depletion is too great for fish to live, although the river is still aerobic. *Curve C:* All oxygen has been used, and the river has become anaerobic.

require more oxygen. Organisms that are food for fish are affected; for example, water fleas *(Daphnia)* have a shorter life span at higher temperatures. The hatching of fish eggs, spawning, and migratory habits of fish are also related to temperature.

One of the primary aims of pollution control programs is to reduce the discharge of oxygen-depleting wastes into rivers and to maintain the desired level of oxygen in the waters. Waste heat discharges also have to be regulated so that the temperature of the river is not elevated very much at any one point. Cold water holds more oxygen in solution than hot water, so even sublethal rises in water temperature can reduce the available oxygen.

Chemical Wastes Chemical wastes affect water in several different ways. They can alter the pH of a river and impart color, taste, and odor. Wastes such as cyanide can poison a river. If liquid wastes from several different sources are emptied into rivers, they sometimes react to produce other compounds. Such reactions or changes in pH can cause materials to precipitate in the river. If the pH becomes exceedingly acid, as happens where acid mine waters drain into a river, the river will be devoid of life.

Nutrients in wastewater and land runoff stimulate algae growth in natural waters. The process of eutrophication is speeded, and the oxygen in the water can be depleted with the death and decomposition of the algae. Obnoxious tastes and odors are sometimes imparted by algae. Sources of nutrients include **phosphates** in laundry wastes and **nitrogen** in domestic wastes and fertilizers. Runoff from urban areas and agricultural lands is often responsible for high concentrations of these nutrients in streams (Chapter 15).

As exploitation of offshore petroleum deposits increases and oil continues to be transported in large tankers, the incidence of oil spills will probably increase, too. Oil spillage in the coastal waters of the United States has devastating effects on marine life and birds and ruin beaches for recreation as well. Some of the hydrocarbons settle to the bottom, where organisms in sediments are destroyed. This can diminish the fish and bird populations for years to come until the area recovers, if ever. In 1989 the largest oil spill in U.S. history occurred in Alaska's Prince William Sound. The spill had devastating effects on the sensitive ecosystems of this unique and biologically diverse area. The spill resulted in the loss of thousands of waterfowl and marine mammals and seriously disrupted the commercial salmon fisheries in the area.

The Environmental Protection Agency is monitoring other forms of ocean pollution in an experimental "Mussel Watch Program." Mussels are used as sentinel species to monitor pollution in U.S. coastal areas. Mussels are filter-feeding shellfish; they extract and accumulate contaminants along with their food. Samples of the mussels, collected and analyzed regularly for pollutants, have thus far provided data identifying several problem areas for corrective action—points along the west coast indicate high levels of the pesticide DDT, and concentrations of heavy metals and PCBs have been located along the east coast.

In Chapter 14, we will discuss the concentration of chemical pesticides in the aquatic food chain and its effects on bird populations that feed on fish and shellfish. Effects of acid rain are discussed in Chapter 8.

The Great Lakes

The five great lakes—Superior, Michigan, Huron, Ontario, and Erie—are the world's largest reservoir of fresh water and represent a tremendously important resource to Canada and the United States. The Great Lakes watershed is densely populated and highly industrialized, so it is not surprising that there are serious pollution problems.

At first, water quality management in the Great Lakes focused on bacteria contamination associated with municipal sewage. As early as the 1940s, serious eutrophication problems were apparent. Recently, controls on organic wastes, especially those containing phosphorus, have improved this problem. In the 1960s and 1970s, alarmingly high levels of PCBs, along with DDT and other pesticides, were detected in fish and other organisms. Mercury levels were also high in some areas. Recent data show that these problems are also improving. The improvements have been won at the cost of billions of dollars by the EPA, not to mention additional billions spent by Canada, state and local governments, and industry.

Today the emphasis is on reducing contamination by toxic materials such as dioxin and toxaphene. Toxaphene is used as an herbicide and a pesticide on cotton and sunflower crops in the southern United States and is carried via the atmosphere to the Great Lakes region.

Attempts to improve the water quality of the Great Lakes have emphasized chemical pollution. Ecological factors are also important and are reflected in changes in fish populations. The lake trout, an important commercial species and sport fish, has been virtually eliminated from all of the Great Lakes. Another native Great Lakes species, the blue pike, has become extinct. Exotic (nonnative) species, including rainbow trout, Pacific salmon, carp, and smelt, have been intentionally introduced to the Great Lakes. Other exotic species have invaded the lakes through the canals built to improve navigation for oceangoing ships. These invaders include alewives, white perch, and sea lamprey, a parasitic species that attacks large fish. Another invader, the zebra mussel, was introduced into the Great Lakes system when ballast water taken on in Europe was released into the Great Lakes. In 1990 zebra mussels were reported for the first time in the Hudson River. Populations of these organisms have increased rapidly, and the species has become a serious pest. Zebra mussels compete with native species for food and become encrusted on intake pipes, disrupting the water supplies of shoreline communities, power plants, and industrial facilities.

The changes in fish populations have been caused by a combination of chemical pollution, changes in habitat, overfishing, introduction of exotic species, and other human activities such as canal building. The Great Lakes illustrate the complexity of water quality management and the importance of considering ecological factors in this process. The large number of political entities that depend on the Great Lakes makes the situation even more complex.

DRINKING WATER QUALITY

Bacteriological Aspects

In 1854 Dr. John Snow dramatically demonstrated that an epidemic of typhoid and cholera in London was caused by a Broad Street well that was polluted by nearby sewers. He removed the pump handle and the epidemic ceased. We can now isolate disease-causing organisms from persons suffering from the disease, but it is virtually impossible to demonstrate their presence in polluted water. It is like trying to dip a net in the water to catch a very small minnow at the very moment it swims by. Consequently, even though we can reasonably expect that polluted water can transmit intestinal disease, we have no practical test to positively prove or disprove the presence of the disease organism.

As a substitute, a bacteriological test has been devised which will demonstrate the presence of the **coliform** group of bacteria. Coliform bacteria are normally found in animal and human wastes, but they can exist elsewhere. This group includes some bacteria that cause intestinal illness, although not all coliforms cause disease and their presence does not necessarily mean that their origin was intestinal. Rather, coliforms are used as an indicator of pollution that *could* contain intestinal organisms. Coliforms produce gas when grown in lactose broth. Thus, if a sample of water is cultured in lactose broth and gas is formed, coliforms are present and we presume disease bacteria could be also. Other tests are then performed, using media in which only certain kinds of bacteria grow, to determine if the bacteria that produced the gas were of fecal origin.

The test for coliform organisms has been used for over 50 years to judge the bacteriological quality of drinking and swimming water. To properly interpret results, however, this test should be used with field surveys to reveal if there are possible sources of sewage pollution (Figure 6.6). Testing involves the use of three groups of indicator bacteria found in mammals: total coliforms (e.g., *Escherichia*, *Citrobacter*, *Enterobacter*, and *Klebsiella*); fecal coliforms (e.g., *Escherichia coli*); and fecal streptococci (e.g., *Streptococcus*). According to the EPA, water supplies with fecal coliform densities greater than one

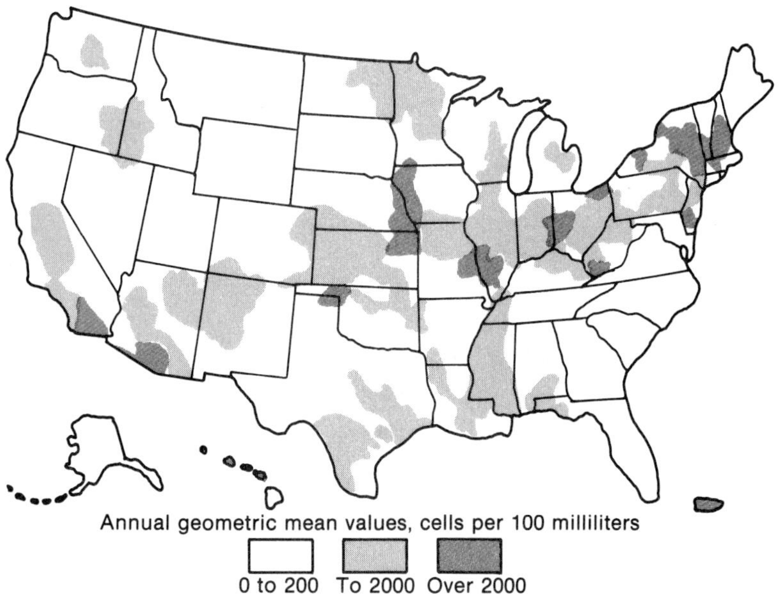

Figure 6.6
Fecal coliform bacteria in U.S. streams.

FACULTATIVE SLUDGE LAGOONS: WASTEWATER SOLIDS TREATMENT

Early in 1980, the new Sacramento Regional Wastewater Treatment Plant was able to take over twenty-one outlying plants because its design takes advantage of the area's long, hot, dry summers to aid evaporation. This method, which treats wastewater already processed past the acid phase of digestion, concentrates solids through natural evaporation. The area needed for final disposal is thus reduced to about one-thirtieth of that formerly needed for the old agricultural reuse system, and the land can be used for 40 to 60 years and possibly much longer if it is stripped and reused. After 5 or 6 years of storage in a lagoon, 5 to 6 percent of the solids sludge will be piped to a permanent land site and injected 1.2 to 1.8 meters (4 to 6 feet) beneath the soil surface.

Leaching into nearby water supplies—a great concern for waste disposal of any type—does not occur from facultative sludge lagoons. Initially, a 5- to 15-centimeter (2- to 6-inch) layer of clay is needed until the lagoons seal themselves with suspended and colloidal materials. This occurs even in sandy soils in 2 to 3 months.

The problem of odor is minimal when a 30- to 91-centimeter (1- to 3-foot) level of aeration is maintained. Algae living in this layer appear to consume most of the odor-causing gases produced in the underlying anaerobic layer. During inversions or on days with no wind, a 75-horsepower wind machine controls remaining odors.

The engineers involved in developing this method feel it can be adapted for use in many areas of the country, not just those with long, hot, dry summers. The reduced use of land, labor, and energy makes facultative sludge lagoon treatment an attractive alternative to conventional methods.

colony per 100 milliliter are generally unacceptable for drinking. A high ratio of fecal coliforms to fecal streptococci (0.7–4.0)/1 indicates that bacterial contamination is at least partly of human origin. A ratio lower than 0.6 indicates that the contamination is from mammals other than humans.

Many public water systems add chlorine to the water to ensure safety. Chlorine has bactericidal qualities and also reacts rapidly with organic material. Viruses are destroyed at higher concentrations of chlorine. Protozoa can enter a protective state known as a cyst, and normal chlorine doses do not kill protozoa in that stage—larger concentrations of chlorine are required. Because of concern about the products of reaction between chlorine and organic compounds in river water, this practice is being reviewed. In Europe, ozone is used for bactericidal treatment. Ultraviolet radiation may also be used.

Chemical and Physical Aspects

The EPA sets national drinking water standards, but states are responsible for implementing these standards. The standards apply to water systems serving an average of at least twenty-five individuals at least 60 days out of the year. They assume that a person will consume about 2 liters (about 2 quarts) of water per day. Standards cover inorganic chemicals, organic chemicals, turbidity, and microbiological quality. To be acceptable for domestic use, water must also be clear (free from turbidity and color) and have a pleasant taste and odor.

State health or environmental protection agencies usually oversee the operation of the public water systems and may be assisted by local health agencies. Sampling and testing of water for compliance to standards are performed according to the methods prescribed by the EPA. In some instances, the EPA accepts methods recommended in *Standard Methods for the Examination of Water and Waste Water.*

Trends in Water Quality

Groundwater Fifty percent of all residents of the United States rely on groundwater. There is increasing concern among investigators that groundwater is becoming seriously polluted (Figure 6.7) by industrial waste landfills and lagoons, municipal landfills, mining and petroleum production activities, septic tanks, and onsite sewage disposal (Table 6.2). Studies of the effects of industrial landfills on groundwater have shown organic chemicals present at many

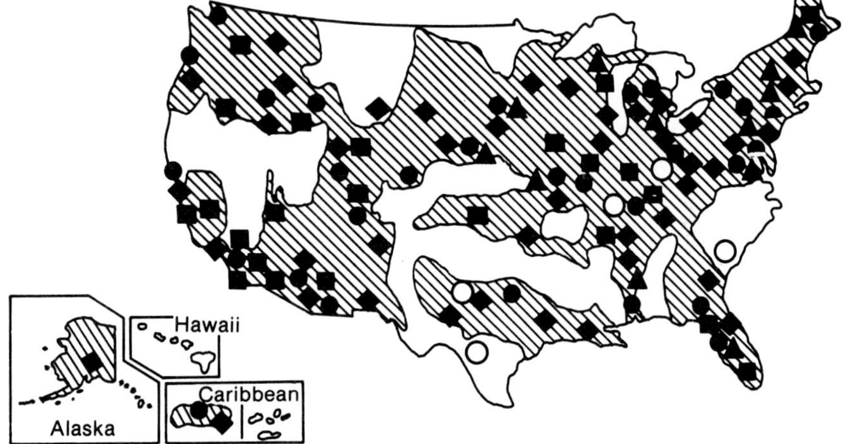

Figure 6.7
Groundwater contamination in the United States. (From U.S. Water Resources Council, *Preliminary Water Resources Problem Statements.* Washington, D.C., 1977.)

Table 6.2
Sources and Causes of Groundwater Contamination

Sources Designed to Discharge Waste or Wastewaters	Sources Not Specifically Designed to Discharge Wastes or Wastewaters	Sources Which May Discharge or Cause a Discharge of Contaminants That Are Not Wastes	Causes of Groundwater Contamination Which Are Not Discharges
Domestic on-site waste disposal systems	Sanitary sewers	Highway deicing and salt storage	Airborne pollution
Sewage treatment plant effluent	Landfills	Fertilizers and pesticides	Water well construction and abandonment
Industrial waste discharges	Animal wastes	Product storage tanks and pipelines	Saltwater intrusion
Storm water basin recharge	Cemeteries	Spills and incidental discharges	
Incinerator quench water		Sand and gravel mining	
Diffusion well			
Scavenger waste disposal			

Source: From Geraghty & Miller, Inc., *Ground Water Conditions*, Interim Series 4, 1977. Prepared for the Nassau–Suffolk Regional Planning Board, Hauppage, New York.

sites and chemical migration from the site commonly taking place. Chemical contaminants included PCBs, chlorinated phenols, benzene and its derivatives, and organic solvents. Heavy metals were present at nearly all industrial sites, and chemical migration was occurring at 80 percent of the locations studied. Selenium, arsenic, and cyanide were common at these sites.

Surveys have been conducted of cities with groundwater supplies that contain volatile chlorinated solvents. One of the contaminants most frequently present was trichloroethylene (TCE), a chemical shown to produce cancer in mice. In January 1980, California closed thirty-seven wells supplying 400,000 people in thirteen cities in the San Gabriel Valley because of excessive TCE. Problems with TCE in groundwater have become widespread, materializing in New York, Massachusetts, New Jersey, and Pennsylvania.

Other carcinogens were also found in groundwater. Besides industrial sources, septic tank cleaning fluids available to the public contain TCE, benzene, or methylene chloride. About 20 million housing units in the United States are using on-site disposal systems.

Surface Water The criteria the EPA uses for surface water quality are shown in Table 6.3. Since the quality of the water at the source determines the treatment required to produce safe drinking water, measurements have been made at sampling stations throughout the United States. Statistics indicate that there has been little or no change in water quality from 1975 through 1979 and that the standards are being exceeded by pollutants such as fecal coliform (35 percent), total phosphorus (50 percent), cadmium (10 percent), lead (25 percent), and mercury (50 percent). Dissolved oxygen is below the standard in 5 percent of the measurements.

In 1980, the EPA examined data on 62,000 community water supplies serving 200 million people. Eighty percent of these communities used groundwater as a source and 20 percent used surface water. Results of these tests are in Figure 6.8. Over 10 percent of the sites tested

Table 6.3
Thresholds Used in the Analysis of National Surface Water Quality

Pollution Indicator	Symbol	EPA Threshold Level
Fecal coliform bacteria[a]	FC	200 cells/100 ml[b]
Total phosphorus	TP	0.1 mg/l[c]
Dissolved oxygen	DO	5.0 mg/l[d]
Total cadmium	Cd	4.0 μg/l for soft water[e] 10.0 μg/l for hard water[f]
Total lead	Pb	exp [(1.51 times natural log of hardness— 3.37)][g]
Total mercury	Hg	0.05 μg/l[h]

1 = liter; ml = milliliter; mg = milligram; μg = microgram.
[a]A measure of water pollution produced by the feces of warm-blooded animals found in improperly treated human sewage, street runoff, combined sewer overflows in urban areas, feedlots and grazing lands in agricultural areas, and wildlife.
[b]EPA criteria level for "bathing waters." There is no uniform national standard for FC concentrations in water used for swimming; standards vary with use and locality. State standards sometimes differ from nationally recommended criteria.
[c]Value discussed by EPA for "prevention of plant nuisances in streams or other flowing waters not discharging directly to lakes or impoundments."
[d]EPA criteria level for "good fish populations."
[e]EPA criteria level for preservation of aquatic life less sensitive than cladocerans and salmonid fishes for water with $CaCo_3$ concentrations of up to 75 mg/l.
[f]Because the EPA criteria level for preservation of the less sensitive aquatic life for water with over 75 mg/l $CaCO_3$ concentration is 12 μg/l, this table uses the more stringent criteria level "for domestic water supply (health)."
[g]A 1979 EPA proposed criteria level for preservation of aquatic life.
[h]EPA criteria level for preservation of "freshwater aquatic life and wildlife."

Source: From U.S. Environmental Protection Agency. *Quality Criteria For Water.* Washington, DC: U.S. Government Printing Office, 1976.

exceeded the microbiological standards. In a more recent nationwide study conducted between 1988 and 1990, the EPA sampled 1300 community water systems and rural domestic wells for pesticides, pesticide derivatives, and nitrates. Based on this study, the EPA estimates that 1.2 percent of the community water systems and 2.4 percent of the rural domestic wells in the United States contain excessive levels of nitrates. They also concluded that 0.8 percent of community water systems and 0.2 percent of rural domestic wells in the United States have unacceptably high levels of at least one pesticide.

SUPPLY AND DISPOSAL

Wells, Springs, and Cisterns

Many suburban, rural, and vacation homes do not have access to public water systems, so water has to be obtained on the same lot as the home. Where water seeps through soil consisting of fine sand and loam, bacteria are usually filtered out in 15 meters (50 feet) or less. Very coarse gravel and some sandstone will not filter out bacteria because the relatively large open spaces between particles will not impede their passage. Neither does water following fissures in rocks or flowing in limestone solution channels receive any filtering.

Naturally filtered groundwater from wells and springs will be bacteriologically safe as long as no surface pollution enters. To prevent surface pollution from entering, wells are located so that surface drainage is away from the well and any barnyards, privies, or septic tanks are situated at least 15 meters (50 feet) away and downhill. The well is covered to divert any pump leakage or surface water away from the well. Similar protection is required for springs. If there are objectionable chemicals in the groundwater, such as dissolved iron, the water might have to be treated to remove them.

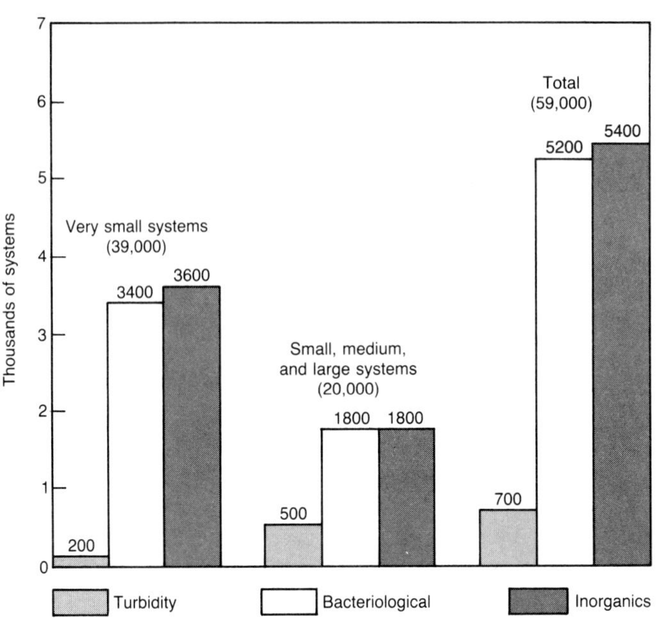

- 6745 community water systems violated the microbiological standard in 1979
- 760 surface water systems violated the turbidity standard in 1979
- 1007 systems violated at least one of the chemical-radiological standards in 1979

Figure 6.8
Estimated number of water supply systems in violation of federal drinking water standards, 1979. (From U.S. Environmental Protection Agency, "Community Water Systems: Financial Aspects of Compliance with Interim Primary Drinking Water Regulations," draft [January 31, 1980].)

Rainwater falling through the air is usually not contaminated with disease organisms or hazardous chemicals and is safe to drink. Water can be collected in roof gutters and conducted to a holding tank *(a cistern)* frequently made of concrete. Since the roof could be contaminated by bird dung and dust, rainwater collected during the first 15 minutes should not be used for drinking or household purposes. Even then, the water should be treated with chlorine before use.

On-Lot Disposal of Human Wastes

About one-third of homes in the United States are located in unsewered areas. Where there is no sewer, human wastes should be disposed of so that they are inaccessible to flies, rats, chickens, or other animals and do not pollute surface or groundwater.

If there is no running water, a **sanitary pit privy** may be used. It is simply a hole dug in the earth about 1.2 to 1.5 meters (4 to 5 feet) deep and covered with a reinforced concrete slab and riser equipped with a seat and lid. A small shed provides privacy. When the pit becomes filled, it is treated with lime and backfilled with earth.

Where running water is available, a **septic tank** and **drain field** may be used. Twenty-five percent of the U.S. population depends on septic disposal systems. A concrete tank about 1.8 meters long, 0.9 meter wide, and 1.5 meters deep (6 by 3 by 5 feet) allows solid material to settle to the bottom, where it is digested by anaerobic bacteria. The volume of solid material is thus reduced, but a residual material called **sludge** is left. As the tank becomes full of sludge, after 5 to 7 years of use, it is emptied and the sludge is buried or otherwise disposed of safely. Septic tanks remove about 40 percent of the organic material in sewage.

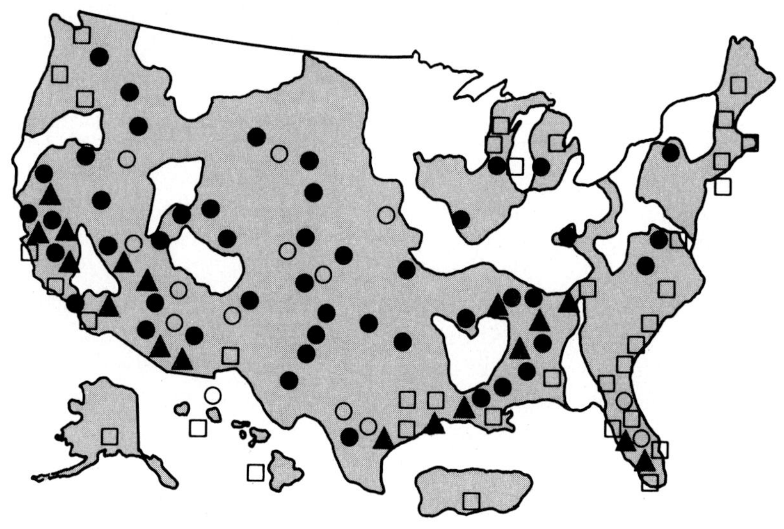

Figure 6.9
Groundwater depletion. (Data from U.S. Water Resources Council, *Preliminary Water Resources Problem Statements.* Washington, D.C., 1977.)

Because the liquid leaving the septic tank can contain disease organisms or have a foul odor, it cannot be released on the surface or to a ditch. Instead, the liquid is conducted to an underground drain field.

Liquid in the drain field seeps into the ground, so **percolation** tests have to be performed to measure the capacity of the soil to absorb liquid. The area of the drain field is determined accordingly. However, if the septic tank is not emptied regularly, solid particles will flow through to clog the drain field. Persons accustomed to living in homes connected to city sewers may not know the tank has to be cleaned regularly or may not even be aware that the house is connected to a septic tank until liquid from the drain field begins seeping to the surface. Furthermore, if the soil's clay content makes it impervious, it may not be practical to use septic tanks and drain fields for sewage disposal. Where they are practical, this system does help to recharge the groundwater.

WATER TREATMENT

Public Water Supplies

Small towns and some cities can secure water from wells and springs that require little or no treatment because natural filtration removes bacteria. Memphis, Tennessee, is a very large city that obtains water from wells, but most large cities turn to rivers and reservoirs for an adequate quantity of water. Although many communities have satisfactory groundwater supplies, some areas have experienced severe water shortages because of uncontrolled development and because water is being extracted faster than it is being recharged (Figure 6.9). Montgomery County, Pennsylvania, and Galveston, Texas, are examples of such areas.

There is developing concern that chemical pollution, not removed by soil filtration, is endangering groundwater supplies. Treatment to remove hazardous chemicals will be needed in such cases.

Well water can contain certain dissolved gases such as carbon dioxide, which lowers pH, and hydrogen sulfide, which gives a bad odor. Simple **aeration** accomplished by spraying the water into the air or running it over a cascade allows the gases to escape. Dissolved iron is oxidized by aeration and precipitates settle to the bottom.

Surface water from rivers, lakes, and reservoirs always has the potential for being polluted by human wastes and is considered unsafe to drink without treatment. The earliest treatment methods were adapted from nature. For example, **sedimentation** occurs in large lakes when heavy suspended material settles as the velocity of the water slows. Some color, bacteria, and other material adhere to the heavy particles and settle with them. Sedimentation is therefore used in water treatment. Plain sedimentation requires very large basins and a lot of time. Treatment proceeds more rapidly and is more effective in clearing the water if cheap chemicals—commonly alum and lime—are mixed with the water in a process called **chemical precipitation.** The ensuing reaction produces a gelatinous material *(floc)* to which color and bacteria adhere. The floc settles readily in relatively small basins, leaving the water clear (Figure 6.10).

Another natural system adapted to water treatment is **filtration.** Filtration was first applied in water treatment in London in 1829. In 1892 an epidemic of cholera occurred in Hamburg, Germany, but Altoona, Germany, was spared because its water was filtered. Both cities obtained water from the Rhine. These early installations were **slow sand filters.** An underdrain piping system was placed in a shallow basin and covered with crushed stone or gravel. Above this was spread a layer of fine sand. Settled water was introduced on top and slowly filtered through the sand to come out through the underdrain system. Mud and other fine material collected on top of the sand and aided in straining out bacteria. Although they were effective, these systems required extensive areas of land to supply large cities and mud had to be cleaned off the filter with a hand shovel. Some slow sand filters were used in Philadelphia, Pennsylvania, until the 1950s.

As land and labor costs rose, another filtering system replaced the slow sand filter. A **rapid sand filter** (Figure 6.11) uses the same principle, but a coarser sand allows water to flow through more rapidly. Some of the floc from the sedimentation process does not settle but is carried over to collect on top of the filter, where it strains out bacteria. The filter is equipped to reverse the flow of water, expanding and agitating the sand to remove the material that has accumulated on top.

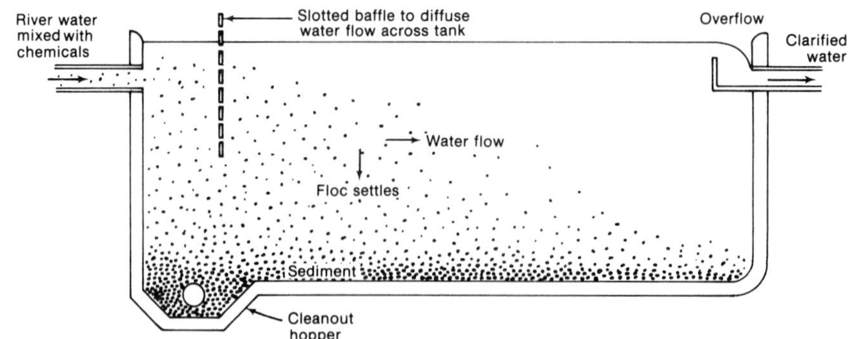

Figure 6.10
Sedimentation tank for water clarification. Chemicals (frequently lime and alum) are mixed with water to form floc. Mud particles and bacteria adhere to the floc and settle to the bottom.

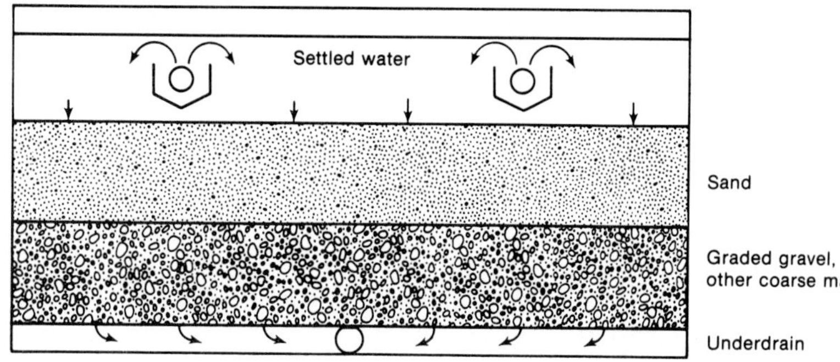

Figure 6.11
Rapid sand filter. Clarified water enters through the trough, overflows, and filters through the sand. To clean the filter, the flow is reversed, expanding and agitating the sand to remove floc and material that accumulates on top.

Disinfection is used in most public water supplies, even those from wells and springs usually considered safe. As we discussed earlier, chlorination is the form of disinfection generally used in the United States. In Europe, the addition of ozone and exposure to ultraviolet light are used. Chlorination is preferred in the United States because the residual chlorine reduces bacterial growth in the water system.

One of the chlorine's disadvantages is that it reacts with trace organic chemicals in river water to form a number of compounds, some of which are suspected cancer agents. **Trihalomethanes (THM)** are some of these compounds. The EPA has established an "acceptable" level for THMs, but no level of a carcinogen can be considered absolutely safe.

Treatments to reduce THM levels include oxidizing agents other than chlorine (e.g., ozone), aeration-improved coagulation, ion exchange, granulated activated carbon (GAC), powdered activated carbon (PAC), and others.

Some areas do not have access to adequate supplies of fresh water but have abundant supplies of brackish or salt water nearby. If the area has potential for settlement and development, it could be worthwhile to remove the salt. Evaporation and subsequent condensation is one relatively economical method of **desalination,** especially if solar energy is used for some of its energy requirements. Corrosion and disposal of salts are problems for these installations. A method called *reverse osmosis* has been used in Florida, Iowa, and elsewhere for treating brackish and highly mineralized water. Reverse osmosis involves forcing water under pressure through a membrane so fine that the salt molecules cannot pass.

Water Pollution Control

Early cities built systems of sewer pipes to conduct sewage to rivers so that slop would not be thrown into the street. By following natural slopes, flow was accomplished by gravity. Because the only consideration was to remove sewage from the city, there was no treatment and no concern about what happened downstream. However, when the river was fouled in the immediate area, treatment systems were developed to prevent pollution problems. These efforts were initially prompted by aesthetics; objectionable floating material had to be removed and odors caused by anaerobic decomposition reduced. Added to these concerns were fish kills caused by depletion of oxygen in the river.

The initial treatment processes were designed to remove settleable solids. Sedimentation with adaptations for handling sewage was one effective method. Some small communities at one time used large septic tanks, but bad odors were frequent and removal of solids not very efficient. Treatment efficiency improves when the sludge is removed from the settling chamber to another

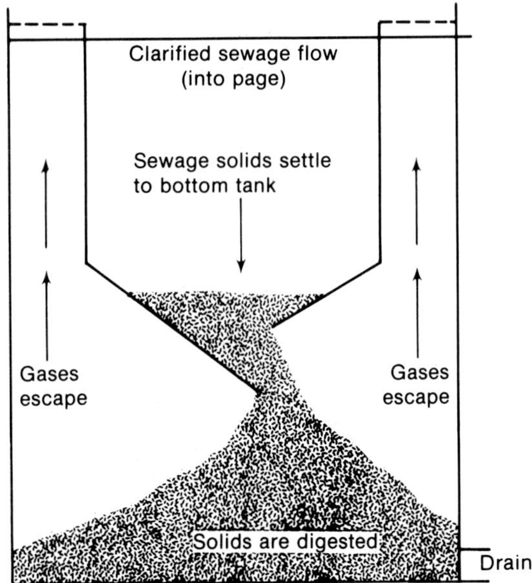

Figure 6.12
Imhoff tank.

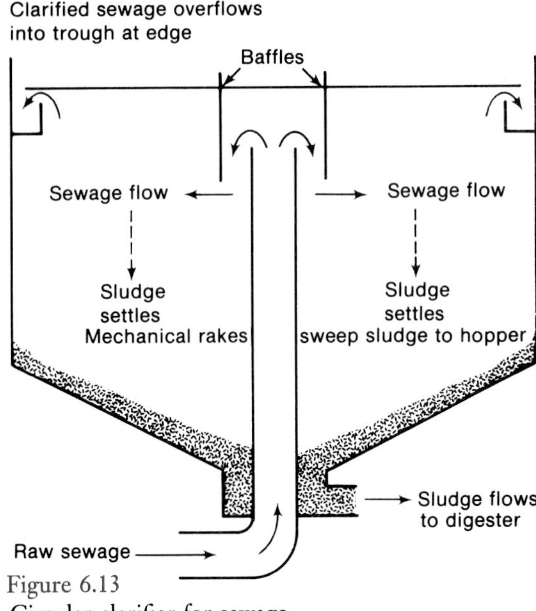

Figure 6.13
Circular clarifier for sewage.

compartment for digestion, as was done in the **Imhoff tank** (Figure 6.12). These tanks were two stories high, with the bottom chamber used for digestion. Imhoff tanks were soon replaced by constructing separate **sedimentation** and sludge **digestion tanks.** By removing the sludge, the sedimentation tank is kept in an aerobic condition, eliminating bad odors and improving its efficiency. Skimming removes floating solids, and sedimentation removes about 50 to 70 percent of suspended solids.

Sedimentation is called **primary treatment,** even though it may be preceded by facilities which remove grit or rags or preliminary treatment to control odors. Sedimentation of sewage solids is made more efficient by adding chemicals to produce a floc, as in water treatment. However, this process is costly and considerably more sludge is produced.

Some sedimentation tanks are long, rectangular basins with sewage entering one end and leaving at the other end. The settled solid material is raked mechanically to a hopper and pumped to a digester. The more recent circular design introduces sewage at the center and causes it to flow to the outside (Figure 6.13).

In some cases, the waste load on a river depletes the oxygen so much that fish are killed. Primary treatment reduces biochemical oxygen demand (BOD) about 25 to 30 percent. This might not be enough, so **secondary treatment** processes might have to be employed. The purpose of secondary treatment is to reduce the BOD, that is, the demand for oxygen put on the river by the waste. Secondary treatment is required in all U.S. municipal sewage treatment plants, but some persons question its need in every case.

The amount of oxygen in a river at the point of discharge as well as the BOD of the waste influence the type and capacity of the secondary treatment required. The amount of oxygen in a river depends on its concentration in the water and the quantity (flow) of water. Oxygen depletion is likely to be most critical during droughts unless the low flows are augmented by releasing

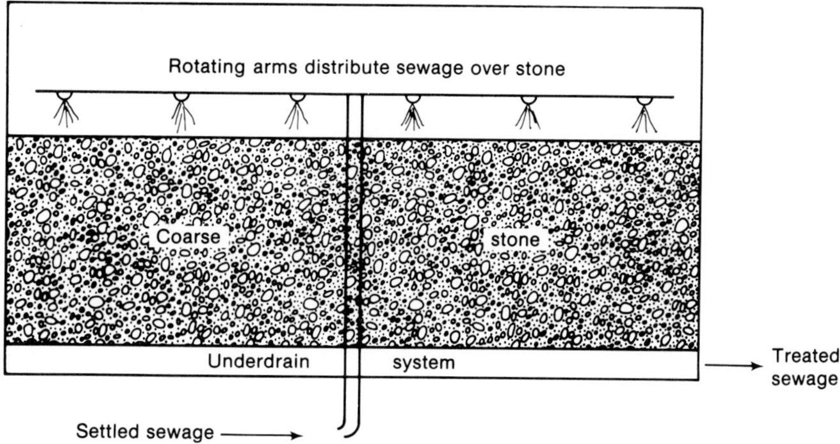

Figure 6.14
Trickling filter for sewage treatment. Sewage flowing over coarse stone is not filtered as with sand. Instead, a biological community grows on the stone and feeds on the organic material from the sewage. This action reduces BOD.

water impounded in a reservoir. Oxygen is sometimes bubbled in water in deep reservoirs to increase the oxygen concentration.

Because the wastes stimulate the growth of biological communities that consume oxygen, what could be more natural than to use these very same communities to reduce the BOD? This can be done in a treatment plant under controlled conditions where additional oxygen or air is supplied as needed.

Two biological treatment systems are commonly used in secondary treatment. One is a **trickling filter,** which is quite different from the filtering process previously described. The trickling filter consists of a bed of large stones or similar material with plenty of open space between individual stones (Figure 6.14). The sewage, distributed over the stone by rotating arms, trickles over the stone. A natural biological community becomes established on the surface of the stones and uses the organic material in sewage as food, thus reducing the BOD. Air circulating through the open spaces supplies needed oxygen.

The other common biological treatment is **activated sludge.** Previously settled sewage is introduced at one end of a long tank and air or pure oxygen is pumped into the tank. In some instances, primary sedimentation is eliminated and sewage flows directly into this aeration tank. As the air mixes with the sewage, biological communities grow on the sewage particles and form visible clumps of particles that resemble the chemical floc used in water treatment. Again the bacteria and other organisms use the sewage as food. Following the aeration tank is a second sedimentation tank, where the floc settles. Some of the sludge and sewage is returned to the entrance to the activated sludge tank to *seed* incoming sewage. The excess sludge is pumped to the primary tank and from there to a digestion tank.

Trickling filters following sedimentation remove about 70 percent of the BOD. If the sewage is recirculated (that is, a large portion of the effluent is pumped back to flow through the trickling filter to be treated again), up to 90 percent can be removed. A conventional activated sludge system should remove 90 to 95 percent of BOD. Although these systems may be properly designed, they may not be as efficient as planned because of poor maintenance and operation (Table 6.4).

While EPA regulations require all municipalities to install at least secondary treatment, each sewage treatment system must be designed to remove enough BOD to maintain the oxygen concentration at desired levels in the river. The dissolved oxygen (DO) in the river is measured,

Table 6.4
Municipal Water Treatment Plant Compliance with BOD and TSS Secondary Standards

	Trickling Filter		Activated Sludge		Total	
Rating	Number of Plants	Percentage	Number of Plants	Percentage	Number of Plants	Percentage
Satisfactory	188	30	527	55	715	45
Unsatisfactory	177	28	178	18	355	22
Poor	269	42	256	27	525	33
Total	634	100	961	100	1595	100

BOD = Biochemical oxygen demand
TSS = Total suspended solids

and the amount of BOD the river can assimilate is computed. Then, a treatment system is designed to remove sufficient BOD from the sewage so that the remaining BOD discharged to the river does not exceed the acceptable amount. Obviously, if a discharge upstream has already reduced the DO in the river, a downstream municipality or industry will have to provide more treatment. This is one reason why plans for waste treatment should include the entire river basin.

With the popular use of home laundry equipment, detergents containing phosphates were added to domestic wastes in large quantities. Fifty to 70 percent of the phosphorus in domestic sewage is attributable to detergents. Human excreta, ground garbage, and food processing wastes add nitrogen and thus supply nutrients that stimulate algae growth in rivers and lakes. Conventional treatment does little to remove these. An activated sludge system may remove up to 30 percent of the phosphorus and 40 percent or less of the nitrogen. So, a **tertiary treatment** system may be added to the series where these problems exist. Tertiary treatment is any form of special treatment used when chemicals are contaminating the water. Tertiary treatment is used where eutrophication of lakes is a question. Phosphates can be removed by adding chemicals such as lime. These chemicals react with the phosphates in the water to form precipitates that will settle in sedimentation tanks.

Conventional secondary treatment removes some nitrogen. Although special nitrogen removal processes are not used for large cities discharging to large rivers, they may be desirable in special situations such as Lake Tahoe. If sewage is passed through shallow lagoons where algae grow, the algae remove the nitrates as they would in the river. Other systems involve mechanical and chemical processes.

If primary or secondary treatment systems fail to remove sufficient BOD, a rotating biological contactor (RBC) may be added to the system. The RBC is a perforated rotating disk, partially in the water and partially out. As the treated sewage flows through the disk, a biofilm containing organisms similar to those found in trickling filters and activated sludge tanks develops on the surface. As the biofilm rotates out of the water, a liquid film keeps it moist. RBCs also remove ammonia nitrogen.

An experimental treatment system involves introducing sewage, effluent from treatment plants, and the liquid overflow from sludge digesters into ponds where water hyacinths grow. During testing of the system, water hyacinth ponds removed 97 percent of BOD from raw municipal sewage and 83 percent from secondary effluent. They also removed 72 percent of the nitrogen and 31 percent of the phosphorus. The system appears to be a promising, cheap, energy-efficient, and effective treatment process, well suited for small communities as a polishing process after secondary treatment. It is sensitive to freezing, however, so enclosure of the ponds may be necessary in some climates.

Chlorination may be used to disinfect sewage plant effluents. The concern for the formation of THMs, however, could raise questions about this process. Because of the possible adverse effects of chlorine on trout. Smithsburg, Maryland, switched to ultraviolet radiation for disinfection.

The Council on Environmental Quality (CEQ) reports that in 1988 slightly more than two-thirds of the population of the United States was served by a municipal wastewater treatment system. Most of these people (84 percent) were served by systems that included secondary or greater than secondary treatment processes.

Sludge

Sludge resulting from sewage treatment is costly and sometimes troublesome to handle. When sludge is pumped from the primary sedimentation tank to a separate digester, it undergoes anaerobic decomposition. Initially, large organic compounds are broken down by facultative bacteria into organic acids, carbon dioxide, methane, and a little hydrogen sulfide. Anaerobes then convert the organic acids to methane and carbon dioxide. The methane can be recovered to heat buildings (including the digester itself). Solid material, now relatively inert, settles to the bottom and is removed periodically to be dried for further disposal. Heating and agitating the digester speeds the digestion process. Where agitation is practiced, a second digester is necessary for sedimentation of the residual solids.

In activated sludge processes without primary clarifiers, the excess sludge is in an aerobic condition and has a low solids content. This causes anaerobic digestion to fail; however, long-term aeration of the digester allows aerobic decomposition to proceed. In aerobic digestion, the problem of separating water from the residual sludge is increased because the sludge is so dilute.

Sludge obtained exclusively from domestic sewage after digestion and drying can be used as a conditioner for sandy soils or soils with high clay content. The added humus helps the soil retain moisture. Sometimes sludge is used as a base and chemicals are added to produce commercial fertilizers.

Land Treatment Systems

In some instances sewage is applied directly to forest and agricultural land. A variety of systems have been used, some of them for more than 50 years. Slow rate systems employ sprinklers or ridge and furrow irrigation approaches to apply treated wastewater to croplands. Other systems use a basin flooding approach to achieve rapid infiltration or overland flow to distribute the wastewater. In most cases, the wastewater is pretreated by standard primary or secondary methods, or by aeration. In rare cases raw sewage is used.

Sludge parasites and pathogens have never been directly incriminated in the transmission of a specific disease, but guidelines for application of sludge to agricultural land have been developed (Table 6.5). More than 50 percent of municipal sludge is applied to land.

When the physical and chemical character of the soil is improved by application of municipal sludge to poor soils, the crop yield increases. Application of treated municipal wastewater to cultivated soil helps remove nitrates and phosphates present in the effluent.

Heavy metals in the sewage will also be present in the sludge. Metals accumulate in the soil and may, in turn, accumulate in plant tissue at concentrations that make the plant unfit for food consumption. The soil, species of plant, and metal element influence accumulation and uptake, but pH is the dominant factor. Cadmium, nickel, and zinc reduce crop yield in areas where the soil is acidic. Where heavy metals are a concern, metal concentrations in soil and plants should be monitored.

Industrial Wastes

Many industrial wastes can be treated by the conventional primary and secondary processes used for municipal sewage. If necessary, coagu-

Table 6.5
Suggested Prevention Guidelines to Reduce Exposure or Contact to Sludge Parasites and Pathogens

Aerosol Inhalation Prevention Guidelines	Potable Water Ingestion Prevention Guidelines	Incidental Physical Contact Prevention Guidelines
1. Sludges should not be spread during periods of strong winds. 2. Low-pressure equipment should be used to spread sludges, and large droplet size formation should be maintained during sludge spreading operation. 3. Isolation distances, downwind buffer zones, and the use of vegetative screens should be practiced. 4. Public access should be controlled through the use of posted notices, barriers, fences, or patrolling.	1. Sludges should not be spread near streams, water supplies, or land which experiences flooding, erosion, or soil movement. 2. Sludge should not be applied on soils with less than 20 inches of depth or on sites having slopes greater than 15 percent. 3. Sludge application sites should not contain areas that are underlaid with highly porous, fracture or stratified formation, fault zones, open bedding places, or sink holes. 4. Sludges should be applied uniformly over the field.	1. Public access should be controlled. 2. Sludges should not be applied near property line or occupied dwellings. 3. Sludges should not be used on land where leafy vegetables are to be grown. 4. Sludge should not be applied to growing or mature crops. 5. Livestock should not be allowed to graze in pastures which have sludge residues remaining on vegetation, and food to be eaten raw should not be grown in sludge at least 3 years after the last land application of sludge.

Source: From Michael H. Gerardi, "Application of Sewage Sludges on Agricultural Lands," *Water Pollution Control Association of Pennsylvania Magazine* 14(1981):25.

lants may be used in conjunction with sedimentation to remove some of the dissolved chemicals in the waste. Even the biological treatment processes will remove some metals and organic compounds; however, some pretreatment is necessary if the waste is toxic to the microorganisms. This would be a simpler process if the toxic materials were not mixed with other wastes.

To discharge new industrial wastes, a company must comply with standards and obtain a National Pollution Discharge Elimination System (NPDES) permit from the EPA. If the state's pollution program is approved, the state may perform this function.

Industries are required to remove sixty-five of the most dangerous chemicals from their wastewater before discharge to municipal sewers. Among the substances are cadmium, lead, mercury, benzidine, PCBs, cyanides, and vinyl chloride. Removal of these substances will make municipal sewage sludge more acceptable for agriculture.

The effects of industrial waste discharges remain for many years. The James River is contaminated from kepone discharge and fishing is restricted because of remaining residues. Three decades of discharge contamination in the upper Hudson River have resulted in the river's sediments testing about 10 parts per million (ppm) of PCB, between 1 and 2 orders of magnitude greater than other large rivers and estuaries contain.

Even so, the Council on Environmental Quality reports examples of dramatic recovery on the Pennigewasset River in New Hampshire, the Hackensack in New Jersey, and the Neches Tidal Basin in Texas. Although there are other examples of improvement, some rivers remain deteriorated or are in the process of becoming so.

Thermal Pollution: Hot Water Discharges

Water is used for cooling in many industrial processes, including thermal generation of electricity. In the simplest cooling systems, water passes through the plant once and is returned to its source. After having passed through the system and absorbed large quantities of heat energy, the cooling water is apt to raise the temperature of the receiving body of water to unacceptable levels. Some regulations permit no more than a 0.35°C (1°F) rise in the river water temperature after mixing with the hot water discharge. Since a temperature of 34°C (94°F) is lethal to fish, no discharge is permitted which results in that temperature, except in situations where the river is naturally above that temperature. To satisfy this requirement, a cooling pond or cooling tower is usually used. In both cases, evaporation of large quantities of water is involved—a consumptive use. There can also be a slight increase of fogging and icing in the vicinity.

Aside from heat problems, at first glance it might appear that using water for cooling adds no pollutants. But evaporation of water will concentrate any minerals and metals in the intake water. Where water is recirculated, the effect is even greater. Sometimes the natural level of chemicals in a river is so high that concentration by evaporation causes unacceptable levels downstream. Furthermore, chemicals used for corrosion control add to those already in the cooling water.

Recycling

For centuries communities upstream have discharged sewage to rivers used as water supplies by communities downstream. Forty percent of the flow of the Ruhr River in West Germany is treated wastewater. Treatment is available to ensure bacterial safety of recycled wastewater, but the main concern is the concentration of toxic inorganic, metallic, and organic compounds by recycling. There is also some question about how effectively the treatment systems remove viruses. Because of these and other aesthetic concerns, recycled wastewater is used generally for purposes other than drinking.

In Windhoek, South West Africa, drinking water has been reclaimed from sewage since 1969. The sewage receives conventional primary and secondary treatment, then goes to a maturation pond where algae remove phosphates and nitrates in the water. Chemicals added to the water cause the algae to float to the top, where they are skimmed off. Detergents are removed by foaming, and then sufficient chlorine is added to "burn out" any remaining impurities and leave a residual of 0.5 milligram/liter. After passing through sand filters and granular carbon columns, the water from this system is blended with water from other sources. At times, reclaimed water has constituted up to 13 percent of the supply.

In 1956 a severe drought threatened to exhaust the water supply reservoir of Chanute, Kansas, on the Neosho River. The effluent from a trickling filter sewage treatment plant was pumped back to the reservoir and directed through the water treatment plant. This emergency system, with a recycling time of 20 days, continued for 5 months. Bacteriological quality was maintained by extensive chlorination, but there were other problems, including a persistent amber color.

Since 1959, effluent from the secondary sewage treatment plant at Santee, California, has been allowed to seep through about 1200 meters (4000 feet) of natural ground and a series of four artificial lakes. In the last lake in the series, boating, swimming, and fishing are permitted and the water is chlorinated. Careful monitoring of the population shows no increase in disease rate.

As discussed earlier, experiments and demonstrations with the use of sewage are proceeding on forest and farm lands. Usually treatment plant effluents are used rather than raw sewage. However, there is concern that metallic salts will build up, as happens with sludge. Also, the rate of application is carefully monitored to avoid clogging the soil and exceeding the infiltration rate. This application of sewage has potential for recharging groundwater and returning nutrients to land, but care must be taken to prevent disease transmission.

River Basin and Regional Planning

Because of the interrelations among wastewater discharges, upstream and downstream uses for

LEGISLATION

Federal Water Pollution Control Act of 1972, The Clean Water Act of 1977, and subsequent amendments. Under these laws the Environmental Protection Agency (EPA) sets national water quality standards and delegates enforcement to states through EPA-approved programs. The EPA also cooperates with the Corps of Engineers to protect the water quality of wetland areas and with the Coast Guard in dealing with oil spills. The law establishes fines and penalties for violators and includes a provision for citizen lawsuits.

Safe Drinking Water Act of 1974, and subsequent amendments. This law is similar to the Clean Water Act in that it directs the EPA to set national standards for safe drinking water and to delegate enforcement to states with approved programs. It also has provisions for fines, penalties, and citizen lawsuits.

drinking water supplies, fish, recreation, agriculture, and industry, allocations of waste discharges are permitted and water uses are planned on a river basin and regional basis. The degree of treatment required downstream is also influenced by waste loads received upstream. Section 208 of the Federal Water Pollution Control Act provides financial assistance for such regional planning efforts.

Those who develop regional plans must consider not only all point sources of pollution, such as municipalities and industries, but such nonpoint sources as agricultural runoff as well. They must also take into account the substances from air pollution that reach streams and the pollutants from urban storm water.

SUMMARY AND CONCLUSION

Rain, as it falls, is usually considered safe to drink. After contact with the surface of the ground, water may become contaminated by contact with animal and human wastes. Bacteria are strained out as water filters through soil, but water may dissolve some chemicals as it percolates through the ground, and water that flows directly into a limestone sinkhole or seeps through coarse sandstone and gravel is not filtered.

Most water pollution problems are caused by human activities—sewage from municipalities, industrial wastes containing chemicals, seepage from mines and waste deposits, urban storm water runoff, agricultural runoff (containing pesticides and fertilizers), and air pollutants in rainwater. These pollution loads can make water unfit to drink, kill fish, accelerate eutrophication, and cause other harm to ecosystems by changing pH and temperature.

The environmental control principles outlined at the beginning of Part Three have been successfully applied to enhance water quality. The Greeks and Romans sought isolated water supplies free from pollution and brought the water into cities by aqueducts. Such sources are still sought, but few surface sources are so isolated that they are safe without treatment. Conventional treatment for public water supply was originally designed to remove bacteria (sedimentation and filtration) and destroy them (chlorination). Now metallic ions and synthetic organics present concerns, and special treatment is required for their removal. Other treatments, such as pH and hardness and fluoride adjustment, are also designed to enhance water quality.

Human waste disposal practices were developed to isolate the wastes, thereby preventing the transmission of intestinal diseases. Sewers removed wastes from the cities and emptied them into streams, protecting people in the cities but polluting the river. Wastewater treatment systems were not designed to produce an effluent free of disease organisms: rather, the treatment was developed mainly to remove BOD so that oxygen levels in the receiving river would be sufficient to sustain fish populations and avoid the bad smells associated with anaerobic decomposition. Most municipal systems now have secondary systems that employ some form of biological treatment such as trickling filters, activated sludge, and biocontractors to reduce the BOD in the effluent. More recently, nutrients in wastes are of concern because they stimulate algae growth and speed eutrophication.

Although technology is available to treat most water pollution problems, substitution of raw materials or preventive processes at the source is a preferred approach. The ban on PCBs is an example. If substitution is not feasible or sufficient, waste recovery at the source and subsequent recycling is desirable. Many communities are experimenting with land application of treated sewage and sludge from removal processes, while some communities are reclaiming water for reuse in recreational lakes and drinking water. The presence of metallic ions and synthetic organics complicates recycling and reuse efforts and may require treatment for removal. These treatment processes create residues that make further disposal processes necessary—either incineration that may produce air pollution (Chapter 8) or solid waste disposal (Chapter 15).

The most efficient use of technology occurs when water supply and wastewater disposal problems are handled on a regional or river basin basis. However, all areas have limited water resources and rivers have a limited capacity to assimilate wastes. When that capacity is reached, maximum waste loads may have to be allocated for various discharges and further industrial and residential development restricted around that stretch of the river. In subsequent chapters we will see that land use and density of development are not just problems for water consumption and wastewater disposal; they also have an impact on air and land quality. The solution of environmental problems requires total consideration of human activities interacting with the environment.

FURTHER READINGS

Ashworth, W. 1986. *The Late, Great Lakes: An Environmental History.* New York: Alfred A. Knopf.

Colborn, T. E., et al. 1990. *Great Lakes, Great Legacy?* Washington, DC: The Conservation Foundation, and Ottawa, Ontatio: The Institute for Research on Public Policy.

Council on Environmental Quality. 1990. *Environmental Quality.* Twentieth Annual Report. Washington, DC.

Council on Environmental Quality. 1982. *Mussel Watch Program.* Environmental Quality—1982. Washington, DC.

Erickson, D. C., H. L. Gary, S. M. Morrison, and G. Sanford. 1982. *Pollution Indicator Bacteria in Stream and Potable Water Supply of the Manitou Experimental Forest, Colorado.* USDA Forest Service Research Note RM-415.

Frederick, K. D. 1990. Water Resources. In R. N. Sampson and D. Hair (eds). *Natural Resources for the 21st Century,* Chapter 7. Washington, DC: Island Press and the American Forestry Association.

Mauritz la Riviere, J. W. 1989. Threats to the World's Water. *Scientific American* 261 (3):80–94.

Potter, L. D., J. R. Gosz, and C. A. Carlson, Jr. 1984. *Water Resources in the Southern Rockies and High Plains. Forest Recreational Use and Aquatic Life.* Eisenhower Consortium Institutional Series, Report 6. Albuquerque: University of New Mexico Press.

World Resources Institute. 1990. *World Resources 1990–91.* New York: Oxford University Press. Chaps. 10 and 11.

STUDY QUESTIONS

1. Discuss the pros and cons of using chlorine for water disinfection if carcinogenic compounds form on contact with some organic pollution.
2. Should we require activated carbon filtration for public water supplies obtained from polluted rivers or simply prohibit organic pollution?
3. Discuss whether we should rely on the "natural assimilative capacity" of a river or use the highest degree of treatment available in treating wastewater.
4. Is it practical to make all surface waters fishable and swimmable?
5. Should we permit ocean dumping of sludge and other wastes?
6. If 34°C (94°F) is a lethal level for fish, should we permit warm water discharge at or higher than that temperature when the river water is already 34°C?
7. Comment on the following: The use of fertilizer should be restricted in the interest of water quality.
8. How do you feel about treating sewage to remove pollutants and recycling the treated water directly for drinking, bathing, and cooking?
9. Can natural filtration remove all pollutants?
10. Do rivers purify themselves naturally?
11. What are the benefits and disadvantages to constructing dams to augment low flows?
12. Should water supply reservoirs be used for recreation?

SUGGESTED ACTIONS

(Adapted from National Wildlife Federation's Citizen Action Guide and other sources.)

1. Pesticides, herbicides, and other toxic lawn care chemicals might eventually contaminate lakes, streams, rivers, or wetlands. Limit your use of these products.
2. Don't dump toxic chemicals down the drain. Find out how to dispose of various toxic household products and use the suggested methods. One source of information is the National Wildlife Federation's Citizen Action Guide published for Earth Day, 1990.
3. Have your drinking water tested. Find out how to do it by contacting your local public health office, state environmental quality agency, or community Cooperative Extension Office. Also, arrange to have the drinking water fountains you use regularly tested for toxic substances, such as lead, and try to correct any problems that you discover.
4. Contact the EPA's Safe Drinking Water Hotline to get information about water quality studies. The number is 1-800-426-4791. The American Ground Water Trust can also be contacted for information about well water. The number is 1-800-423-7748.

7

THE ATMOSPHERE AND GLOBAL CLIMATE CHANGE

Imagine what the earth would be like without an atmosphere. There would be no oxygen to support animal life. There would be no water vapor, and hence no water for any kind of life. The earth would be intensely hot where there was sunshine and extremely cold where there was none. Small meteors from space would bombard the earth's surface instead of burning up in the atmosphere. Deadly ionizing radiation from the sun could reach the surface of the earth.

The earth's atmosphere has a chemical composition that is unique in our solar system. What is its origin? How does vegetation govern its composition? What changes can human activities make in this atmosphere? Will such changes make the earth uninhabitable? In this chapter we will discuss the characteristics of the atmosphere and how they relate to these questions.

THE EARTH'S ATMOSPHERE

Composition of the Atmosphere

The earth's atmosphere contains gases essential to plant and animal life. When we compare the planets in our solar system, we see that the planets more distant from the sun have atmospheres containing methane and hydrogen while those nearer the sun have little or no atmosphere.

Why is the earth's atmosphere unique? Scientists speculate that at one time, hundreds of millions of years ago, the earth and planets closer to the sun had atmospheres similar to those of the more distant planets. The planets nearest the sun then lost their original atmospheres. This probably occurred because the pull of gravity was not great enough in the early stages of their formation, and the lighter elements escaped into space or formed heavier compounds with other elements. On earth, subsequent volcanic activity released gases to form a primitive atmosphere composed largely of methane and ammonia plus water vapor. This was quite different from the present atmosphere, which is rich in nitrogen and oxygen (Table 7.1).

Table 7.1
Composition of the Earth's Atmosphere

Gas	Percent by Volume (Dry)
Nitrogen	78.09
Oxygen	20.95
Argon	0.93
Carbon dioxide	0.03

NOTE: Sometimes water vapor (0.00–3.0 percent) and carbon dioxide are shown as variable constituents of the atmosphere.

Transformation of the Atmosphere

The transformation of the earth's atmosphere has been attributed to the atmosphere's continuing interaction with living organisms. Condensation of water vapor released from volcanic rocks combined with the early surface water to form oceans. Because of the composition of the earth's primitive atmosphere, this water must have contained dissolved carbon dioxide (CO_2) and ammonia (NH_3). All living organisms require oxygen in some form for their life processes, but if the primitive atmosphere contained no free oxygen, the earliest organisms were probably anaerobes that had the capability to break down carbon dioxide, ammonia, and sea water producing organic molecules and thus releasing nitrogen and oxygen into the air. Study of life at the ocean vents may give us some insights into early life.

Eventually, plants evolved that could use the energy from sunlight to separate the carbon from carbon dioxide (see photosynthesis in Chapter 2). By combining carbon with water, these plants were able to make organic compounds necessary for plant life.

Respiration and the decomposition of plant and animal remains release carbon dioxide into the atmosphere. Active volcanoes and combustion of fuels also produce abundant quantities of carbon dioxide. In this chapter, we will consider how carbon dioxide balance is maintained, the influences of human activities on the atmospheric system, and other characteristics of the atmosphere that are important to life on earth.

Thermal Profile of the Atmosphere

The layer of the atmosphere next to the earth where the temperature decreases with elevation is the **troposphere** (Figure 7.1). Above that is the **stratosphere,** where the temperature of the air increases with height. Between the two is a transition zone, the **tropopause,** where there is no change in temperature with elevation. Absorption of the sunlight's short-wavelength radiation by the ozone and oxygen in the stratosphere causes the temperature to rise.

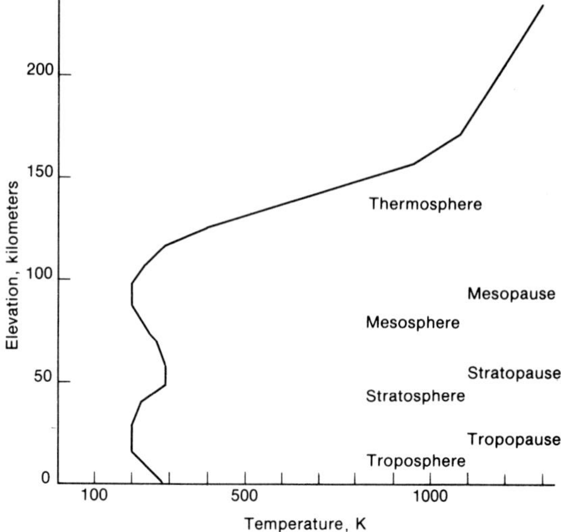

Figure 7.1
Temperature profile of the atmosphere.

The **mesosphere** is the layer above the stratosphere, and the temperature declines with height in that layer. Above the mesosphere, cosmic radiation causes ionization and results in an increase in temperature from absorption of the energy.

The Earth's Radiation Balance

Sunlight Energy Our solar system has the sun at its center and nine planets orbiting about it. The earth, the third planet, is about 150 million kilometers (93 million miles) from the sun. Thermonuclear reactions (fusion of atomic nuclei) in the sun provide the light and heat for our solar system, resulting in a surface temperature on earth ranging from −88°C to 58°C, with a mean at 10°C. This temperature range is significant because, over much of the earth, water is in a liquid state, a necessary condition for life as we know it.

How does the composition of the atmosphere affect solar and earth radiation and the earth's heat balance? To answer this question, we must examine some of the physical properties of sunlight and heat energy.

The temperature of an object indicates its potential to transmit some of its heat through space to another object at a different temperature. The scales we use to measure temperature are **Celsius** (°C, formerly called centigrade) and **Fahrenheit** (°F). Water freezes at 0°C (32°F) and boils at sea level at 100°C (212°F).

Heat will flow from a hotter to a colder body. There is a temperature so low, however, that no heat can be transmitted. This temperature is called *absolute zero*, which is measured as −273.16°C (−459.688°F). To measure the temperature above absolute zero, or the *absolute temperature*, the **Kelvin** scale (K) is used. Absolute zero, then, is represented as 0K. The temperature of an object in kelvins (K) is equal to its temperature in degrees Celsius (°C) plus 273.16.

A hot body, such as the sun, emits heat in the form of radiant energy that can be absorbed by a colder body, such as the earth, causing its temperature to rise. Any body above absolute zero can emit some **radiant energy.** Radiant energy has a wavelike form and is also known as electromagnetic radiation (Figure 7.2). **Electromagnetic waves** have been compared to the waves on the surface of water that are caused by dropping a rock into the water. The ripple or wave spreads in all directions on the surface of the water. An electromagnetic wave, however, moves in all directions like an ever-expanding sphere. Radiant energy as electromagnetic waves can travel through space or a vacuum. In fact, that is the only way energy can travel through a vacuum.

The distance between adjacent wave crests is called the **wavelength.** The number of wave crests passing a fixed point in 1 second is the **frequency** of the wave, which is expressed in cycles per second or in *hertz* when referring to radio waves. The wavelength multiplied by the frequency is the speed of travel, or **velocity.** The speed of electromagnetic waves through space is

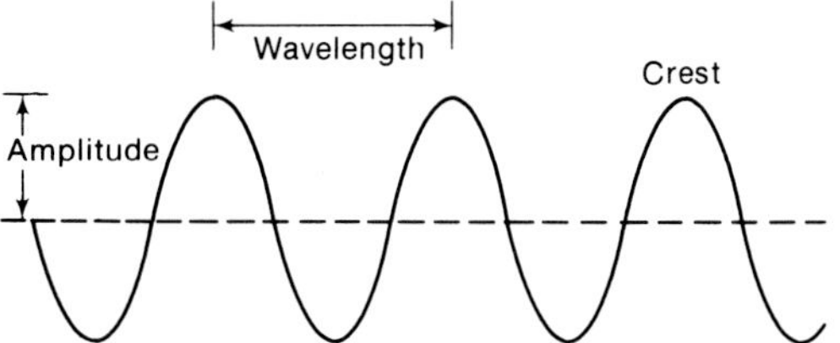

Figure 7.2
The wavelike form of electromagnetic radiation helps describe its characteristics: wavelength, frequency, and amplitude. Wavelength is the distance for a complete wave cycle (distance between wave crests). Frequency is the number of cycles in a certain time period, usually seconds. Amplitude is the height of the wave crest above a mean value, shown as a dashed line. Although it cannot be shown here, an electromagnetic wave moves in all directions from the source.

3 billion meters per second, sometimes written as 3×10^9 meters per second. Not all electromagnetic waves are the same length, and a hot object will emit radiation covering a band of wavelengths. Much of the radiation, however, will be distributed in a relatively narrow band around the wavelength showing the maximum amount of radiation. The wavelength of this maximum radiation varies inversely with the temperature of the radiating body; that is, the hotter the body, the shorter is the wavelength emitted. This feature influences the absorption and radiation of heat by the earth, as we shall see shortly.

The range of all possible wavelengths is called the **electromagnetic spectrum** (Figure 7.3). This spectrum contains radiation of short and long wavelengths and includes visible light. Visible light is an emission of radiant energy with an electromagnetic wavelength in the range detectable by the human eye (0.4–0.7 micron; μ is the abbreviation for micron. One micron equals 0.0001 centimeter). Red is the longest wavelength seen by the human eye; violet, the shortest. Black is an absence of light, and white is a mixture of all detectable wavelengths. The longer wavelengths (over 0.7 micron) include **in-**

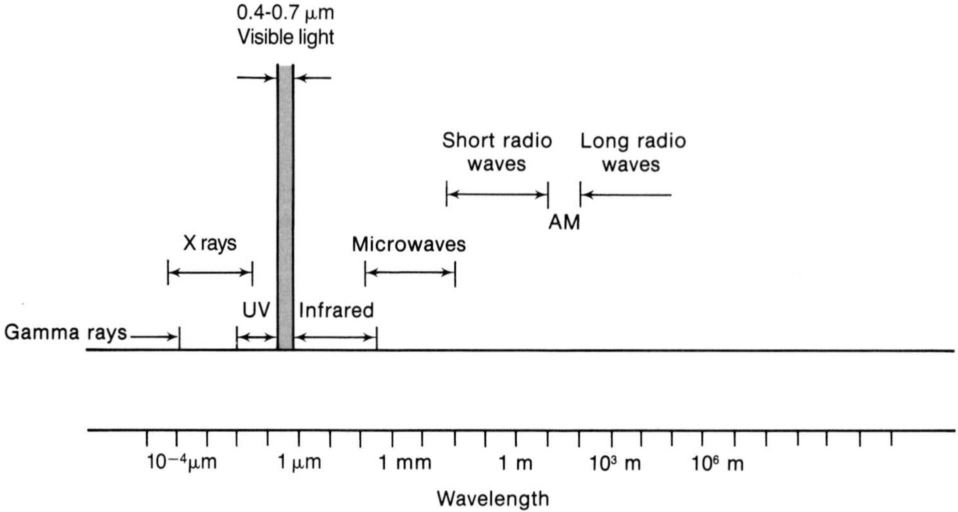

Figure 7.3
The electromagnetic spectrum. Wavelengths seen by the human eye are from 0.4 to 0.7 micron (μ). The shortest wavelengths are violet, and the longest wavelengths are red.

frared radiation (like the heat from a stove that can be felt but not seen); radio waves; and radiation from radar and microwave ovens. The shorter wavelengths (less than 0.4 micron) include **ultraviolet** (UV) light, X rays, and gamma rays.

Persons exposed to short-wavelength radiation may experience undesired effects to their health. A limited exposure to ultraviolet light can cause sunburn. Prolonged exposure to sunlight for many years may cause skin cancer in persons living in tropical areas. Ultraviolet radiation is sometimes used to kill microorganisms in air and water, and X rays and gamma rays can damage and destroy all kinds of cells. Intense exposure to the short wavelengths, as from an atomic bomb blast, is lethal (see Chapter 15). In 1954, 70 percent of the Japanese fishermen working within the area of an accidental atomic fallout pattern received ulcerous skin lesions 21 days after exposure.

Absorption and Radiation of Energy

An object exposed to electromagnetic radiation of any wavelength will absorb energy and warm up. Of course, depending on the characteristics of the material, some radiant energy is reflected. White snow reflects most of the radiation in the visible range. A *black body* is an almost perfect absorber and emitter.

A warmed object, in turn, reradiates electromagnetic energy. Since the electromagnetic spectrum of the radiant energy emitted depends on the absolute temperature of the radiating body, more of the radiation emitted from extremely hot bodies will be in the shorter wavelengths; colder bodies will emit more radiant energy in the longer wavelengths. A wood-burning stove emits infrared radiation even though it may not be obvious just by looking at it that it is hot. An incandescent lamp is so hot at the filament that it emits visible light as well as invisible radiant energy. Most of the solar radiation is in the visible light spectrum, 0.4 to 0.7 micron.

Ozone Protection

If all the short wavelength radiation reached the earth's surface, it is doubtful that higher life forms, including human beings, could survive. In the upper atmosphere, about 15 to 50 kilometers above the earth's surface, the energy in sunlight causes oxygen molecules (O_2) to dissociate to form atomic oxygen (O). Atomic oxygen in the presence of sunlight and other particles combines with oxygen molecules (O_2) to form ozone (O_3). The ozone selectively absorbs the portion of solar radiation in the short-wavelength range, thus preventing most of the ultraviolet, X, and gamma radiation from reaching the earth's surface. The radiation reaching lower levels is thereby limited to those wavelengths longer than 0.3 micron.

Oxygen and ozone in the upper atmosphere absorb 1 to 3 percent of the incoming solar radiation (the ultraviolet and shorter wavelengths). At lower levels, dust particles suspended in the air, clouds, and the earth's surface reflect about 30 percent of the incoming radiation back into space. About 20 percent is absorbed by water vapor, water droplets, and dust in the air. Approximately 50 percent reaches the surface and is absorbed by the earth's land and water surfaces (Figure 7.4). Land with vegetation reflects 20 percent of solar radiation, calm seas 5 to 10 percent, and snow-covered land and ice-covered seas 60 to 80 percent.

Ozone Depletion

When products discharged into the atmosphere are stable—that is, they do not readily degrade—their effects are cumulative. So while it is difficult to imagine a little squirt of hair spray changing the world, billions of squirts can make a difference if the propellant is a stable chlorofluorocarbon (CFC). Over time the CFCs diffuse over the earth and up to the stratosphere. At altitudes above 25 kilometers the CFCs will degrade as they absorb solar energy, releasing some chlorine atoms. The chlorine in turn reacts with ozone, depleting the ozone concentration. The chain reaction that occurs can result in the removal of 100,000 molecules of ozone for every chlorine atom released. The ozone layer in the upper atmosphere is irreversibly depleted by this reaction.

If the global release of fluorocarbons continued at the current rates, some scientists estimate the concentration of ozone in the stratosphere could decline by as much as 16 percent by 2050. The effects of ozone breakdown would be widespread. A 16 percent decrease would increase damaging ultraviolet (UV) radiation reaching the surface of the earth in middle latitudes by 44 percent, resulting in several thousand additional cases of melanoma per year in the United States and several hundred thousand more other skin cancers. A 30 percent depletion of ozone would allow practically all UV radiation to penetrate, producing catastrophic results for living systems. In early 1991 the Environmental Protection Agency reported that the thinning of the ozone

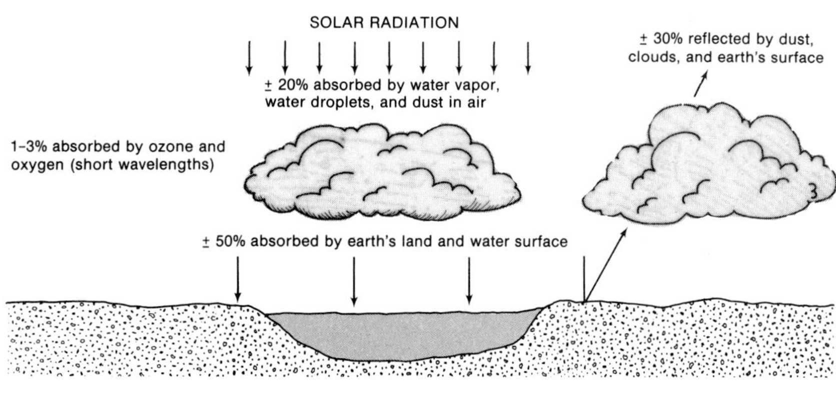

Figure 7.4
Ozone and oxygen in the upper atmosphere absorb radiation of very short wavelengths, protecting the earth from the lethal effects of gamma, X, and ultraviolet radiation. Other degrees of absorption and reflection of solar radiation are also shown.

layer appeared to be about twice as rapid as previously believed and that deaths from skin cancer could rise to even higher levels.

CFCs still have important industrial uses (Figure 7.5), such as foam packaging materials, refrigeration and air conditioning, sterilants, and solvents for cleaning circuit boards. Intense efforts are underway to find alternatives for these CFC-producing industrial processes.

As with other global issues, an international strategy is necessary to deal with this problem. Sweden, Canada, Norway, and the United States have taken actions to control the use of fluorocarbons as aerosol propellants. Other propellants or simple atomizers can be used for spray cans.

In 1987 an international agreement called the Montreal Protocol was negotiated at a meeting in Canada. This agreement resulted in a 1989 treaty signed by 81 nations. The protocol calls for a phaseout of ozone-depleting chemicals by the year 2000. The protocol establishes a framework for regulation, but it is up to individual countries to decide how to control the uses of CFCs within their borders. Some developing countries have been reluctant to sign the protocol because they feel it could slow their economic development. Consequently, considerable effort is directed toward finding substitutes for the ozone-depleting chemicals that developing countries feel they need for industrialization.

There was concern for a while that the discharge of nitrogen oxide (NO) from supersonic airplanes (SSTs) would deplete the ozone layer by chemical reaction. A fleet of seventeen Concordes operating 7 hours per day at an elevation of 17 kilometers was computed to cause a 1-percent reduction of ozone concentration. Even larger fleets could cause a 20-percent reduction by the year 2000. However, this fear has not materialized because SSTs have yet to prove their economic feasibility and the extent of their use has been less than anticipated.

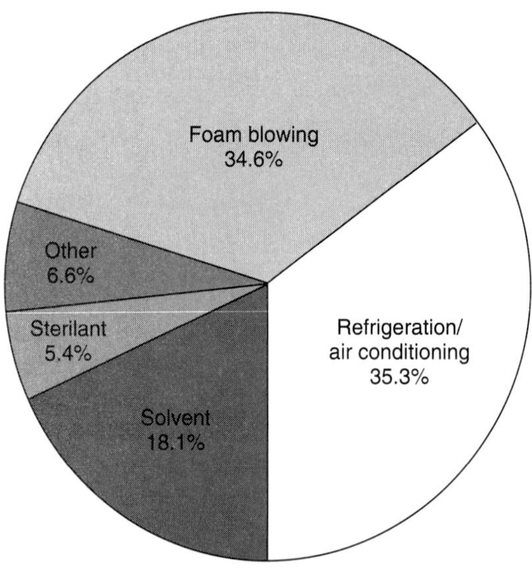

Figure 7.5
Chlorofluorocarbon use. CFCs are used intensely in U.S. industry for a range of purposes. EPA says substitutes can be adopted for most purposes within 10 years.

HUMAN ACTIVITIES AND CLIMATE

When Alfred Wegener proposed a theory of continental movement, he thought that apparent changes in climate (revealed by fossils and other signs found on continents) could be explained by continental migration to higher latitudes. He believed it was more likely that the continents moved than that the basic air circulation patterns had changed.

The forces which control air movement, on first appraisal, seem to be beyond society's ability to control. Even though the amount of energy generated by the sun is beyond our control, we can change the composition of the atmosphere in ways that will determine the amount of sunlight energy that will reach the earth and be retained.

Cooling—Particle Interference

Particles discharged from volcanoes or from industrial smokestacks can restrict the amount of sunlight reaching the earth, possibly leading to a cooling of the earth's atmosphere. The eruption of Tambora in 1815, which killed 12,000 people, discharged 80 cubic kilometers (19 cubic miles) of rock fragments into the atmosphere. Some of the small particles remained in the stratosphere and circled the earth. The following year, known as the "Year Without a Summer," affected crops in northern parts of the United States, causing reduced yields.

Table 7.2
The Greenhouse Gases and Their Potential Contributions to Global Warming

Carbon dioxide	50%
Methane	16%
Tropospheric ozone	8%
Nitrous oxide	6%
CFCs and others	20%

Warming—Carbon Dioxide

Gases in the earth's atmosphere are responsible for maintaining the heat balance of the earth. The temperature of the earth remains relatively constant because the solar energy absorbed by the earth is radiated back into space. The earth, however, radiates at a much lower temperature than the sun. Since the spectrum of the electromagnetic energy is governed by the temperature of the radiating body, the lower temperature of the earth (200–300 K) causes the spectrum of the radiated energy to change to a range of about 4 to 20 microns. Maximum emission is around 10 to 12 microns.

Water vapor absorbs radiation strongly at 5 to 7 microns and above 12 microns. Carbon dioxide absorbs best between 4 and 5 microns and above 14 microns. Most of the incoming solar radiation is able to pass through air containing carbon dioxide and water vapor because most of its electromagnetic energy is in the shorter wavelengths. However, as the earth radiates heat back into space, the shift to the longer wavelengths puts the radiation in a range readily absorbed by water vapor and carbon dioxide. Therefore, any increase in the carbon dioxide content of the atmosphere may result in more of the outgoing earth radiation being absorbed in the air and a warming of the earth's atmosphere.

Other gases also have the potential to cause global warming. The World Resources Institute has estimated that human activities associated with energy development account for almost half of the gases responsible for global warming, while deforestation, agriculture, and industry are responsible for the rest.

This warming effect has become widely known as the **greenhouse effect,** even though a greenhouse behaves quite differently. A greenhouse lets in sunlight through the glass roof to warm the surfaces and air inside and physically confines the warmed air in the glass enclosure, keeping it from rising. The water vapor and carbon dioxide in the atmosphere let the short-wavelength sunlight pass through the air to the earth's surface but absorb the long-wavelength heat radiated by the earth.

Patterns of climate change and its effects are being explored through use of climate change models. These models indicate that as a result of climate change various areas of the earth will be warmer or cooler, or wetter or drier than they are today. Some projections indicate an average rise in temperature of 2 to 5°C in 100 years.

Because variations in the carbon dioxide concentration in the atmosphere have the greatest potential to generate climatic change, attention has been focused on the increasing rate of combustion of fossil fuels (oil, coal, and gas) in the United States and elsewhere. Combustion of fuels, volcanic activity, and decomposition of vegetation put carbon dioxide into the air. Some of the infrared portion of the earth's radiation into space is absorbed by carbon dioxide, leading to a warming of the atmosphere. While the actions of individuals in burning fuels and other activities seem to be insignificant on a global scale, we learned that over a geological time period green plants were apparently able to convert an atmosphere of carbon dioxide, ammonia, and meth-

PREDICTING THE WEATHER

Weather forecasters, also known as meteorologists, have a very complex task. Working at National Meteorological Centers, they use complicated sets of data gathered from satellites, surface and upper-air weather maps, radar images, and other sources to make short-term and long-term weather forecasts. Because of the complexity of the data, computer models are widely used to help meteorologists study large amounts of information and to find patterns that will assist them in making forecasts. Special centers have been established to forecast hurricanes and severe storms, including national hurricane centers in Florida and Hawaii, and the National Severe Storms Forecast Center in Kansas City, Missouri.

While it is often important to be able to predict weather or climate changes over broad regions, most people are especially interested in the day-to-day local weather patterns that affect them directly. For these people the most important weather predictions are the short-term forecasts based on observations at one location. Known as single-station forecasts, they use some of the principles of weather behavior discussed in this chapter. Because they are so localized these forecasts tend to be somewhat general and tentative, and local conditions are often modified by changes occurring elsewhere. Even so, amateur meteorologists can try their hand at local weather prediction by applying certain rules of thumb found to be generally applicable to midlatitude weather in the Northern Hemisphere.

1. At night, air temperatures will be lower if the sky is clear than if it is cloud covered.
2. Clear skies, light winds, and a fresh snow cover favor extreme nocturnal radiational cooling and very low air temperatures by dawn.
3. Falling air pressure may indicate the approach of stormy weather, whereas rising air pressure suggests that fair weather is in the offing.
4. The appearance of cirrus, cirrostratus, and altostratus clouds, in that order, indicates that less dense air is flowing up and over denser air ahead of a warm front, and precipitation is a possibility.
5. A counterclockwise wind shift in wind direction from northeast to north to northwest is usually accompanied by clearing skies and the movement of a cold air mass into the area (i.e., cold air advection).

6. A clockwise wind shift from east to southeast to south is usually accompanied by clearing skies and the movement of warm air into the area (i.e., warm air advection).
7. A wind shift from northwest to west to southwest is usually accompanied by warm air advection.
8. If radiation fog lifts by late morning, a fair afternoon is likely. Radiation fog is a ground level cloud that forms when very humid air is cooled.
9. With west or northwest winds, a steady or rising barometer, and scattered cumulus clouds, fair weather is likely to persist.
10. Towering cumulus clouds by midmorning may indicate afternoon showers or thunderstorms.

The accuracy of local weather forecasts can be increased by also considering historical weather patterns for a particular location and taking the general climate for the region into account. Sometimes geographical characteristics of the landscape can have important effects on weather patterns. For example, the weather of coastal areas is affected by land and sea breezes, while the presence of mountains causes rain shadow effects. It is also common for mountains to have downslope winds like the spring chinooks that occur on the east slope of the Rocky Mountains.

It is possible to learn about weather forecasting by studying the daily weather maps in local newspapers and watching weather reports on television. Sometimes local newspaper and television stations offer guides to weather forecasting.

Most of the information for this case study has been adapted from J. P. Moran and L. W. Morgan, *Meteorology. The Atmosphere and the Science of Weather*, 3rd ed. New York: Macmillan Publishing Company, 1991.

ane to one rich in nitrogen and oxygen. As the earth's population has increased and industrial processes have become more extensive, it appears that relatively small individual actions can collectively cause vast worldwide changes.

The carbon dioxide concentration in the air has been rising (Figure 7.6). The increase in carbon dioxide concentration is about one-half the amount expected from the rate of burning fuel. Several natural processes seem to be moderators. The ocean acts as one **sink,** where carbon dioxide is dissolved as such and combined as carbonates and bicarbonates. An increase in carbon dioxide in the air acts as a stimulant in plant growth and thus more carbon is fixed in plant parts. Carbon fixed in marine organisms is deposited in the ocean when the organism dies, effectively removing that carbon from any cycle.

We might think that the carbon dioxide content of air would not be influenced by national strategies for meeting energy needs. Such is not the case. Proposals for a significant increase in the use of coal to meet energy needs of the United States and to reduce reliance on imported oil would add considerable amounts of carbon dioxide to the atmosphere. Combustion of coal produces 1.8 times more carbon dioxide than combustion of natural gas and 1.2 times that of oil!

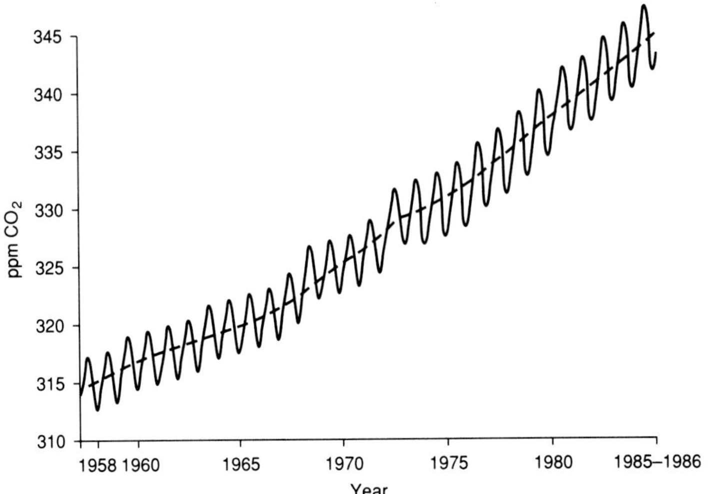

Figure 7.6
Concentration of atmospheric carbon dioxide at Mauna Loa Observatory in Hawaii. The yearly oscillation is due primarily to seasonal cycles of photosynthesis and respiration by plants in the Northern Hemisphere.

The combustion of coal also increases the emission of particulates, but the Committee on Health and Environmental Effects of Increased Coal Utilization did not think the effects would be as significant as the increase in carbon dioxide concentration. There is also some uncertainty as to whether the particulates would cause cooling or warming. The committee, established by the National Institute for Environmental Health Sciences, noted projections of an increase of two- or threefold in carbon dioxide concentration in 100 years. They expressed concern that climatic changes after 50 years could induce sociopolitical disruptions because of changes in climate and agriculture. Even a rise to a level of 375 to 400 parts per million by the year 2000 (as forecast by some scientists) could cause the earth's temperature to rise 0.5°C, although the exact temperature increase is impossible to predict.

Some scientists do not think combustion of fossil fuels is as significant to stability of the carbon dioxide concentration in the atmosphere as the abundance of green trees in tropical rain forests and green plants in the oceans. These scientists believe that converting tropical forests to agriculture or destroying plankton in oceans would have a much greater influence because the dominant means of removing carbon dioxide from the atmosphere would be destroyed.

A warming of the earth could cause a shift of air circulation cells toward the poles, changing patterns of rainfall. The Midwest could become a desert, and farming could move to the Hudson Bay area in Canada. The ice caps could also melt, causing an estimated 13 to 38 centimeters (about 5–15 inches) rise in sea level by 2025. If this sea level rise continued, it would result in the inundation of coastal lowland areas including important wetlands and heavily populated areas such as Bangladesh and urban centers near the coast.

Global warming could have severe effects on the biodiversity of the earth as plants and animals adjust to the changing climate. For example, Kirtland's warbler (see Chapter 10) could face extinction if global warming occurs. The bird might be able to move with the shifting jackpine zone it depends on. However, a northward shift of the jackpine zone would take the bird from the sandy soils it also requires.

The rate of climate change is also an important factor affecting how plants and animals might respond to global warming. Paleoecologists have shown that tree species in eastern hardwood forests were able to migrate northward as the climate in North America began warming 15,000 years ago. However, these same tree species may not be able to keep up with the much more rapid climate change projected by some climate change models.

Scientists estimate that up to 80 percent of the oxygen produced by photosynthesis is from green plants in the ocean (algae), especially in the coastal areas. Water pollution in coastal areas could destroy these plants or reduce their numbers.

Warming or cooling of the atmosphere is important because it could influence vast climatic changes, such as glacial recession or expansion. In recent history the temperature of the earth increased from the 1800s to the 1940s, but it has decreased since the 1940s by 0.3°C. There is no indication that this fluctuation is associated with pollution.

Even though variations in the concentration of greenhouse gases in the atmosphere have the potential to generate climate change, not all scientists agree that a long-term warming trend is actually underway.

Short-term increases or decreases in temperature of about 2°C over a period of a decade have been experienced without great ecological changes. We know that glacial periods are associated with accumulation of snow over thousands of years and that there appear to be cyclical interglacial periods when glaciers melt and retreat. The cause of these cycles is not clear. At present, some scientists think we are nearing the end of an interglacial cycle. Perhaps a warming of the atmosphere will delay the next glacial period. However, more cloud cover and more particles in the air could reflect heat, increasing cooling tendencies. These trends and their influences will be debated for some time.

Like the problem of ozone depletion, the potential warming of the earth's atmosphere requires a global response. The World Resources Institute reported that in 1987 more than 50 percent of the annual warming potential of the atmosphere was contributed by only five countries—the United States, the U.S.S.R., Brazil, China, and India. Efforts to reduce global warming in these countries are especially important, although all countries must contribute to solving the problem. Several actions to reduce global warming have been suggested:

1. Reduce or slow the rate of population growth.
2. Find ways to meet energy needs without burning fossil fuels.
3. Slow the rate of deforestation and institute tree-planting programs.
4. Find substitutes for the CFCs which contribute to global warming as well as to ozone depletion.

Heat Islands

Human activities can modify the weather in unintentional ways. On a small scale, brick, concrete, and other construction materials in urban areas absorb and hold the sun's heat. In addition, heating, air conditioning, and generation of electricity result in waste heat. Consequently, **heat islands** develop in large urban areas (Figure 7.7). The air circulation pattern between urban and rural areas is similar to the sea breezes induced by convection currents when the atmosphere is relatively stable. When there is a wind,

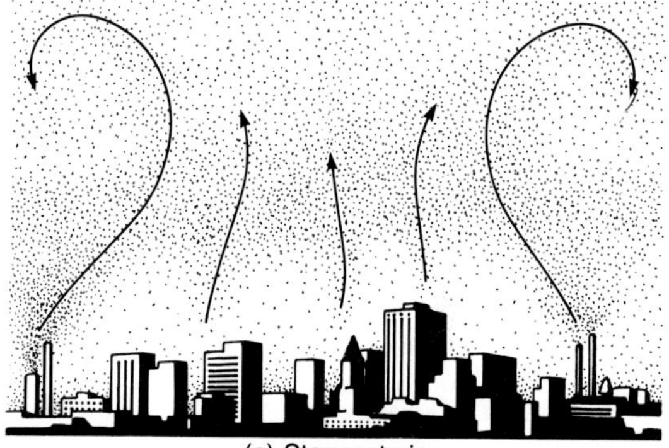

(a) Stagnant air

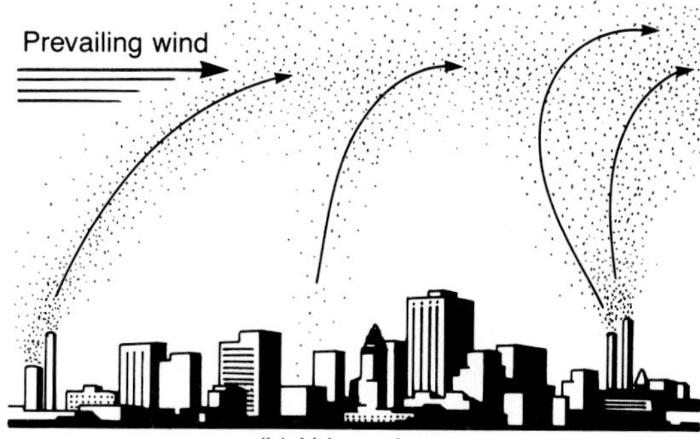

(b) Urban plume

Figure 7.7
Heat island effects.

the pattern resembles a plume moving downwind from a huge smokestack. This localized pattern of air circulation transports urban air pollutants, such as auto exhaust and the resulting ozone and other oxidants, to rural areas.

In addition to the heat island effect urban areas also experience increased fog and precipitation associated with particulate matter discharged into the air, especially from burning coal. As London has eliminated air pollution, the weather has changed to more days of sunshine and fewer of dense fog.

Cloud Seeding

Intentional efforts to modify the weather are directed at dispersing fog, suppressing hail and lightning, inducing rainfall, and reducing the intensity of tornadoes and hurricanes. Rain is induced in moist air when drops or ice crystals become sufficiently large to begin falling. Cloud seeding with silver iodide crystals is a method which provides a nucleus for freezing supercooled cloud droplets, which then leads to the growth of precipitation-sized particles.

Cloud seeding efforts have been controversial. The effectiveness of seeding is debatable, although some investigators think that 10 percent or more additional precipitation may be induced. In addition, there are serious legal and political questions. Efforts to see clouds over South Dakota in 1972 were followed by 0.3 meter (1 foot) of rain that caused a flash flood in Rapid City, killing many people. While that flood was almost certainly not a result of that seeding, the potential hazard is real. Other efforts to dissipate the force of hurricanes by seeding have raised questions about the desirability of interfering with one of the ways heat accumulations at the equator are rapidly dispersed toward the poles. Some meteorologists fear drastic climatic changes if hurricanes are suppressed.

Some applications of cloud seeding for dispersal of fog (a cloud on the ground) have been successful. The Salt Lake City airport is sometimes closed because of fog. A small plane takes off in the fog, seeds the fog cloud, and shortly lands in the *hole* created in the cloud. The hole is open long enough for commercial airliners to land. Seeding is also practiced to prevent severe hailstorms.

AIR CIRCULATION AND WEATHER PATTERNS

Air circulation patterns are influenced by a complex array of factors. A major influence is the absorption and release of heat by the atmosphere, land masses, and bodies of water. When the sun shines, the temperature of oceans and lakes rises more slowly than that of land. When there is no sunshine, as at night, the temperature of the water bodies falls more slowly. Wind and wave action may act to mix the water so that heat absorbed at the surface is distributed downward in the water to a depth of as much as 100 meters. However, heat penetrates the ground to very shallow depths. Because land heats and cools more rapidly than water and the heat storage in land is limited by the shallow depth of penetration, adjacent bodies of water tend to moderate the local temperature on shore by absorbing large quantities of heat in hot weather and releasing it in cold weather.

Land temperatures change rapidly while ocean temperatures are relatively constant, with the result that the ocean is cooler than the land in the daytime and warmer at night. These conditions favor the movement of air associated with sea breezes in the morning and land breezes in the evening. Furthermore, prevailing winds induce ocean currents, which carry warmed water from the equator toward the poles. At the poles their stored heat is released, thereby distributing heat over the earth. Also, the heat stored in the oceans in the summer is released to the atmosphere in the winter. Both of these actions moderate the climate.

Places where circulating air warms and rises tend to be areas of low pressure (lows), and where the air cools and subsides, areas of high pressure (highs). Winds occur when air moves from high pressure to low pressure areas. Be-

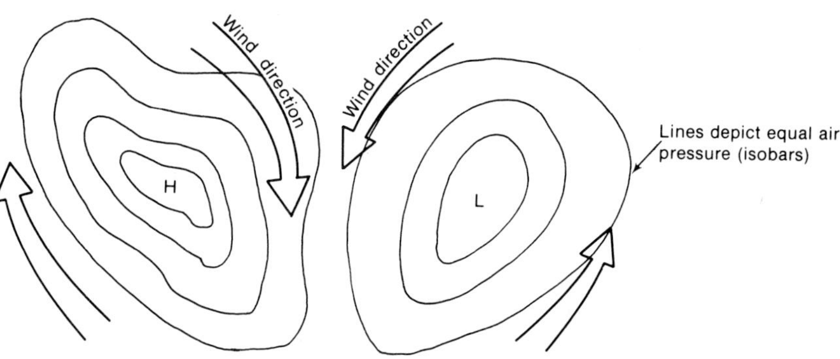

Figure 7.8
High- and low-pressure areas. Instead of flowing directly from high-pressure to low-pressure areas, winds in the Northern Hemisphere circle highs in a clockwise and outward direction; they circle lows in a counterclockwise and inward direction.

cause of the influence of the earth's rotation (Coriolis effect), air flows clockwise around highs (anticyclones) and counterclockwise around lows (cyclones) in the Northern Hemisphere (Figure 7.8). Near the ground, surface friction tends to reduce the Coriolis effect, resulting in a spiral flow outward away from the anticyclone (high-pressure system) and inward toward the center of cyclone (low-pressure system).

These basic systems of air motion not only influence weather patterns, but also affect the distribution of air pollutants. For example, when a high-pressure system is stationary over Bermuda, the basic clockwise air motion will tend to sweep air pollutants along the eastern seaboard of the United States from Richmond, Washington, Baltimore, Philadelphia, and New York City to the New England area.

Regional Influences on Weather

Systems of cyclones and anticyclones govern the general weather patterns, but other conditions can cause regional winds. The phenomena of land and sea breezes have already been discussed. These are due to diurnal (twice a day) cycles of convection currents resulting from the difference in heating and cooling of the land and the sea. Similar mountain–valley currents are established by air rising up warmed slopes in the daytime and down cooled slopes at night. High mountains, such as the Himalayas, often act like a dam to deflect or obstruct a surface flow of cold air. They can also serve as a temperature barrier. For example, in the winter the temperature on the north side of the Himalayas in Siberia might be −60°C, while at the same time on the south side in India it might reach 30°C.

Some regional winds are the result of gravity flow of cool, dense air down the slope of mountains. Examples are the *mistral* from the Alps down the Rhone Valley and the *bora* in Yugoslavia toward the Adriatic Sea. Sometimes a high-pressure wind descending a mountain slope will heat up because of compression. Recall that if the pressure of a gas increases, the temperature will increase. Examples of hot winds are the *foehn* of the Alps, the *Santa Ana* of southern California, and the *chinook* east of the Rockies.

Warm and Cold Fronts

The development of cyclonic air circulation systems and the movement of high- and low-pressure systems across the country may cause relatively warm (less dense) air to move over cooler (denser) air. The intersection of these dissimilar air masses is a *warm front.* If the warm air is uplifted, rain may be produced. Warm air is also uplifted at ground level where the leading edge of relatively cooler air moves in under warmer air, as with a *cold front* (Figure 7.9). This also causes shower activity.

As rising air masses become cooler cloud formation may occur. The clouds that resemble huge puffs of cotton are called **cumulus** clouds.

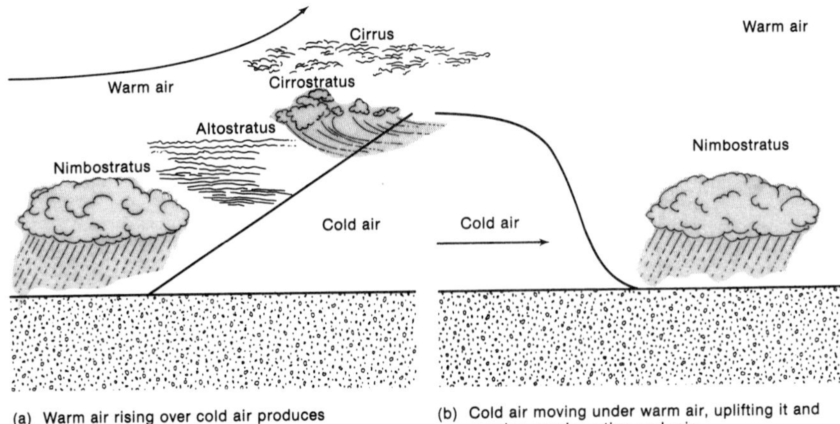

Figure 7.9
Rain caused by (a) warm front; (b) cold front.

Clouds associated with rainstorms or approaching storms are formed in layers and are called **stratiform** clouds. Thin, feathery clouds known as **cirrus** clouds appear at a high level (6–18 kilometers) and are often an indication of an approaching storm. **Cirrostratus** clouds form a thicker layer at a high level. **Altostratus,** a middle-level (2–6 kilometers) cloud, appears as a uniform white or gray layer and is thick to the point of obscuring the sun. **Nimbostratus** clouds are low (0–4 kilometers) clouds in a uniform gray layer from which rain falls.

Jet Streams

During World War II, U.S. and German pilots flying at high altitudes encountered very strong winds. The United States abandoned high-altitude bombing flights over Japan because the 350-knot (400-mph) winds interfered with accurate bombing. We now recognize these air currents as **jet streams.**

Jet streams are normally thousands of kilometers long, more than 100 kilometers wide, and a few kilometers thick, and they move at a speed in excess of 30 meters (100 feet) per second. In both hemispheres, the surface air temperature is relatively uniform over wide areas except for a sharp change in horizontal temperature gradient at the boundary between cold polar air and warm tropical air, which is called the *polar front.* Blowing from the west, the two polar front jet streams meander around the earth above these polar fronts. Another jet stream flows above the subtropical high-pressure belt during the winter season only. In the hemisphere experiencing winter, a polar-night jet stream develops at a height of 32 to 48 kilometers (20–30 miles) at middle to high latitudes due to equatorial and polar temperature differences in the stratosphere. Jet streams tend to narrow at places where their speeds are extremely high, becoming **jet maximums.**

Jet maximums are always associated with strong fronts in the lower atmosphere. In the Northern Hemisphere, areas of low-pressure air will be to the left and areas of high-pressure air to the right of a jet stream. Air tends to ascend in the low-pressure area and descend in the high-pressure area (Figure 7.10). Thus, the jet streams are involved in horizontal transport and vertical mixing of ozone and air pollutants.

Closer to the earth's surface, stormy weather is associated with jet maximums. Sharp temperature differences between air masses tend to induce a cyclonic circulation pattern around the low-pressure areas. The jet maximums accelerate this tendency and thunderstorms and tornadoes may develop. The jet stream may also help initiate the development of cyclonic air flow.

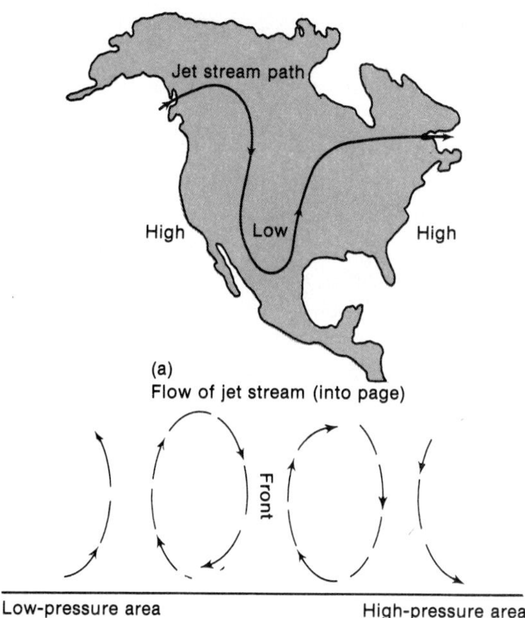

Figure 7.10
Jet streams. (a) A typical jet stream path. (b) Cross section of a jet stream path.

Topographic Uplifting

Where air in motion approaches a mountain barrier, a general uplifting of the air produces clouds and often rain as moist air is cooled. Such rain is usually of low intensity and is called **orographic** from its relationship to mountains. Because of the uplifting over the mountain, there is more precipitation on the windward slope and light precipitation, or *rain shadow*, to the leeward (Figure 7.11). Areas in rain shadows are arid and deserts may form, as found in the western part of the United States. While rain on the ascending windward slope may be of relatively low intensity, some of the highest annual precipitation rates occur in these areas. The Sierras, Cascades, Rockies, and Himalayas experience these high rates of precipitation because prevailing winds bring moist air over oceans to these mountain slopes. Of course, unstable air associated with fronts may produce torrential rains in these areas.

STORMS AS HAZARDS

Hurricanes

A **hurricane**, sometimes called a typhoon, is an intense cyclone that develops over tropical oceans. For example, in the South Atlantic and the Caribbean, heat from the summer sun warms the surface of the ocean and evaporates water. Each gram of water evaporated represents 580 calories of heat waiting to be released when the vapor condenses. The buildup of heat in the late summer makes the area *ripe* to produce thunderstorms. When this tendency for storms is combined with a cyclonic weather system moving through the area, the thunderstorms may be so intense as to release sufficient heat to warm the air over a large area. The rising warm air creates a local low-pressure area, and air flows around and inward toward the low-

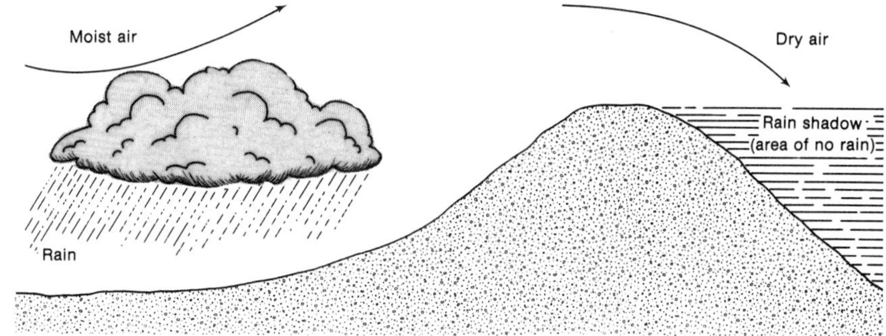

Figure 7.11
Rain shadows. Movement of moist air up a mountain slope cools air and causes condensation and rain on the windward side. No rain falls on the leeward side, creating a rain shadow.

pressure center. The spiraling of air inward increases the speed. The center of the hurricane, the *eye*, is calm, and there the skies are nearly clear. As the moist air rises, condensation releases energy that further warms the air, increasing the speed of circulation. This brings in more moist air which releases more energy, and in this manner the system expands (Figure 7.12).

Hurricanes can reach 800 kilometers (500 miles) in diameter, with winds over 256 kilometers (160 miles) per hour. Interaction between the wind and water can produce waves 6 to 10 meters (20 to 30 feet) high as well as cause water to run up on shores, flooding coastal areas. When hurricanes move over land, the moist air is not readily available to supply energy, so winds dissipate and the hurricane dies out in rainstorms. Even as hurricanes dissipate over land, rain may be so intense as to cause flooding far away from the shore. Surges from storms are greatest when they coincide with maximum tides.

Since 1900, barrier islands on the Atlantic and Gulf coasts have been crossed by more than 100 hurricanes. Six thousand people were killed in 1900 when a hurricane hit Galveston, Texas, and 2000 in Florida, in a 1928 hurricane. While early warning reduced loss of life in later storms, Hurricane Camille caused $1.4 billion property damage in 1969, Agnes $2 billion in 1972, and Frederick $2.3 billion in 1979. Still, fewer than 20 percent of the barrier island residents along the Atlantic and Gulf coasts had experienced a major storm. The Atlantic coast had not had a major hurricane since Donna in 1969. This changed in 1985, when eight tropical storms struck the U.S. coastline. Among them was Hurricane Gloria, an Atlantic coast storm that killed eight people and caused $1 billion in damage. Total damage from the eight storms was $4 to 5 billion, and the death toll was thirty. The last time the United States experienced eight tropical storms in one year had been in 1916, when damage was $33 million and 107 people were killed. The potential for damage was much greater in 1985 because of extensive coastal development and a coastal population that had increased from 12 million in 1916 to 42 million.

In 1988 Hurricane Gilbert killed 318 people, 202 of them in Mexico, the others in the Caribbean islands, Central America, and the United States. Total property damage was estimated at $5 billion. In 1989 Hugo, caused $10 billion in damages as it ripped through the northeastern Caribbean islands and struck the Carolina coast of the United States, killing forty-nine people. The 1990 hurricane season was rather quiet, with most of the storms remaining at sea. Another damaging hurricane, Andrew, struck Florida and Louisiana in 1992.

Tornadoes

A **tornado** is another type of cyclone, but is much smaller in diameter than a hurricane (Figure 7.13). Tornadoes are about 100 meters to 1 kilometer (330 feet to 0.6 mile) in diameter, with wind speeds exceeding 200 meters (660 feet) per second. In addition to damage caused by wind speed or by objects propelled through the air, the very low pressure in the center of a tornado causes buildings to explode. Tornadoes usually form over land when cooler, drier air moves rapidly over warm, moist air. This creates an unstable condition, and the rising warm air creates a turbulent situation. The condensation of the water vapor releases energy to warm the air more and to speed up the circulation. Although tornadoes may occur at any time, they are frequently spawned in the spring when there is a maximum temperature differential between the cold and warm air masses. They are also associated with severe thunderstorms and may be produced on the fringes of large hurricanes.

Not all cyclonic systems are as violent as hurricanes and tornadoes. The vast majority of cyclones are the low-pressure systems that produce a great deal of the *weather*. High-pressure systems (anticyclones) are usually characterized by fair weather because the descending air becomes

(a)

(b)

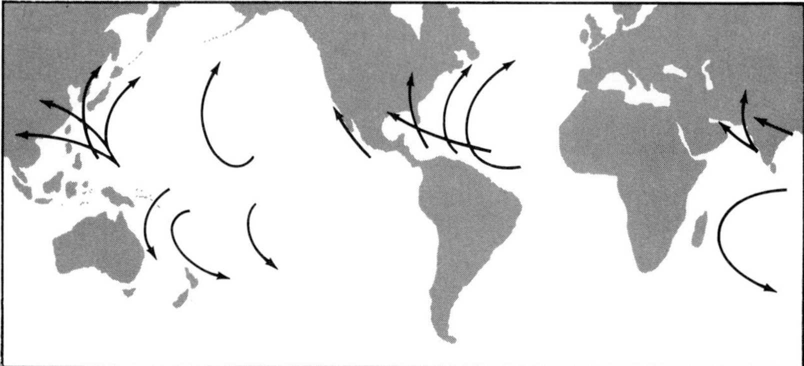

(c)

Figure 7.12
(a) Satellite photo of hurricane Ione in the Pacific in 1974 (Courtesy of NOAA). (b) Cross section of a hurricane showing cloud distribution. (From Richard A. Anthes et al., *The Atmosphere*, 3rd ed. Columbus, OH: Merrill, 1981.) (c) Map of regions of hurricane formation indicating paths of movement. (Data from NOAA.)

Figure 7.13
Tornado as it entered Union City, Oklahoma, on May 24, 1973. (Courtesy of NOAA.)

warmer as air pressure increases and any droplets evaporate. Wind speeds and mixing depths are usually lower in anticyclones. Because there is less rapid dispersion and diffusion, higher concentrations of pollution are usually associated with the highs.

Because intense storms like hurricanes and tornadoes often endanger human life and property, hazard management systems have evolved to deal with them. These management systems involve procedures for forecasting and warning, hazard protection, and disaster relief and recovery. (We discussed the flood hazard management system in Chapter 5 and systems for geophysical hazards—volcanoes, earthquakes, landslides, etc.—in Chapter 4.)

SUMMARY AND CONCLUSION

The earth's unique oxygen-rich atmosphere, which is necessary for mammal respiration, evolved as a result of plant photosynthesis. Now human activities, particularly the combustion of

LEGISLATION

Montreal Protocol on Substances that Deplete the Ozone Layer (1987). This protocol is a formal agreement, similar to a treaty, designed to protect the ozone layer by taking precautionary measures to control global emissions of substances that deplete it.

Disaster Relief Act of 1974. This law authorizes the president to provide assistance to an area suffering a natural catastrophe such as a severe storm or an earthquake. After declaring a disaster the president delegates most dispersements to the Federal Emergency Management Agency (FEMA). FEMA provides disaster relief funds to state and local governments, and loans or grants to individuals.

fossil fuels and the destruction of forests, may increase the carbon dioxide concentration enough to alter the earth's heat balance. Other human activities may deplete ozone in the upper atmosphere, with the result that ultraviolet and ionizing radiation normally absorbed by ozone could reach the earth. Increased skin cancer or worse results are possible consequences.

The basic air circulation disperses and transports pollutants. During periods of temperature inversions, however, pollutants are not dispersed but are concentrated in a particular area. Pollutants, by absorbing or reflecting solar energy, could alter weather patterns drastically.

So far, attempts to modify the weather, such as cloud seeding, have been controversial. On the other hand, especially in urban areas, some human activities have produced unintentional small- and large-scale effects. Examples are urban heat islands, an increase in carbon dioxide, and other greenhouse gases.

The future of human beings is related to the condition of the atmosphere because life, food supplies, and ways of life are bound to the weather. Consequently, we must carefully evaluate the effect of human activity on the atmosphere to determine what global changes are being wrought. Thoughtful action now may avoid irreversible disaster later. (The role of the atmosphere in the hydrological cycle is covered in Chapter 5.)

FURTHER READINGS

Brookes, Warren T. 1989. The Global Warming Panic. *Forbes* 144 (14): 96–102.

Dunlop, Storm, and Francis Wilson. 1987. *Weather and Forecasting* (Macmillan Field Guides). New York: Collier Books, Macmillan Publishing Company.

Flavin, Christopher. 1990. Slowing Global Warming. In Lester R. Brown (ed.). *State of the World 1990.* Washington, DC: Worldwatch Institute.

Moran, Joseph M., and Michael D. Morgan. 1991. *Meteorology. The Atmosphere and the Science of Weather.* New York: Macmillan Publishing Company.

Schneider, Stephen H. 1989. The Greenhouse Effect: Science and Public Policy. *Science* 243: 771–81.

Shea, Cynthia Pollock. 1989. Protecting the Ozone Layer. In Lester R. Brown (ed.) *State of the World 1989.* Washington, DC: Worldwatch Institute.

Weatherwise. A magazine published quarterly by the Helen Dwight Reid Educational Foundation in association with the American Meteorological Society.

White, Robert M. 1990. The Great Climate Debate. *Scientific American* 263 (1):36–43.

STUDY QUESTIONS

1. What are the implications for society of the changes in the atmosphere from its original composition?
2. What is the significance of carbon dioxide and ozone to plant and animal life?
3. What are the mechanisms that moderate climate? What scientific principles are involved?
4. What would happen to life if the sun's heat were not distributed over the earth?
5. Why do inversions aggravate the effects of air pollution?
6. Is it socially desirable to modify the weather? What can be done about human activities that inadvertently modify the atmosphere?

SUGGESTED ACTIONS

(Adapted from National Wildlife Federation's Citizen Action Guide.)

1. Use a coffee mug or thermos instead of styrofoam cups, many of which are made from CFCs.
2. If you have an auto air conditioner with an ozone-depleting refrigerant, have it repaired or recharged at service stations that recycle the refrigerant.
3. Trees use carbon dioxide, a major greenhouse gas. Support programs that attempt to slow the rate of deforestation, especially in tropical rain forests. Also, become involved in tree planting programs. If planted near houses, trees can also reduce the amount of fuels needed for heating and cooling.

8

AIR POLLUTION

In the previous chapter we saw how air pollution can affect the heat balance and chemical equilibrium of the earth's atmosphere. Wind can transport pollution many miles from its source, causing such effects as acid deposition, and inversions can cause pollutants to accumulate in an area in dangerous concentrations. We will examine the effects of air pollution on people, animals, and vegetation and explore ways to minimize pollution.

TYPES OF AIR POLLUTION

We have already noted how air pollution can affect the earth's heat balance and modify climate and weather patterns. Atmospheric conditions also influence the severity of air pollution. Sunlight produces photochemical smog, and atmospheric inversions can cause pollutants to accumulate in dangerous concentrations.

Air pollution is not a new problem. In 1273 King Edward I banned use of sea-coal in London, and Parliament established other controls in 1306. A 1661 report, Fumifugium, attributed one-half of all deaths in London to pulmonary conditions brought on by polluted air. Great Britain's Public Health Act of 1848 included provisions for smoke abatement.

With the use of coal for power in industry and locomotives, smoke became a nuisance in many cities. After conditions became so bad you could barely discern the sun at midday, Pittsburgh pioneered in attacking the smoke problem, mainly by restricting the burning of soft coal (Figure 8.1).

Near the beginning of this century, acid fumes from a smelter at Copper Hill, Tennessee, completely denuded the land in the area, creating a "devil's playground" of barren eroded hills (Figure 8.2). Other forms of air pollution have since been found to affect all kinds of vegetation.

In the 1940s another type of air pollution was recognized in the Los Angeles area. This pollution was most severe on sunny days, when there were also temperature inversions (see next section). Vegetation damage was observed in 1944, and as the problem became more severe, eye irritation caused tearing. In the 1950s it was demonstrated that auto exhaust gases, when irradiated by sunlight, formed compounds that had oxidizing and irritating qualities. This chemical soup was called photochemical smog (Figure 8.3).

(a)

(b)

Figure 8.1
(a) Air pollution near Pittsburgh's Liberty Tunnel in 1946, prior to restriction on burning of soft coal. (b) Liberty Tunnel in 1957. (From Arthur C. Stern, Henry C. Wohlers, Richard W. Boubel, and William P. Lowry. *Fundamentals of Air Pollution.* New York: Academic Press, 1973. Photos from Allegheny County Health Department, Pennsylvania.)

Effects of Air Pollutants

Human and Animal Health In December 1930, a mysterious irritating fog formed in the Meuse Valley in Belgium. People became ill on the second day and men donned gas masks to search vainly for some unknown poison gas cache, perhaps left over from World War I. Not until 1948, in Donora, Pennsylvania, did people recognize that air pollution confined by a temperature inversion could be deadly. Four thousand deaths during a 1952 air pollution episode in London attracted the world's attention.

Air pollution exposes humans to poisoning by toxic materials in the environment. Toxic substances enter the human body by ingestion, by absorption through the skin or eyes, by means of a puncture or injection, or by inhaling a dust or gas. Air pollutants enter the body through the respiratory system. The cleansing mechanisms for the lungs bring some particles up to where they are swallowed or expelled.

In general, four factors influence how a toxic substance will affect an individual: concentration, duration of exposure, toxicity, and individual susceptibility.

Concentration and Duration of Exposure Major air pollution episodes resulting in human deaths have involved a complex interaction of high pollution levels, stagnant air, a concentration of pollutants within valleys or other topographic basins, and a condition of atmospheric stratification referred to as an **inversion,** which places a "lid" over the polluted area and prevents dispersion of the pollutants.

Under normal conditions, the temperature of air declines with height, but under conditions of zero or low wind an inversion might occur—a condition where a band of warmer air overlays cooler air. Such inversions may be caused by any of several conditions. When the sun sets, the ground cools more rapidly than the air. Nocturnal cooling of air next to the ground frequently leaves a layer of warmer air immediately above. A nocturnal inversion breaks up when the early morning sun warms the earth, heating the air next to it. On a larger scale, the overrunning of cooler air by a warmer air mass or the subsidence of a warmer air mass over a layer of cooler air causes an inversion that will not break up until the weather system changes. Also, a wedge of cooler air may move in underneath warmer air.

Figure 8.2
Devil's playground. Acid fumes from a copper smelter at Ducktown, Tennessee, killed nearby vegetation and caused rapid erosion. (Courtesy of A. Keith, U.S. Geological Survey.)

(a)

(b)

Figure 8.3
Smog is caused by chemical reactions between hydrocarbons, oxides of nitrogen, and other pollutants when they are irradiated by sunlight. The St. Louis skyline is readily seen when there is no smog (a), but is not visible when there is smog (b). (Courtesy of George Arnold, Southern Illinois University at Edwardsville.)

When no inversion exists, a discharge of warm gases from a smokestack will be buoyed upward and dispersed by the wind. However, when there is an inversion, there is little if any horizontal air motion. The stack discharge rises until it encounters the inversion layer, whereupon it ceases rising. This condition may be observed in a mountain valley after the sun goes down. Smoke from a cabin will rise to the inversion layer and then spread outward.

Inversions lead to a concentration of pollutants discharged into the atmosphere. The pollutants accumulate at the inversion level, gradually spreading out horizontally in all directions and diffusing downward. Since the concentration of pollutants will be a maximum at the inversion layer, this bodes ill for persons who live on upper floors in high-rise buildings if that is the level of the inversion layer. The concentrations at ground level can also exceed safe limits during an inversion with deadly results.

Where mountain valleys or other physical features confine the air, inversions cause more acute concentrations of pollutants than a flat terrain, where relatively unlimited horizontal diffusion occurs. When nocturnal inversions break up, vertical mixing takes place that brings the pollutants accumulated near the inversion elevation to the ground level. Thus, the highest concentration of pollutants at ground level may be at the time of the breakup of the inversion.

Toxicity Two different substances can cause different effects even though the concentrations and duration of exposure are identical and the test animals exposed are as nearly alike as possible. The difference in effect is caused by toxicity. Toxicity is commonly measured by how much of a substance kills 50 percent of exposed animals, a quantity called LD_{50}. The National Institute of Occupational Safety and Health publishes a list of toxic substances and their known toxic effects.

In some cases people do not seem to react to a toxic substance until some level of exposure, significantly above zero, is reached. The level where physiological reactions of humans or test animals begin to be observed is called the **threshold level.** As we will see in Chapter 15, the concept of a threshold level is also used to examine the effects of radiation. Some substances do not appear to have a threshold. In other words, any exposure—no matter how small—causes some reaction. Substances that have no threshold are considered most hazardous because even one molecule at a vital place could cause trouble. Those that emit ionizing radiation are considered to be in this category, and asbestos, and perhaps some toxins, may be.

Individual Susceptibility Individual susceptibility depends on the person's health history. People with lung and heart ailments are most affected by air pollutants, as are the very young and the aged. Individuals who are allergic to certain substances are sensitive to lower exposures and have worse reactions when exposed to those substances.

We are also aware that greater reaction is produced when a person is exposed to two or more of certain substances simultaneously than to either substance alone. This effect is called **synergism.** Such combined effects have been noticed in community air pollution episodes where the concentration of a pollutant in the air appears to cause adverse reactions at levels below those observed in laboratory experiments.

Gases are readily carried to the depths of the lungs, but there are natural defenses to keep out particles. Some particles impinge on mucus in the nasopharynx area (Figure 8.4) and are washed out. The nasal and bronchial passages are lined with hairlike *cilia* that tend to sweep particles out. As a result, particles of dust, carbon, and pollen larger than 10 microns are usually kept out of the lungs (Figure 8.5). Particles less than 1 micron readily pass to the lungs and become lodged. Coughing dislodges and expels some, but not all, such particles. Particles less than 0.01 micron will act like chemical molecules and can be absorbed through cell walls in the lungs.

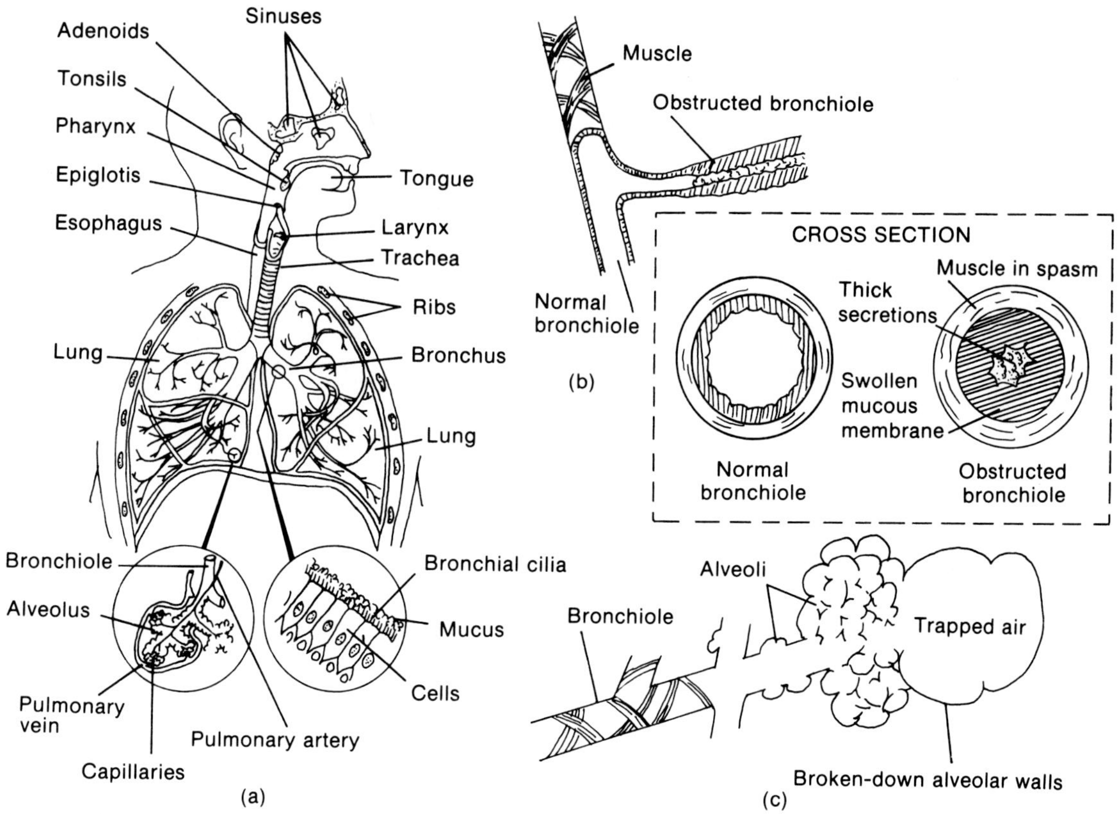

Figure 8.4
(a) The respiratory system. (b) Obstruction of a bronchiole during an asthma attack. (c) In pulmonary emphysema, loss of elasticity and deterioration of alveoli walls deter exhalation of carbon dioxide gas. (From the American Lung Association, *Air Pollution Primer.* New York, 1974.)

Figure 8.5
Sizes of selected particulates. (From the Public Health Service, *Air Quality Criteria for Particulate Matter.* Washington, D.C: U.S. Department of Health, Education, and Welfare, 1969.)

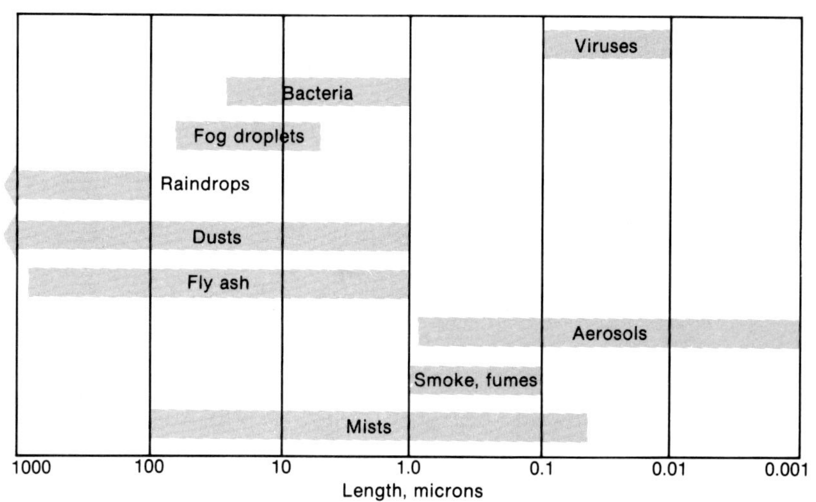

Pollutants entering the body affect specific organs. Figure 8.6 shows the primary targets of some major pollutants. Gases such as chlorine, ammonia, and ozone are pulmonary irritants. Silica and asbestos are notorious among irritating dusts for causing fibrosis in the lungs. Some substances (beryllium, for example) are granuloma-producing agents, and some metals produce fever. Carbon monoxide (CO) and hydrogen sulfide (H_2S) are asphyxiating pollutants. Lead, other metals, and some pesticides cause systemic poisoning.

Perhaps the people most aware of distress from air pollution are the *hay fever* sufferers. They are sensitive to natural pollens of flowers, trees, and grasses. These people and others can also become allergic to dusts and other pollutants or contract diseases through some airborne fungi.

As scientists link some 60 to 90 percent of cancer to environmental stimuli, air pollutants

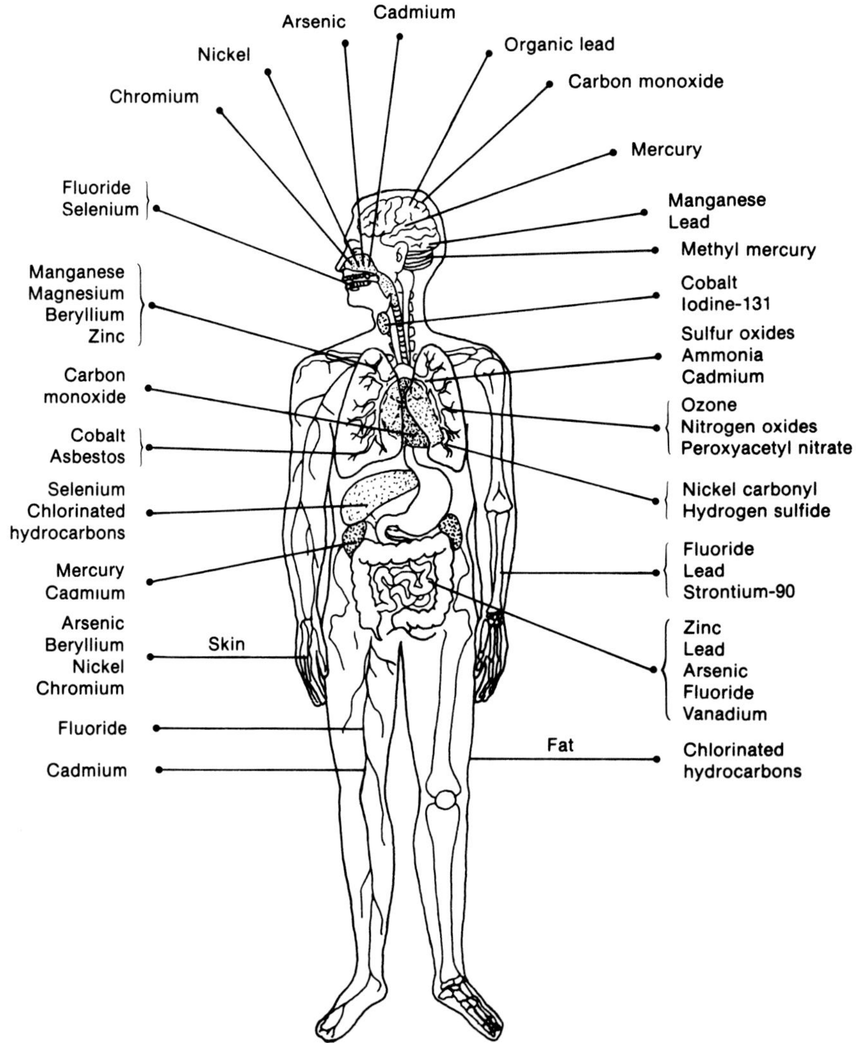

Figure 8.6
Main targets of major air pollutants in the human body. (From George L. Waldbott, *Health Effects of Environmental Pollutants*, 2nd ed. St. Louis: C.V. Mosby, 1978.)

become suspect. More lung cancer occurs among persons living in urban and industrial areas than among those in rural areas and among persons raised in areas of high air pollution than among those in areas of low pollution. However, cancer cannot definitely be attributed to air pollution on this basis alone.

Years of research have preceded the identification of certain carcinogenic air pollutants. Most lung cancer deaths have been attributed to smoking cigarettes. Exposure to asbestos, especially when accompanied by smoking, has been associated with *mesothelioma*, a relatively rare tumor. *Chimney sweep's disease* is now recognized as a form of cancer (cancer of the scrotum). Organic compounds found in diesel exhaust and other combustion emissions are also suspected carcinogens. In addition to possibly causing cancer, some pollutants are suspected to be *mutagenic* (causing mutations in genes) or *teratogenic* (causing birth abnormalities).

Plants and Material Goods

Air pollution effects on plants can best be seen near the source of pollution. For example, tree foliage along turnpikes is damaged in a band where fumes from diesel truck exhaust touch the leaves. Cement dust deposited on leaves, when moistened, will form incrustations; other dusts plug the leaf openings, or *stomata*. Where ozone levels are high, pine needles turn brown and die (*necrosis*). Sulfur oxides can cause acute injury, resulting in tissue drying to an ivory color or darkening to a reddish brown. Chronic injury leads to pigmentation of leaf tissues and a gradual yellowing called *chlorosis* (Table 8.1). Just as

Table 8.1
Examples of Pollutants Toxic to Plants

Pollutant	Toxic Level (parts per million)	Indicator Plants	
		Sensitive	Resistant
Sulfur dioxide	0.1–3.0	Pumpkins	Potatoes
		Barley	Onions
		Squash	Corn
		Alfalfa	Maple
		Cotton	Most trees
		Wheat	
		Apples	
Fluoride	0.0001	Gladioli	Alfalfa
		Tulips	Roses
		Prunes	Tobacco
		Apricots	Tomatoes
		Pine	Cotton
Ozone	0.15	Tobacco	Mint
		Tomatoes	Geraniums
		Muskmelons	Carrots
		Beans	Gladioli
		Spinach	Peppers
		Potatoes	Beans
Oxidant smog	0.2	Petunias	Cabbages
		Lettuce	Corn
		Oats	Cotton
		Pinto beans	Wheat
		Bluegrass	Pansies

From Waldbott, George L. *Health Effects of Environmental Pollutants*, 2 ed., 1978. St. Louis: The C. V. Mosby Co.

PROTECTING THE VIEW—VISIBILITY IN THE NATIONAL PARKS

Besides providing enjoyment and protecting natural and cultural resources, national parks play an important symbolic role in our society. E.P.A. Administrator William Reilly recently commented that, "the Grand Canyon is to the United States what Notre Dame is to France. It is a worldwide symbol of our country's spirit and beauty." (*National Parks* March/April 1992). Despite their status as treasured landscapes, however, air quality in some of our national parks is deteriorating. During a recent television news interview, a national park visitor, disgusted by the haze that spoiled his family's enjoyment of the park, said he felt like demanding a refund of the money his family spent to visit the park.

As a result of the Clean Air Act Amendments of 1977, many national parks and wilderness areas were designated as Class I air quality regions. These Class I regions were also referred to as Prevention of Significant Deterioration (PSD) regions because they were areas where the air was relatively clean and no deterioration of air quality was to be permitted. Despite the best intentions of Congress, Class I areas have not been effectively protected and several national parks and wilderness areas are experiencing serious air quality problems. As Charles Howe points out (*Environment*, September 1991), there are two main reasons why the air quality of wildland areas is deteriorating. First of all, pollutants are often carried long distances from source areas and air quality regulation operates primarily at the point of emission. In addition, some pollutants such as sulfur dioxide and nitrogen oxides undergo chemical transformation in the atmosphere to form pollutants that cause acid deposition and produce the haze that impairs visibility.

The National Park Service has established a nationwide network to monitor air pollution. The network employs a combination of automated cameras, devices called nephelometers that continuously measure light scattering by haze particles, and samplers that are used to determine the chemical composition of haze.

One of the parks in the monitoring network, Great Smoky Mountains National Park, is noted for scenic views of its dense forests and beautiful, fog-draped ("smoky") mountains. Surveys of park visitors have shown that scenic vistas are one of the most important reasons people come to the park. However, in 1990 scientists reported that in the last 30 years visibility has deteriorated about 30 percent in the air quality region that includes the park. During the summer, haze is responsible for 87 percent of the visibility impairment and only 13 percent is due to precipitation, clouds, and fog. While 5 percent of the haze is from naturally occurring organic compounds, the rest of it originates as human-generated pollution.

Grand Canyon National Park, also known for its spectacular scenic vistas, experiences its own visibility problems. Pollutants carried long distances, possibly from as far away as southern California and Mexico, are responsible for the haze that, at times, makes it impossible to see across the canyon. In 1991 environmental groups and representatives of a coal-burning utility signed an agreement that will reduce smog problems in the park. Under the agreement operators of the Navajo Generating Station, located 80 miles northeast of the park headquarters, will achieve a 90 percent reduction in sulfur dioxide emissions by 1999. One estimate of the total cost of this reduction is $1.8 billion. In addition, the Grand Canyon Visibility Transport Commission was established to recommend actions that will reduce air pollution on the entire Colorado Plateau. This region encompasses the Grand Canyon, along with several other national parks and wilderness areas.

Visibility problems plague many other scenic wildland areas and we must find ways to deal with the complex air quality issues that contribute to these problems. Prevention of significant deterioration of air quality in Class I areas will require a regional approach to pollution regulation. This could result in the adoption of controversial measures such as requiring expensive pollution control devices for automobiles and industrial facilities, locating potential polluters away from sensitive areas, restricting industrial development, and reducing automobile traffic in and near Class I areas. It will be necessary to find solutions that are compatible with the rapidly growing tourism industry and with the desire of communities for the employment and prosperity that industrial development might bring.

some pollutants can damage plants, deposits of arsenic, lead, and fluoride on leaves or grass can poison grazing animals.

Pollutants also reach plants through the soil. Suppression of plant growth has been attributed to magnesium oxide deposits on soil. Cement dust makes soil more alkaline and acid rain makes it less so. According to the type of soil each needed, different plants would be retarded or favored by changes in pH. However, a soil that is too acid will be completely barren.

Vegetation injury in the Los Angeles area was first attributed to photochemical smog in 1944. Since then, ozone has been identified as the major culprit. Ozone possibly causes more damage to vegetation in the United States—trees, flowers, and crops—than any other air pollutant. Bleaching chlorosis, necrosis, and pigmentation are symptoms of ozone injury. Care must be taken to distinguish damage caused by air pollution from damage caused by other factors such as low temperature, lack of moisture, nutritional deficiencies, pests, pesticides, and disease.

Buildings, fabrics, and cars are also affected by particulate matter. Did you ever notice the soil and dirt on the first two or three floors of a building on a well-traveled street? Acids absorbed on these particles accelerate corrosion in humid areas. Sulfur oxides speed the deterioration of building materials, especially marble and limestone. Statues and other art objects that withstood centuries of exposure in the relatively dry, pollution-free atmosphere of Egypt began to deteriorate rapidly when moved to London, with its humid and formerly smoky, sulfur oxide–polluted air. Fabrics, leather, and steel all deteriorate upon exposure to sulfur oxides. Ozone cracks rubber in auto tires and reduces the life of fibers. Nitrogen oxides can fade sensitive dyes.

Community Air Pollution

Many point sources of air pollution cause hazards or concern to immediate neighbors: metal fumes from a smelter, smoke from the boiler of a dry cleaning plant, smoke from burning leaves, and smoke and odors from backyard barbecues. An accumulation of pollution emitted from many point sources, or a few large ones, can affect the air quality of an entire region.

A serious problem since the 1940s has been pollution from increasing numbers of autos, trucks, buses, and other *mobile sources.* Also contributing to regionwide air pollution are *line sources*, such as heavily traveled highways, and *area sources*, such as shopping centers and sports and amusement centers.

Government programs aimed at controlling these sources of air pollution in the United States began with cities and states. With passage of the Clean Air Act in 1963, the federal programs became dominant. Under the 1970 amendments to that act, the EPA issued standards for major community air pollutants.

The Clean Air Act was finally amended in 1990 after 13 years of controversy and debate over how best to regulate air quality. The new law emphasizes control of the emissions of sulfur dioxide and nitrous oxides that contribute to acid deposition. The act also has strong provisions for controlling tailpipe emissions from automobiles and airborne toxic chemicals. These provisions call for a 90 percent reduction in the release of the 189 toxic chemicals that the Environmental Protection Agency identified as the most hazardous.

Pollution control is based on two kinds of standards: *ambient standards* for the overall quality of surrounding air, and *emission standards* that limit the amount of pollution a particular source can release (Table 8.2). A major objective of federal legislation is to prevent a significant deterioration of air quality anywhere. In wild and recreational areas, virtually no increases in pollution are permitted. In most areas, a moderate increase in pollution may be permitted as long as ambient standards are not exceeded. In areas where ambient air quality standards are already exceeded (nonattainment areas), no new sources may be added unless the amount of increase is offset by reducing pollution emissions. For example, a West Coast oil company installed additional pollution control equipment for a utility's power plant to offset the pollution from importing and refining Alaskan oil.

Standards are set for particulates, sulfur oxides, carbon monoxide, nitrogen oxides, hydrocarbons, ozone, and lead. The EPA is empowered to establish emission standards for new installations that are possible sources of air pollution and other sources that release such hazardous substances as asbestos, beryllium, and mercury. Federal law also sets the emission standards for autos, and some states prescribe standards for other pollutants. For example, California has more stringent regulations than the federal government.

Federal agencies control community air quality by setting primary standards that protect health, with a reasonable margin of safety, and secondary standards that protect other aspects of public welfare. In most instances, the primary and secondary standards are identical, as we saw in Table 8.2. Some environmental health profes-

Table 8.2
National Ambient Air Quality Standards (established pursuant to the Clean Air Act of 1970)

Pollutant	Period of Measurement[a]	Primary Standard		Secondary Standard	
		μg/m[3b]	ppm[c]	μg/m^3	ppm
Carbon monoxide (CO)	8 hours	10,000	9.00	Same	Same
	1 hour	40,000	35.00	Same	Same
Hydrocarbons (HC) (nonmethane)	3 hours	160	0.24	Same	Same
Nitrogen oxides (NO_2)	1 year	100	0.05	Same	Same
Ozone (O_3)	1 hour	240	0.12	Same	Same
Sulfur oxides (SO_x)	1 year	80	0.03	None	None
	24 hours	365	0.14	None	None
	3 hours	None	None	1,300	0.5
Total suspended particulates (TSP)	1 year	75	—	60	—
	24 hours	260	—	150	—
Lead	3 months	1.5	—	None	None

NOTE: Primary standards are designed to protect public health, with a margin of safety; and secondary standards protect other aspects of public welfare.
[a]Concentrations are averaged over each period of measurement. The annual TSP concentration is a geometric mean of 24-hour samples: all other concentrations are arithmetic mean values. Standards for periods of 24 hours or less may not be exceeded more than once per year, except ozone may use a 3-year statistical average to determine if the standard is exceeded.
[b]Micrograms per cubic meter.
[c]Parts per million.

sionals have suggested that a standard be set for asbestos. Sources of major pollutants and their estimated emissions are shown in Table 8.3. While installation of controls will alter these emissions over the years, notice that the major source of carbon monoxide is transportation and that sulfur oxides come principally from stationary fuel combustion. Nitrogen oxides come mainly from combustion, stationary and mobile.

Sources and Control

Suspended Particulate Matter Burning coal produces unburned or incompletely burned particulates of carbon (smoke and soot) and solid residue (fly ash). More complete combustion results when there is adequate air and mixing of combustion gases above the burning coal bed. Oil also produces particles, but not to the same extent as coal. However, an inadequate air supply to burning oil produces soot, a process sometimes used intentionally to make lamp black.

Substituting natural gas for coal and oil would eliminate particulate emission problems from combustion, but because it is scarce it is reserved for home heating and manufacture of synthetic materials. Similarly, nuclear plants do not have these air pollution problems, but people have other concerns about using nuclear fuel. The social and economic costs and benefits, including environmental impacts, need to be evaluated in order to choose the best fuel system.

Dusts, another form of particulate matter, are produced from a variety of manufacturing processes which involve grinding and crushing or the handling of dusty materials: cement plants, asphalt plants, foundries, and so on. Construction and demolition make dust which can blow away (fugitive dust).

Particles from combustion and dusts from manufacturing and processing can be captured by simple air cleaning equipment. Among the most used are cyclones, scrubbers, bag houses, and electrostatic precipitators.

In a **cyclone,** dust-laden air is introduced at the top outer edge to whirl around and around inside the cylinder. Particles in the airstream are

Table 8.3
Sources of Air Pollution Emissions

Source	Percentage by Weight	Major Pollutants
Transportation	55.4	Carbon monoxide Nitrogen oxides Hydrocarbons
Fuel combustion by stationary sources	20.9	Sulfur oxides, Nitrogen oxides
Industrial processes	15.5	Hydrocarbons, Carbon monoxide
Solid waste disposal	2.1	Carbon monoxide
Miscellaneous	6.1	Carbon monoxide

thrown against the outer wall and settle down to the bottom, where they are withdrawn from time to time. The air that is freed of particles escapes through a duct in the center of the cylinder (Figure 8.7). Depending on particle size and the power used, conventional cyclones will remove about 50 to 80 percent of particulate matter and most of the particles will be 10 microns or larger. The cyclone is not efficient in removing smaller particles, which are in the respirable size range.

By putting a high electric charge on a wire, **electrostatic precipitators** create a charge on particles in the airstream. A grounded plate is placed between the wires to collect charged particles, which slide down the plates to a hopper at the bottom for withdrawal. This system works very well on particles that become electrically charged, like carbon. Its efficiency depends on the amount of energy put in the charge, the number of precipitators in the series, and the temperature of the airstream. Sometimes electrostatic precipitators are used after cyclones. Under some conditions, overall collection efficiencies as high as 99 percent can be achieved. Again, the collection is more efficient for larger particles than for those in the respirable range.

Bag houses consist of long sleeves or bags made of fabric that will withstand high temperatures. As the airstream enters the sleeve and coats the fabric with particles, particles in the airstream are filtered out as the air passes through the fabric to the other side and escapes

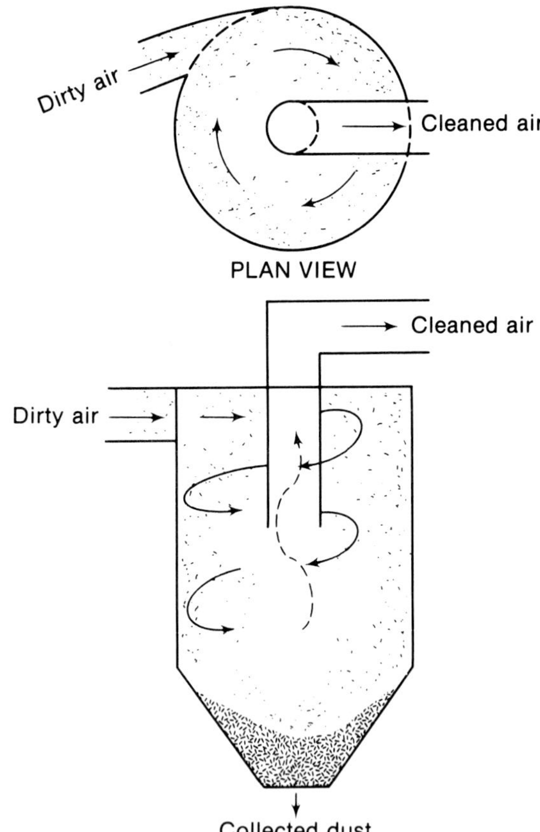

Figure 8.7
A cyclone dust collector depends on centrifugal force to fling dust against the outside wall, where gravity causes the dust to slide to the hopper at the bottom.

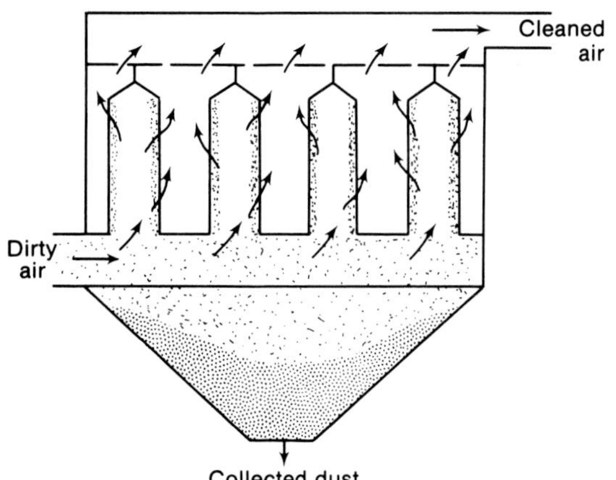

Figure 8.8
In a bag house, dirty air is forced by air pressure through fabric sleeves, much like a tank-type vacuum cleaner used in the home. As particles collect on the fabric, filtering of fine particles becomes more efficient, but resistance to air flow increases. The sleeves are then shaken to cause particles to fall into a hopper.

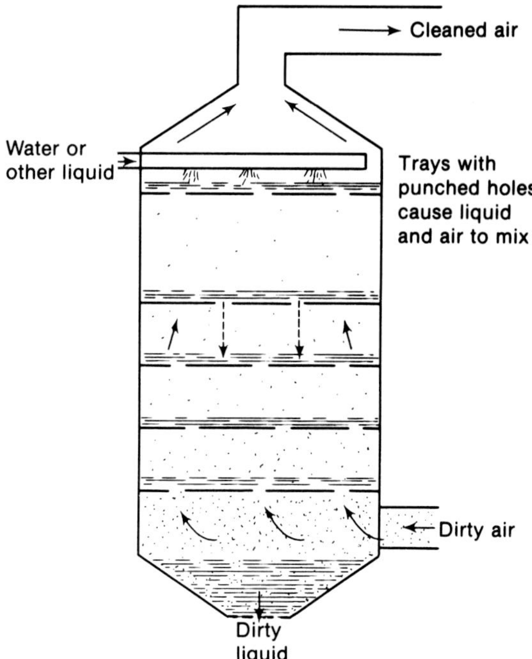

Figure 8.9
Schematic of a scrubbing tower that may be used to remove particles and gaseous pollutants. Water may be used if the gas is soluble in water; otherwise, another liquid in which the gas is soluble must be used. The dirty liquid has to be treated before it can be reused or discharged.

(Figure 8.8). As the coating builds up, the system becomes more efficient, but resistance to air flow increases. Ultimately, the sleeve has to be taken out of service momentarily to be blown or shaken clean. The temperature in the bag house has to be higher than the dew point of water or any vapors in the airstream to prevent condensation. Bag houses are used in conjunction with other systems, like cyclones. Efficiencies approach those of electrostatic precipitators, but bag houses are more effective in removing very fine particles from the air.

Scrubbers can be simple screens of water spray that will knock some of the large particles out of the air or towers with trays of packing material to mix liquid and air (Figure 8.9). For high efficiencies, the particles and water have to be thoroughly mixed. This is accomplished in a *venturi scrubber*, where the airstream is forced into a restricted portion of a pipe (the *venturi*) where water is flowing at a high velocity. The problem with this system is cleaning the water before discharge or recirculation.

Oxides of Sulfur Because the combined effect of sulfur oxides and particulate matter is thought to be a principal cause of deaths in air pollution episodes and a continuing cause of bronchitis and emphysema in lesser concentrations, there is great concern about sulfur oxides. Some oxides of sulfur are emitted by smelters, oil refineries, and paper mills, but community pollution is usually caused by burning fuel that contains sulfur. When such fuel is burned, most of the sulfur converts to sulfur dioxide and a small fraction converts to sulfur trioxide. Sulfur dioxide is partly oxidized to sulfur trioxide by photochemical processes, and moisture will convert sulfur trioxide to sulfuric acid.

Sometimes you can see issuing from a stack a white plume that quickly disappears. That is most likely water vapor. In other cases, there is no plume visible at the top of the stack, but a short distance away a white plume develops and persists. That could very well be due to the formation of sulfuric acid aerosols.

Sulfur dioxide can be controlled by switching to a low-sulfur fuel such as natural gas. But, because it is scarce, natural gas is no longer practical for power plants. Furthermore, low-sulfur coal and oil (with less than 1 percent sulfur) is in limited supply. Nuclear energy, desulfurization of fuel, or stripping sulfur oxides from stack gases are possible alternatives. Sulfur can be stripped from oil, but with declining reserves of oil in the United States and elsewhere a sound energy policy suggests use of coal fuel for power plants.

Some of the sulfur in coal is in the inorganic form of iron pyrite. By crushing and washing, up to one-third of the iron pyrite can be removed. The rest of the sulfur is chemically bound in organic form to the coal. Only through other processes, such as coal gasification, can this sulfur be removed. Consequently, any large-scale use of coal will require flue gas desulfurization. While the EPA considers such processes feasible, some engineers point to problems of corrosion, maintenance of removal equipment, and disposal of the sludge that most of these processes generate (Figure 8.10).

Carbon Monoxide The major source of community pollution is the internal combustion engine, which is designed to run on rich mixtures of gasoline and air. In the past automobile engines operated at an air/fuel ratio (the weight of air used in combustion compared with the fuel) of 13 to 1. This mixture, known as a fuel-rich ratio, produces more power. Complete combustion of the fuel would produce CO_2. However, at this ratio there is not enough oxygen from the air to complete combustion processes and carbon monoxide (CO) is produced instead (Figure 8.11). Introducing more air creates a leaner ratio and thus produces less carbon monoxide. If the design of combustion equipment and engines were changed to accommodate an air/fuel ratio of 15 to 1, more complete combustion would

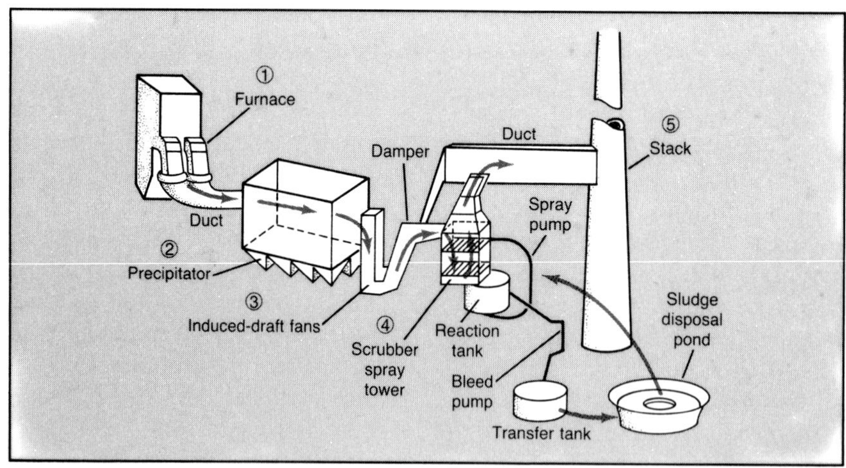

Figure 8.10
Inside a coal scrubber. (Copyright © 1981 by The New York Times Company. Reprinted by permission.)

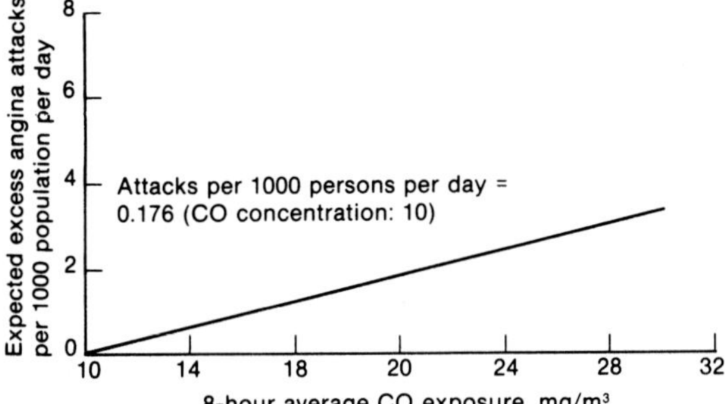

Figure 8.11
Carbon monoxide levels and expected angina attacks. Exposure to carbon monoxide may cause heart attacks in persons with *angina pectoris*, a kind of heart disease. (Data from J. Knelson and K. H. Jones. From Council on Environmental Quality, *Environmental Quality–1977*. Washington, D.C., 1977.)

result. A change to a ratio of 18 to 1 would reduce emissions of all kinds. A diesel engine provides more complete combustion and therefore less carbon monoxide than a gasoline engine.

Another approach to reducing carbon monoxide levels would be to complete combustion as the exhaust gases leave the engine. However, this could produce a hazardous amount of heat in the exhaust system. Sometimes these processes can be performed at lower temperatures in the presence of special substances that do not enter the reaction themselves (**catalysts**). Most autos using internal combustion engines now come equipped with *catalytic converters* that reduce the amounts of both carbon monoxide and hydrocarbons exhausted. Because catalysts can be poisoned by lead and made ineffective, lead-free gasoline is required. Other approaches include substituting electric power systems for internal combustion engines. Of course, by walking, riding bicycles, or using public transportation we can also reduce generation of carbon monoxide.

Carbon monoxide is absorbed through the lungs and reacts with the hemoglobin of the blood to form *carboxy hemoglobin* (CO Hb). Hemoglobin has an affinity for carbon monoxide that is 200 times stronger than for oxygen, preventing oxygen from reaching body tissues. Absorption increases with carbon monoxide concentration, duration of exposure, and ventilation rate of lungs, which varies with exercise. However, carbon monoxide is expired and an equilibrium is reached after a period of exposure. Impairment of performance is associated with carboxy hemoglobin levels of 5 percent, and cardiovascular changes occur at higher levels. Cigarette smokers have a carboxy hemoglobin level of about 5 percent, contrasted with nonsmokers at 0.5 percent. Exposure to high levels of carbon monoxide produces death quickly. Exposure to lower levels (6 weeks at 58 grams per cubic meter) produces changes in the hearts and brains of test animals (see Figure 8.11). Nonsmokers exposed to similar concentrations had visual acuity impairment and had trouble estimating time. From this evidence it appears that there might be no safe threshold for carbon monoxide exposure.

Carbon monoxide is a special problem for street traffic and tunnel police officers and for underground garage personnel. Ventilation can help persons who work in tunnels and garages. Traffic police in Tokyo have been equipped with gas masks and oxygen supplies.

Oxides of Nitrogen, Hydrocarbons, and Ozone The oxides of nitrogen (NO_x) and hydrocarbons (HC) will undergo chemical reactions with each other and other compounds when exposed to sunlight. These photochemical reactions produce strong oxidizing agents or oxidants, chiefly ozone. While ozone in the upper atmosphere is beneficial, high concentrations near the ground are undesirable. The complex mixture of pollutants generated in this way is called *photochemical smog*. The operation of internal combustion engines is an ideal producer of nitrogen oxides, although this is not the only source.

Formation of nitrogen oxides could be controlled by using pure oxygen for combustion, but this is not practical. Using fuels other than gasoline and changing air/fuel ratios reduce emissions. Other approaches include designing engines to operate at lower temperatures, using specially designed exhaust control systems, and of course, using autos less.

Gasoline evaporation is a major source of hydrocarbons. Placing floating roofs on storage tanks cuts losses substantially. Auto losses are reduced by eliminating evaporative emissions from tanks and crankcases. Positive crankcase ventilation (PCV) systems take the vapors from the crankcase and recycle them to the engine to be burned.

Lead Sources of airborne lead are primarily smelters and automobile exhaust. Smelters may be a problem for nearby residents. Lead concentrations due to auto exhaust will be greatest near heavily traveled streets and highways.

The EPA conducted a study in California to assess the health effects of airborne lead from automobiles. Children living near a major freeway in Los Angeles showed substantially higher lead levels than those who lived in the rural desert town of Lancaster, California. The blood lead levels of 10 percent of those tested in the Los Angeles study were high enough to impair human health, while levels at Lancaster were much lower.

In March 1976, the National Resources Defense Council won a suit to compel the EPA to establish air quality standards for lead. A criteria document was issued in December 1977, and a standard of 1.5 micrograms per cubic meter on a 3-month average was adopted a year later. To achieve this standard, the EPA has reduced the allowable lead content in U.S. gasoline to a very low level (0.1 gram per gallon). Most of this reduction was achieved between 1985 and 1986, when the standard was lowered by 90 percent. As a result, emissions of lead in the United States declined from 203,000 metric tons in 1970 to 8100 metric tons in 1987.

Emergency Controls Control of various pollutants is designed to be effective under normal variations of atmospheric conditions. Where temperature inversions persist for several days in highly industrialized areas, extraordinary measures might have to be taken to protect lives. Particularly susceptible people (especially persons with heart and lung disease) can be moved from the area or placed in air-conditioned rooms. Emission sources can also be reduced by curtailing traffic and closing industrial plants.

Emergency controls should be initiated when weather forecasts indicate that a severe inversion will persist for several days. The first measure is to notify mass media (newspapers, television and radio stations) and to alert heart and lung patients, their physicians, and hospitals of the danger. The next phase is to ask people not to drive autos into cities and request industries to curtail operations that emit pollutants where possible. Following this, traffic could actually be restricted to essential use and selected polluting industries, or certain operations, could be shut down. Finally, all industry could be shut down and only emergency and police vehicles allowed to travel. A system for comparing pollution concentrations and health effects for precautionary action has been developed by the EPA (Table 8.4).

Trends in Community Air Quality

The control measures we have described have proven effective in reducing emissions in many cases (Figure 8.12). Sulfur oxide levels in cities have been reduced by limiting the sulfur content in fuel, usually by burning natural gas or low-sulfur fuel oil. Switching to coal that has a higher sulfur content could reverse this trend unless sulfur is removed from stack gases. Nevertheless, even with improved reduction of sulfur oxide emissions, acid rain continues to be a problem in the northeastern United States.

Carbon monoxide and hydrocarbon emissions from autos have been reduced largely by exhaust controls (Figure 8.13). But in many areas, air quality standards for pollutants associated with auto exhaust will not likely be achieved

Table 8.4
Comparison of Pollutant Standards Index (PSI) Values with Pollutant Concentrations and Health Effects

Index Value[a]	Air Quality Level	Health Effect Descriptor	General Health Effects	Cautionary Statements
500	Significant harm		Premature death of ill and elderly. Adverse symptoms that affect normal activity of healthy people	People should remain indoors, and keep windows and doors closed. Everyone should minimize physical exertion and avoid traffic.
400	Emergency	Hazardous	Premature onset of certain diseases; significant aggravation of symptoms; decreased exercise tolerance in healthy persons	Elderly and people with existing diseases should stay indoors and avoid physical exertion. General population should avoid outdoor activity.
300	Warning	Very unhealthful	Significant aggravation of symptoms, decreased exercise tolerance in persons with heart or lung disease; widespread symptoms in the healthy population	Elderly and people with existing heart or lung disease should stay indoors and reduce physical activity.
200	Alert	Unhealthful	Mild aggravation of symptoms in susceptible persons; irritation symptoms in the healthy population	Persons with existing heart or respiratory ailments should reduce physical exertion and outdoor activity.
100	NAAQS			
50	50 percent of NAAQS	Moderate		
0		Good		

[a]Index values are based on National Ambient Air Quality Standards (NAAQS) for total suspended particles (TSP), SO_2, CO, O_3, and NO_2.

by exhaust controls only. In such cases less use of gasoline-powered autos for personal transportation might be necessary. Gasoline consumption declined 12 percent between 1978 and 1982. Consumption has increased slightly since then, but by the end of 1985, it was still 8 percent lower than in 1978.

More than 100 cities have never achieved the federal standards for ozone or carbon monoxide emissions. The National Wildlife Federation has

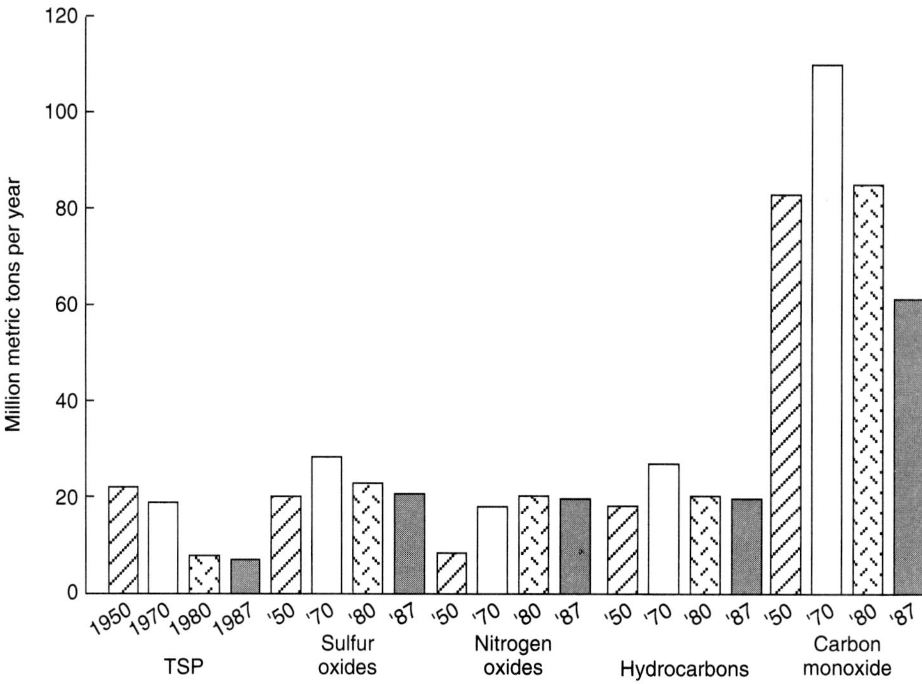

Figure 8.12
National air pollution emissions. (Data from Council on Environmental Quality.)

pointed out that this is primarily due to pollution from automobiles. Even though today's cars are less polluting, there are more of them. There was one car for every 2.5 citizens of the United States in 1970 and one car for every 1.7 people in 1988.

Hazardous Air Pollutants While most efforts in recent years have been directed at meeting ambient air quality standards, the Clean Air Act also covers air pollutants that are considered to present special health hazards. The EPA has proposed or issued standards for seven hazardous air pollutants: asbestos, mercury, beryllium, vinyl chloride, benzene, radionuclides, and arsenic (Table 8.5). Other substances under consideration for designation as hazardous air pollutants are polycyclic organic matter, cadmium, ethylene dichloride, perchloroethylene, acrylonitrile, methylene chloride, methyl chloroform, toluene, and trichloroethylene.

At present there is no system for monitoring airborne carcinogens. In some locations, benzopyrene has been recorded at concentrations ranging from 0.5 to 4 nanograms per cubic meter. With the burning of fossil fuels and wood, the values are higher in winter. They are also higher in the vicinity of coke ovens. A general reduction of benzopyrene concentrations occurred in the 1960s and 1970s as a result of less coal consumption, bans on open dump burning, and reduced coke-oven emissions. In future years, however, coal burning may increase.

Energy Conservation Hazards Under most circumstances energy conservation results in conservation of the environment. Certainly, burning

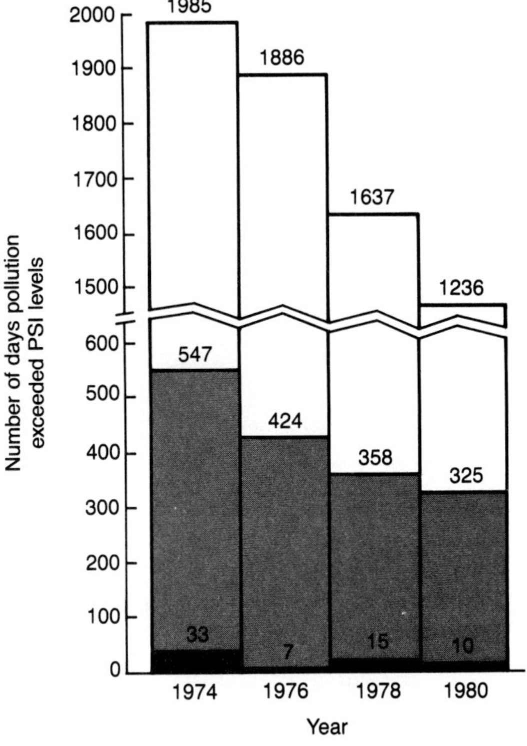

Figure 8.13
National trends in urban air quality. Graph shows combined total of number of days that air quality was poor in twenty-three metropolitan areas during the period 1974–1980.

less fossil fuel and driving an auto less reduce pollution emissions from these sources. There are adverse aspects to conservation efforts, however, that need to be considered. Foremost is **indoor air pollution.**

As houses are made more secure to prevent heat loss, air pollution generated inside the house accumulates. Nitrogen dioxide (NO_2) levels can be several times higher than outdoors and above the recommended standards for NO_2 emissions, particularly where gas stoves are used without adequate ventilation. Higher than average levels of radon gas have been found in homes in phosphate mining areas. This radioactive substance (radon) is produced by the decay of radium, which occurs in trace amounts in soil and rock. It also comes from other sources. Prolonged exposure to higher than normal levels of radon can induce lung cancer. Because smoking is a major source of respirable particles and poor ventilation leads to its concentration in homes, home installation of electrostatic filter units can assist in the removal of smoke particles and dust.

In older uninsulated homes, urea formaldehyde foam insulation has been used to fill the cavities in exterior walls. Under conditions of high moisture and temperature, the chemical reaction that formed the foam insulation may be reversed, leading to the release of formaldehyde gases. Some people are very sensitive to formaldehyde and cannot remain in houses with high levels of this gas. Urea formaldehyde resins are also used in the manufacture of plywood and particle board and may be used with some rugs and upholstery fabrics. Research is now underway to show whether plants can aid in filtering air.

As the prices of other fuels rise, many people are burning wood in stoves or fireplaces. Stoves are thermally more efficient than fireplaces, but fireplaces may provide more complete combustion. Emissions of respirable particles (that is, less than 2 microns) from wood combustion can easily exceed all other same-size particle sources in winter months, as almost all emissions are in the respirable range. These emissions also contain toxic and priority pollutants, carcinogens, cocarcinogens, cilia toxic, mucus-coagulating agents, and other respiratory irritants.

Faulty flues from wood stoves are a frequent cause of fires and carbon monoxide poisoning. To avoid depletion of oxygen and build up of carbon dioxide, all combustion facilities—wood stoves, furnaces, kerosene and gas heaters, and

Table 8.5
Sources of Hazardous Pollutants with Established Emissions Standards

Pollutant	Source
Asbestos	Asbestos mills
	Road surfacing with asbestos tailings
	Manufacturers of asbestos-containing products (fireproofing, etc.)
	Demolition of old buildings
	Spray insulation
Beryllium	Extraction plants
	Ceramic manufacturers
	Foundries
	Incinerators
	Rocket motor manufacturing operations
Mercury	Ore processing
	Chlor-alkali manufacturing
	Sludge dryers and incinerators
Vinyl chloride	Ethylene dichloride manufacturers
	Vinyl chloride manufacturers
	Polyvinyl chloride manufacturers

fireplaces—should be provided with an outside air supply to obtain oxygen and a vent to discharge combustion gases out of the house. Venting requirements are usually covered by local building and fire codes.

ACID DEPOSITION

In the 1950s, observers noticed an increase in the acidity of water in lakes in various parts of the world. By 1959 Norwegians had established a connection between the increase in acidity and a decline in fish populations. Presently, in the Adirondacks there are 180 lakes with no fish; in Ontario, 140. Salmon have disappeared from rivers in Nova Scotia. The Boundary Waters Canoe Area and Voyageurs National Park of Northern Minnesota have been damaged. If the trend continues, life in thousands of lakes could disappear in 10 to 20 years. These effects are attributed to acid deposition (Figures 8.14 and 8.15).

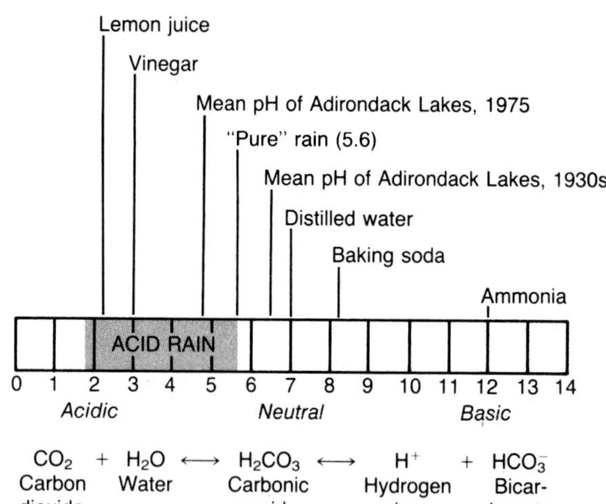

Figure 8.14
The slight natural acidity of normal rain is due to the presence of carbonic acid (H_2CO_3) formed by the reaction of atmospheric carbon dioxide (CO_2) with water.

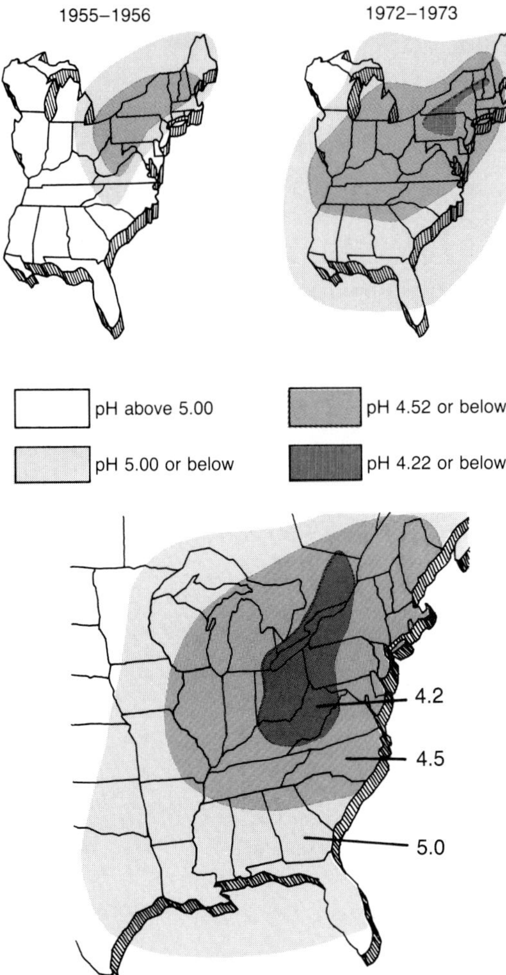

Figure 8.15
Trend in rain acidity. Comparison of the weighted annual average of pH of precipitation in the eastern United States shows a decided increase in the number of states affected. (Adapted with permission from Gene E. Likens and C. V. Cogbill, "Acid Precipitation," *Chemical and Engineering News 54* [1976]:31. Published 1976 by the American Chemical Society.)

The major components of acid deposition or acid rain are the polluting sulfur oxides (mainly SO_2) and oxides of nitrogen (NO_x). SO_2 and NO_x are transformed (oxidized) in the atmosphere, particularly in the presence of sunlight, to sulfate and nitrate aerosols. If moisture (in the form of rain, snow, dew, or mist) is present, sulfuric and nitric acids form. The acid particles can be deposited as dry deposition or fall with precipitation. If the particles fall with snow, the effect is most pronounced during the spring thaw, when the environment receives a "pulse" of acidified meltwater. The primary sources of SO_2 and NO_x are burning of fossil fuels by electrical utilities, smelters and other industrial facilities, and transportation (Figures 8.16 and 8.17).

As we discussed earlier, air circulation patterns often transport pollutants over great distances. Acid deposition in Scandinavia comes largely from pollutants carried there from industrialized areas of Europe. The summer surface air flows in North America carry pollution into Canada, as shown in Figure 8.18, and occur when the highest values of acidity are observed in the eastern and northeastern United States.

Ecosystems vary in their sensitivity to acid deposition (Figure 8.19). Sensitivity is related to the ability of bedrock and soil to neutralize acidity, temperature, amount of precipitation, steepness of terrain, and the acid-neutralizing capacity of surface waters.

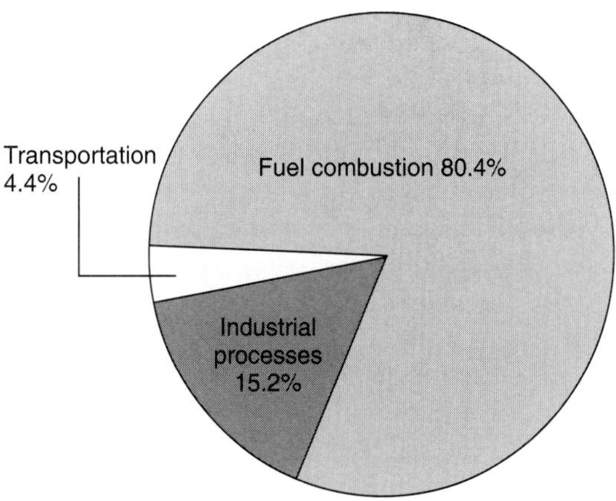

Figure 8.16
Percentage of 1987 national SO_x emissions by source category.

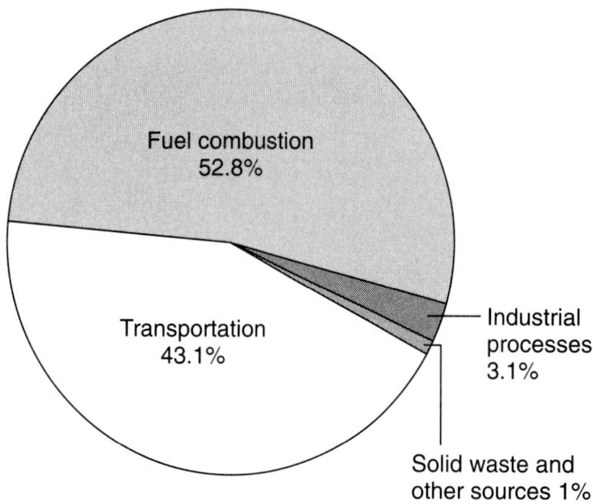

Figure 8.17
Percentage of 1987 national NO_x emissions by source category.

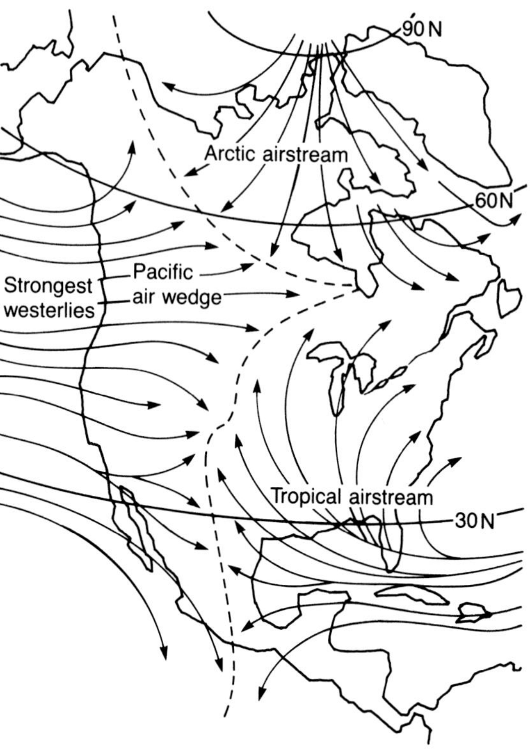

Figure 8.18
Schematic representation of the surface flow across North America based on July resultant surface winds. (From R. A. Bryson and F. K. Hare, eds., *Climates of North America*. Amsterdam: Elsevier Science Publishers, 1974. Used with permission.)

Normal rain around the world has a pH of 5.6, as we saw in Figure 8.14. Rain in the Adirondacks has a pH of 4.2 and a storm in West Virginia had a pH of 1.5. The effect of acid deposition in the Great Lakes area is buffered by alkaline soil delivered from carbonate rocks such as limestone. In areas where soil is derived from decomposed granite and basalt, as in New England and the Canadian Shield, the soil has a tendency to be naturally acidic; therefore the area is much more sensitive to acid rain (Figure 8.19).

The first victims of acid deposition are eggs of amphibians and fish, especially trout. The spring thaw drops pH in Nova Scotia rivers from 6 to 4 in a few days. Salmon eggs and larvae are killed. The river returns to normal, but the damage to salmon is permanent. By 1970, no fish were caught in small rivers, where in the 1950s, 100 to 200 were caught annually.

As lakes become more acidic, bacteria disappear, frogs die, and leaves and other litter pile up on the lake bottom. Below 5.5 pH, traditional plants are replaced by mosses, fungi, and algae. Adult fish are probably the last to die (Figure 8.20).

There is also some evidence that acid deposition is responsible for damage to forests in Europe and North America. Even if pollution doesn't kill or damage trees directly, it places forests under stress and probably makes them more vulnerable to diseases and insect attack.

The 1970 Clean Air Act (and 1977 and 1990 amendments) emphasizes the control of the quality of ambient air. These are some of the measures suggested for reducing emissions of SO_2 and NO_x, precursors of acid deposition:

Emission reduction techniques

Precombustion "washing" of coal to remove sulfur

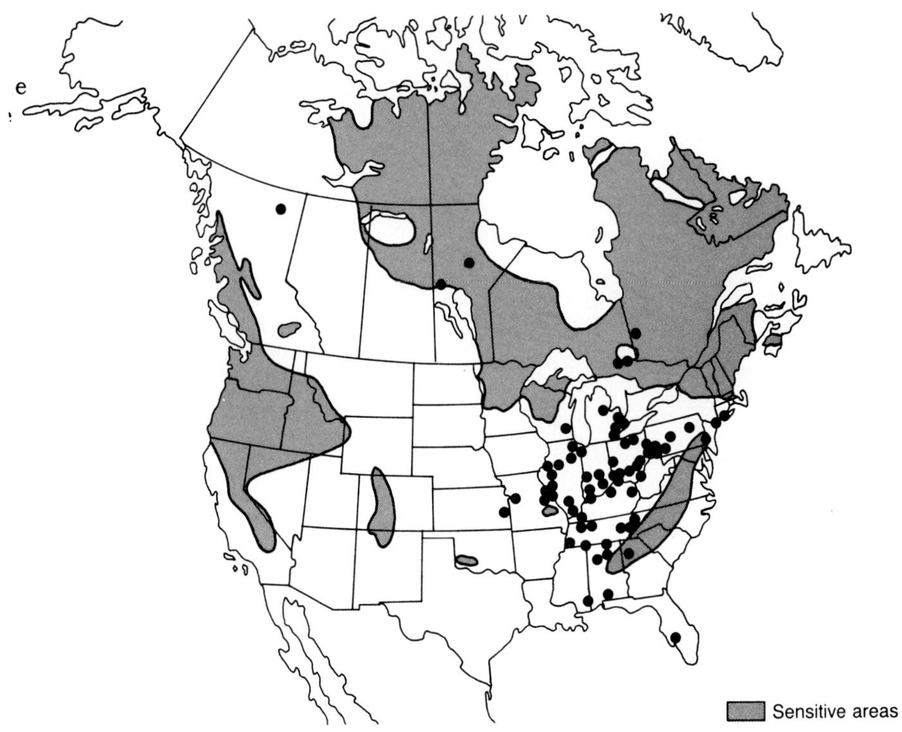

Figure 8.19
Shaded areas have few natural buffers and are especially susceptible to acidification; dots designate areas with the greatest concentrations of SO_2 emissions (more than 100 kilotons per year). (Courtesy of Friends of the Boundary Waters Wilderness.)

Using "scrubbers" (flue gas desulfurization)
Emission controls on automobiles
New, developing technologies

- Specialized methods of burning coal, such as fluidized bed combustion
- Coal gasification (see Chapter 13)
- Coal liquefaction (see Chapter 13)

Switching to low-sulfur coal
Switching to nonpolluting energy sources
Energy conservation (see Chapter 13)

- Reducing energy use
- Increasing efficiency of energy use

To meet sulfur dioxide standards, new coal-burning electric generating stations were built in rural areas and equipped with tall smokestacks. While local urban air quality improved, pollution now traveled hundreds of kilometers from the source.

Between 1970 and 1979, 429 smokestacks more than 61 meters (200 feet) high were built, mostly in the eastern United States. Prevailing winds sweep pollution from these stacks into New England and Canada. About one-half of the sulfur deposited in eastern Canada originates in the United States, causing Canadian officials to urge the U.S. Congress to initiate better sulfur emissions controls. Most of the tall smokestacks are in Pennsylvania, Ohio, West Virginia, Indiana, and Illnois.

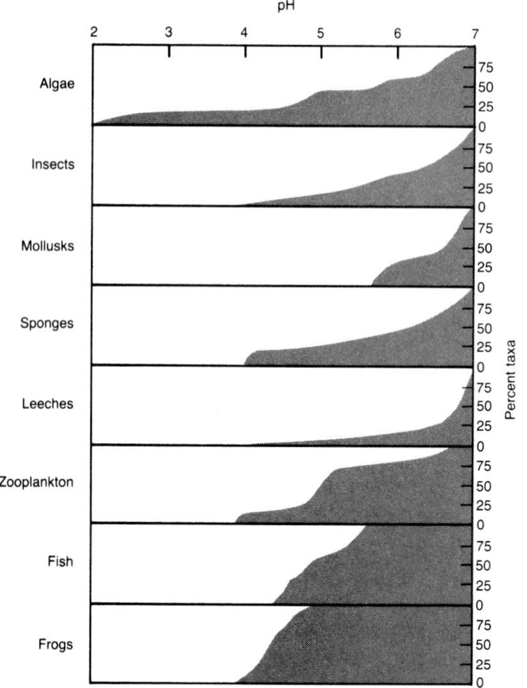

Figure 8.20
As the pH of a lake goes down, the diversity of aquatic life is reduced until very few species survive. (Reprinted with permission from Bette Hileman, "Acid Precipitation," *Environmental Science and Technology* 18[1981]:1122. Copyright © 1981 by the American Chemical Society.)

Another control measure involves removing SO_2 by scrubbing stack gases (see Figure 8.10). Flue gas desulfurization is 95 percent efficient. In 1967 Japan began a program to reduce sulfur emissions. Between 1970 and 1975, its SO_2 levels fell 50 percent while energy consumption rose 120 percent. As a result, Japan set new goals to limit SO_2 emissions even more.

Other studies indicate that acid deposition is also a serious problem in the western United States, although it is not yet well understood. The causes of acid deposition in the west are different from those in the east. Nitrogen oxides from transportation sources contribute more significantly to acid deposition in the west. Smelters and power plants presently account for the sulfur oxide emissions. Western power plants are likely to become a major source of SO_2.

In 1991 the Environmental Protection Agency ordered the coal-burning Navajo Generating Station to reduce sulfur dioxide emissions by 90 percent between 1995 and 1999. This action was taken to protect the once relatively clean air of Grand Canyon National Park, which is only 16 miles away from the power plant, and now likely to be polluted by the station.

It is important to aggressively control acid deposition in the West before it causes the serious damage to ecosystems that has been identified in the East. Acid deposition could threaten eleven national parks in the West, and human populations in urban areas could also be threatened—highly acidic fog has been reported around Los Angeles and San Francisco.

Some progress is being made in understanding and controlling acid deposition. A comprehensive, multiagency monitoring program called the National Acid Precipitation Assessment Program (NAPAP) was created in the late 1970s. NAPAP established a nationwide system of regional monitoring networks and a long-term research program. The network has proven very helpful to researchers and decision-makers who are trying to understand and control acid deposition.

NOISE POLLUTION

Noise may be considered a form of air pollution. Since sound pressure waves move in every direction, their energy is spread over a larger and larger surface area as the sound moves away from the source. Consequently, the amount of sound energy spread over a square meter decreases in relation to the square of the distance from the source. Distance is therefore an important factor in reducing the effect and perception of noise (Figure 8.21). While the effects of noise usually decrease with distance from the source, nearby surfaces such as building walls reflect noise and increase the perceived intensity. If noise is channelled between buildings or be-

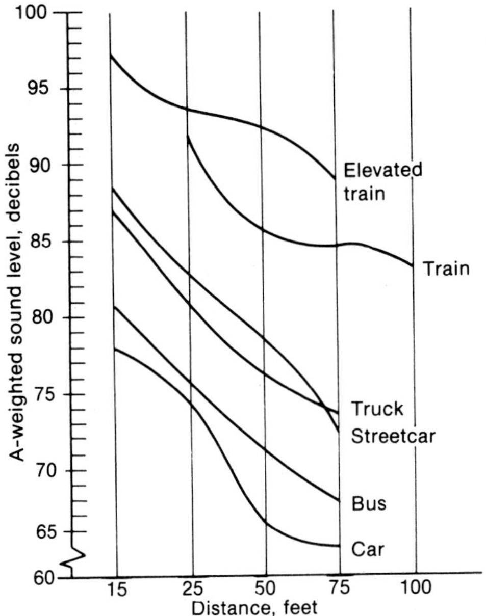

Figure 8.21
A survey of specific noise sources shows that distance is a factor in noise intensity. (From Clifford R. Bragdon, *Noise Pollution.* Philadelphia: University of Pennsylvania Press, 1970.)

tween other surfaces such as hot air ducts, the intensity is also increased. Some surfaces—curtains, rugs, acoustical tile, and others—absorb noise and decrease its effects.

Industrial Noise

A number of industrial operations generate noise. Stamping metal into auto fenders, punching holes into metal plates, riveting plates together, and crushing different materials all produce impact noise, and grinding and drilling metal produce continuous noise. Rapid air motion caused by jets of air, blowers, and fans, and vibration of equipment also cause noise. Although industrial noise mainly affects workers in the industry, some of this noise also reaches nearby homes.

Community Sources of Noise

Most community noise originates from transportation. Transportation noises are generated by a vehicle's power unit, such as a jet's engine or a truck's motor, and from the contact between the tires or wheels and the road or rails. The greatest amount of aircraft noise is produced upon landing and takeoff because more power is used at these times. Homes and businesses under and near the landing and takeoff path receive most of this noise (Figure 8.22).

Greater vertical distances, like greater horizontal distances, reduce sound intensity. For example, the higher an airplane flies upon leaving or approaching an airport, the less the noise is perceived on the ground. However, because

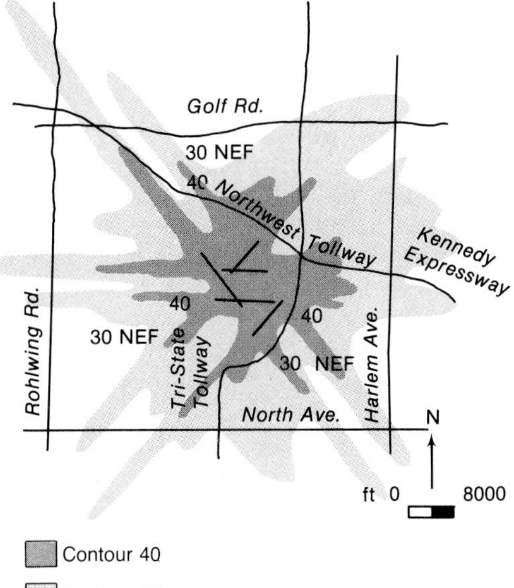

Figure 8.22
Noise intensity for Chicago's O'Hare International Airport. The Noise Exposure Forecasts (NEF) contours for airports are determined by the number of daytime and nighttime jet aircraft operations, major runway locations and flight paths, and whether supersonic aircraft will use the airport. Outside contour 30, land is normally acceptable for residential housing, but hospitals, schools, and churches may need special construction to shield aircraft noise. (Reprinted with permission from Peter A. Franken and Daniel G. Page, "Noise in the Environment," *Environmental Science and Technology* 6 [1972]: 125–27. Copyright by the American Chemical Society.)

street noises are channelled up between buildings on each side of the street, taller structures up to three stories high receive more intense noise from street sources at their upper floors (Figure 8.23).

Aside from the effect vehicle noise has on the surrounding community, the noise also has an impact on the vehicle's operator and occupants. The operators of large earth-moving machines frequently suffer hearing loss. Dangerous situations arise when vehicle noise masks the sounds of sirens and the driver is unaware of the warning sound of an emergency vehicle.

A residential community abounds with its own sources of noise that add to those from industry and transportation. The variety is almost endless: lawn mowers, loud radios and televisions, motorbikes, banging metal garbage cans, power tools, construction projects, amplified rock music, loud conversations, children's screams, barking dogs, roosters, and so on. The familiarity of a sound as well as its characteristics determine how annoying an individual will find the noise.

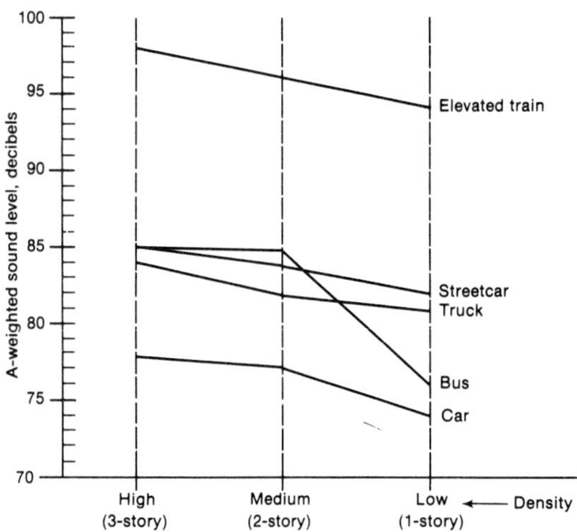

Figure 8.23
Building height as a factor in noise propagation. Data suggest that ground-level noise is higher on streets with taller buildings. (From Clifford R. Bragdon, *Noise Pollution*. Philadelphia: University of Pennsylvania Press, 1970.)

People are more annoyed by noise in the evening and at night than at any other time. People on daytime work schedules use this time for rest and relaxation.

Residents of multiple-family dwelling units frequently complain of noise transmission between units. It is not just a matter of annoyance but affects the privacy of conversation and other activities. Mechanical services and equipment such as elevators, air conditioners, and refuse handling also make noise that disturbs tenants.

Noise Control Methods

Noise is amenable to the same principles of environmental control advanced for other environmental problems. Substitution of a quieter machine design, process, or material may be an easy and effective means of eliminating or reducing a noise problem. This principle can be or has been applied in several ways. For example, a low-speed propeller fan makes less noise than a high-speed one, and a squirrel-cage blower makes less noise than a propeller-type fan. Because of improved design of the complete system, the new subway trains in San Francisco come into the station in a *swoosh* rather than a roar. Installing devices that resist motion and thereby damp vibration or using stiffer materials that resist vibration, can reduce noise from vibrating surfaces. Forces that cause vibration can sometimes be isolated by mounting the equipment on springs or resilient materials such as rubber and by using flexible connections. Jet exhausts can be modified to produce less air turbulence, and mufflers reduce noise from engine and air exhausts.

If noise cannot be reduced to acceptable levels in industry, personal protection or special design considerations may be necessary. Fitted earplugs sometimes protect the ear more effectively, but the visibility of earmuffs makes their use easier to supervise. Workers may also

PLATE 1 Lake Tahoe is an oligotrophic lake. (P. W. Purdom)

PLATE 2
The coastal ecosystem at Cape Foulweather, Oregon, shows a variety of habitats, from forests to tidepools. (Ron Beiswenger)

PLATE 3
Sage grouse have traditional *leks,* or dancing grounds, where mating occurs. (Mark Boyce)

PLATE 4
Lake vegetation is one food source for moose. (Grant Heilman)

PLATE 5
Lava flowing from Kilauea Caldera, Hawaii. (R. B. Moore, U.S. Geological Survey)

PLATE 6
Advancing lava flow in the Royal Gardens subdivision, Hawaii. (J. D. Griggs, U.S. Geological Survey)

PLATE 8
Aerial view of a large sinkhole that formed in Winter Park, Florida, in May, 1981. (George Remaine, Orlando Sentinel Star)

PLATE 7
Sequence of events in the eruption of Mount St. Helens, May 18, 1980. From left: Before May 18, minor eruptions covered the winter's snow and made the volcano appear dark; a minor earthquake causes a bulge on the north slope to slide toward Spirit Lake (some landslide material is visible in the lower right corner), beginning the eruption; within seconds, the ash cloud begins to expand vertically (darker material), but with even greater lateral force (lighter material); the lateral blast races out, destroying nearly everything in its path. (Photos courtesy of Keith Ronnholm)

PLATE 9

Sour cherry sprayed with oil. (Michael Treshow, Department of Biology, The University of Utah, Salt Lake City)

PLATE 10

A satellite photo of the lower Mississippi shows meanders winding eastward. Natchez, Mississippi, is in about the center of the right third of the photo.

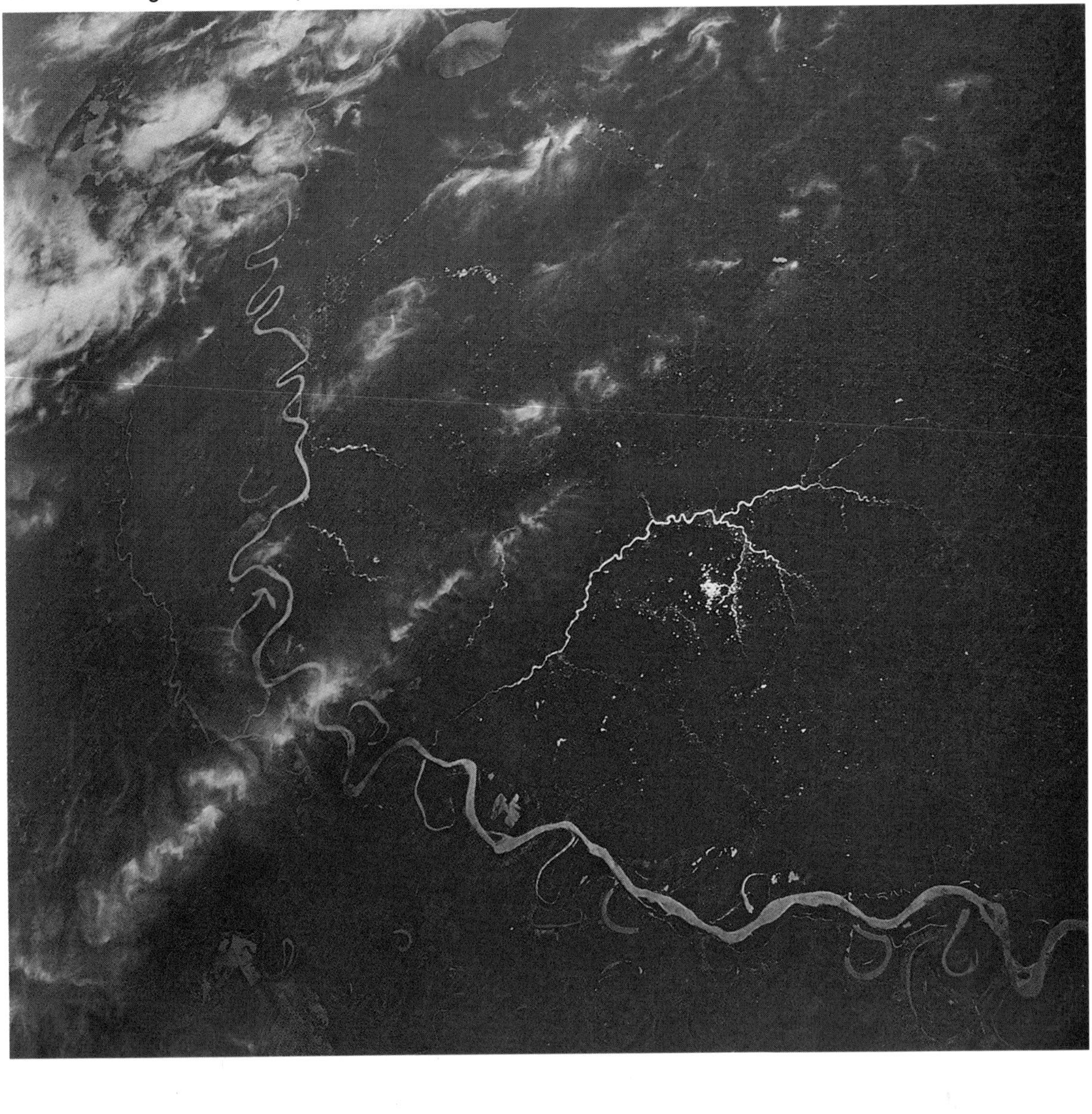

PLATE 11
Ozone injury to ponderosa pine. (A. Clyde Hill, Environmental Studies Laboratory, University of Utah Research Institute, Salt Lake City)

PLATE 12
Temperature inversion near San Diego acts as a lid, trapping pollutants below. (James E. Patterson)

PLATE 13
High plains grasslands are habitat for antelope. (Stanley Anderson)

PLATE 14 Prairie falcon are cliff nesters. (Warren Garst)

PLATE 15
Dam constructions on the upper Mississippi River have interfered with the movement patterns and populations of the paddlefish. (Patrice)

PLATE 16
Biologists learn about the movement patterns of birds by banding and telemetering, as in the case of this endangered Yuma Clapper Rail, found in the lower Colorado River marshes. (Vincent Semonsen)

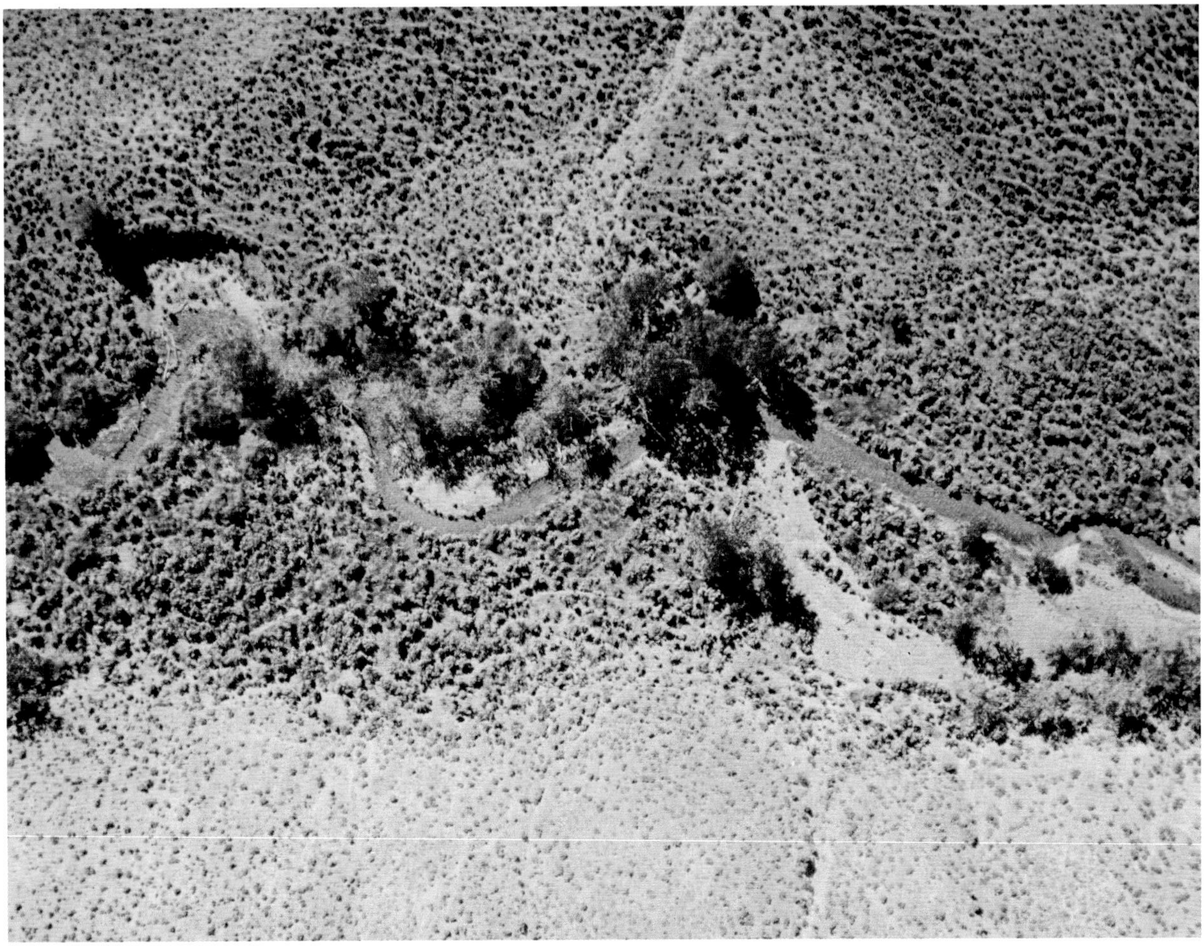

PLATE 17
Infrared photography helps examine habitat inventory. Slight overexposure of the film reveals water clarity; dark magenta indicates a clear stream. Infrared photos also show stream shade, intensity of grazing near the stream, stream bank stability, and sedimentation of the bed. (Paul Cuplin, U.S. Bureau of Land Management)

PLATE 18
Infrared photo shows heat loss around windows of the National Bureau of Standards Building. (U.S. Department of Commerce)

LEGISLATION

Clean Air Act (amended in 1990). Already listed in energy chapter. *National Environmental Education Act of 1990.* The act establishes an Office of Environmental Education in the Environmental Protection Agency. This office will oversee environmental education training, grants to schools and organizations, internships, and a national awards program.

Noise Control Act of 1972 (amended by the Quiet Communities Act of 1978). Under this law the Environmental Protection Agency sets noise pollution standards and approves state enforcement programs. Occupational Safety and Health Adminitration (OSHA) enforces the law in the workplace.

be shielded by enclosing or partly enclosing the noisy operations. If reverberation is a problem, surfaces can be treated with sound-absorbing materials that do not reflect sound. In offices and homes, rugs, curtains, and acoustical ceiling materials are used to absorb sound. Because heating and air conditioning ducts provide a path for sound transmission, they can be lined with sound-absorbing material or baffled to reduce noise.

Community noises such as those produced by transportation, lawn mowers, and construction equipment can be reduced by modifying the noise generator. The EPA has taken steps in this direction by establishing noise performance standards for the manufacture of new equipment such as lawn mowers. If garbage collection were scheduled for normal daytime hours instead of late at night, there would be less annoyance even though the sound level would remain the same. Substituting plastic bags for refuse storage prior to collection would eliminate the clanging of cans and lids; however, the advisability of using plastic bags has to be evaluated with respect to the hazard of rats as well as to noise.

Other approaches to noise abatement involve land use controls. By providing open space or locating noisy operations remote from residences, the noise source can be separated from the receptor. This approach is particularly suited to the control of noise near airports. Some noise problems will always exist near airports because the power used in takeoff cannot be reduced too much or the plane will stall, endangering the passengers' lives. However, land use controls could be used to keep residential areas from creeping in to surround the airport. But, if the airport expands or new planes are developed that make more noise (like the SSTs), the original buffer zone might not be adequate to protect homes from noise. When Los Angeles found itself in this position, the city had to purchase an extensive area of homes at an enormous cost in order to continue operating its airport.

The influence of roadway design on traffic noises is important. Trees and shrubbery that conceal the roadway reduce noise complaints but do little to reduce noise levels. The design of other types of facilities must also consider the relationships between design, noise level, and the land use category of the site.

While external community noises can be kept from intruding into a building with the use of appropriate materials, the reduction of noise generated inside buildings usually requires treatment at the source to stop the production of noise or reduce it to acceptable levels. In addition to traveling through air ducts, noise inside buildings is transmitted mechanically through the structural members and materials. Structural treatment may therefore be required to impede transmission of vibrations.

Attention to layout and careful location of functions within an apartment building can help alleviate the annoyance of noise. For example, if the kitchen of one unit contains a garbage grinder, it should not be located adjacent to the bedroom of another unit. Kitchens and toilets placed back-to-back would be better from the standpoint of noise and might be cheaper in plumbing installation costs. Rooms in which quietness is desired could be located on the side of the building away from the street. Hotels and motels built adjacent to airports demonstrate that it is possible to prevent excessive external noise from entering dwelling units through proper design and use of materials.

SUMMARY AND CONCLUSION

Community air quality standards have been set for suspended particulate matter, sulfur oxides, carbon monoxide, hydrocarbons, nitrogen oxides, ozone, and lead. The last five of these are associated especially with auto exhaust. Upon exposure to sunlight, hydrocarbons and oxides of nitrogen undergo photochemical reactions to form ozone and other oxidants. Carbon monoxide is produced when there is insufficient oxygen to complete combustion. Particulate and sulfur dioxide emissions are associated with stationary combustion, principally processes using high-sulfur coal.

Acid deposition is a problem of global proportions. It has damaged forests, lakes, and other ecosystems in Europe and eastern North America. More recently, evidence of acid deposition has been found in the western United States.

On a smaller scale, we must also try to improve the individual dwelling unit and residential environments. By considering mental and physical needs, residential areas can be developed to safeguard health and also provide a stimulating, satisfying quality of life. Providing for urban transportation is crucial to relieving the stress and tedium of congested traffic, conserving energy, and reducing air pollution, but residential areas need to be protected from the hazards and noise of heavy traffic.

The interaction of all these factors is complex, and many problems still need to be resolved. Land use decisions and environmental quality regulation are basic to the implementation of any plan to preserve the character of the environment and the quality of life.

STUDY QUESTIONS

1. Should community air quality standards be based on (a) first physiological response; (b) undesirable physiological response; or (c) proven detrimental health effect?
2. To what extent should air pollution control standards consider effects on (a) vegetation; (b) animals; (c) materials; and (d) visibility?
3. How can we distinguish between air pollution's effects on people and other environmental exposures?
4. If it takes 15 to 30 years for cancer to develop, what information would we need to determine if cancer is caused by air pollution?
5. Community A is burning high-sulfur fuel, and community B uses low-sulfur fuel. Wind blows sulfur oxides from A to B. Community A meets SO_x standards; B does not. Should A or B be required to burn fuel with lower sulfur content?

6. Do you agree with these federal policies?
   (a) Areas with clean air cannot increase air pollution significantly even if standards would continue to be met.
   (b) Areas exceeding standards may add new pollution if there is equal offset.
7. How can Norway and Sweden be protected from effects of acid rain caused by sulfur oxide emission in England and Germany?
8. Should carbon dioxide be considered an air pollutant?
9. Should community air quality standards be based on inversion conditions or normal conditions?
10. Should individual communities determine air quality standards for their areas?
11. How should we go about changing our mode of transportation from the auto to a less polluting system?
12. How does the noise of a jet plane's landing and taking off affect nearby people and homes? Could jetports be designed to reduce noise?
13. What materials in buildings reduce noise?

SUGGESTED ACTIONS

(Adapted from National Wildlife Federation's Citizen Action Guide.)

1. Find out about companies in your area that pollute the air by calling the EPA's Emergency Planning and Community Right-to-Know Information Hotline at 800-535-0202. The hotline will put you in touch with state officials who can tell you if any companies in your area are releasing toxic air pollutants.
2. Buy a low-pollution, fuel-efficient car to reduce the emissions of air pollutants and, at the same time, reduce your gas consumption.
3. Walk, bicycle, carpool, or use public transportation whenever you can.

THREE

•

ENVIRONMENTAL MANAGEMENT

9

FORESTS AND GRASSLANDS

When European explorers arrived in the world that was to eventually become the United States of America, they discovered a land of beautiful and diverse landscapes and a rich store of natural resources awaiting someone with the ingenuity to make them useful. Early settlements were in the eastern forests, and local economies developed around forestry and agriculture.

Settlers moving west in the late 1700s found vast prairies and deserts that presented a different challenge to those who wished to make a living from the land. As the human wave of migration and resource development spread across the country, competition for land and resources intensified, and resource management became an imperative.

Today, one of the most challenging aspects of resource management is the need to balance conflicting demands on the development of our forests, grasslands, shrublands, and deserts. When is it appropriate to manage an area intensively for a commercial product? How do we identify and protect unique wildland areas? How do we balance immediate needs against the needs of future generations? What compromises will ensure that various interests are treated fairly?

LAND AND RESOURCES

The Evolving Idea of Conservation

Attention to resource conservation in the U.S. began with George Washington, Thomas Jefferson, Patrick Henry, and George Perkins Marsh, but their concerns about the depletion of resources fell on deaf ears. The first real effort was made in 1908 when President Theodore Roosevelt invited governors, congressional leaders, scientists, sportsmen, and foreign experts to a White House conference on natural resources. The National Conservation Commission, under the leadership of Gifford Pinchot, was created, and its first task was to make a national inventory of resources. During this time, the systems of national parks and national forests were begun.

The resulting public awareness of conservation continued until the outbreak of World War I, then reappeared during Herbert Hoover's administration. Interest exploded during Franklin Roosevelt's administration, partly because of the many jobs he created under the Public Works Administration and the Civilian Conservation

Corps during the Great Depression. These organizations improved access roads and trails into wilderness areas, aided flood control projects, and established wildlife refuges.

The Taylor Grazing Act established a system for managing grazing lands. In 1946, the Bureau of Land Management was created to manage the vast rangelands of the West. The Soil Conservation Service was also established in the 1930s to help landowners with management problems.

World War II again diverted attention from these matters. During the early 1960s, concern about the deteriorating environment revived. President John Kennedy convened a White House conference on conservation in 1962 to discuss the status of American resources, including the preservation of wilderness areas. He established the Youth Conservation Corps to supplement some existing programs. The Multiple Use–Sustained Yield Act became law during this period, giving direction to the management of national forest lands.

In the later 1960s, the environmental movement gained momentum, stimulated by the campus unrest resulting partly from the Vietnam War. In the spring of 1970, colleges and conservation organizations celebrated the first Earth Day, which focused on pollution, population, and conservation problems. While many emotions were originally tied to the environmental movement, organizations began to focus seriously on causes of and cures for environmental problems. Legislation such as the Clean Air Act, the National Environmental Policy Act (NEPA), and the Wilderness Act proved that resource conservation had a stiff price tag. Now resource conservation is a way of life for many. As our educational system provides more and more information on environmental interactions, people have come to realize the place human beings have in the natural system and the public's involvement in environmental decision making has increased considerably. This public interest was given a boost with the twentieth-anniversary celebration of Earth Day in 1990. Concerns about global climate change, the loss of biodiversity, and toxic waste disposal, and a renewed interest in recycling and reuse of resources dominated the discussions in 1990.

Land Ownership

The United States has complex patterns of land ownership. Some land is privately owned by individuals or companies, but 40 percent of the total land area of the United States is publicly owned. Some land is controlled by state and local governments, but 85 percent of public land is controlled by the federal government. Publicly owned land really belongs to all citizens.

Resource economist Marion Clawson traces the history of U.S. public lands through six major eras:

1. The era of *acquisition* began when the original thirteen states ceded their western frontier lands to the new federal government of the United States. These lands extended westward to the Mississippi River, forming the nucleus of what we call our *public domain.* Other lands were acquired through purchase (as in the case of the Louisiana Purchase), by war, or by treaties with other nations (1776–1867)
2. The era of *disposal,* during which homesteading, land sales, grants to railroads and states, and other means were used to transfer public domain lands to state and private interests (1776–1934)
3. The era of *reservation,* which established national forests, national parks, wildlife refuges, and other lands that were to be under permanent federal management (1872–present)
4. The era of *custodial management,* when public domain were protected, but demands for resources from these lands were low (1934–1950)
5. The era of *intensive management,* during which demands for resources from public domain increased, as did conflict among var-

ious interest groups and users of public lands (1950–present)

6. The era of *consultation and confrontation*, resulting from public participation in land and resource management as required by laws such as the National Environmental Policy Act and others (1970–present)

As Clawson points out, the role of federal lands in the national life has changed dramatically over the past 200 years. We will consider some of the principles involved in understanding and managing forests and grasslands, especially those that are publicly owned and should concern us all as responsible "landowners."

At the federal level several agencies are involved in managing public land. The national forests are under the control of the Forest Service, the Fish and Wildlife Service manages wildlife refuges, the national resource lands (primarily western grazing lands) are managed by the Bureau of Land Management, and national parks and monuments are managed by the National Park Service. Some land associated with federal reservoirs is managed by the Bureau of Reclamation, the Army Corps of Engineers, and the Tennessee Valley Authority.

Forests and Grasslands: Multiple-Use Resources

National forests are managed according to provisions of the Multiple Use and Sustained Yield Act, which directs the U.S. Forest Service to consider timber harvest, wildlife habitat, livestock grazing, watershed management and recreation on its land. Grasslands and other rangeland areas are also managed for multiple uses. The concept of multiple use requires seeking maximum benefit from lands by using them for several purposes at the same time. This can mean planning a logging operation to provide needed timber and to improve wildlife habitat and water runoff patterns at the same time (see Chapter 10). In grassland areas, livestock grazing, wildlife habitat, watershed management, and recreation can coexist compatibly with careful management.

Table 9.1
Proposed Distribution of Forest Uses in the Management Plan for Medicine Bow National Forest

Use	Percentage of Land Area
Recreation	15
Wildlife habitat	18
Grazing	35
Timber harvest	16
Wilderness	5
Water yield	8
Special situations	3

Multiple use can also mean that some areas within a region are for recreation and wildlife habitat while adjacent areas are dedicated to timber production and water resource management. The areas dedicated to single use must be of sufficient size to support that use. Table 9.1 illustrates the multiple-use concept as applied to the Medicine Bow National Forest in Wyoming. The plan emphasizes grazing, wildlife habitat, recreation, and timber harvest in this forest. Other national forest plans also employ the multiple-use concept, but specific land uses vary from region to region because of the unique ecological characteristics of the nation's forests and the particular human needs of the region.

FOREST ECOSYSTEMS

The world's diverse forest ecosystems range from the open forests of arid regions to the dense tropical forests in warm, humid regions. Even the United States has a variety of forest ecosystems (Figure 9.1). Effective forest management requires understanding forests' ecological characteristics.

"Trees are renewable." This statement is often used by the timber industry to emphasize that even though trees have to be cut down to provide lumber and other wood products, they

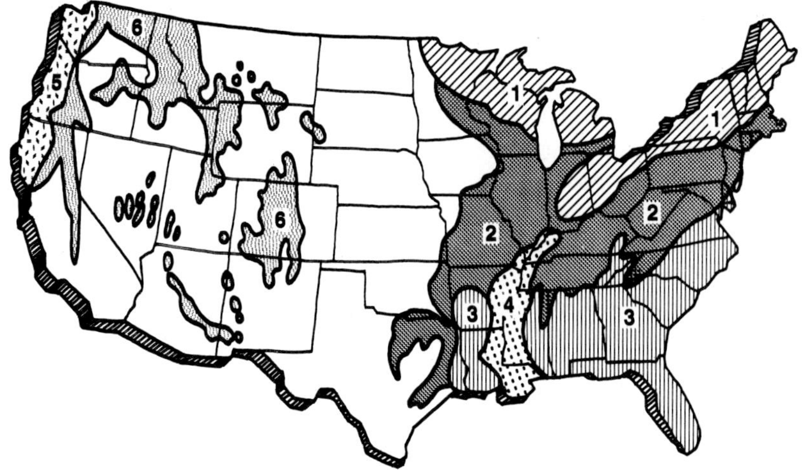

Figure 9.1
Major forest types of the United States (F.C. Hall, "Forest Lands in the Continental United States," in *Using Our Natural Resources*, 1983 Yearbook of Agriculture. Washington, DC: 1983.)

1. Northern forests of white-red-jack pine, spruce-fir, aspen-birch, and maple-beech-birch groups.
2. Central forests of oak-hickory.
3. Southern forests of Oak-pine, loblolly-shortleaf pine, and longleaf-slash pine groups.
4. Bottom land forests of oak-gum-cypress.
5. West coast forests of Douglas-fir, hemlock-sitka spruce, redwood, and some western hardwood groups.
6. Western interior forests of Ponderosa pine, lodgepole pine, Douglas-fir, white pine, western larch, fir-spruce, and some western hardwood groups.

can reproduce, and with proper management the forest will eventually grow back. Ecologically sound management schemes require understanding of the biological characteristics of trees and the structure and function of forest ecosystems. For example, we now understand that vegetation management is essential to maintaining nutrient cycling patterns and protecting forest soils from erosion.

Improper forest management can have wide implications. A study conducted in northern hardwood forests shows that minerals such as calcium, potassium, and nitrate disappear from an area following deforestation. A fertilization program is required to renew the land. As a result, stream flow out of an area increases considerably, partly because of the reduction of water loss through vegetation **(transpiration)** and partly because water does not percolate into the soil after logging. Forests also help buffer and detoxify environmental urban and industrial pollutants that reduce soil fertility.

The Life and Death of Trees

A major topic of current research is the long-term ecological effects of cutting mature trees for processing into wood products. Much of this research has centered on the role of dead and decaying woody material in forest ecosystems of the Pacific Northwest.

A tree may live for 300 years or more before it dies and falls to the ground. It may then take another 300 years to decompose (Figure 9.2). Throughout its life the tree is involved in photosynthesis, respiration, energy flow, the cycling of nutrients and water, and other ecological processes. It provides food, nesting sites, shade, wind protection, and concealment for wildlife. After it falls and begins to decay, stored nutrients are released, and the tree continues to provide protection and food for many forms of plant and animal life (Figure 9.3).

Forests in Europe and Asia that have been intensively managed for hundreds of years may

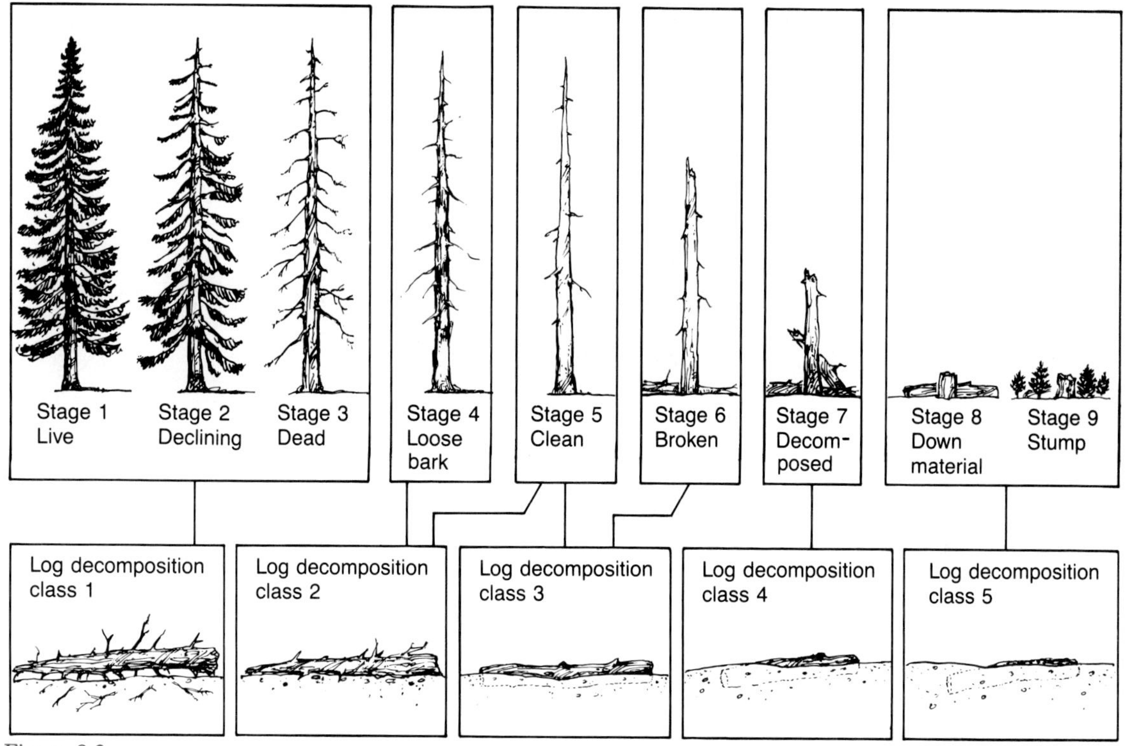

Figure 9.2
When they fall, trees and snags immediately enter one of the first four log decomposition classes. (From J. W. Thomas [ed.]. *Wildlife Habitats in Managed Forests—The Blue Mountains of Oregon and Washington*, Agricultural Handbook no. 553, U.S. Dept. of Agriculture, Forest Service.)

be experiencing loss of soil productivity. The years of removing wood products from the land may have depleted the necessary nutrients for continued forest growth. If this is the case, it is important to learn how to avoid this problem in North American forests, which have been intensively managed for a shorter period of time.

Protection of Wildland Areas

The management approach we have described recognizes the value of maintaining some old growth areas in a wild or wilderness condition. The Wilderness Act of 1964 created the National Wilderness Preservation System. In 1989 the Wilderness System included 490 areas in 44 states—areas kept in a wild state for human appreciation and enjoyment (Table 9.2). By law, these areas must be managed to preserve natural conditions and remain unmodified and uncontrolled so as to provide opportunities for solitude or primitive recreation. Wilderness Areas are federal lands that must be designated by an act of Congress.

There are other systems for preserving nature in a wild state. The federal government has a Research Natural Area program under which federal lands can be set aside for research. Wildlife refuges and national parks are other means for protecting wildland areas. Aquatic ecosystems can be protected under the Wild and Scenic Rivers Act, and many states have their own systems for protecting wild nature. A private conservation group, The Nature Conservancy, has been especially successful in developing land

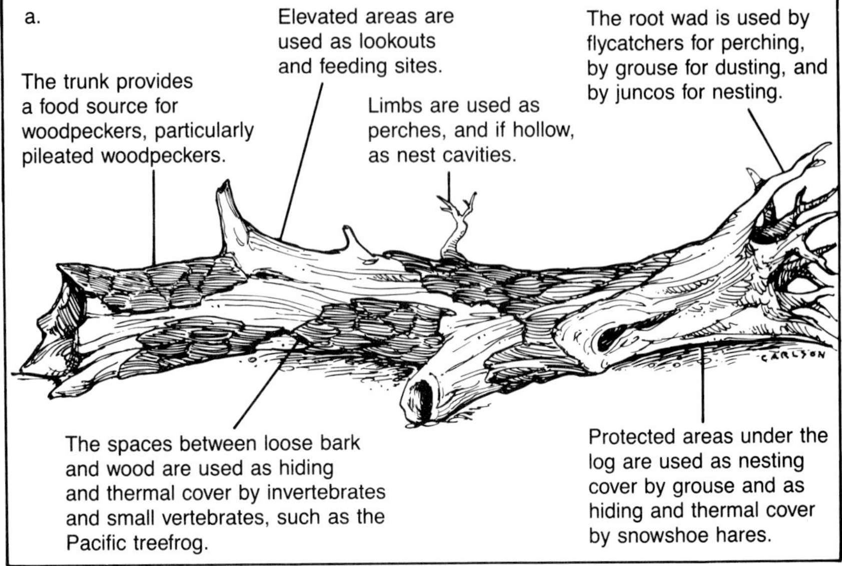

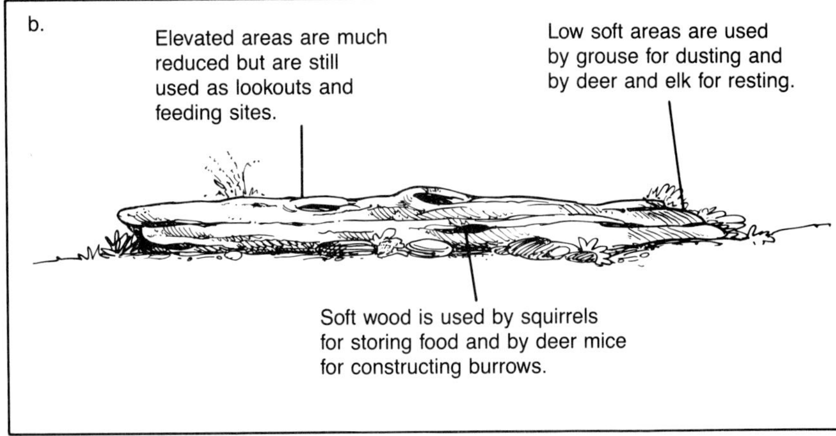

Figure 9.3
(a) A decaying log showing some of the structural features important to wildlife; (b) A log in an advanced stage of decay. (From J.W. Thomas [ed.], *Wildlife Habitats in Managed Forests—The Blue Mountains of Oregon and Washington*, Agricultural Handbook no. 553, U.S. Dept. of Agriculture, Forest Service.)

Table 9.2
National Wilderness Preservation System, March, 1989

Agency	Number of Areas	Percent of Areas	Million Acres	Percentage of Acres
National Park Service	43	9	38.50	42
Fish and Wildlife Service	66	13	19.30	21
Bureau of Land Management	27	6	0.47	1
Forest Service	354	72	32.46	36
Totals	490	100	90.76	100

protection systems. This organization devises ways to identify unique natural areas and protect them through land purchases, acquiring special easements, or providing incentives to landowners who agree to keep their land in a wild state.

Forest Ecological Succession

Tree species vary in their responses to environmental conditions. For example, some species, like the Ponderosa Pine, are fire-adapted; because of their thick bark and other characteristics, they can withstand a forest fire of moderate intensity. Other species, like the Engelmann Spruce, are quite susceptible to fire. Some species become established quickly in open sunny areas (shade intolerant species); other species grow only in the shady understory of an established forest (shade tolerant species).

The generalized pattern of forest ecological succession (see Chapter 2) shown in Figure 9.4 illustrates how shade tolerance and other characteristics might interact to cause successional change. An opening may occur in the forest as a result of wind damage, fire, an insect outbreak, disease, or a logging operation. At first the forest opening will be invaded by grasses and forbs (herbaceous plants). Within a year or two, seedlings of shade intolerant trees will appear. As they grow in height and diameter, the canopy of the forest becomes dense enough to shade the forest floor, and shade tolerant species become established and eventually become the dominant forest species (Figure 9.4). The sequence of succession usually occurs over hundreds of years, and the particular pattern of change varies from site to site. Table 9.3 gives examples of successional patterns for different regions of the United States. As the sequence of succession unfolds, patterns of water and nutrient cycling, plant and animal species that live in the forest, and other factors also change.

Forest Succession and Multiple Use

Research is yielding an understanding of these successional patterns that will help predict forest change so it can be manipulated for more effective management. One interesting approach that has emerged is a planning strategy that views a forest as a kind of "patchwork quilt" of old-growth forest intermingled with other patches of forest at various stages of succession. The idea is to apply understanding of forest ecology to provide the mixture of forest habitats our society needs for its diverse interests. This concept has been developed in the Pacific Northwest as a means of preserving "islands" of natural areas in

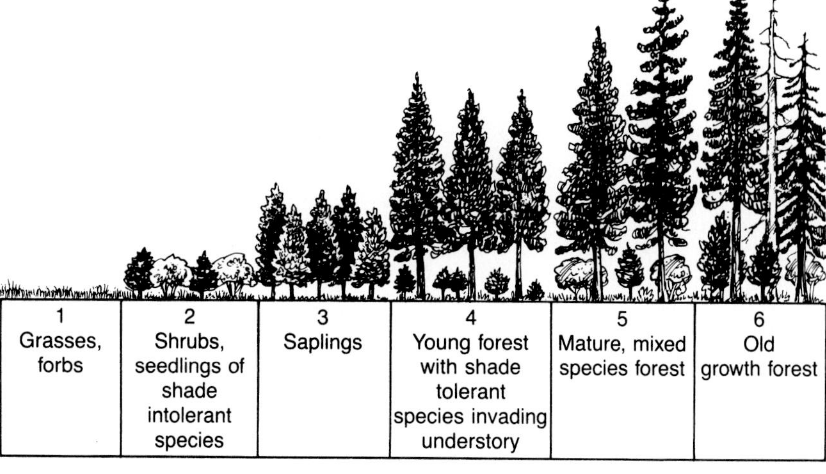

Figure 9.4
Ecological succession in a coniferous forest. (From J.W. Thomas [ed.], *Wildlife Habitats in Managed Forests—The Blue Mountains of Oregon and Washington*, Agricultural Handbook no. 553, U.S. Dept. of Agriculture, Forest Service.)

Table 9.3
Selected Patterns of Forest Succession for Various Regions of the United States

Region	Pioneer Tree Stage	Young Forest	Mature Forest	Old-Growth Forest
Northern	White pine	Pine with oak understory	Oak with beech and maple understory	Beech and maple
Southern	Southern pine	Pine with oak and hickory understory	Mixed pine, oak, and hickory	Oak and hickory
Rocky Mountains	Lodgepole pine	Pine with understory of Engelmann spruce and subalpine fir	Mixed pine, spruce, and fir	Spruce and fir
Pacific Northwest	Mixed Sitka spruce, red cedar, western hemlock, Douglas fir		Western hemlock	
Rocky Mountains	Mixed Douglas fir and Ponderosa pine		Ponderosa pine (with fires of moderate intensity)	
Pacific Northwest			Mixed Douglas fir and Ponderosa pine (with no fire)	

a "sea" of forestland that is more intensively used.

Figure 9.5 shows an example of this idea, an area of old-growth forest that has taken 200 years or more to develop. In this plan, the central core area of old growth would be left alone, and the surrounding area divided into six stands for harvesting in a sequence of cuts at 20-year intervals (Figure 9.5a). At the end of the 120-year cutting cycle, the forest would have stands of diverse ages (Figure 9.5b) that could provide a variety of potential uses, ranging from grazing in the younger stands to enjoyment of the aesthetically pleasing old-growth areas. Examples of some of these uses are given in Table 9.4.

Urban Forests

An urban forest is that part of an urban–suburban area covered by trees, associated vegetation, soil, water, wildlife, and open space. In recent years more attention has been given to managing these urban forests. Trees cover about

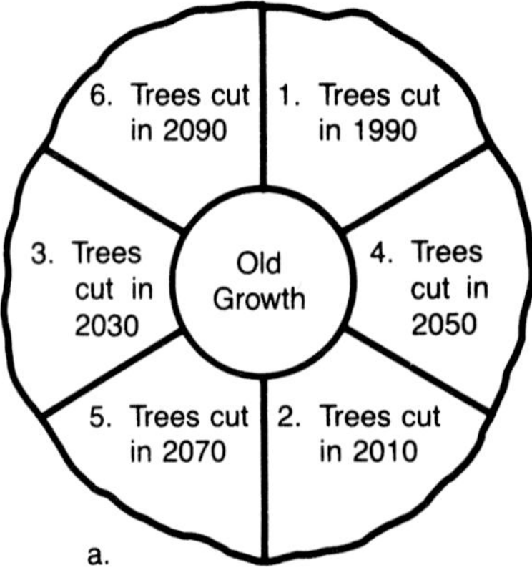

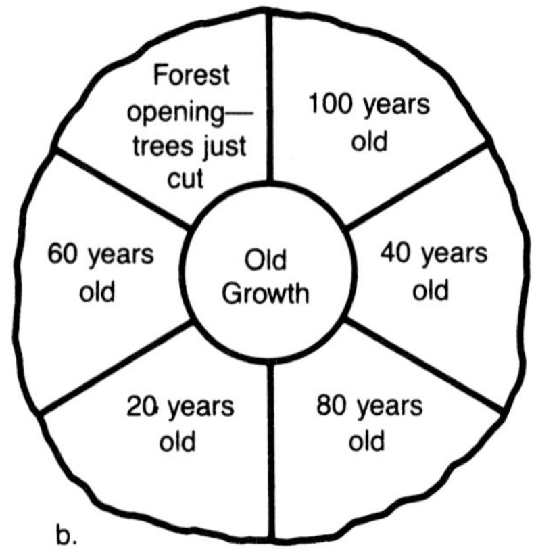

Figure 9.5
Patterns of harvest designed to produce a diverse forest with stands of various ages. (a) Planned cutting sequence; (b) Forest in the year 2090. (Adapted from L. D. Harris, *The Fragmented Forest: Island Biogeographic Theory and the Preservation of Biotic Diversity.* Chicago: University of Chicago Press, 1984.)

Table 9.4
Forest Use at Different Stages of Succession

Stage of Forest Succession	Forest Uses
Grass/forb stage	Grazing areas for cattle and elk Habitat for nongame wildlife Snow accumulation areas—increased runoff
Shrub/seedling stage	Forage area for deer
Sapling stage	Hiding cover for deer and elk Snow accumulation areas Habitat for nongame wildlife and small game animals Fence posts, firewood, pulpwood for paper
Mature forest stage	Thermal cover for elk and deer Habitat for nongame wildlife Firewood, lumber products Campgrounds and picnic areas
Old-growth stage	Aesthetic enjoyment and primitive recreation experiences Habitat for nongame wildlife Ecological "services" such as stable nutrient cycling patterns, weather modification, erosion control, etc.

30 percent of the surface area of an average city in the United States. Urban forests are valued at $25 billion, and it is estimated that cities spend about $300 million a year to maintain them.

Urban forests provide many benefits, such as enjoyment of the wildlife that depends on forest habitats (see Chapter 10). Other benefits include noise reduction, improvement of air quality, recreation, and aesthetic values. Trees also provide direct economic benefits by increasing property values (by as much as 20 percent in some areas), and by reducing the costs of air conditioning and space heating.

FORESTRY

Forests have always been a major resource for wood for homes, furniture, fuel, and paper; bark for waterproofing material; nuts, fruits, and sugar for food; and land for grazing livestock. In recent years, forests have become recognized as important recreational resources for camping, hiking, fishing, or just enjoying the view. We have discovered forest lands to be parts of watersheds which release water slowly and prevent flooding, as well as areas that contain energy fuels and minerals.

Forests in this country have been cut at rates unequaled in the world. In 1975 the average American used over 272 kilograms (600 pounds) of paper products annually—five times as much as in 1925. Europeans used 61 kilograms (135 pounds) per person and Asians 10 kilograms (23 pounds) per person. To meet this demand for forest products, approximately 202 million hectares (500 million acres), or one-quarter of the continental United States, are either forested or thought to be suitable for commercial timber production.

The lumber industry developed first in New England, expanded to the Great Lakes states, then followed the Appalachian Mountains southward into the southwestern stages. Later the trees of the Pacific Northwest were logged, followed by the mountain states (Figure 9.1). The early practice was to cut down an entire forested area. Some parts of Minnesota, Michigan, Wisconsin, and eventually the Pacific Northwest were left without commercial trees, forcing the loggers to leave.

In 1891, President Benjamin Harrison created a forest reserve around Yellowstone Park. President Theodore Roosevelt's adviser, Gifford Pinchot, added land and brought the U.S. forest reserve to 60 million hectares (148 million acres). More important, Pinchot established a system of national forests from what were formerly called forest reserves and implemented forest management practices that made these areas permanent sources of timber.

If we consider forests in the United States only, we are ignoring the importance of the world's timber. Forests in some parts of the world, particularly the lesser developed countries, present difficult management problems. Currently, forests in Africa, Asia, and South America are disappearing at a very rapid rate. In many cases, commercial timber companies operate on a cut-and-get-out basis—much as in the northeastern forests of the United States—making no effort to manage for the continuity of the resource. In less developed countries forest removal for agricultural purposes is often done by burning down the trees, with little or no use being made of the wood. Although many areas of the world, particularly Central America, have enacted some kind of management regulations, underdeveloped nations have insufficient funds and personnel to enforce them. In addition, nations in Central and South America allow poorer families to exercise "squatting rights." If they live on the land for more than one year, they cannot be expelled by the government, thereby eliminating the land from alternative use.

Timber Management

To supply society's demand for forest products, effective management as well as conservation measures must be instituted to make forests yield more high-quality timber and to prevent massive forest losses. As we discussed in Chapter 1, forests should be managed on a sustained-yield basis, meaning that timber harvest plans must ensure that tree removal will be balanced by regrowth of new trees. Sustained yield is a difficult concept to apply to forests because it takes a long time for trees to grow and it is not always clear what time period and how large an area to use to calculate the balance between tree cutting and tree growth.

Two common timber management systems are **uneven-age management** and **even-age management.** In uneven-age management, stands of trees differing markedly in age are planned. A type of uneven-age management is *selective cutting*. Individual trees or small groups of trees are selectively removed, resulting in a forest that consists of trees of various ages.

Even-age timber management results in large stands in which all the trees are about the same age. This is usually cheaper than selective cutting and creates a more uniform forest that is easier to manipulate. Entire stands are cut at different times, as in western Douglas fir forests where the rotation cycle is about 120 years. Consequently, blocks of different successional ages are found throughout the larger forest. In a comparable setting in the southern part of the United States, many pine plantations are maintained in even-aged stands; however, they have a shorter rotation cycle of about 30 to 40 years.

The practice of harvesting an entire stand of trees in a block or strip is called *clear-cutting*. It is most effective in forests such as Douglas fir, where seedlings need the full sun for growth in the debris-covered ground. This form of management has been used extensively in the Pacific Northwest forests (Figure 9.6).

Figure 9.6
Clear-cutting of Douglas fir in the Pacific Northwest. (U.S. Forest Service.)

TROPICAL FORESTS: OUR GREATEST MANAGEMENT CHALLENGE

Wet tropical forests are among the earth's most important ecosystems. Tropical forests provide valuable hardwoods, and many people depend on tropical forests as a source of fuelwood. In addition, some tropical forests have been cleared (deforested) for growing crops or to allow livestock to graze. The Amazonian forest in Brazil is so expansive that it influences weather and climate patterns by producing large amounts of water through transpiration from the lush vegetation and by its use of carbon dioxide in photosynthesis. Many of our domesticated food plants and drugs originated in the tropics. Scientists believe that tropical plants and animals are among the most important natural resources because of their potential for practical use. There are millions of species of plants and animals in tropical forests.

Despite their obvious value, tropical forests are being deforested at an alarming rate. A combination of conversion to cropland, cattle ranching, fuelwood gathering, and commercial forestry are responsible for most deforestation. Hugh Iltis, University of Wisconsin botanist, says, "The destruction of tropical forests in the world today is so extensive, so devastating, so irrevocable that humanity may soon lose its richest, most diverse, and most valuable biotic resource."

Management of tropical forests is clearly an issue of global significance, complicated by the fact that most tropical forests are in countries that need to develop economically to improve their peoples' well-being. At the same time, experience has taught us that clearing a tropical forest for a banana plantation, a cattle ranch, or commercial hardwood production can provide short-term benefits; however, without careful planning, these activities can lead to long-term problems and permanent loss of unique resources.

Governments and corporations in the U.S. and other countries contribute to deforestation through development projects related to energy, mineral extraction, lumbering, agriculture, and road building. In a sense, all of us contribute to tropical deforestation. We demand tropical products such as hardwoods, fruits, and medicines that come from tropical areas. Norman Myers wrote, for example, of the "hamburger connection," describing how our desire for fast food created a demand for beef from cattle raised on cleared tropical forest land. In response to this knowledge, some fast-food restaurant chains in the United States now purchase beef from other sources. International lending institutions, such as the World Bank, have also changed their policy to require environmental assessments before approving loans for development projects.

Resource development in tropical forests is also affecting native human cultures, especially in Central and South America, much as the American Indians were affected by the western expansion of European settlers in North America. For example, indigenous peoples in Brazil existed for centuries at low population densities, making a living by slash and burn agriculture, hunting, and fishing. These people numbered over 500,000 as late as 1940, but today have dwindled to fewer than 50,000. Is this a price of development that must be paid, or is there a way to preserve this ancient lifestyle?

Local rubber tappers are also affected by deforestation. They extract latex, Brazil nuts, and other products from the forest, but their livelihood depends on leaving the forest intact. World public opinion was aroused in 1988 when Chico Mendez, a leader of the Brazilian rubber tapper's union, was murdered. Just before his death he had been honored with an international award for his efforts to prevent deforestation in the Amazon. Two cattle ranchers were found guilty of the crime. The Brazilian government recently established protected tropical forest areas called extractive reserves so that rubber and other products will be available for extraction on a sustainable basis.

The diverse biota of tropical forests represents a rich storehouse of potential future resources. Can we practice enlightened forest management so that local people can benefit economically from the resources and still protect the natural diversity of tropical forests and the primitive human cultures that depend on them? It is essential that any economic development project be carried out in a way that is ecologically and culturally sustainable. Following are some proposals to help answer these questions:

- Establish a global system of tropical forest preserves financially supported by a tax on countries that are economically well-off. In some cases "debt-for-nature swaps" can be made. A large debt owed by a tropical country is forgiven if the country agrees to establish nature reserves to protect tropical forests.
- Encourage local support of nature reserves by finding ways for local people to benefit from them. Compatible human uses, such as small-scale agriculture, can be permitted within their borders, and local people can be hired to help manage the reserves.
- Concentrate on improving the fundamental well-being of the people who need tropical forests for food and fuelwood so they will be motivated to manage their forests more carefully.
- Provide foreign experts to advise local officials on proper forest management.
- Help tropical countries train experts on forest ecology and build local natural history museums to encourage ecological research.

- Launch an extensive public education campaign to convince people in tropical countries that good forest management is an important goal.
- Modernize tropical countries to provide local peasants with more conveniences without the need to clear vast forest areas to make a living.
- Fund large-scale research projects that provide a fundamental understanding of tropical forest ecology.
- Fund large-scale research projects that emphasize applied research.
- Help developing countries formulate population policies that will reduce the need to clear forests to accommodate rapidly growing populations.
- Encourage land-use policy changes to put more land into the hands of peasant farmers who will need to clear only small patches of forest.
- Slow deforestation by working to reduce consumption of products from tropical forests, particularly by people in developed countries.
- Establish reservations for indigenous peoples who are displaced when their homeland is deforested.
- Begin extensive reforestation programs to restore tropical forest ecosystems.

For further information, see Hugh H. Iltis, "Tropical Forests: What Will Be Their Fate?" *Environment 25*, 10 (1983):55–60; Peter T. White, "Tropical Rain Forests: Nature's Dwindling Treasures, " *National Geographic* 163, 1 (1983):2–46; and Philip M. Fearnside, "A Prescription for Slowing Deforestation in Amazonia," *Environment 31*, 4 (1989):16–20, 39–40.

If improperly carried out, clear-cutting can leave unsightly scars and large amounts of debris, including parts of unwanted trees. When debris is removed or burned, some seedlings will not grow well in the exposed sunlight. If cuts are poorly made in relation to the wind, the edge trees can be severely damaged by wind. If erosion is not prevented after clear-cutting, water runoff can remove valuable topsoil. All result in habitat changes.

Another harvest method for even-age management is the *shelterwood system*, in which a stand is removed in a series of cuts. This leaves a partial forest to provide protection for the seedlings that try to establish themselves or are planted in the newly cut areas. A third kind of even-age timber harvest is the *seed tree method*, in which a clear-cut leaves standing a few of the more desirable trees to provide a source of seed.

Timber management is the key to managing other forest resources. When ecological principles are considered, timber harvest need not be destructive; it can improve wildlife habitat, enable more effective management of watersheds, and create new recreation areas. A clear-cut mimics a natural event such as a fire, insect outbreak, or a destructive tree disease. Properly carried out, clear-cuts can be beneficial and result in a forest with stands of diverse ages.

Recently, construction of logging roads has become particularly controversial. Although new logging roads make the forest more accessible for recreation, it is sometimes necessary to obliterate roads after the logging operation is over to protect an area that cannot withstand heavy human use.

New Forest Management Perspectives

A series of events in the late 1980s and early 1990s caused the Forest Service to critically examine its management philosophy. In November 1989, sixty-five national forest supervisors told their boss, the Chief of the Forest Service, that their agency was out of touch with the values of the public and many of its own employees. At the same time some employees organized the Association of Forest Service Employees for Environmental Ethics (AFSEEE), which grew from two members in 1988 to more than 5000 members by 1991. In response to these events the Forest Service launched a new program called New Perspectives for Managing the National Forest System. According to the Chief of the Forest Service, the program emphasizes four items:

1. Enhancing recreation, wildlife, and fisheries resources. Recreation facilities will be improved and the quality and quantity of recreation opportunities will be increased. Also, the protection and management of biodiversity will receive a greater emphasis.
2. Producing commodities in ways that are environmentally acceptable. This includes a reduction in timber harvest levels and a steady decline in the number of timber sales that are referred to as below-cost. Below-cost timber sales have been criticized because their administrative cost is higher than the payment the Forest Service receives for the timber. Also, the condition of rangelands in national forests will be improved and sensitive riparian areas will be given special protection. Water quality will also be emphasized.
3. Improving the base of scientific knowledge about natural resources.
4. Responding to global resource issues such as climate warming. In general, the New Perspectives approach tries to emphasize the management of forests as ecosystems as opposed to forests as sources of commodities. The program also calls for greater cooperation with landowners to improve the management of private forest lands.

Forest Management Actions

Reforestation When trees are removed from a large area, immediate revegetation is essential to prevent soil erosion and water loss (Figure 9.7). In many areas this is easily done by aerial seeding; other locations require ground planting by hand or machine. Frequently tree seedlings are planted with grass seeds. Land damaged by pollutants or erosion might need to be treated before any regrowth of vegetation can occur.

Fire To most people, fire is something to be feared and eliminated. In the early history of logging, fire used to burn out of control because of poor management practices, destroying thousands of acres and whole towns. Public policy beginning in the 1930s and continuing in the

Figure 9.7.
Soil erosion following loss of vegetation on a hilly slope. (U.S. Forest Service.)

1940s and 1950s dictated that fires be suppressed. Smokey Bear became the symbol of this campaign.

Although suppression of fire is imperative where human life and property are concerned, we know that some species are fire-adapted and that fire is necessary to maintain some forest ecosystems. Without fire, ponderosa pine seedlings cannot develop because they can take hold only on open ground in the pine forests. Thus, in the year following a fire, many seedlings appear. When fire is suppressed, dry pine branches form on the lower parts of tree trunks and a great deal of dry matter accumulates on the ground, creating a severe fire hazard and preventing new trees from growing.

In some of our northern bog forests, fires help maintain plant and animal diversity. Without fire, these areas reach a mature state with relatively uniform structure. Fire is also used to destroy disease-producing organisms, as a means of keeping hardwoods out of pine forests, and as a way of releasing nutrients into forests, and as a way of releasing nutrients into forests. Normally, fire can be used only in limited areas with great care on days of very low fire hazard.

Pest Control Insects and disease-causing organisms destroy many trees in our forests. Frequently, the lack of fire alters the habitat to favor these organisms. Bark beetles have now invaded a major portion of our forest reserves. Chestnut blight, introduced into the United States from China around 1904, has made the once abundant chestnut tree almost extinct, and Dutch elm disease is a constant threat to elm trees.

Control is a two-stage process. First, the disease must be understood and suitable controls instituted depending on the tree species. The second step is to develop ways of keeping the forest pest within acceptable levels. Biological control, development of resistant tree species, and proper management of the ecosystem through biologically sound practices are some effective control methods.

Figure 9.8
Tree farm. (U.S. Forest Service.)

Tree Farming Tree farming, which is more of an agricultural than a forest management practice, results in a monoculture that can be harvested at prescribed intervals. Most of the more than 29,000 tree farms in the United States are maintained by paper companies for pulpwood (Figure 9.8). Since the trees are of uniform size, the plant and animal diversity of these systems is usually low, and invasions of insects and other pests can easily occur.

FORESTRY AND THE FUTURE

The Outlook for Timber in the United States

Because timber is an important product in the U.S. economy, economic planners try to estimate patterns of future supply and demand. The timber supply projections for 1990 (16.5 billion cubic feet), shown in Table 9.5, were published in 1983. They turned out to be slightly lower than the actual harvest of 17.9 billion cubic feet. Even with this margin of error it is important to make these projections, because timber is one commodity that the United States can export. Unless the value of imported oil, automobiles, and other products is counterbalanced by exports, a balance of trade deficit disrupts the economic system. Table 9.5 also shows a high demand for wood products within the United States. These demands for domestic supply and timber for export will undoubtedly increase the pressure for more intensive forest management.

Intensive Forestry

As the prices of forest products increase, there is a desire to control and manipulate forests. Intensive forestry practices, such as genetic hybridization which allows trees to grow more rapidly, are combined with management practices to reduce the length of the growth cycle. Foresters believe that Douglas fir can be regenerated in the western United States every 40 rather than 120 years, and southern pine every 35 years or less. Cloning and tissue culture techniques along with advanced irrigation and fertilization methods also appear to be promising methods of increasing growth rates of commercially important tree species.

However, a number of problems result from shorter rotation cycles. As discussed earlier, decaying trees provide nutrients for the soil. In the detritus food web (see Chapter 2), energy and minerals are released which can be used by young trees, other plants, and wildlife. When this cycle is broken, the system requires an energy subsidy in the form of fertilizers to produce trees. Each link in the system is altered. Since wildlife use downed wood for food and shelter, some rodents are likely to decline in numbers. As rodents are food for some hawks and owls, those populations decline in turn and the species composition changes.

Deforestation

Because of the increased demand for timber products, the world's forests are decreasing in size at the rate of approximately 18 to 20 million hectares (44 to 49 million acres) per year. About one-fifth of the world's land surface is now covered by forest. If the deforestation rate continues, experts predict forests will cover only about one-sixth of the land's surface by the year 2000.

Deforestation is especially severe in tropical areas where forests are being cleared for agriculture or to provide hardwoods for export. In other parts of the world, forests are being destroyed for fuelwood.

Ancient Forests

A major controversy in recent years has involved the management of old-growth forests in the Pa-

Table 9.5
Demand, Exports, Imports, and Supply of Timber Products from U.S. Forests (billions of cubic feet)

Item	1952	1962	1970	1976	Projected[a] 1990	2010	2030
Total U.S. demand	11.9	11.6	12.5	13.4	18.8	22.8	25.5
Exports	.1	.5	1.5	1.8	1.5	1.5	1.3
Imports	1.5	1.9	2.4	2.8	3.8	4.2	3.8
Demand on U.S. forests	10.8	10.2	11.6	12.4	16.5	20.1	23.0
Supply from U.S. forests	10.8	10.2	11.6	12.4	16.5	20.1	23.0

[a]Projected based on relative prices rising from their 1970 level (inflation).

Source: F. C. Hall, "Forest Land in the Continental United States," in *Using Our National Resources, 1983 Book of Agriculture* (Washington, DC: U.S. Dept. of Agriculture, 1983), pp. 130–39.

cific Northwest. These forests, called ancient forests by those who wish to preserve them, are economically very important because they are a source of valuable timber and the employment that accompanies timber production. At the same time these forests are highly valued natural ecosystems because of their aesthetic values and the many ecological services they provide. For example, they are linked to the greenhouse effect because their photosynthetic activity makes them a "sink" for carbon dioxide. They also play a major role in watershed conservation and the maintenance of biodiversity.

A controversy arises because these old-growth forests are habitat for the spotted owl, a threatened species protected under the Endangered Species Act. People concerned about the economy of the region fear that protecting old-growth forests to preserve the spotted owl will result in the loss of jobs and the economic benefits society derives from timber harvest and a healthy economy. On the other hand, those advocating preservation of old-growth forests argue that, at current rates of timber harvest, the old-growth forests will soon be gone anyway and we will need to develop new economic activities to replace the lost timber-related jobs. In April 1991 the Fish and Wildlife Service proposed logging restrictions on 11.6 million acres of old-growth forests in Washington, Oregon, and California. Anyone who wishes to cut trees in these restricted areas will be required to obtain special permission from the Fish and Wildlife Service. However, logging will be totally prohibited only in specific locations where it would threaten the continued survival of the spotted owl. This decision was later modified by a committee called together by the Secretary of the Interior. More of the restricted area was opened to logging.

GRASSLAND

Because grasslands cover approximately 20 percent of the earth's surface and produce much of the world's food, knowledgeable management of these areas is of great importance. Grasslands are biotic regions (see Chapter 11) with low to moderate erratic rainfall. Some grasslands have been converted to croplands. Often, however, the topography is rough, the drainage poor, and cold, unsuitable temperatures hinder cultivation. These areas, which provide forage for free-ranging native and domestic animals, are referred to as *rangelands* or *range*. Although rangelands have traditionally been thought of as grazing lands, they are also managed for multiple uses. They produce wood products, water, and wildlife, and provide recreational opportunities. Whether the land is used for crops or grazing, we need to understand and respect the ecosystem and its characteristics.

A Brief History

Before the English settled at Jamestown, Spaniards brought sheep, horses, and cattle as they settled in the Southwest (especially the middle Rio Grande). Settlers quickly found the vast grasslands ideal for allowing cattle to roam freely and graze. As more and more people moved west, barbed wire fences went up, confining the increasing numbers of cattle to smaller areas. In the north, the westward movement brought pioneers who soon discovered the grasslands to be excellent farmland. The ranchers overgrazed some grassland, while farmers depleted the soil nutrients in other areas.

The U.S. government tried to dispose of its land as quickly as possible during the nation's early history. For example, the Homestead Act of 1862 gave 160 acres of land to settlers who resided on it for at least 5 years. Since not all land was suitable for crops or grazing, some of these plots had to be abandoned.

While the Forest Service managed timbered and some mountain grazing land in the early part of the twentieth century, much rangeland did not come under the jurisdiction of any major government agency. Local agencies (loosely organized into the Grazing Service) had this function, but their control was limited. In 1946, the

Bureau of Land Management was created and given the responsibility of holding over 65 million hectares (160 million acres) of federal rangeland for management or for later disposal to private citizens. The Bureau did not receive full legal authority to administer grazing, recreational, or other activities on a permanent basis on those public lands until 1976.

RANGELAND ECOSYSTEMS

Rangeland Vegetation

As with forest management, the key to successful range management is to manage the soil/vegetation system. Grass plants grow and reproduce in a relatively short growing season. In the grasslands of the United States, cool-season grasses do best in the spring and are replaced by warm-season grasses later in the summer. Most grass species have fibrous root systems that efficiently hold soil and prevent erosion. Because grass plants grow from the base of the leaf blades, they recover quickly from grazing or from a fire of moderate intensity.

It is generally thought that the upper 50 percent of the grass shoot (the metabolic reserve) can be removed without damaging the plant. When more than 50 percent is used, the plant must spend all of its energy to synthesize and store food, leaving no energy for reproduction. In overgrazed land, the livestock or wild herbivores (such as antelope, deer, elk, and buffalo) in excess of the land's carrying capacity not only eat into the metabolic reserve but clip the vegetation to the bare ground, causing starvation and death of the root system.

Grass species also vary in their palatability and nutritional value to grazing animals. All these features are important to consider when developing grazing management systems.

Range Productivity

Lands in different regions vary in their capabilities for growing forage plants for grazing animals. Table 9.6 shows the variation in the amount of grazing land needed to support one animal unit (a cow or a cow and a calf) in different regions of the United States, ranging from 0.65 hectares (1.6 acres) in Iowa to 13.9 hectares (34.4 acres) on the less-productive grasslands and desertlands of Arizona. Measured in relation to the land's natural potential to produce vegetation, a fair range in Iowa may be more productive than a good range in Arizona.

Range Condition and Trends

Rangeland quality is measured by how closely a site's vegetation and groundcover approach their natural potential for a particular climate and soil type (the climax ecosystem of grassland succession). Quality is classified as excellent, good, fair, or poor. A good or excellent range is dominated by climax grass species and has a stable, healthy groundcover. A fair or poor range is dominated generally degraded and dominated by less desirable, weedy plant species.

Concern about the condition of rangelands in the western U.S. has prompted nationally organized efforts to manage public rangelands, beginning with the Taylor Grazing Act in 1934. By

Table 9.6
Amount of Land Needed for Grazing Animals in Selected States

State	Area in Hectares (acres) per Animal Unit
Iowa	0.65 (1.6)
Wisconsin	1.1 (2.6)
Alabama	1.5 (3.7)
California	2.0 (5.0)
Louisiana	2.0 (5.0)
Texas	4.1 (10.2)
Oregon	4.6 (11.3)
Colorado	5.1 (12.6)
Wyoming	9.4 (23.3)
Arizona	13.9 (34.4)

Source: *America's Soil and Water: Condition and Trends* (Washington, DC: U.S. Dept. of Agriculture, Soil Conservation Service, 1980).

that time, however, an estimated 85 percent of the rangelands had already been severely depleted. Range condition improved somewhat between 1936 and 1966, but most of the improvement was on lands classified as fair or poor. A 1988 assessment (Figure 9.9), the Bureau of Land Management (BLM) classified 38 percent of the public rangelands in the United States as good or excellent, and 72 percent as fair or poor, indicating a need to manage grassland and shrubland resources carefully. Forested rangelands managed by the Forest Service also require good management practices. Despite 50 years of effort, U.S. rangelands have been slow to recover, although recent trends are encouraging. Between 1975 and 1988 the proportion of rangelands in good and excellent condition increased from 17 to 38 percent.

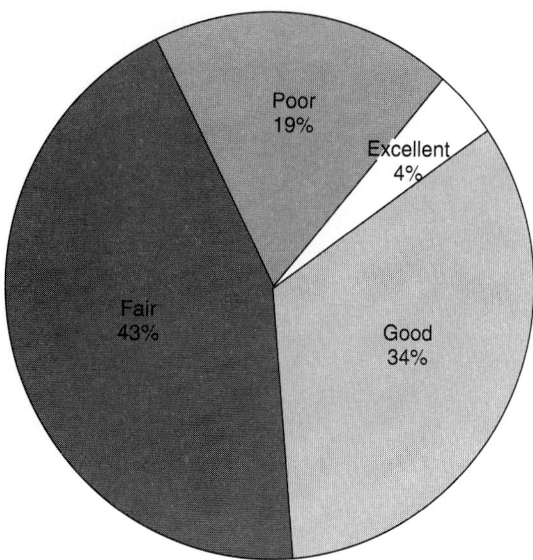

Figure 9.9
Condition of public rangelands in the United States in 1988.

In the early 1990s public concern about the quality of public rangelands led to a movement calling for the removal of livestock from public lands. The people advocating this approach think the BLM should place more emphasis on recreation, wildlife habitat, and other multiple uses. They also believe that the grazing fees ranchers pay to use these lands are too low and represent a unnecessary subsidy of the ranching industry. "Cattle free by '93" is the rallying cry of this movement. Proponents of public land grazing on the other hand, point out that recent ecological research shows that many ranges would change little, if at all, with livestock removed. Furthermore, productive rangelands with improvements such as watering holes benefit wildlife as much as they do livestock. Proponents also argue that livestock grazing is very compatible with other multiple uses of rangeland areas and may even prevent the conversion of these lands to housing subdivisions incompatible with recreation and wildlife habitat.

This controversy shows us there are some complex management questions that the BLM and other agencies should consider. For example, what is an appropriate grazing fee for using public lands? Should some unique areas be declared off limits to grazing? When are fences appropriate on public lands, and should these fences be designed to permit passage of wildlife through or over them? What kinds of predator control activities are appropriate on public lands?

Effects of Overgrazing

When rangeland is overgrazed, the natural vegetation is destroyed and undesirable species take over. If cattle herds graze on rangeland, they trample young plants, preventing their growth. They also compact soil so that seeds cannot sprout. This results in a reversion to an earlier stage of succession in which colonizing plants (usually undesirable) enter an area. In addition, overgrazing accelerates erosion as topsoils are washed away, along with minerals and other nutrients.

RANGE MANAGEMENT

In 1949, range management was defined as the science and art of planning and directing range use so as to obtain the maximum livestock production consistent with the conservation of range resources. More recently, the definition was broadened to include the management of rangeland for all animals and for a variety of other multiple uses. Range management, therefore, often involves controlling grazing pressures, manipulating wildlife, and sometimes using the tools of fire and reseeding. It involves the application of principles from the biological, physical, and social sciences. Managers must understand the response of vegetation to cropping or cutting and the response of animals that harvest the crop. Physically, management deals with climate, topography, and moisture factors; and socially, it considers people's desires along with the need to use rangeland as a source of food and recreation.

In management, it is important to recognize that the grasslands throughout the United States do not receive a dependable supply of rain. Although we talk of the average annual rainfall, precipitation is highly variable. In drought years the productivity is lower, so a rancher may need to reduce the number of cattle he grazes on an acre by as much as 75 percent or more or rotate stock from one pasture to another to allow needed recovery time. Since moisture is so important to effectively maintain grassland, managers must first find the percentage of net productivity that can be used annually by grazing animals.

Grazing Systems

A number of different techniques can be used in managing rangeland. **Rotational grazing,** the movement of livestock from one pasture to another according to a schedule, can reduce grazing intensity, allowing vegetation time to recover and grow. Long rotations can be achieved by longer grazing periods, more pastures, or both, while short rotations, though seldom employed on extensive rangeland, are widely used for harvesting forage on improved pastures.

Deferred rotation occurs when a pasture is not grazed until all seed and grass maturity is ensured. This method enhances plant vigor and seed production. A pasture that is not grazed at all in a given year is called *rested.*

Selection of Livestock Breeds

Choosing the right kinds of domestic animals to graze on rangeland has an important effect on the system. Since all animals do not have the same forage preferences, selecting an animal that likes the plant species present in the area makes wiser land use possible. Sheep and goats are best adapted to thick grass and shrubs, while cattle are chiefly grass grazers although they consume some shrub plants. Horses, the most selective domesticated animals, are primarily grass eaters, taking relatively small amounts of other forage.

Competition with Native Herbivores

Grasslands feed a variety of animals, such as buffalo in North America, kangaroo in Australia, camels in Africa, and cattle in Asia. Many other animals use the range for food including prairie dogs, rabbits, and small rodents. As more and more domestic animals are grazed, however, their direct competition with the wild animals that normally eat the vegetation must be evaluated.

When grazing animals are first placed in a range environment, the effects on the land are intensified, resulting in a reduction of the native animals. Problems arise when some of the larger game animals are sought as hunted species. Although state game departments require a license fee for hunting these species, with revenues going to a fund for game management, ranchers complain that native animals are utilizing forage that would otherwise be available to their domestic animals. This results in competition among the ranchers, the conservation agency, and the hunter who is interested in recreation.

LEGISLATION

Federal Land Policy and Management Act of 1976 (FLPMA). This law serves as an organic act for the Bureau of Land Management and directs the agency to inventory their resources and to provide for multiple use and sustained-yield planning and management. It also calls for the BLM to conduct a study to determine which of its lands are suitable for wilderness designation.

Land and Water Conservation Fund Act of 1964 (LWCFA or LAWCON). This law provides funds to the states and federal agencies to acquire public lands for recreational purposes and to support state and local projects related to outdoor recreation. States must provide matching funds and submit a State Comprehensive Outdoor Recreation Plan (SCORP) to be eligible for funding. A major source of funds for this program comes from leasing Outer Continental Shelf petroleum reserves.

Multiple Use–Sustained Yield Act of 1960 (MUSYA). This law directs the Forest Service to manage the national forests for multiple use and for a sustained yield of forest benefits.

National Forest Management Act of 1976 (NFMA). This law requires the Forest Service to carry out long-range planning which must include special consideration for biodiversity of the forest, and requires management guidelines for timber harvest, especially the practice of clear-cutting.

Wilderness Act of 1964. Described in an earlier chapter.

Livestock Management

Effective range management requires determining the number of livestock a rangeland can support. Distribution of livestock is also important. Livestock movements can be controlled by several means, such as the use of herders, fencing, and strategic placement of saltlicks or water sources.

Controlling Pests

Undesirable Plant Species When undesirable plant species such as unwanted shrubs, weeds, or poisonous plants become established in an area, they can be removed chemically with herbicides or mechanically. In some cases, controlled fires can effectively destroy unwanted plant species, however, following the fire, the desired plant species must be able to grow rapidly in order to overtake reproduction of the less desirable plants. Reseeding and fertilization can increase the grass production.

Animal Pests Many small animals such as rabbits, rodents, and even insects, forage on grasses and other range plants. In years when rodent populations are particularly high, the forage available to domestic animals is considerably reduced. This condition amplifies the dilemma existing between predator control and biological food chain. For example, ranchers often seek to have animals such as coyotes and eagles destroyed because they use young domestic livestock as a source of food. However, when populations of predators are reduced to very low levels, rodents multiply rapidly, thereby reducing the amount of food available to domestic animals. Because of nature's delicate balance, biologists who manage rangelands must understand all of the principles of food chain dynamics, succession, and competition to adequately manage this complicated resource. (This is discussed further in Chapter 10.)

Rangelands of the World

Rangeland degradation is a serious problem in various parts of the world, to which years of overgrazing and other abuses, often in combination with climatic change, have contributed. Desertification, discussed in Chapter 5, often has improper range management as one of its root causes.

SUMMARY AND CONCLUSION

Forests and grasslands can be managed for multiple use. A tree cutting operation can be planned to provide for wildlife habitat, watershed management, and other forest uses, in addition to providing lumber products. Good grassland management yields recreational opportunity, healthy watersheds, and forage for grazing animals.

Forest and grassland resources are renewable; trees can be harvested and livestock can graze the grassland vegetation, and, with proper management, these resources will regrow. Society should be able to expect a continuous flow of forest and range products. As wise managers, we seek to conserve these resources so they will continue to be available to future generations. To be able to do so, we need to understand the physical and biological principles of ecosystem interactions and realize that people subjectively judge plant and animal species as desirable or undesirable.

Although we cannot correct some situations, such as the loss of a unique old-growth forest or a stand of climax prairie grasses, other damage can be repaired. We can to a great extent rehabilitate Appalachia—its land, wildlife, and people—where it is devastated by coal mining. But more important, we must learn not to repeat our past mistakes. We also need to learn how to supply our needs without further damaging the resources. Many people believe in reaping everything possible for the present, hoping that competition will force optimal utilization of resources and that the cost factor of supply and demand will eventually force conservation. When people realize their own enjoyment, environment, or pocketbooks are threatened, they demand action to protect them. Unfortunately, many of us do not see the end results of resource use and thus rationalize them as unimportant, irrelevant to human life, or so negligible that they do not warrant consideration. To help us with future decisions, we must consider why we approach life with this attitude.

FURTHER READINGS

Chandler, W. J. (ed.) 1990. *Audubon Wildlife Report, 1989/1990.* San Diego: Academic Press (also published in 1985, 1986, 1987, and 1988/1989).

Clawson, M. 1983. Reassessing Public Lands Policy. *Environment* 25: 6, 8–17.

Dasmann, R. F. 1984. *Environmental Conservation,* fifth edition. New York: John Wiley & Sons.

Dregne, H. 1985. Aridity and Land Degradation. *Environment* 27: 16–20, 28–33.

Dwyer, J. F, F. J. Deneke, G. W. Gray, and G. H. Moeller. 1983. Urban Forests, Where Trees and People Go Together. pp. 498–507. In *Using Our Natural Resources, 1983 Yearbook of Agriculture.* Washington, DC: U.S. Department of Agriculture.

Gup, Ted. 1990. Owl vs. Man. *Time Magazine* 135 (26): 56–63.

Hall, F. C. 1983. Forest Lands in the Continental United States. pp. 130–39, In *Using Our Natural Resources, 1983 Yearbook of Agriculture.* Washington, DC: U.S. Department of Agriculture.

Harris, L. D. 1984. *The Fragmented Forest: Island Biogeography Theory and the Preservation of Biotic Diversity.* Chicago: University of Chicago Press.

Hendee, J. C., G. H. Stankey, and R. C. Lucas. 1990. *Wilderness Management*, second edition. Golden, CO: Fulcrum Publishing.

Holechek, J. L., R. D. Pieper, and C. H. Herbel. 1989. *Range Management Principles and Practices.* Englewood Cliffs, NJ: Prentice Hall.

Maser, C., and J. M. Trappe (eds.). 1984. *The Seen and Unseen World of the Fallen Tree.* USDA Forest Service, General Technical Report PNW-164.

Postel, S., and J. C. Ryan. 1991. Reforming Forestry. Chapter 5. In L. R. Brown, *State of the World 1991.* New York: W. W. Norton and Company.

Young, R. A., and R. L. Giese (eds.). 1990. *Introduction to Forest Science*, second edition. New York: John Wiley & Sons.

Yuskavitch, J. 1985. Old Growth: A Question of Values. *American Forests* 91: 22–25, 46–49.

STUDY QUESTIONS

1. "Forests are renewable." What does this mean?
2. What ecological problems might result from two or three cycles of forest removal and regrowth in a particular area?
3. Can fire act as a tool of conservation? Explain.
4. What ecological factors determine whether an area can support forest or grassland?
5. How can foresters decide on the best forest management plan?
6. Why does the manager need to evaluate the impact of wildlife when considering the number of cattle a particular plot of grassland can support?
7. Why would the number of animals that rangeland can support vary from year to year?
8. List some advantages and disadvantages of maintaining publicly owned land.
9. Give an example of how forest ecological succession can be managed to provide for a particular human need.
10. Outline a rangeland management plan that includes a grazing system and a way to control cattle distribution.

SUGGESTED ACTIONS

1. Recycling one ton of paper saves seventeen trees. It also saves energy, water, and landfill space. Reuse and recycle paper whenever you can, and encourage your local newspaper to use recycled newsprint (adapted from National Wildlife Federation's Conservation 90 publication).

2. Take reusable cloth shopping bags with you to grocery stores instead of using paper or plastic bags.
3. Reduce paper consumption by limiting the amount of "junk mail" you receive. You can have your name taken off most mailing lists by writing to the Direct Marketing Association, 6 East 43rd Street, New York, NY 10017 (adapted from National Wildlife Federation's Conservation 1990 publication).
4. Practice low-impact camping and hiking. Find out how to minimize environmental damage as you enjoy outdoor recreation activities, especially in wilderness and other wildland areas, and encourage others to join you in your efforts.
5. Watch for opportunities to comment on national forest plans, wilderness proposals, or environmental impact statements. Read a section of the report that interests you and submit written comments to the appropriate agency.
6. Develop a position paper that explains how you feel about a complex land management issue such as a wilderness proposal, below-cost timber sales, or the dilemma concerning an appropriate grazing fee to charge for public land grazing. Try to use sources of information (i.e., articles and interviews) that will give you a broad range of perspectives about the issue. Interest others in the issue by discussing your position paper with them, or by submitting it to your local newspaper as a guest editorial or a letter to the editor.

10

WILDLIFE AND FISHERIES

Wildlife and fishery management has evolved into an important means of conserving our natural resources. Early settlers in the New World found an abundance of wildlife for their taking. Most settlers removed enough to satisfy their food and shelter needs; others began a program of exploitation. Commercial use of wildlife expanded without controls. As settlers became more affluent, sports hunting and fishing became more common. Efforts to control wildlife take were initiated as people, particularly hunters and fishermen, noticed a decline and sometimes loss of species. Wildlife management became the focus of many private conservation groups, which spurred interest on the part of the federal and state governments to develop sound management practices. From the beginning, wildlife managers found their profession tied closely to the political arena.

WILDLIFE MANAGEMENT

Wildlife management can be defined as "the science and art" of changing the characteristics and interactions of habitat, wild animal populations, and people to achieve specific human goals by means of managing wildlife resources. Thus, wildlife management is the manipulation of populations or habitat to achieve a desired goal. The goal may be one or more of the following:

1. To increase population
2. To remove individuals from a population on a continuing basis yet leave enough individuals to reproduce to replace those that are removed
3. To stabilize or reduce population

In one form or another, everything done in wildlife management is done for people. Because hunters demand game species, conservationists are concerned about endangered species, and bird watchers want a diversity of bird species, managers must decide which species or groups of species are to be managed. They then work to provide the habitat so those species can prosper or be controlled. Propagating the habitat often requires extensive knowledge of ecological principles (see Chapter 2), population dynamics, animal behavior, and a good sense of the public's desires.

Management Approaches

A manager can use one of three basic approaches to satisfy hunting, conservation, and other public desires and simultaneously preserve wildlife habitat for future generations. One possibility involves allowing the area to proceed in natural succession. In other words, let nature take its course, creating change in wildlife populations as succession occurs. Often proposed by preservationists, for the most part this type of management is not feasible because most areas are influenced to some degree by the human population.

A second form of management is aimed at achieving and maintaining the diversity of wildlife in a community through a variety of management techniques. In managing for species diversity, the objectives are to ensure that the largest possible number of wildlife species are maintained in a particular area. Generally, habitats such as a forest, grassland, or riparian community (community near water) are managed with the idea of keeping a diversity of species. In this form of management, all species are important. The general approach is to provide a broad array of habitat conditions including different seres, types of plants, and plant communities. In addition, it is important to know the total space requirements of wildlife communities. This approach can be used on land when wildlife management is a secondary objective. For example, foresters can maintain an area for timber production as well as wildlife. This can be achieved by allowing a few snags to stand and clear-cutting in such a way that large tracts of land remain uncut instead of leaving long strips of timber. Special hunts can be prescribed where wildlife is expanding beyond the carrying capacity of the land. Managers must gather information on the habitat needs of wildlife species and communities in order to formulate a comprehensive management plan, in some cases, covering hundreds of years.

A third approach is featured or single species management, which is commonly used for endangered or hunted species. A thorough knowledge of the managed species, habitat requirements, and relationships with other species is needed to be successful. Featured species management can be undertaken on land being used for other primary objectives.

For centuries most wildlife management has emphasized single-species management. Hunters and fishermen have prevailed on the powers of government to protect their source of food and sport. Conservationists seized the single-species management concept to enact endangered species legislation to protect declining species, often through habitat preservation. Today conservationists are recognizing that no species, including humans, can be viewed as a single entity outside the natural system. They see the need for protecting the diversity of life throughout our planet. In this chapter, we examine some of the traditional wildlife techniques. In Chapter 11, the questions and problems of managing for diversity are examined.

Management Techniques

Whether the objectives are to manage for a single species or for a community of wildlife, management entails some modification of the plant community, alteration of the successional sequence, or population manipulation. Since wildlife responds to community structure, managers must understand the relationship of wildlife communities and species to habitat structure. Some forms of habitat management have already been discussed in regard to forestry management practices; other examples follow.

Fire Fire largely determined the composition and structure of presettlement vegetation in such areas as Michigan's Upper Peninsula, a good portion of northern Wisconsin, Minnesota, and large parts of the central and western United States. These fires burned in irregular patterns through an area, opening up understory (vegetation below the forest canopy) and creating additional edge habitat for many birds and mammals.

Species such as the endangered Kirtland's warbler are presently maintained by prescribed burning. The Kirtland's warbler in the upper portion of Michigan's Lower Peninsula requires a perpetual supply of jack pine 1.8 to 2.4 meters (6–8 feet) high.

Although foresters tend to seek a *hot burn* to eliminate as much dead debris as possible, wildlife managers use a *cool burn* for habitat manipulation. Cool burns destroy only some of the surface litter. They are used early in the season so that new grass can grow.

Vegetation Manipulation Because all forms of vegetation change affect wildlife populations, vegetation can be manipulated to accomplish wildlife management goals. Thus, applying herbicides, clearing for roads and transmission line rights-of-way, and even trampling down ground cover result in habitat modification. In some cases, desired species are associated with a certain stage in succession. Deer, for instance, prefer the earlier stages of succession; therefore, their population increases following brush clearing projects and removal of dense forests.

Selective logging, clearing understory, and removing small stands of timber all have a major impact on wildlife communities. Opening the forest canopy promotes herbaceous and shrubby vegetation, which draws many birds and small mammals that require this habitat. Generally, logging is most effective when some logs and debris are left as refuge for wildlife. Small mammals and birds attracted to these sites revegetate by transporting seeds and nutrients into the cleared area.

When vegetation is cleared, the amount of *edge*—the zone where two plant communities meet—often increases. The area along the side attracts edge species, thus providing a mixture of forest and edge species. Flycatchers, woodpeckers, and hawks use the more exposed branches, and deer, moose, and elk feed in edge areas. In New Mexico the dense canopies in pinyon-juniper woodland reduced the amount of midstory, browse, and understory herbage, which in turn reduced deer and elk use of the area. When small clearings were made in the woodland, populations increased; however, when extensive clearings were made, herbage increased but deer and elk did not (Figure 10.1).

The size and shape of clear-cuts and their position in relation to uncut timber are important to improvement of wildlife habitat. One study revealed that deer and elk in New Mexico made use of logged areas adjacent to uncut timber. Circular openings of approximately 8 hectares (19 acres) in spruce and subalpine fir and 18 hectares (43 acres) in ponderosa pine were most beneficial.

Thinning to improve tree growth can also be beneficial. Herbaceous vegetation generally decreases as forests age and canopies close. The production of forage is usually inversely related to the base area of the remaining trees. Increasing forage production through forest management practices encourages use of areas by deer and elk. In some parts of the eastern United States, thinning, together with other changes in land use practices, had led to major increases in white-tailed deer population.

A study of bird populations in an eastern deciduous forest showed that clearing for transmission line corridors reduced the nesting area for migratory bird species but increased the nesting potential for resident species. The additional edge and open area also brought in quail and woodcock as well as predators such as hawks, which could hunt along the corridors.

Wildlife biologists are finding that habitat size is an important environmental factor. For example, when eastern deciduous forests are subdivided below 810 hectares (2000 acres), the population of nesting migratory birds, particularly neotropical migrants, begins to decrease.

Habitat improvement techniques can be used to attract wildlife. In a number of cases, particular plant species are associated with certain an-

FIRE: A BENEFIT TO WILDLIFE

Fire has often been thought of as a destructive force for wildlife populations. An ideal opportunity to evaluate its effects on wildlife occurred in 1976. On July 30, lightning ignited a fire on the Seney National Wildlife Refuge on the Upper Peninsula of Michigan. By the time the fire had been contained on September 21, it had burned through close to 28,329 hectares (70,000 acres) of wildlife habitat.

Wildlife biologists initiated a 3-year study to evaluate the fire's effects on wildlife and wildlife habitat, which included censusing mammals, birds, fish, amphibians, and reptiles in burned and unburned study sites on the refuge. Vegetation changes were monitored on each site and correlated with wildlife changes each year. Soil and water analyses were conducted to determine the fire's effects on soil properties and water chemistry that might influence wildlife habitat.

Vegetation structure was the wildlife habitat component most affected by the fire. Structural changes occurred throughout the refuge. Trees were killed because the root systems were destroyed; some of these were left standing as snags, but most toppled over. Throughout the refuge, the fire burned in very erratic patterns, destroying most of the vegetation in some areas but only burning around the bases of trees in others. Quaking aspen and jack pine seedlings as well as blueberries and bracken ferns were common in the burned forest stands in the spring following the fire. Grasses became firmly established in open areas.

Analysis of the bird and mammal populations showed that no species were eliminated. Because a greater edge habitat was opened up, additional species of animals (such as the black-backed three-toed woodpecker, formerly uncommon in the area) began to inhabit the area. Prime waterfowl nesting habitat near some of the dikes was destroyed, reducing the population of nesting waterfowl in the years immediately following the fire. This was balanced to some degree by newly created grassy edges of burned bog, which provided new nesting habitat for ring-necked ducks, mallards, black ducks, and green- and blue-winged teals.

Larger mammal populations were affected in several ways. The white-tailed deer, a common animal on the Upper Peninsula of Michigan, was not benefited because good winter habitat was lacking. However, the dense crop of berries that followed the fire provided ideal feeding areas for black bear. Thus, their populations began to increase. Alder thickets that grew in the third year following the fire were used extensively by beaver.

Fire is apparently an important occurrence in natural ecosystems in the Upper Peninsula. As human intrusion retards fire, we find that succession proceeds to a later stage. Fire, which plays an important role in maintaining diversity of wildlife and wildlife habitat, needs to be used as a management tool.

For further information, see S. H. Anderson, *The Effect of the Seney Wildfire on Wildlife and Wildlife Habitat.* U.S. Fish and Wildlife Service Publication RP 146, 1982.

imals. The sage grouse, for instance, is found only around sage that has nearby edge. Planting grass improves grassland communities while the construction of nest boxes attracts cavity nesters. Through management, nesting cover can be maintained along the edge of fields or in hedge rows. Brush piles are attractive to some species while natural and artificial roosts can be constructed for others. Recent studies show that eagles, ospreys, and some colonial nesting birds occasionally use platforms to nest. In Wyoming's Powder River Basin, golden eagles' nests hampered strip-mining activities in the area. However, the nests could not be destroyed because golden eagles are protected by federal law. By evaluating the type of terrain, exposure, and vegetation around successful nests, however, biologists erected nest platforms and got the birds to nest successfully in new sites (Figure 10.2).

Figure 10.1
An elk.

Figure 10.2
Young golden eagle in nest on artificial platform.

Most wetland habitats, including standing water, vegetation along the edge of streams, rivers, ponds, and lakes, as well as communities along the ocean coasts and bays, have a very high diversity of wildlife. These communities are also very desirable for human use. Thus, our wetlands are declining rapidly due to development, agriculture, and recreation. Since the Declaration of Independence, the lower forty-eight states have lost an estimated 50 percent of their original wetlands. Twenty-two states have lost 50 percent or more. California has lost 91 percent of its wetlands. These tremendous losses have a major impact on the wetland wildlife and adjacent communities.

As people deposit waste material and destroy the habitat, many wildlife species can no longer live in these areas. Efforts are underway to preserve communities that are important to wildlife. Preservation appears to be an effective technique to maintain wetlands for wildlife use. Habitat manipulation must be undertaken only after careful evaluation of the impact on the wildlife species.

Riparian communities (communities along lakes, streams, and marshes) are also important wildlife habitats, particularly in arid regions of the West. Riparian communities are attractive to people, who often construct businesses or homes in these areas. Ranchers allow cattle to graze in riparian communities because the vegetation is often dense and water is available. Controlled grazing does not always harm the communities, but excessive grazing removes enough vegetation to cause severe siltation of the water and soil erosion, destroying wildlife and fish habitat.

Snags, which are dead trees still standing, are traditionally felled for either aesthetic or safety reasons. Many wildlife species utilize snags for nesting and foraging sites. In fact, we know of some eighty-five species of North American birds that use them for nesting. *Primary cavity nesters* such as chickadees, nuthatches, and woodpeckers excavate holes in snags. *Secondary cavity nesters* such as bluebirds, saw-whet owls, and tree swallows as well as mammals such as flying squirrels, bushy-tailed wood rats, and some bats use the abandoned holes.

Seeding In areas where grass and shrubs are common, seeding native or introduced grasses or other plants can result in early greening following winter. Species that supply high-quality plant food help animals recover faster in the spring following stressful winters. In many areas seeded species can supply significant amounts of green vegetation for wildlife food. Sometimes big-game winter areas are seeded when native grass species are dormant or unavailable. Wildlife, particularly big-game species, can benefit from range seeding when livestock and other cattle are present.

Water Impoundments Most water impoundments created specially for wildlife are found on wildlife refuges. These ponds and lakes supply food and habitat for waterfowl and other animals. Water impoundments made for livestock and recreational activities also provide wildlife habitat: marsh animals live around the edges; osprey and eagles hunt over large impoundments of water; raccoon, deer, and snakes find food along the shore; and amphibians lay eggs and pass through their larval stages in the water.

Water impoundments often require specialized features to make the area more attractive for wildlife. In Kansas, an impoundment for waterfowl (Lovewell Project) is surrounded by 21 kilometers (13 miles) of fencing to prevent cattle from damaging plants and harming nesting birds. At the Canyon Ferry Reservoir in Montana, islands made from dredge materials resulted in a threefold increase in the Canada geese population over a 3-year period.

Figure 10.3
Concentration of tundra swan attracted to corn stubble. (Courtesy of R. Munro, U.S. Fish and Wildlife Service.)

Hunting Hunting is a form of recreation which serves as a major stimulus for wildlife management. When habitats are manipulated to maintain an abundant supply of game animals, hunting becomes a means of raising revenues through the sale of licenses and permits.

Some populations, such as the North American mallard, are reduced more than 25 percent annually by hunters. Game biologists are constantly trying to increase the mallard population by creating marshes for nesting and water impoundments for feeding, controlling predators, and reducing diseases that destroy birds which hunters otherwise could take. Other forms of habitat manipulation are used for different games species. Corn fields attract swans and geese (Figure 10.3), and shrub cover provides habitat for pheasants. Unfortunately, not all birds shot by hunters are killed. Biologists estimate that 15 percent of living ducks have shot in their tissue (Figure 10.4).

When hunting is permitted on natural populations—that is, those that, are not maintained specially for hunters—it becomes a means of keeping the population within the carrying capacity of the habitat. This is a form of population management. Managers must be keenly aware of this limit to prevent damage from an excessive number of animals. Deer populations, for example, can more than double every second year. At maximum productivity, one doe can produce fifteen or more fawns in an average life span of 8 years. If all her young and their young survive to the same age and breed successfully,

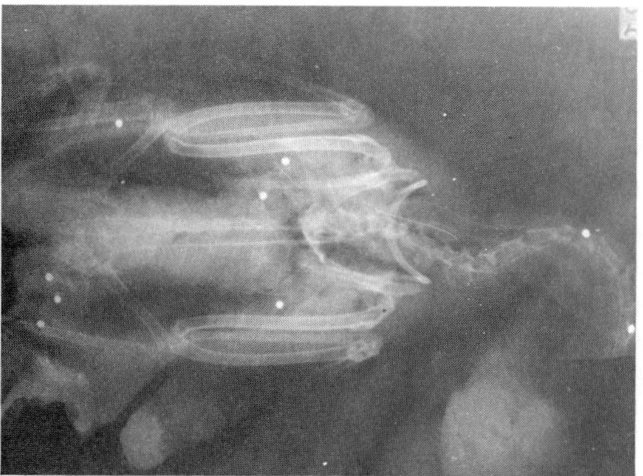

Figure 10.4
X ray of a living canvasback duck showing shot imbedded in muscle tissue. (Courtesy of M. Perry, U.S. Fish and Wildlife Service.)

there would be more than 150 animals before the mother's death. This is true of many other animals such as rabbits, squirrels, quail, and ducks. A particular plot of land cannot support this type of population growth. Hunting, then, substitutes for natural mortality factors to keep the populations within the habitat's carrying capacity. If the population is reduced to such a level that it can no longer reproduce and maintain itself, then the harvest pressure is too great.

Animal rights groups have focused on hunting in many parts of the United States (see Chapter 3). It is quite clear that battle lines are being drawn. Unfortunately, members of both groups have ignored some of the concerns of management. Hunters must be environmentally conscious, taking only animals that they can use and not leaving others wounded to die. Antihunters must examine the concept of carrying capacity and the biology of the animals. After changing our landscape much of our hunted population is without natural predators.

Evolution of Management in the United States

The first settlers exploring the undisturbed habitat in the United States reported an unlimited supply of wild animals. Bison grazed the plains and could be found eastward into New York State in such numbers that Buffalo, New York, was named for them (Figure 10.5). Trains moving west stopped so passengers could shoot buffalo, only to leave them where they died. Grizzly bears and wolves were common throughout much of the West and in some parts of the north central United States. Meriwether Lewis and William Clark reported California condors in the Willamette Valley, Oregon, where the proper habitat was available.

As people removed forests, planted crops, built roads, and hunted, habitats were disrupted. Grizzly bears, wolves, martins, and fishers decreased in numbers while coyotes, quail, rabbits,

(a)

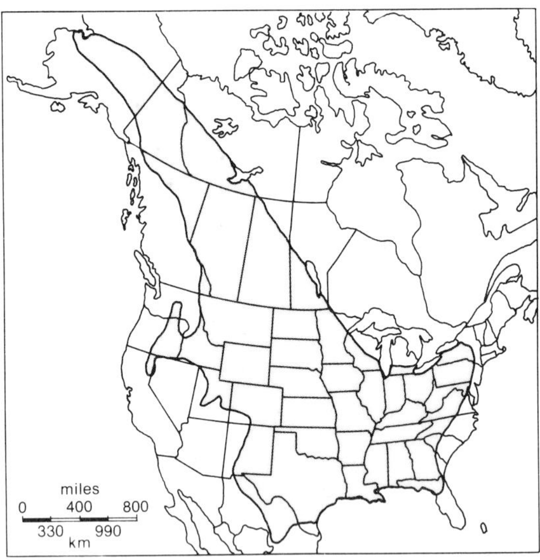

(b)

Figure 10.5
(a) Bison. (U.S. Fish and Wildlife Service. Photo by Wesley D. Parker.) (b) Original distribution of bison in North America. (From E. R. Hall and K. R. Kelson, *The Mammals of North America.* Copyright © 1959 by Ronald Press. Reprinted by permission of John Wiley & Sons, Inc.)

and deer became abundant. Robins increased with lawns, barn swallows with building eaves, and cliff swallows with bridges and other structures.

Through the years, people deliberately changed some populations. Ground squirrels competed with cattle by eating grass, so eradication programs were undertaken. Wholesale slaughter of passenger pigeons occurred so they could be sold as delicacies to restaurants for two cents apiece. This slaughter, coupled with habitat destruction through logging practices, caused the birds' extinction.

State and Federal Roles The responsibility of administering wildlife as a trust of the people was assumed by the states. California and New Hampshire were the first states to establish fish and game commissions in 1878 for the purpose of conserving their states' wildlife. In 1896, the Supreme Court upheld the state as the public owner of wildlife. States are thus responsible for managing wildlife within their borders. They set hunting and fishing limits. An exception are migratory birds, including all our waterfowl. Because the federal government is responsible for migratory bird management through a series of treaties with foreign countries, states and the federal government jointly set migratory bird seasons within the guidelines of the treaties. These seasons are set by flyway councils. There are four in the United States corresponding to the four flyways: Pacific, Central, Mississippi, and Atlantic (Figure 10.6). These flyways are the major migration corridors of migratory birds in the spring and fall.

The federal government also manages endangered species through legislature acts. Much of the work is done jointly with the states. Several specific laws give federal government responsibility for managing special groups of wildlife. Examples are the Bald Eagle Protection Act (1940), Marine Mammal Protection Act (1972), and the Fisheries Resource Act (1976).

Agencies of the federal government affect wildlife through habitat management, specifically through wildlife conservation on publicly owned land such as national parks and forests. Agencies designated to administer large areas of public land are the U.S. Forest Service, National Park Service, U.S. Fish and Wildlife Service, Bureau of Land Management, Department of Energy, and Department of Defense.

Figure 10.6
Administrative waterfowl flyways in North America.

Public Versus Private Land Management There is a great deal of difference in managing public and private lands. Public land management agencies are usually established with specific objectives. The U.S. National Park Service was established to preserve unique areas of the country for their wildlife, geological, historical, and vegetation values. The U.S. Forest Service is responsible for managing forests for multiple use. Wildlife is one component of multiple use, but generally receives lower priority than timber production. The U.S. Bureau of Land Management manages grassland, primarily for grazing, but multiple use has become an important concept of their management practices. The U.S. Fish and Wildlife Service manages national wildlife refuges often for specific animals or groups of animals. Many refuges were bought primarily to allow waterfowl populations to nest, winter, or stop over during migrations. States have also set aside land for wildlife; some state wildlife lands are managed for sports use.

The first federal wildlife refuge was established in 1904 on Pelican Island, Florida, where 1 hectare (2.5 acres) was reserved for brown pelicans and egrets. Egrets, which nested near the pelicans, were the targets of plume hunters. In the first half of the century, waterfowl refuges were established to aid propagation and provide rest areas along migratory routes. Some big-game refuges, such as the National Bison Range in Montana, were established in the western U.S. during this time.

Many of the initial refuges were created by designating public land specifically for wildlife management, so as to protect threatened habitats, migratory birds, marine mammals, and resident species. Currently there are more than 375 refuges (Figure 10.7).

Although most refuges have been created by withdrawing land from public domain, some occupy land purchased by citizens' groups. For example, members of the Boone and Crockett Club raised funds to buy private land in northwest Nevada that became the heart of what is now the Charles Sheldon National Antelope Refuge. In Jackson Hole, Wyoming, local concern for the elk herd that wintered in the Teton Valley prompted people to purchase land for the elk. This effort, spearheaded by the Issac Walton League, resulted in creation of the National Elk Refuge, under the auspices of the U.S. Fish and Wildlife Service. It provides winter habitat for more than 7000 elk. The Nature Conservancy is very active in purchasing land for wildlife. Frequently their properties are later purchased by the federal government.

Generally, public lands are maintained by both federal and state laws. Managers operate within the confines of these laws to manage the land for the prescribed purpose. Some of the laws relate to multiple use, so that wildlife must be considered along with grazing, timber production, recreation, and other uses. Producing products such as timber on public lands has sometimes led to conflicts in managing wildlife. The goals of public land management change with economic conditions, public demand, legislation, land capacity, and politics.

Individual owners have a great deal of say in private land management; for example, the number of cattle grazed or trees removed are under the owner's control. Efforts must be made to educate private landowners in maintaining wildlife species that can be aesthetically pleasing and beneficial. This is sometimes difficult, especially since some cattle ranchers think certain species of wildlife (such as prairie dogs) are better off dead. In fact, poisoning prairie dogs can be detrimental to burrowing owls, black-footed ferrets (Figure 10.8), and other desirable species. As we will discuss, too many predators (such as coyotes) can cause economic hardships by killing livestock, so that sometimes a control program is necessary.

Thus, the manager's role changes when dealing with private landowners. Only through effective education and personal contact can the manager persuade private landowners to cooperate in developing and implementing management plans.

When corporations own land for timber harvest and other uses, wildlife managers must work with the companies and with their private biologists to help develop the best possible wildlife habitat for the animals that live on private properties, while still maintaining corporate objectives. For example, timber management practices can be manipulated to enhance game habitat without seriously hindering economic objectives.

Figure 10.7
Wildlife refuges in the United States.

Figure 10.8
Black-footed ferrets.

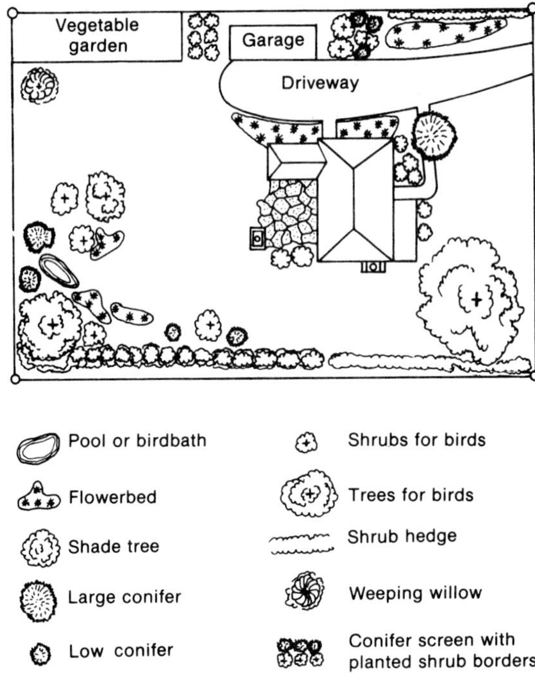

Figure 10.9
Garden planning to attract birds around the home.

Planning

People are confronted by wildlife management decisions daily, but effective management requires a plan based on desired goals and specific methods. For example, many homeowners want birds around their homes (Figure 10.9), and small landowners may want to attract wildlife to their property for their enjoyment. Those working to reclaim strip mines dictate what wildlife can live in the area by the reclamation procedures they choose.

Land management differs greatly between the eastern and western United States. In the East, much of the land is privately owned and therefore managed by individuals. The West contains vast areas of public land governed through legislation and pressure groups. Alaska is similar to the West, with politicians now making major decisions at the state and national levels. We must remember that what we do now determines options for the future.

Planning is, basically, establishing objectives. *Objectives* might be to maintain diverse wildlife species in a particular area, or to maintain hunting. Objectives can be specific, such as to produce trophy animals, or more general—to produce a large number of animals. After objectives are established, *strategies* are developed to meet the objectives. Thus, if an objective is to enhance waterfowl population, the strategy might be to develop more waterfowl impoundments and produce better habitat (Chapter 1). If the objective is to increase the big game species, hunting might be reduced or habitat improved to maintain that species. Objectives regarding most endangered species are to increase the number of individuals; to do so, it is necessary to find out why the species have declined and reverse that action. For example, in the United States pesticides were discovered to have an impact on certain bird and fish species. By reducing the number of pesticides in the environment, some populations of species that were previously declining have now begun to increase.

All planning requires public and private landowners to work together. Public input must be a component of the planning process. Planning also requires an understanding of wildlife legislation. The states and the federal government look to the Constitution in establishing wildlife law. The states generally have authority over wildlife that reside within their boundaries. States enforce hunting regulations, but must abide by all federal legislation that applies to wildlife.

Funding Licensing the hunting public, usually a result of wildlife conservation programs, proved to be a primary source of funding until about 1960. By congressional act in 1934, the federal government had also authorized the sale of migratory bird hunting stamps to raise money for the acquisition of refuge land. In addition, excise taxes on guns, ammunition, and fishing tackle were levied to support wildlife conservation efforts.

Today, conservation has spread beyond hunting circles. In 1985, approximately 135 million people spent 1.6 billion days watching wildlife and 29 million people spent 175 million days photographing wildlife. Nonconsumptive users spent $14 billion on wildlife. In contrast, 17 million Americans spent 334 million days hunting wildlife. Their total expenditures were $10 billion. Legislatures have now recognized the need for conserving all wildlife and have designated monies from general revenues to maintain our wildlife resources. States and the federal government are responding to nonconsumptive user groups.

Legal Restrictions A number of legislative acts govern the operation of wildlife management. Under the discussion of endangered species, we will talk about the Endangered Species Act. Another example of controversial legislation is the Wild and Free-Roaming Horse and Burro Act of 1971, which states that the management of burros and wild horses is an integral part of public rangeland policy. According to this law, wild horses cannot be killed. If they are creating a problem for ranchers, the horses must be moved to new locations. Under the federal Adopt-a-Horse Program, private individuals, for a fee, can receive title to a wild horse after a 1-year stewardship.

Since the federal policy provides limited funds for relocation, a controversy has arisen. Ranchers claim the horses and burros are eating plants that livestock could use, thus limiting the number of cattle that can be grazed on the land. As no immediate solution appears in sight, the conflict between legislative moderates and ranching interests continues.

Some legislation affects wildlife indirectly. The National Wild and Scenic River Act, while not directed at wildlife management, has the effect of preserving wildlife habitat.

SPECIES MANAGEMENT

Big Game

In terms of wildlife management, one is likely to think first of the big-game species. These include a variety of animals from bear and mountain lion through ungulates such as elk, deer, and moose, as well as goats, buffalo, and pigs (Figure 10.10). Every effort has been made to maintain adequate summer and winter ranges for these species. Based on some form of census techniques, managers determine the number of individual animals that can be harvested each year, then hunting licenses are issued for hunt areas. In some areas hunting emphasizes trophy hunts, others, just removal of wildlife. Today, many people are concerned about maintaining big game simply for aesthetic purposes. People visit

Figure 10.10
Moose feeding on aquatic plants at lake's edge.

national parks and other areas to see and photograph these animals. Thus, management of big game has taken on a slightly different perspective; a manager must maintain animals not only for the hunter, but for the public as well.

Waterfowl

Habitats are managed throughout the United States for waterfowl production and harvest. Wetlands must be protected for breeding, migration, and wintering. Habitat loss through a good portion of the United States has been responsible for extensive waterfowl decline. Predation, disease, and toxic materials have also reduced waterfowl populations. Populations of waterfowl declined 25 percent from 1981 to 1990. Extensive methods have been initiated to increase the waterfowl populations, including reducing the number of birds that hunters can take and changing the hunting season. Generally, waterfowl management is controlled by the federal government because of international treaties. Local hunts are set in collaboration with the states through four flyway councils. As with big game, nonconsumptive use includes better viewing of waterfowl species.

Nongame Wildlife

Until the public became ecologically aware during the 1960s and early 1970s, most funding dictated that, except for endangered species, wildlife management be directed at consumptive animals. A change in public attitude now dictates that more effort be turned toward management of nongame. Nongame is actually an administrative unit for a subset of wildlife species that are not hunted, harvested, or intentionally removed by humans. Obviously nongame can also include game species that must be managed as nongame, such as those in the national parks. Federal government surveys indicate that people in the United States are very interested in nongame species. Results of a national survey showed that during the late 1980s, 23 percent of the people fished, 9 percent hunted, and 35 percent spent some time bird watching. Approximately 42 percent of the population spent time observing and photographing wildlife. Clearly the public is becoming more aware of our wildlife resources and spending some time and money on wildlife activities. As a result, managers are expected to answer more questions about and spend more time on nongame.

It is difficult to manage a nongame species; therefore, most management is directed at achieving and maintaining the diversity of wildlife in a community. Each community's combination of wildlife species results from the region's abiotic and biotic characteristics. The animals form an intertwining network that is altered when people encroach on the system. Managing nongame means maintaining a community so as not to disrupt its wildlife. Most approaches to managing nongame thus involve maintaining a reasonable habitat size to promote a diversity of animals. This may involve burning programs to recreate different successional stages. Protecting the nests of large birds-of-prey as well as providing walkways in swamps are management approaches that allow the public to view wildlife without harming the habitat. The general level of funding for nongame is considerably lower than that for all the game species.

Endangered Species

Why do species become extinct? As we discussed in Chapter 3, species extinction is a natural phenomenon, but human encroachment often greatly accelerates the process. Extinction occurs when species fail to replace their number of the population. Extinction can be grouped into several categories for wildlife:

- A reduction in the population is created by habitat loss, so that nest sites or cover are not available for the population to reproduce. Each species generally has a minimum critical area in which it can survive; when this is no longer available, species needs cannot be accommodated. The possible response by a species: adaptation (possibly through natural selection), migration, or extinction.
- Ecological stability of a species can be disrupted. Disruption can include changes in the food web or loss of food sources. This often occurs when new species are introduced into an area, causing an ecological imbalance that may destroy existing populations. As land bridges form between habitats, predator/prey cycles and competition may change. The coyote, for example, now interbreeds with red wolf, which changes the genetic makeup of the wolf.

When populations decrease, their members tend to inbreed and become less fit (Chapter 3). Biologists have calculated how small a population can get and still survive, but the figures must be interpreted with care. Behavioral factors and habitat factors such as isolation may prevent mates from meeting. Sex ratio can be a factor; for example, currently all surviving dusky seaside sparrows are males. When a population is declining, efforts must be made to reverse that trend.

Stress is also a problem for some species. As animals are moved out of their natural habitat because of habital change or people encroaching on the area, they may be unable to survive or reproduce. Sometimes the signs are obvious; other times they are not. The manatee, found in Florida, is a large, seal-shaped, gray or brown aquatic mammal with a flat, spatulate tail (Figure 10.11). The forelimbs are paddle-shaped and there are no hind limbs. The animals survive on vegetation in bays and slow-moving streams. They can reach 2.03 to 6.08 kilograms (4.5 to 13.5 pounds). People often scuba dive out to see the animals. They can move up and "pet" them.

Figure 10.11
Manatee. (Courtesy of the U.S. Fish and Wildlife Service.)

Boat propellers are a major cause of mortality for these large animals, which move sluggishly near the water's surface. It is important to know how animals like the manatee respond to increased pressure from humans, in order to maintain these animals.

The Endangered Species Act The Endangered Species Act, a series of legislative actions that began in 1966, designates species that have decreased in number and initiates plans to allow their recovery. The act calls for cooperation among states and federal agencies, particularly the U.S. Fish and Wildlife Service, to develop recovery plans for restoring the species to nonendangered status. The public as well as an agency can nominate species to the endangered status.

A species can be listed as **endangered** (threatened with extinction) throughout all of a significant part of its range, or **threatened** (likely to become endangered in the foreseeable future). The recovery plan is set up to prescribe possible methods of increasing the numbers of individuals such as the California condor, grizzly bear, and the Kirtland's warbler. Once a viable population level is achieved, the species is removed from the list. For example, due to excessive hunting and poaching, the American alligator was first classified as endangered in 1967. Subsequent protection allowed the alligator to recover in many parts of its range and thus be removed from the endangered classification. Unfortunately, the California condor and other species' populations have been reduced to the size where extinction appears likely.

Most endangered species recovery plans involve habitat manipulation. For example, water level control in Florida is crucial to the survival of the apple snail, the principal diet of the endangered Florida Everglade kite. In some ref-

MARKERS REDUCE SANDHILL CRANE COLLISIONS WITH POWER LINES

South central Nebraska, around the Prairie Bend region of the Platte River, provides a staging ground for some half a million sandhill cranes each spring. Cranes arrive from the southern United States in early March and depart for northern nesting areas by mid-March. During their stay in Nebraska the cranes roost nightly on shallow sandbars in the Platte River to protect themselves from predators. At dawn, large flocks fly to the surrounding fields to feed on grain stubble left from the previous fall's harvest. The flocks return to the roosting sites at dusk.

The whole area has transmission lines of all sizes and heights, going in many directions. Numerous crane collisions occur at some of the transmission lines.

In the late 1980s, a study was conducted by the University of Wyoming to locate areas where there was high crane mortality from collisions with lines. At nine such locations, yellow, fiberglass aviation marking balls 30 centimeters in diameter (1 foot) with black stripes were installed at approximately 100-meter (109-yard) intervals on alternate spans. A span is the section of wire between two support structures.

After the installation, observers made daily observations of cranes found dead under the spans. A significantly larger number of crane carcasses were found under unmarked spans. Observation of flocks of cranes flying indicated that nearly half of the flocks reacted to the presence of the balls by increasing their altitude when approaching the marker. Factors such as visibility and wind contributed to the reaction of cranes. The study showed that in areas of high bird mortality from collisions with power lines, markers can be effective.

uges, dams and levees are required to hold water at a suitable level for the snail to survive.

A number of other approaches are used to increase endangered populations. To prevent the pearly-eyed thrasher from taking over the nests of endangered Puerto Rican parrots, scientists make artificial nest structures for the parrot. The new nest has a deep cavity that keeps eggs and chicks out of the sight of predators and is protected from flooding by a rain shield. These nests are hung near thrasher-occupied parrot nests, allowing both species to live in the same area. Also, the restriction on the use of DDT and other pesticides has assisted in the recovery of the brown pelican and bald eagle populations.

By clearing streams of fish that were crowding out greenback cutthroat trout and then transferring artificially raised greenbacks to their old habitat, biologists have restored this species to Colorado streams. This procedure may have to be repeated if less desirable fish are reintroduced. An egg relocation technique has been used to aid recovery of the bald eagle population by moving eggs to incubation by nonproductive eagles in Maine, Virginia, and New York.

Because of the difficulty of breeding wild birds in captivity, the U.S. Fish and Wildlife Service has perfected artificial insemination techniques in order to breed endangered species such as whooping cranes and Aleutian Canada geese. The young are then released into the wild.

Preservation of the black-footed ferret, a weasel-like predator on prairie dogs in the plains states, is now underway. Because prairie dogs are destroyed by poisoning and shooting, the black-footed ferrets have very small areas of habitat remaining. Captive breeding of ferrets has been successful. Now as they are released into the wild, we will have a chance to see if these captive-reared animals can find food and shelter, avoid predators, and successfully reproduce to maintain their population.

If people are not successful in maintaining natural habitat so species can survive, zoological gardens may be the only places some animals can survive. Although today's zoos provide valuable research for preservation of some endangered species, natural habitat protection must be a key in maintaining our wildlife resource.

When reviewing endangered species programs, we must remember that changing a habitat is likely to change the species composition. Thus, if we wish to encourage the continuation of all species, we must recognize and attempt to maintain natural habitat conditions.

Several techniques can be used to maintain endangered species, but funds are necessary to identify the species and implement the processes. As we will see in Chapter 11, there are difficulties in identifying all endangered species. Meanwhile, government sources say funds are far from inadequate to protect endangered species. In 1990, the U.S. Department of Interior's inspector general issued a report that the U.S. Fish and Wildlife Service needs an additional $4.6 billion annually to recover all *known* endangered species. Their total agency budget, for all wildlife, is just under $1 billion. Conservation efforts are going to be very expensive in the future. At the same time, we cannot afford to ignore our natural communities.

Urban Wildlife

Since people living in urban areas appreciate outdoor surroundings, increased emphasis is being placed on urban wildlife management. Wildlife populations found in urban areas contain relatively few species because urbanization tends to have profound effects on species' environments. Due to changes in food and habitat conditions, the populations of nesting birds, mammals, and reptiles decrease dramatically. Therefore, habitat management that depends on maintaining a diversity of corridor-connected habitats, located along natural drainages and free of land development, is the basic tool used to attract urban wildlife.

Natural areas around many of our population centers are now being developed as wildlife refuges. This assists in maintaining wildlife populations and allowing people to appreciate the animals in their natural settings. Large urban wildlife programs in most states encourage people to landscape and plant to attract wildlife. Food shrubs can attract many birds to people's homes. Bird feeders are also popular ways to encourage people to appreciate our wildlife resources.

Urban wildlife management also involves special presentations to involve people in wildlife work. Such efforts can mean more pleasant parks and waterways in cities. Some animals, such as the endangered peregrine falcon, have been encouraged to nest on building ledges of large cities. These magnificent birds can create a lot of interest in wildlife.

People and their habits often create habitat for less desirable wildlife. Thus, rats, mice, and snakes will be attracted to garbage. Gulls can be a problem when dumps are placed too close to living quarters. Sometimes poor planning results in wildlife problems. When new developments are placed in snake migration routes, homeowners have difficulties. Lack of control of geese populations in parks results in their numbers expanding to golf courses and into swimming pools.

Urban wildlife management is a fine art. It requires the understanding of habitat that will attract wildlife. It involves planning and a great deal of public relations. Most of all, urban wildlife management is a form of integrating people and wildlife. Protected or secluded sites for wildlife are important. Some species require complete seclusion. Other species are more adaptable to people. All animals retain a behavior of the wild. People must understand and respect this behavior and not attempt to domesticate wildlife around their living quarters.

Predator Control

Wildlife species have always had some effect on human activities. In rare instances, wild animals (such as the grizzly bear in the United States or the tiger in India) attack and kill humans. More commonly, wildlife species cause damage to human commodities, such as agriculture, livestocks, gardens, homes, and buildings. Wildlife managers are called on to prevent or control damage, usually by reducing the population to a size that will not cause damage, or perhaps by altering animal behavior. Predator control has an emotional element, especially regarding sheep predation by coyotes, mountain lions, bear, bobcat, and other animals. Government animal control groups recommend methods of predator control.

When people move into an area and disrupt the natural system, new predatory behavior often sets in. The removal of wolves from large parts of North America eliminated natural control on the coyote, resulting in an increase of coyote populations. Coyotes then began preying on domestic sheep (Figure 10.12). Similarly, removal of large expanses of native prairie vegetation and replacement with monoculture disrupts the habitat in such a way that insects, rodents,

Figure 10.12
Coyote.

and some large ungulates begin to use farmers' crops for food. When acreages are converted into man-made ecosystems, wildlife species utilize the man-made as well as the natural system. Introduction of starlings and English sparrows resulted in extensive damage to crops and buildings because there are no natural predators.

Most forms of damage control programs have been directed at the coyote, including hunting, use of baits, and target devices to kill the animals. The extensive poisoning program has perhaps caused the most controversy. A toxic compound, 1080, appears to be effective in killing coyotes, but it requires sacrificing lambs. The 1080 poison is placed on the sacrificed animal, often in the form of a neck collar. Coyotes attack, ingest the poison, and die shortly after. Unfortunately, 1080 also causes a number of other animals to die.

Managing to control wildlife damage includes altering the habitat. Theoretically, if the habitat of animals that cause damage can be altered so that the species no longer finds the habitat satisfactory, the damage should cease. Rats, for example, use ground debris for cover, as well as garbage in alleyways and between buildings. Removal of garbage and debris eliminates the food source and renders the habitat less suitable.

Many forms of control involve some method of population management. Bounty hunting, in which a sum of money is paid to anyone who brings in an undesirable animal, has been used for many years but has not been highly successful in reducing populations. Successful harvest programs require knowledge of the biology of the animal—what time of day it is active, when it breeds, how many offspring it produces, and how the young disperse.

Harvest methods include trapping and removal. For example, one reason for the decline and endangerment of the Kirtland's warbler is that a high proportion of its nests are parasitized by cowbirds. The brown-headed cowbird lays one or two eggs in the nest of the warbler. The cowbird chicks generally hatch sooner than the warbler's, pushing the eggs or young warblers out of the nest. The cowbird is then raised by foster parents. In the early 1970s, when the warbler population had declined to fewer than 300 birds, the Fish and Wildlife Service and the Forest Service introduced a program to remove cowbirds. Cowbird decoy traps were erected in 1971, and in the period from 1975 to 1981, a total of 24,158 cowbirds were removed from Kirtland's warbler nesting areas. Between 1971 and 1981 the number of nests parasitized by the cowbirds dropped from close to 60 percent to less than 5 percent.

FISHERIES

More than 70 percent of the earth's surface is covered by salt water, and fresh water in lakes, streams, and marshes covers much land surface. Thus, widespread aquatic habitat destruction and overfishing in these waters affect many people throughout the world. Peru and Norway are economically dependent on ocean fish, and people living on the Asian coast or in the Pacific islands use fish as a dietary staple. Although commercial fisheries in the United States provide a living for a small segment of our population, they supply a food base for many people.

Fishery Economics

The economics of fisheries can be classified as commercial or recreational. In commercial fishery people fish for their livelihood, from the individual who owns one boat and hires a crew member to bring in a daily catch for sale to a local processor, to the large canning companies that own or lease many ships and hire captains and crews to obtain a huge supply of fish. Commercial fishing is based on supply and demand; if the demand for a fish product goes down, commercial fisheries and canners lay off workers and let their shipping fleet lie idle until the economy picks up. The individual who owns one ship may be more severely affected by economic recession or changes in demand. There have been recent shifts in the use of individual boats, however; along the East and West coasts, a number of former fishing boats are used for wildlife-watching excursions.

The so-called delicacy foods, such as lobster and crab, are apparently not affected by the supply-and-demand curve; the people who buy them are not concerned with price increases. As more people demand these delicacies, exploitation of the resources increases, and eventually the only way to supply the luxury demand may be through aquaculture.

Recreational fishing has increased in the last several years. In 1980, there were 250 million fishing days; each person who fished spent an average of $27 per day. The economics of recreational fishing includes more than expenditures on the sport itself. States with recreational fishing have greater tourism, with expenditures at hotels, motels, and campgrounds. Local businesses want their state game and fish department to provide adequate fish stocks to continue to attract the sport fishing enthusiast. This industry is not concerned about the total amount state or federal agencies spend as long as the funds keep local businesses flourishing. Thus, fisheries must be managed not only to maintain fish as a part of the natural ecosystem, but as part of people's livelihood.

Freshwater Fisheries

Managers in the United States work with natural and artificial habitats. Both require knowledge of such habitat requirements as spawning grounds, water temperature, movement and depth, and life history patterns.

Natural Habitats Managers place special emphasis on enhancing streams for sports fisheries, particularly for trout, smallmouth bass, and catfish. As habitat conditions improve, more fish survive and reproduce.

State laws generally protect fish in lakes, reservoirs, and ponds, unless the lake crosses state boundaries or in the case of the Great Lakes, which operate under the Great Lakes Fishery Commission. Lakes have been subjected to a great deal of human and chemical waste pollution. Reservoirs, which biologists treat as half a lake, contribute considerably to freshwater fisheries and require management similar to that of other types of fisheries resources. Generally, management aims to optimize recreational fishing.

Management techniques to rehabilitate or improve a waterway include manipulating a population by selectively removing undesirable species or stocking only a desirable species. Carp, suckers, gar, and gizzard shad populations can be reduced in small bodies of water by poisoning, netting, or trapping, but these fish are harder to control in large lakes. When undesirable fish become overabundant and difficult to thin, the only alternative is to destroy the entire fish population and restock with desirable fish.

A number of habitat management programs have been initiated to maintain fish populations in reservoirs and lakes. During construction of reservoirs, the bottoms are cleared of trees and other vegetation. If some trees are left around the sides, fish and other wildlife can find protection from predators when the area is inundated by water. Brush, tires, and concrete or other permanent structures placed in the bottom of reservoirs provide areas where fish can find food and shelter.

Control of water flow in reservoirs to prevent excessive fluctuation helps regulate some species. Using dams to regulate water flow each month prevents excessive flooding or drought conditions and consequent damage to the fisheries resource. Small impoundments in reservoirs upstream from dams help avert sudden changes in reservoir depths when hydroelectric facilities begin to operate.

Streams have a variety of habitats, from open, flowing water to small riffles. Streams or rivers are generally characterized by the speed at which water flows through them. Open, slowly flowing water can be similar to a lake. The different physical conditions of each of these waterways provide habitats for different fish.

In Libby Dam, Montana, eroding stream banks were seeded with deep rooted grasses to reduce stream velocity and retard erosion. Other devices used to stabilize banks include rocks, fences, logs, old automobiles, and wire mesh. Overhanging banks have been effective forms of fish habitat improvement, as trout prefer to hide under the protection of the overhangs (Figure 10.13).

Spawning grounds are of particular importance because they are vulnerable to pollution, siltation, and predation. Dams impound water, human waste material uses up oxygen, and siltation destroys the gravel beds that fish need for spawning. Open water is a suitable spawning environment for some species, but others use estuaries or marsh areas, which are highly susceptible to pollution. By placing small check dams constructed of concrete, rock, or metal pieces in some smaller streams, we can control the currents to improve the habitat for some species.

Spawning riffles occur over gravel beds in shallows of flowing streams and rivers. Trout and salmon deposit their eggs in depressions carved out of spawning beds. There small pieces of gravel protect eggs and newly hatched fry while water percolating through the gravel provides oxygen.

Artificial Habitats Some forms of artificial habitats are also used for fisheries. The construction of fish ponds is effective for sports fishermen in some parts of the United States, as are new lakes created by dams.

Fish pond management considers such factors as water temperature, oxygen content, vegetation control, and available food. Species best suited to the water temperature must be selected. Trout generally need ponds with a surface temperature that seldom or only briefly exceeds 18°C (64°F) on the hottest summer day. In warmer water, bass, bluegill, and catfish are possibilities. The food source, of course, begins with nutrients and phytoplankton, but competition is also an important consideration. The best fishing occurs where several species are present and predator–prey interactions develop. Hiding places should be provided, and debris and water

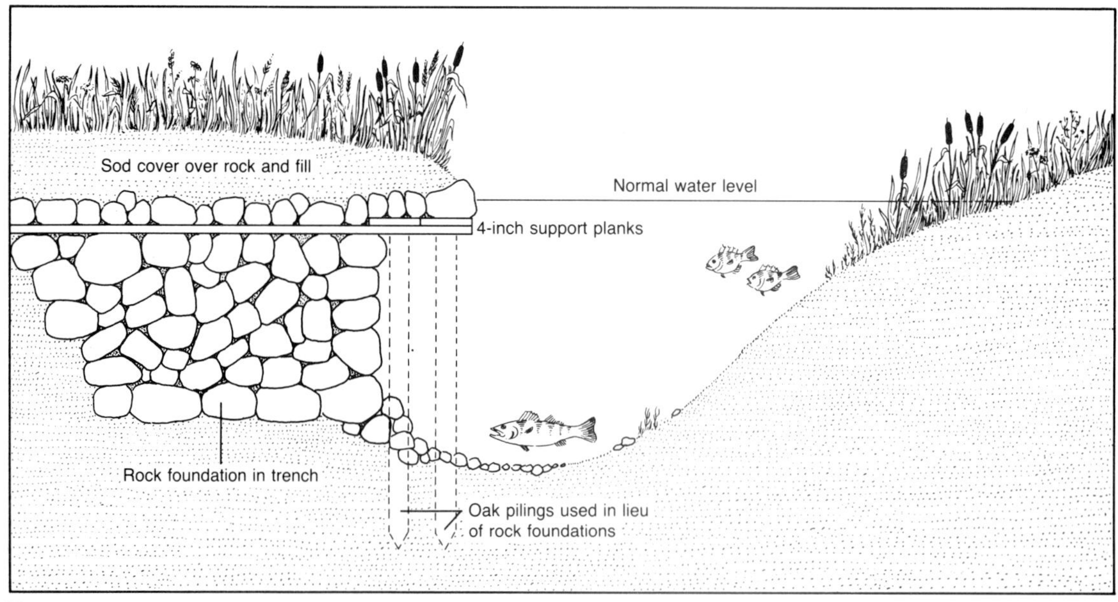

Figure 10.13
Overhanging bank cover to improve fish habitat.

plants must be controlled to prevent oxygen depletion. Harvesting fish ponds on a regular basis prevents other mortality factors from becoming dominant.

Effective managers also consider disease. Fish are attacked by fungi, algae, parasitic protozoa, tapeworms, roundworms, leeches, bacteria, and viruses. Fish diseases can be controlled by habitat manipulation, genetic improvement of the stock, and the use of disinfectants and antibiotics.

Fish Culture

Fish culture involves raising fish in captivity for release into the wild. Many principles of fish culture are similar to those of aquaculture (Chapter 12).

Livingston Stone traveled to California in 1871 to establish a salmon egg collection station on the McCloud River, an effort that gave rise to the National Fish Hatchery system, a system of ninety-three national hatcheries, fish health centers, technology centers, and support facilities. Early fish culture efforts were directed at collecting eggs from the wild for spawning species such as salmon or shad in an attempt to reverse the decline of localized fish populations. Large-scale programs developed, some of which encompass entire watersheds or coastwide populations.

Confinement of fish populations in ponds and raceways showed that adequate diets and means to control disease did not exist. Desirable fish species that grew well in the wild succumbed under intensive cultural situations. At the same time, species that grew well in hatcheries often fared poorly in the wild. Fish culture now uses a highly concentrated protein-efficient fish diet, usually of pellets, formulated to meet specific nutritional needs. Some of the food is so concentrated that it converts at a ratio of 1.1 pound of food for every pound of fish weight gain.

In 1990, the U.S. Fish and Wildlife Service produced 2,800,000 kilograms (6,200,880 pounds) of fish in hatcheries. Over half consisted of various species of trout, about a third were anadromous fish, and the rest bass, bluegills, sunfish, catfish, and walleyes.

Disease plagues fish cultures. Although effective control of many external parasites and bacteria has all but eliminated the catastrophic loss of cultured fish, major losses occur as a result of viruses.

Biologists can produce fish that endure many of the stressful conditions at high-production hatcheries and flourish in the wild, but these changes, brought about in part by genetic management, raise questions about what will happen to natural fish populations. Can fish released from hatcheries interbreed in the wild? Will the gene pool of fish in the wild have enough variation so the population can survive under natural conditions?

Hatchery-reared fish have begun to play a major role in fishery management plans (Figure 10.14). Because of physical changes in waterways resulting from pollution and habitat alteration, many areas now have only stocked fish populations. Fish hatchery programs are relatively expensive. Extensive sums of money are devoted to the Trout Hatchery Program, making it a very expensive form of conservation. Cost accounting methods are not generally agreed upon; however, the Wildlife Society estimates that a pound of fingerlings—fish that are 2.5 to 7.6 centimeters (1–3 inches) long—costs about 83 cents. Trout about 20 to 25 centimeters (8–10 inches) long cost about 81 cents a pound. These costs include raising as well as planting the fish.

Figure 10.14
Fish-rearing tank in hatchery. (U.S. Fish and Wildlife Service.)

LEGISLATION

Migratory Bird Treaties. In 1916 the Convention for the Protection of Migratory Birds was signed between the United States and Great Britain (signing for Canada). Subsequent treaties have been signed between the United States and Mexico (1936), Japan (1972), and the Soviet Union (1978). These treaties, later established as laws by Congress, specifically protect migratory birds. They allow for open hunting seasons on game birds and protect nongame birds. They protect nests, eggs, or young birds except by special permits for scientific use.

Other treaties on fish, polar bears, antarctic seals, fur seals, whales, and other animals have been signed. All give our national government responsibility for working with other nations to protect wildlife.

Federal Funding Legislation. In 1937, the U.S. Congress passed the Pittman–Robertson Act for wildlife restoration that levied a 10 percent (later, 11 percent) manufacturers' excise tax on sporting arms and ammunition to support wildlife restoration work by state wildlife agencies. The taxes are collected by the federal government and turned over to the states based on their size and population. The states can then fund projects involving land acquisition habitat development and research with 25 percent of their own monies and 75 percent of these federally collected taxes.

In 1950, the Dingle–Johnson Act was passed to tax fishing equipment and pass the funds to the states for fish restoration work. In 1984, the Wallop–Breau Act amended the Dingle–Johnson Act to generate funds from a tax on additional fishing-generated sales of small boats and fuel for boat motors.

In 1979, Congress passed a nongame funding act similar to the Pittman–Robertson and Dingle–Johnson Acts. Products used by nonconsumptive users, such as binoculars and bird food, were to be taxed. Congress then decided not to tax these items but use appropriate funds from general revenue tax. They have not, however, acted on an appropriation under the legislation.

Saltwater Fisheries

Although the ocean ecosystem may appear to be the least affected by human activities, its fisheries resource has experienced a tremendous impact (see Chapter 12). The best oceanic fishing is found on continental shelves which are often polluted by human activity. Garbage, chemicals, and "junk" are dumped in these productive areas. Oil pollution from ships and oil spills is taking a heavy toll on sea life along the coastlines. Distruction from the Alaskian oil spill by Exxon is likely never to be repaired. There is also a political controversy regarding who has the right to fish where. Most nations now want to extend their continental waters out to 322 kilometers (200 miles), which includes most of the continental shelf.

Oceanic fisheries provide the major food source for some countries of the world. Today, fishing fleets have elaborate systems of radar to detect fish. These fleets have processing ships which cut up the fish and freeze or package them for sale. Such operations deplete the fishery resource faster than it can renew itself. Likewise, behavior patterns are disturbed so that schools of fish might be reduced to sizes that they can no longer return.

Capture methods are also of concern. Nets strung out to capture tuna often pick up dolphins and kill them. Nets themselves are not discriminatory, thus all fish that cannot get through the mesh are captured. Some water species are killed and discarded or used for fertilizer.

In all likelihood, our exploration of oceanic food sources is reaching its limits. Some potential may exist for aquaculture if water pollution doesn't destroy habitat. If people from all nations do not work cooperatively, overexploitation and pollution will lead to a drastic decline in the numbers of species and the abundance of species in our oceans.

Anadromous Fish

Anadromous fish migrate from fresh water to salt water and back again. Salmon are the best known species in this group. In the West, from the Pacific region up through Alaska, salmon hatch in high mountain streams and then make their way downstream and remain in the open ocean for 3 to 4 years. The adults, apparently led by their sense of smell, return to their home streams to spawn.

Dams and impoundments are a deterrent to maintaining anadromous fish populations. Massive dams along the Columbia River, between Oregon and Washington, have caused a decline in salmon numbers. Biologists have worked with engineers to design fish ladders that allow salmon to return to the streams where they were born to spawn (Figure 10.15). Maintenance of stream habitat is thus extremely important; logging, extensive grazing, and dumping waste into these small tributaries can seriously alter the

Figure 10.15
Lower Granite Dam and fish ladder on the Columbia River in eastern Washington. In 1985, nearly 119,000 adult steelhead trout migrated over the dam on the way to their spawning grounds. (Courtesy Dept. of Energy, Bonneville Power Administration.)

habitat. Gravel bottoms that are disrupted by human activities, siltation, or excessive runoffs can also be detrimental to anadromous fish. Managers must encourage people who use habitat around mountain tributaries to create a band of natural vegetation so that water temperature, oxygen content, flow, and stream bottoms remain unaltered. The U.S. Fish and Wildlife Service releases salmon smolt (young salmon) on the East and West coasts to try to restore this resource.

The sea lamprey virtually destroyed the Great Lakes commercial fishing industry in the 1960s by parasitizing the adult fish (Chapter 2). Since then, extensive fish restoration programs have been developed. During 1976, 22 million trout and salmon were planted in the Great Lakes by the U.S. Fish and Wildlife Service and other cooperating agencies. This included lake trout, coho salmon, rainbow trout, steelhead, and brown trout. Poisons, electrical wires, and trapping devices reduced the lamprey population in Lake Superior 95 percent during the period from 1973 to 1977 (Figure 10.16).

Figure 10.16
Electric shocking system that prevents sea lamprey from moving upstream. (U.S. Fish and Wildlife Service)

SUMMARY AND CONCLUSION

Fish and wildlife are important natural resources. Early colonists exploited their abundance, but gradually, wildlife management evolved into laws that allow conservation of the resource. Wildlife management applies to single species, such as those that are harvested, predators, or endangered, as well as to whole communities of wildlife. Managers use an array of techniques, including vegetation manipulation and construction of impoundments for water, to control wildlife habitat. Controlled hunting and fishing are also important management tools. By substituting hunting as a mortality factor, populations can be maintained within a habitat's carrying capacity.

Most of the funds for managing fish and wildlife come from hunting and fishing fees, so most effort is directed toward these species.

Fishery managers primarily work to maintain habitat suitable for the species they are managing. Changes in waterflow, mineral content, or substrate can all have an impact on the fish species present. The hatchery program has become a large operation for stocking many species of fish, of which trout are the most common.

Management of saltwater species has not been a major objective of managers; however, the decline of these species means that pollution control measures as well as harvest control must be instituted for the oceans. Anadromous fish that migrate from fresh water to salt water and back are an important fishery resource in parts of the world, and their continuance depends on unpolluted streams and oceans.

FURTHER READINGS

Anderson, S. H. 1991. *Managing Our Wildlife Resources.* Englewood Cliff, NJ: Prentice Hall.

Everhart, W. H., and W. D. Young. 1981. *Principles of Fishery Science.* Ithaca, NY: Comstock.

Meade, J. W. 1989. *Aquaculture Management.* New York: Van Nostrand Reinhold.

Teaque, R. D. (ed.). 1971. *A Manual of Wildlife Conservation.* Washington, DC: The Wildlife Society.

U.S. Fish and Wildlife Service. 1988. *1985 National Survey of Fishing, Hunting, and Wildlife Associated Recreation.* Washington, DC.

STUDY QUESTIONS

1. What is wildlife management?
2. What is the difference between species management and managing for species diversity? Give examples of wildlife that would be managed by each method.
3. Why isn't there greater emphasis on management of nongame?
4. How are hunting and fishing factors in wildlife management?
5. Explain why extermination of predators can be bad.
6. Should all wildlife be managed?
7. Why are fish hatcheries in operation when they are so expensive to run?
8. Describe management techniques for anadromous fish.
9. What is the difference between the federal and the state governments' roles in managing wildlife?
10. What causes extinction of wildlife?

SUGGESTED ACTIONS

1. *Hunter/Landowners.* If you are in an area that has a lot of hunters or sports people, develop community organizations that involve sports people and landowners. Work to establish working relationships so landowners will allow hunting, fishing, and wildlife viewing. Create a sense of responsibility in the general public to respect the rights and needs of landowners (e.g., leave gates closed, don't dump trash, or drive off roads).
2. *Protection of Migratory Wildlife.* Look at movement patterns of wildlife in your area. Consider migration areas of big game, birds, fish, and other animals. Become an activist in encouraging these routes to be open for wildlife use. Such activity could encourage underpass construction for wildlife under highways, maintenance of wetlands for migratory birds, and freedom of rivers and streams from obstacles to fish migration.

11

BIODIVERSITY

People tend to take changes in the natural system for granted. To be sure, travelers on our interstate system speed through the countryside, often missing the true nature of the living community. As drivers leave the eastern deciduous forest and move out onto the Great Plains, they see few of the native plants or animals. They may see a few ducks at a river crossing or note antelope in the high plains, and people rush through deserts. Masses of people head to our large national parks such as Yellowstone or Yosemite. Here they pack themselves in tents or trailer cities that would be unacceptable as normal living conditions. A bear, deer, or moose generates tremendous excitement, causing traffic jams and providing film processing firms with great amounts of money. Spectacular scenes such as the Grand Tetons or the fall colors along the Blue Ridge Parkway are also the targets of many shutter clicks.

The world, however, is composed of diverse communities of living organisms. Looked at separately, each has its own unique group of plants and animals, each with its own forms of interactions. The overlooked deserts and plains have some of the most varied species and combinations of species.

Looking at the world from a satellite, we see diverse combinations of life. As we move into each community, we can count the number of plants and animals in each species and the number of species to view their diversity. Within each species, every individual varies; thus there is a diversity of chemical and genetic material in a species as a whole. Diversity, then, is the cornerstone of our planet. Today people are beginning to appreciate and manage for biodiversity.

WHAT IS BIODIVERSITY?

In Chapter 2, we described species diversity as a measure of community diversity. Biological diversity, or biodiversity, is a more comprehensive term meaning the diversity or variety of any living material plus how components are structured and how they interact. The term, therefore, applies to *species diversity*, *habitat diversity*, and *genetic diversity*. It includes composition, structure, and function. Composition encompasses the varieties of species; structure is the physical organization; and function is the process such as energy flow, nutrient cycling, and disturbances. We can measure the diversity of plants and animals, but biodiversity is more complex. Biodiversity encompasses all forms of life at all levels, from the genetic DNA to soil microorganisms to large vertebrates to large communities of deserts, forests, and oceans. We therefore need to define the level at which we use the term biodiversity.

Measuring Species Diversity

Biologists use diversity as a measurable characteristic. Diversity has two components, *richness* (number of items, e.g., number of species per unit area or genetic material per individual) and *equitability* (evenness or relative abundance of items in an area—are the majority of individuals in one species [uneven], or are the individuals evenly distributed between all species [even]?). Thus, if we want to know the diversity of trees in a city park, we can go to a park and count all the different kinds of trees. As a result, for example, we may find that one park has primarily cotton wood trees and a few spruce as follows: 78 percent cottonwood and 22 percent spruce. Another park may have 27 percent cottonwood, 33 percent spruce, 21 percent elm, and 19 percent willow. We would say that the second park appears more diverse in terms of trees. This park has more species, thus the richness component of diversity is higher than that of the first park. Likewise, the evenness component is high because the percentage of trees of each species is divided among the four species. The evenness of the first park is low because most trees are cottonwood.

This is an easy example. Actually, many mathematical formulas can be used to measure diversity. Some are found in books listed in Further Reading at the end of this chapter. Diversity is often hard to measure. For example, the diversity of vegetation in your back yard might be possible to measure if you could identify all the species of plants. However, to measure total diversity, you would need to identify all the soil organisms—some requiring high-resolution microscopes to see. You would need to know the names of all insects as well as birds and mammals. Here is another problem: how do you count a bird or insect that flies in and out of your yard? Obviously, to measure diversity you need to define limits, such as how much time you plan to spend, and whether you need a total count, or can get adequate data from a sample.

Why Is Biodiversity Important?

All forms of living material are important to our biosphere and natural system. In a complex system where a predator can find a number of other organisms as a food source, the loss of one may not disrupt the entire system—thus, the notion that the *more diverse systems are more stable.* A tropical system (Figure 11.1) should therefore be more stable than an arctic system, which is much less diverse. This is generally true. Loss of biodiversity in the tropics is occurring at a rapid rate because of logging, burning, and development. Loss of diversity in the arctic is also occurring because small changes can be very destructive. Pipelines and roadways can block animals from seasonal food sources. Removal of one predator through hunting or habitat loss can disrupt vast areas.

Genetic diversity is the total genetic material possessed by a species. Most individuals of a species have genetic material that is not identical to that of another individual in the species. Collectively, species have genetic material that is called a gene pool. When changes in the environment or the introduction or removal of another species occurs, animals with one combination may survive while other combinations may not be suited. A classic example is mentioned in Chapter 14. Flies with DDT-resistant genetic material became dominant with the advent of the pesticide.

Scientists see each species with genes (see Chapter 3) ranging in number from about 1000 in bacteria and 10,000 in fungi to 400,000 in flowering plants. A house mouse has about 100,000 genes. These genes are divided up among all the individuals in the species. Often endangered species have lost a number of genes, and most of the individuals are very similar genetically. This means that conservation efforts to preserve an endangered species deal with a reduced genetic diversity.

Figure 11.1
Hawaiian forests have a high diversity of plants.

When new disease organisms arise, they may destroy all organisms without genetic factors necessary to resist them. Only the organisms with disease-resistant genetic material produce offspring. If a species has no individuals with the disease-resistant genetic material, all individuals in the species may die and so become extinct. If a species has only individuals with identical genetic material, that species is less stable than other species with a high diversity of genetic material.

Today the world has some 5 to 30 million species of living organisms. These diverse organisms inhabit virtually all parts of the globe, from ocean depths to mountain tops and from the polar regions to the tropics. They include groups of living things, such as microorganisms, insects, plants, fungi (Figure 11.2), algae, and vertebrates. Biologists feel that the number of undescribed organisms is quite large. In tropical forests, many new species of insects are found daily. In 1989, biologists found a new species of monkey, never before described. Forest floors, coral reefs, and the bottoms of oceans are poorly explored and undoubtedly have many more undescribed species.

Some areas, such as the tropics, springs, and estuaries, have conditions that are more conducive to life. Thus, these areas have more species than others. The tropics are among the least explored areas in terms of species; so it is likely they have the most undescribed species.

Biodiversity takes us a step beyond strict diversity of species, into the way individuals relate and interact with one another. It can be seen in a hierarchial manner (Figure 11.3). Diversity is much more easily measured than biodiversity since it can be examined from the perspective of how some organisms change (indicators). Monitoring is necessary at each level (Figure 11.3) because changes in one level create changes at other levels.

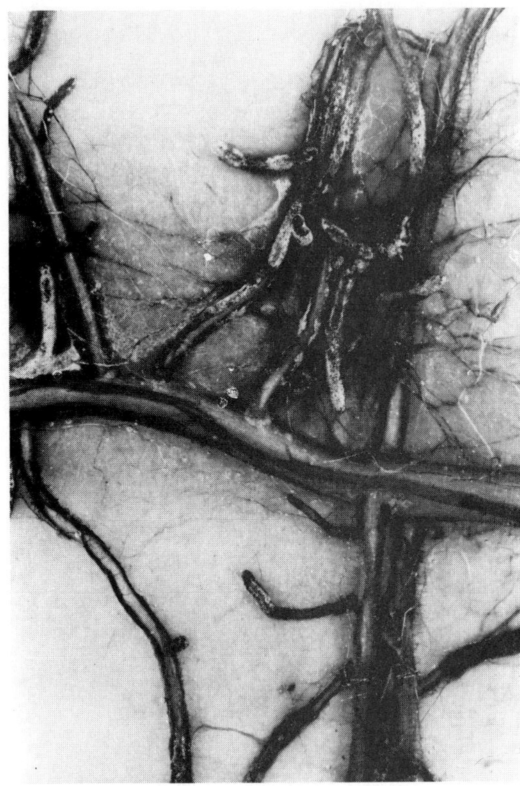

Figure 11.2
Root fungi on trees add to the diversity of forests and aid trees in the uptake of nutrients.

Each area of the world has its own unique combination of living organisms. The living material interacts in such a way as to provide functional organization. If one species or a group of species are destroyed (Figure 11.4), the whole interacting system changes (see Chapter 2). If a number of species are lost, then the number of possible interactions is limited. The more species that are present, the more interaction occurs; therefore, it is necessary to preserve species diversity.

Scientists estimate that more than 1000 species on our earth are becoming extinct each year with the rate likely to reach 5000 per year around the year 2000. Extinction is thus a major problem because we lose genetic diversity, important links in a species, and community stability to interact and withstand stress. Thus, we lose important needs of future generations to control disease and human suffering, and to manage the environment and restore damaged habitat. We must, therefore, develop methods to manage for biological diversity. We must look beyond managing for an endangered species—a deer, duck, or a trout—to consider all life as a global resource necessary for the well-being of all living things and sustaining life.

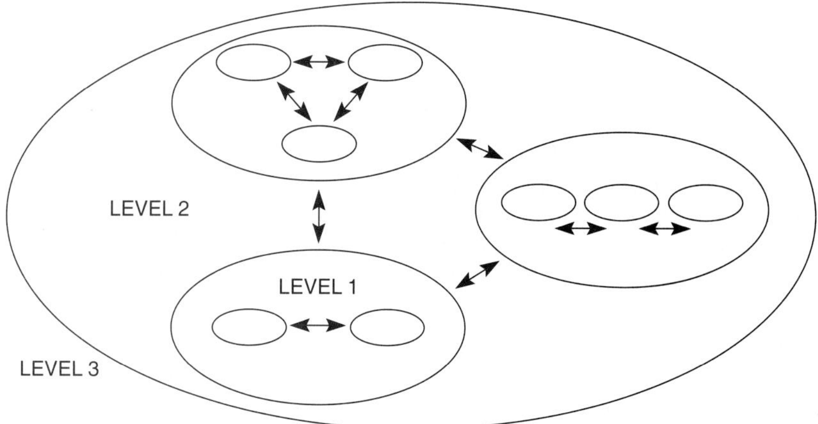

Figure 11.3
Biodiversity needs to be examined in a hierarchical system.

Figure 11.4
Paddlefish found in parts of the Mississippi River are declining rapidly. (Courtsy of John Ramsey.)

BIOMES

The large recognizable communities in different parts of the world are called **biomes.** Biomes are the biological expressions of the interactions of organisms with the physical factors in different regions of the world. Figure 11.5 shows the world as it is subdivided into biomes. Similar environmental conditions create similar biomes in different parts of the world. Climate generally plays a major role in the type of plant and animal life of a region.

People can live almost everywhere, except in a few areas with extreme environmental conditions. Locations with ideal conditions for human living and food gathering are modified and developed to satisfy people's desires. These regions are often altered to such an extent that the natural biotic region no longer exists.

Although each biome is normally named for its climax community, it is composed of life in all developmental stages. We will examine the interactions of abiotic and biotic factors in the environmental regions of North America with some reference to the rest of the world biomes.

Tundra

The northern parts of the North American continent, Europe, and Asia are known as **tundra.** The tundra has a short summer and growing season and a very long, cold winter. The ground below the surface is frozen most of the year. In the summer, the top few inches of soil can thaw;

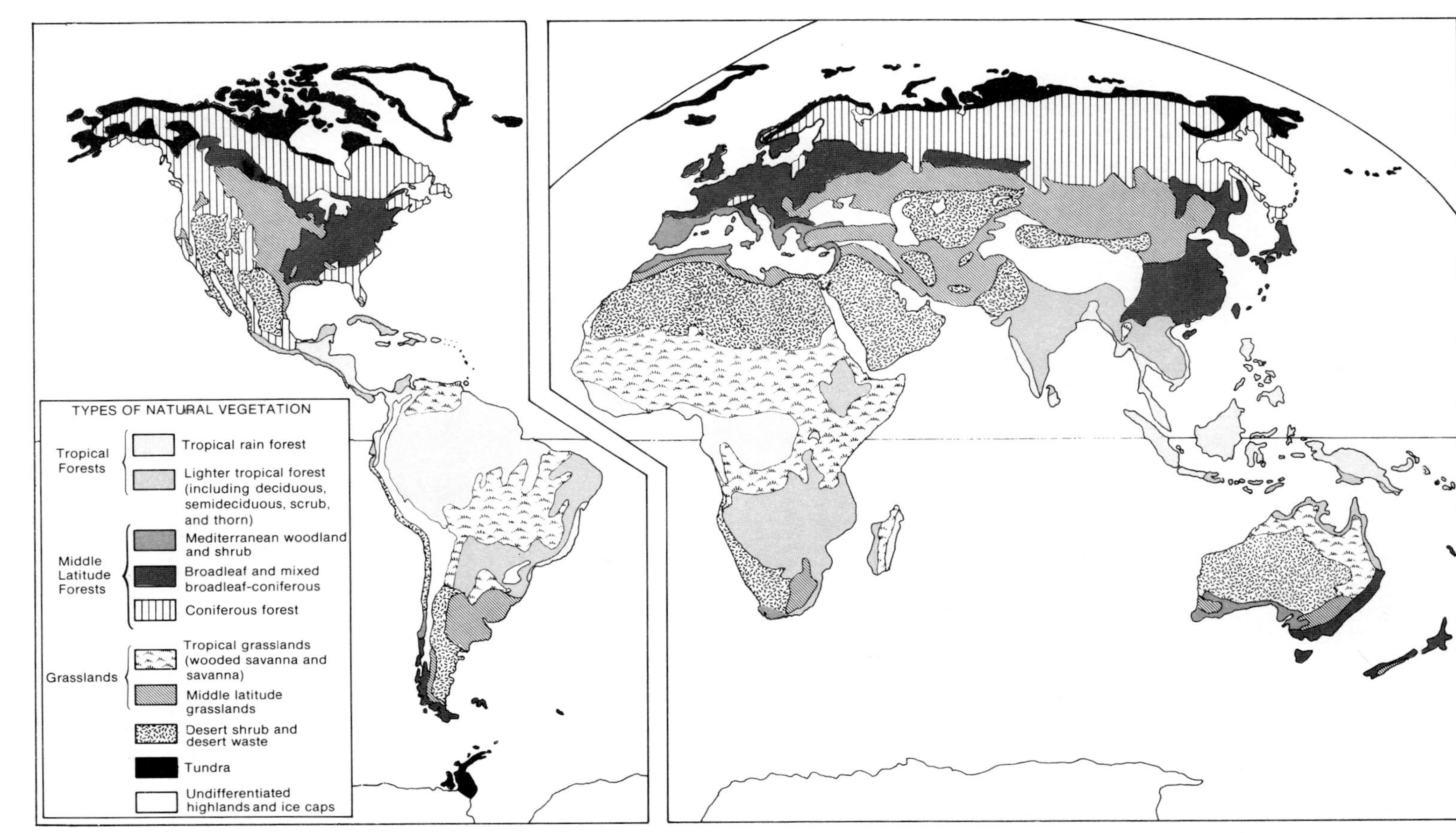

Figure 11.5
Biomes of the world.

however, the rest of the ground remains frozen **permafrost.** During the short summer months, grasses grow quickly and attract a number of species of breeding birds and mammals. Insects also thrive abundantly during this time. Very few woody plants can live here because of the short growing season, so most of the vegetation is in the form of grasses, lichens, sedges, and dwarf willow. All of the plants must complete their life processes very quickly before the cold winter returns. There are two classifications of animals existing in the tundra: those present only in the summer and those resident throughout the year. A number of migratory birds, including many waterfowl, are attracted to these areas. The insects generally pass the winter as eggs or larvae, grow quickly, then reproduce.

Because of the short growing season, the tundra ecosystem is very delicate. Disturbances are likely to take many years to heal. For example, tire ruts in the ground might remain for 50 or more years. This delicate system has been of great concern during the construction of the Alaskan oil pipeline.

Boreal Forest

Moving south from the tundra, we find a large number of coniferous forests—mostly black spruce, white spruce, balsam fir, and tamarack. This biotic region exists throughout most of Canada, Scandinavia, and the northern part of Siberia. Here we find the climate slightly warmer than the tundra with much more precipitation—about 38 to 102 centimeters (15 to 40 inches). The frozen soil melts in the summer, allowing tree roots to penetrate more deeply and the soil to be more fully developed. This region is referred to as the **boreal forest,** or *taiga* in the Soviet Union. Conifers deposit litter on the soil which decays slowly because of the cold climate. The acid products from this decay are carried into the soil by rain or melting snow, making the soil relatively infertile for most crops. The growing season is also very short. Many migratory birds, some of the larger animals (moose, caribou, and wolverine), and the snowshoe hare live here.

Deciduous Forest

Continuing south into the United States, Europe, or the central Soviet Union, we come to deciduous forests (Figure 11.6). In the United States this region extends throughout the East and down into parts of the South. The dominant trees in a deciduous forest are oak, maple, hickory, beech, and other hardwood. Most shed their leaves in the late fall and overwinter in a dormant state. Precipitation is relatively high and distributed throughout the year. Rainfall averages 76 to 152 centimeters (30 to 60 inches) per year, so there is an abundance of plant growth. The warm humid summers and cool winters encourage vegetation to become very dense. Animal species characteristic of this area are white-tailed deer, ruffed grouse, cottontail rabbit, red fox, raccoon, fox squirrel, and wild turkey.

Grassland or Prairie

Between the eastern deciduous forest and the western desert is the grassland of the central United States. Most continents have a similar biotic region in their central area. Grassland climate is intermediate between the forest and desert with a relatively low rainfall. Wet and dry cycles often alternate for periods of several years. The summers are warm and winters are cool.

Several forms of grasses exist in this region. A prairie community maintained by periodic fires is dominant adjacent to forests, while shorter, less dense grass is found in the drier areas further west. The prairie has an ideal climate for crop production. The soil is rich in organic matter because litter from the grass decays in the upper soil. Minerals are not leached out because rainfall is light, under 76 centimeters (30 inches) per year, and the leaching processes are retarded during winter. This is what people refer to as

Figure 11.6
Deciduous forests once dominated the eastern part of the United States.

prime agricultural land. Animal life includes many grazing animals such as bison, pronghorn antelope, and ground nesting birds (Figure 11.7).

Deserts

Deserts are usually found where mountain regions block the flow of water (both as rivers and rain) into an area and where potential evapotranspiration exceeds rainfall. Desert-type areas also exist in continental interiors. The desert region of the southwestern United States has less than 25 centimeters (10 inches) of rainfall per year (Figure 11.8). The two main deserts are the high desert or Great Basin, extending between the Rocky Mountains and the Sierra Nevada; and the low desert, including the Mojave in California and the Sonoran in New Mexico, Arizona, and southern California. The Great Basin vegetation is characterized by sagebrush, low shrubs, and often small conifers and junipers. The low desert has desert shrubs, creosote bush, and a variety of cacti. While most plants in the

Figure 11.7
Sharp-tailed grouse are found in the central and northern Great Plains where grassland is interspersed with a few trees and shrubs. (Charles G. Summers, Jr.)

Figure 11.8
The southwest desert in the United States has many unique species. Shown here are the commonly photographed saguaro cacti.

world reproduce each spring, taking their clues from daylight, desert plant reproduction is triggered by rainfall.

Chaparral

The biotic region along coastal southern California is part of the **chaparral** biome and has a relatively stable climate throughout the year. This is referred to as Mediterranean climate since the area around the Mediterranean Sea is very similar. People consider the climate ideal, and they move in great numbers to this area. The chaparral is interspersed with bushes, trees, and shrubs. Thoroughly dry in the summer, it receives most of its rainfall in the short winter season; thus, fire is common in the summertime. The many people now living in this region often have to fight huge fires, which are spread quickly up the canyons by the strong, dry Santa Ana winds. Some plants have evolved seed structures which open only following fires.

Tropics

Tropical biotic regions exist in southern Mexico, Central America, and the northern part of South America, as well as in Africa and Asia, where rainfall exceeds 229 centimeters (90 inches) per year. Instead of the usual four seasons, most tropical forests have two. They are called wet and dry, even though it rains year-round.

We once thought tropical forests were impenetrable but now know that they are very susceptible to human impact. Soil in many tropical regions is very poor because the vegetation contains most of the minerals and nutrients. When trees die or leaves fall, the minerals are recycled into other living material quickly. Removing vegetation robs these areas of the nutrients necessary for the survival of plants and animals. Extremely varied habitats exist in the Central and South American tropics primarily because of the heavy rainfall, which can vary from 229 to 762 centimeters (90 to 300 + inches) per year.

Nine tropical countries have virtually eliminated their closed forests. Efforts to maintain forests that remain are hampered by the large number of people who live in poverty.

How to Use Biomes in Examining Biodiversity

Biodiversity of individual biomes as well as between biomes can be examined. These data are useful when collected over a period of years, allowing us to relate increases or decreases in diversity to other fluctuations such as changes in air contents of metals (e.g., lead) or chemicals (carbon dioxide—global warming). The biodiversity of communities within biomes can be examined to see how human impact affects one but not another. To date, few studies of biodiversity in biomes have been undertaken. Thus, we lack valuable information that could be useful to help us plan and convince politicians of the existence of worldwide problems in natural resource management.

LOSS OF BIODIVERSITY

As the number of species continues to decline at an alarming rate, we need to focus on why. The primary cause of species extinction, and therefore loss of biodiversity, is habitat alteration or destruction. Today our wetlands are being destroyed at an alarming rate, with 53 percent of this habitat lost in the United States in the past 200 years (Figure 11.9). Wetlands are habitat for many popular wildlife species (Figure 11.10).

Tropical forests are another example of severe habitat loss. Most tropical forests contain more than half of all living species. The forests are found in areas of heavy rain in Southeast Asia, Central and Western Africa, Central America, and northern South America. Only about 3.36 million square kilometers (830 million acres) remain in Indomalaysia, while 60 percent of Africa's rain forests and 19 percent of the American forests are lost. All together we have lost nearly 50 percent of the original tropical forest with its high species diversity. Other habitats in the world are likewise being destroyed at a rapid rate.

With this decline, biodiversity appears to be dropping rapidly. This means that crop yields in many parts of the world will be difficult to maintain as climates change, soil erosion increases, water supplies are depleted, pollinating animals may be unable to survive, and pest species become more common. As these processes continue, productive land for life is converted to wasteland. Deserts, where life is sparse, will expand. At the same time, we will witness an in-

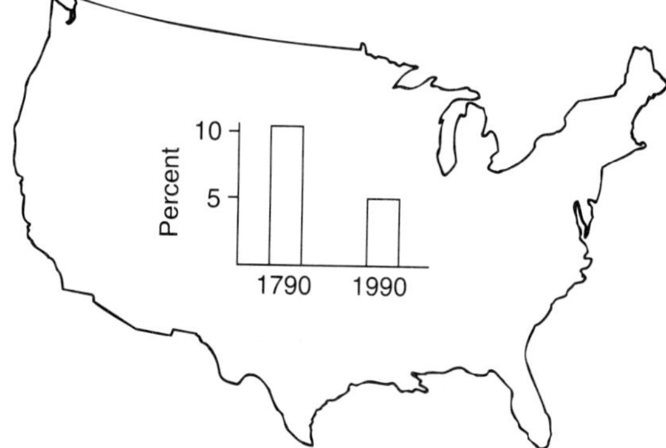

Figure 11.9
Wetlands loss in the United States. Wetlands composed 11 percent of the continuous United States in 1790. By 1990, they had been reduced by 53 percent, to about 5 percent of the country.

Figure 11.10
Wetlands provide food and shelter for many waterbirds. (USDA Soil Conservation Service.)

crease in air pollution and changes in climate as well as water and mineral cycles. People, as a result, will find that the habitat into which they evolved has changed, causing environmental stress. Buffers against disease will be lost. Natural disasters that were once controllable (e.g., fire) will no longer be preventable. Biologist Paul Ehrlich says that civilization will disappear, "not with a bang but a whisper."

Compounding the problems of habitat destruction in biomes are more subtle and difficult to measure changes that occur from human impact at local sites. Many critical species are found in our natural systems. These species, sometimes called *critical link species*, are ones that play a vital role in energy movement (Chapter 2). Regardless of their size, some bacteria and other microorganisms are necessary for the decomposition cycle to operate. These species are examples of *keystone* species because of their importance in ecosystems. As we dispose of oils, pesticides, and other wastes, we may destroy these critical species, making decomposition cycles inoperable. The fact is, we have not identified all these species. Their decline would never result in an endangered species listing.

In Chapter 3, we discussed species interactions such as predation and mutualism. These and other forms of population interaction are essential to keep natural systems operating. Many plants depend on insect pollinators. In some cases, specific plant–insect interactions have evolved in a process of *coevolution* to provide mutualistic survival for both species. Still, little is known about the coevolved relationships in the different biomes. Biologists suspect that the very diverse biomes such as the tropics have an abundance of these relationships. Current management approaches seldom deal with these intricate relationships.

To successfully preserve biodiversity and convince the world of its importance, we must move toward understanding the role various species play in the environment. It is clear that some species play a more critical role than others in maintaining relative abundance. We need to characterize the strengths of our energy and decomposing cycles and the relationships and independence of species.

NATURAL BIODIVERSITY

The earth's living systems have evolved over millions of years and continue to do so. Specialized organisms evolve in stable systems where they can operate without a great deal of environmental change. Within a system, however, the organisms are in a dynamic state. While some biologists feel that stability leads to a state of equilibrium, current research shows that a stable equilibrium seldom exists, but rather a dynamic system which, in human terms, may be changing over the short or long term. When we say that reducing impact causes a return to stability, we really mean a dynamic stability with natural changes directing the community of organisms.

In every natural community, many changes are occurring at the same time. As energy flows from the sun to green plants and then to animals (see Chapter 2), biomass is created and used. The decomposition in turn reallocates the biomass to its original parts, often minerals and gases. Climate influences this process. Drought slows the process down, while moisture may accelerate it.

In stressed ecosystems, plants may respond to the lack of water by growing slower, growing only after rain, and devoting more of their energy to reproduction. Thus, a system may operate in the same fashion but at a different rate depending on the influence of physical factors.

In forests, changes are always occurring. We discussed the dynamics of succession in Chapter 2. As one plant species makes the area unsuitable for itself, one or more others establish themselves. Within the succession sequence many small changes occur. For example, the *gap phase*, in which a tree within a forest falls creating an opening, offers an opportunity for many changes in plant and animals diversity. A greater edge is created, allowing these species to use all the area. A transition stage in forest succession occurs in which different forces operate until a new tree or trees grow large enough to be dominant. This cycle repeats itself many times throughout the forest in each successive cycle. The trees within a forest gap interact by sharing resources, minerals, water, and light. The trees beyond the boundary of the gap have different interactions for the resources with more emphasis on one than the other.

By examining the whole forest, we can see a mosaic of gap size patches, each with its own dynamics, and each contributing to diversity. Each gap's dynamics and species are neither related to nor independent from others or from the forest. The gaps composing a mosaic community share similar species under similar growing conditions and dynamics. They exchange seeds more often within the cove than gap or nearby valleys and ridges. These similarities can be viewed as an area of characteristic size within gaps that interact on a characteristic frequency.

Thus, evolution provides a diversity of life from generalist to specialist over long spans of time, succession in a dynamic sequence provides survival habitat for people, plants, and animals.

In grassland ecosystems, succession also occurs. Open ground gives way to colonizing species of grass, which prepare the ground for still other grasses. In the plains and grasslands, colonies of prairie dogs, a close relative of squirrels, may be found (Figure 11.11). Prairie dogs are herbivorous rodents that build intricate burrow systems underground. Adults can weigh as much as 1 kilogram (2.2 pounds). Currently these species are found in the mountain and great basin states. They have been viewed as competitors of cattle, thus a large-scale poisoning program has been underway to reduce their numbers to less than 2 percent of what they were in the 1800s. Along with the loss of prairie dogs, we have witnessed the decline of raptors, ground predators such as coyotes and badgers, birds of prey, and

Figure 11.11
The prairie dog of the high plains assists in maintaining diversity of the area. (USDA Soil Conservation Service.)

the loss of the black-footed ferret, all depending on the prairie dog for food.

Areas that house prairie dog colonies (which can range in size from a hectare to thousands of hectares) are in a constant state of change. These little creatures are constantly digging burrows and bringing soil to the surface. They tunnel up to 5 meters (16 feet) below the surface, mixing 200 to 225 kilograms (440–495 pounds) of soil per burrow system. Dirt mounds around their burrow entrances show how they loosen, mix, and pile the soil. This process in turn aerates the soil and provides ways for rain and snow to get into the soil, creating a different diversity in the grassland system.

Above ground, prairie dogs constantly clip and graze grasses and other plants. Near their burrow entrances they keep the grass well clipped, changing the successional portion of grasses. The action of the prairie dogs actually changes the competitive interaction of plants, creating a mosaic of the plant community that can be compared to the grass in the forest.

Within a colony of prairie dogs are social systems. Dominant males often have harems of females with young. Young females may remain in the social group but young males are forced to establish their own groups and eventually their own harems. This means that within a colony there are core areas of extensive burrow systems and heavy use of the vegetation separated by areas of taller vegetation.

Research by biologists April Whicker and James Detlingon on the Wind Cave National Park in South Dakota has shown that grazing animals such as bison, elk, and pronghorn preferentially select prairie dog colonies over uncolonized grasslands for feeding. They point out that in areas where prairie dogs have grazed, plant biomes have a higher standing crop and more protein and some nutrients. The grass is more digestible than grasses from uncolonized prairie. As a result, large mammals obtain more nutrients per unit of food consumed, which means grazing bison, elk, and pronghorn need to expend less energy per unit of time to obtain their food. At the same time, some of those large grazing animals forage, opening wider areas for prairie dogs to use. Thus, it is likely that the prairie dog–large grazer relationship is mutualistic.

From the perspective of preserving diversity, poisoning prairie dogs has wide-ranging effects. The prey base for many animals has changed. This has caused some predators to turn to domestic animals for food. The nutritional cycles have been altered, resulting in changes in many microorganism species and the vegetation community. The large mammal population is also impacted.

Natural biodiversity is therefore the result of interaction of all organisms. As changes occur in one, the impact is felt by many other species. Each community's integrated work system is thus subject to change.

BIODIVERSITY DATABASE

The Nature Conservancy is a private conservation organization established to conserve biodiversity by establishing natural area preserves. The Conservancy has established a series of State Natural Heritage Programs which, among other things, collect data and organize it in such a way as to identify the most important areas that should be established as preserves. The program extends into some Latin American countries.

Using data available in each state on plants and animals, and collecting some of their own, Heritage personnel have developed a system that allows evaluation of sites where inventories are needed. They can assist in development by showing where rare or endemic species are located. Candidate species for endangered species listing have been identified. The system has been used to set priorities on conservation, preserve selected species, and monitor population changes.

The Heritage data system accumulates data for university, state, and federal agencies. Files on plants and animals are collected and coded so they can be easily assessed and cross-referenced. From these files they have developed a series of map files which show the distribution of plants and animals. These maps can then be used in an overlay fashion to examine how existing species are in some locations or identify areas of rare species concentration. Regional and national data centers are now being established so that data on species distribution will be standardized and can be viewed on a regional basis.

The National Heritage Program interacts closely with most state game and fish agencies. They work closely with the U.S. Forest Service, National Park Service, and the U.S. Fish and Wildlife Service to identify critical areas on land managed by those agencies.

SINGLE-SPECIES MANAGEMENT

As we learned in Chapter 10, many of our management efforts have been directed toward a single species. The goal of traditional wildlife management has been to sustain the yield of selected species. Whether we are interested in hunting elk, deer, cougar, waterfowl, or rabbits, we emphasize the productivity of these animals for human benefit. Predator control programs are the same, except the emphasis is on removal of "pest" species for the betterment of people and their choice animals and plants. Only in the last 30 years have we shifted our views to look at conservation, but this has been primarily the single-species approach, again through endangered species legislation. Hard lessons have taught us that to maintain desired species, for conservation or harvest or to remove undesirable species, we must manipulate the habitat to maintain the integrity of the ecosystem for all.

For example, the habitat of many fish species has been so disrupted that the competitive interactions of the fish are destroyed and the natural aggregate no longer exists. As a result, a great deal of the sport fisheries are "put and take" operations. (Fish are placed in a stream or lake for immediate removal by fishing enthusiasts.) Game bird and game animal farms are similar types of operations which are nothing more that a form of animal husbandry.

Improved natural fisheries occur where stream habitat diversity is increased to provide suitable spawning, hiding, and wintering habitat for prey fish. Size limitations in some areas allow larger fish to be available. Trophy hunting to satisfy hunters also requires example manipulative techniques. Antler restrictions allow larger animals to survive for the benefit of a few people.

In essence, all our traditional wildlife management techniques are antidiversity. We tend to seek one species to remove or preserve. We are fortunate that habitat preservation is necessary to keep these animals in the wild. It is unfortunate that put and take fisheries as well as game ranches lead us further from the concept of biodiversity. While single-species management leads us away from biodiversity, some of our legislative acts and policies also appear to work against maintaining biodiversity.

Multiple Use

The U.S. Forest Service and the U.S. Bureau of Land Management, two of the largest controllers of land in our country, are mandated to manage for multiple use. This means that recreation, wildlife, and resource removal (logging, grazing, mineral extraction) all are a part of agency management plans. Thus, the natural balance is disrupted as forests are logged, roads are cut, and the grasslands are grazed by livestock. The government is subsidizing the loss of biodiversity through low-cost timber sales and very inexpensive grazing rights on public lands.

The natural balance is altered in the tropics where acid rain and forest practices such as intercropping, cutting, and shifting cultivation occur. Conservation practice in the tropics often pose extreme hardship for the poor people who depend on the forest for their survival. Here, people extract their basic needs from or depend on forests and grasslands for their livelihoods. Some countries, as well as some towns in the United States, could collapse economically if strict laws on forest protection are adopted.

The challenge for maintaining biodiversity, in light of the multiple use question, is great. Conservationists must not only convince the public of the need for maintaining the natural community, but also must provide avenues for these people to survive or continue their livelihood. Both small-scale and national projects must be considered.

In countries where the economy can support national parks, extensive battles over preservation versus development are constant. Yellowstone National Park has been preserved, but it looks like an island. Extensive logging and grazing occur right up to the border of the park. When conservationists want to preserve a natural system for one species, extensive bitter battles often erupt. The spotted owl in the Northwest is an example. While conservationists argue that the old-growth forest must be preserved for the owl, loggers have a very different view. They see jobs lost and therefore argue the multiple-use principle. Unfortunately, at the present rate of harvest, old-growth forests will be gone in 20 years.

Exotics

An *exotic species* is one that has been introduced into a habitat some distance from its natural area. There are numerous examples of exotics. One individual wanted to introduce all animals mentioned in Shakespeare's writings into the United States. This has resulted in the European starling spreading throughout much of our country. These birds often spend nights in large communal roosts from late summer until spring. Their noise and droppings, as well as behavior, make them a pest in city parks, suburbs, and farms.

Some animals are accidentally introduced, such as the Norway rat which arrived on ships. Mongoose populations introduced to control rats have virtually destroyed all ground-dwelling birds on the Hawaiian Islands. Profit has also been a motive in the introduction of exotics. The nutria, a large rodent found in many of our small lakes and marshes, was introduced as a source of fur in the southern part of our country. Their fur was inferior, so the scheme was a failure and they were released into the wild where they have expanded their range widely throughout the country. Nutria, unlike beaver, destroy embankments and dirt dams. They create havoc for many water-dependent organisms. Game ranching poses problems when exotics are brought into an area and escape. The exotic animals can bring new diseases that infect native animals. Sometimes they outcompete the native animals, thereby changing the community diversity. Carp have been introduced into our waterways, creating very silty waters because they spawn and use the muddy shores. The result is the loss of the microbes and invertebrates that support the natural community, including sport fisheries.

Not all exotic introductions have been bad. In Chapter 14 on pesticides we see how organisms that operate as biological control agents can benefit our communities and agriculture. The brown trout that now inhabits many of our streams and lakes is an exotic species that has become a popular sport fish.

The *Eucalyptus* trees in California were introduced from Australia some 135 years ago. They are found in large groves and along major highways. Many *Eucalyptus* trees grow along the side of El Camino Real (the Kings Highway). An attempt was made to remove exotics from California state parks and recreation areas, but many people opposed removing the *Eucalyptus.* They cited studies that showed this exotic had a positive impact on wildlife. In fact, 57 percent of the birds on Angel Island were found in *Eucalyptus.* Salamanders were more dense in *Eucalyptus* stands. Studies also showed that the western population of the monarch butterfly, which winters in coastal California, preferred *Eucalyptus* trees. This behavior may have resulted because the trees provided a source of food with their flower nectar and a source of shelter, whereas native trees provided only shelter. On the Monterey Peninsula, conservationists in large numbers protested when a private landowner decided to remove a grove of *Eucalyptus* trees that contained a large number of wintering monarch butterflies.

The California *Eucalyptus* controversy resulted in editorials in the *San Francisco Examiner* in favor of the trees. The city's supervisor went on record as opposing the removal of the trees. At the same time some biologists protested that the removal of the trees could create severe erosion on some hillsides. The issue in this controversy is whether all exotics should be regarded as equally undesirable and therefore their removal equally justified. How can we view the question of biodiversity in this context? Clearly, community functions and interactions are difficult to manage both with and without exotics. Can we set a priority system and judge "good" and "bad" exotics? We must consider the role exotics play in maintaining both urban and wilderness ecosystem functions. Clearly we as people are setting priorities as we see them.

Exotic species are likely to be a more serious problem in the future as we become a more mobile population and the climate changes. People who travel see plants that they feel might be attractive in their yard, so they try to take samples home. Quarantine measures in California and many nations attempt to prevent this on the basis that exotic plants often bring in pest insects. Some states have laws against removing native plants. Arizona, for example, protects some of its cacti species in this manner.

Global warming creates an opportunity for natural range expansion of plant and animal species. Sometimes native species are stressed and immigrants better adapted to the changing conditions move in. If the climate changes are rapid compared to evolutionary rates, many ecosystems may experience mortality of native species. Very specialized species are likely to be the "unfit" under such conditions, leaving the door open for the generalists or colonizers. Colonizers will likely play a larger role in maintaining stability such as nutrient retention and productivity. If the colonizers in an ecosystem are exotics according to our definition, then their removal could have a disruptive effect.

MANAGING FOR BIODIVERSITY

The obvious goal of managing for biodiversity is to have the greatest diversity of species survive. This is a new way of thinking for many managers who usually consider a single species. What are some of the methods that can be used?

A Hierarchical Approach

Earlier we discussed the changing dynamics of a forest or grassland with the influence of many gaps or openings in the community. If we look at a forest such as the eastern deciduous forest or the western coniferous forest, we find that each ridge, valley, plain, or gully is composed of many interacting stages or gaps. We can take a defined unit such as the south, facing slope of a pine forest and characterize the area as a level of interactive organization with its composite group of plants and animals—the results of the existent stage of succession, the series of gaps, and the physical features that allow the community to exist. Thus, we have at one level a series of gaps that are the result of the surrounding forest, and the physical elements. This gap can be looked at over levels of organization (Figure 11.12). The *stand* of trees, with its composite of gaps all influenced by climate, water, and other physical factors, is a second level of organization. By measuring the physical and biological factors, we can manage a forest stand and maintain one form of diversity.

Several or many stands on different slopes around local drainage basins compose a third level of organization. This level involves a *watershed* that may have stands of different tree species, open areas, and stands of mixed tree species. The trees share a similar resource base and interact among themselves. They differ because solar radiation reaching the north- and south-facing slopes around a watershed are different. Likewise, the ability of trees to retain snow and therefore capture water differs because of sunlight, wind, and plant structure. Logging or grazing in one part of the watershed can have an impact on organisms in other parts of this level of organization.

Moving up to a fourth level of organization we can define *landscapes* as units of similar, interacting watersheds. Here, boundaries might be large geographic features such as mountain ranges or large river systems. Such features gov-

FOUR LEVELS OF A FOREST HIERARCHY

Level	Boundary definition	Scale
Landscape	Physiographic provinces; changes in land use or disturbance regime	10000s ha
Watershed	Local drainage basins; topographic divides	100s–1000s ha
Stand	Topographic positions; disturbances patches	1s–10s ha
Gap	Large tree's influence	0.01–0.1 ha

Figure 11.12
Organization of forests from gap phase to landscape.

ern weather patterns and control movement of some species. More recently we have recognized how all changes in the landscapes along our major drainage systems influence life all along those systems. Timber removal or introduction of exotics can have effects many miles away through erosion, siltation, or the resulting change in the interactions of functions in a system.

Beyond the landscape, or overlaying this level of organization, we find the regional vegetational biomes. Here, again, the major climatic patterns influence the type of plants and animals that can survive.

The hierarchy system allows us to see how, at each level, interactions occur that influence the functional aggregates at the next higher level. The components of each level tend to interact among themselves. We can see that a gap has its own internal dynamics but contributes to the behavior of a stand, watershed, and landscape. Each patch at any level is at once an integral whole and a part of a higher-level component.

Biologists tend to collect data on one level but not consider how interactions with another level may occur. The hierarchical system allows us to examine events on one level and realize that other levels are influenced. Thus, a small forest can create the form of a gap. Its dynamics are influenced by the surrounding plant and animal species that can colonize the area. The changes in the soil, small patterns of erosion, and changes in water flow patterns influence the watershed and landscape. Think about what this means in terms of dispersing of waste material in soils or opening landfills in an area.

If we can develop tools to examine our ecosystems on a hierarchical level, we can develop predictive models to show us the impacts of changing practice in one area or large regions. If we can scale our activities to mimic natural patch dynamics, thereby taking advantage of the natural adaptive mechanism in a local species aggregation, we may disrupt the system less. As a result, rather than destroying functional interactions and reducing biodiversity, we would maintain diversity. As an example, foresters have found that the size of a clear-cut can mimic the size and shape of a natural disturbance. This means that all the regenerative power of the plants and animals surrounding the clear-cut is used to replace the trees removed. Too large a clear-cut can result in massive runoff and erosion.

Fire used in controlled burns prevents large destructive fires which result when massive amounts of debris are allowed to accumulate. Livestock grazing can be made to mimic grazing pressure from native animals. These forms of management form the basis for the discipline of landscape ecology. Landscape ecology, therefore, requires not only the examination of how levels of natural organization exist, but also how they interact functionally. Human-modified landscapes may be developed to imitate the patterns in the natural system.

Biological Preserves

National parks and wilderness areas have long been thought of as natural preserves. A preservative approach to park management allows the natural system, including succession, to progress naturally. Unfortunately, parks are the focus of much public activity because they are usually associated with popular natural features. Thus, parks contain roads, hiking trails, camping, and all sorts of recreational activities. Some have so many vehicles and people in the summer season that they exceed the human population of some cities. Thus, preservation of diversity is only one feature of parks, and one that is sometimes difficult to keep as a major objective.

Proponents of biodiversity management suggest a series of biological preserves managed with effective methods of retaining biodiversity. Preserves provide the opportunity to protect species of interest, rare and endangered species as well as those found in only one location *(endemic)*. Preserves offer the opportunity to maintain diversity by preserving functional communities. Keystone species, along with all community interactions, can be preserved. All of the species can also be preserved—the maximum species rather than first selected species.

Selection of sites for species was discussed under GAP analysis. In addition, cultural, political, and economic factors must be brought into the equation. If the government or a private conservation organization already owns land that could be used as a preserve, it would be less expensive and less controversial to select such an area. Obviously, such sites must have the desired habitat and species. We must protect what we want (e.g., high diversity, rare species sites, endemic species).

Design of Preserves

Here we must examine the size, number, and proximity of one preserve to another. In some areas, human development and the loss of natural areas means preserves cannot be close together.

Biologists conducting research on size and shape have concluded that a few dispersed small sites usually contain as many species as a single site of equal area. Sizes are used here as relative values because preserves will be differently designed in each community type. The fact that several small sites are equal in value for preservation of biodiversity assumes that the habitat is identified in both areas. Preserves have buffer zones to protect them from detrimental effects of human activity. Thus, sites in parks, forests, or in areas distant from human activity might be desirable.

To plan preserves, biologists must know the size of the area necessary to support the desired community. The preserves must provide enough habitat and space that extinction of species does not occur through succession. It must be clear that critical habitat will be protected in the total landscape. Thus, we can refer to the hierarchical system discussed earlier. How much of a stand on watershed, with its buffering, must be protected to maintain the dynamics of the system along with the species diversity?

Corridors also play a role in maintaining diverse preserves. When species movement can occur through riparian zones, fence rows, and other strands of natural vegetation, the total preserve might be more diverse. Corridors can also have a negative influence. Powerline right-of-ways and roadways can open areas of natural vegetation to less desirable or exotic species that may take advantage of the disturbance.

Rapid Assessment

Because we do not know all the plants and animals that live in many areas, we are likely to lose important species when development occurs. Biologists are trying a new technique of *rapid assessment* to evaluate biodiversity in areas such as the tropics on which we do not have adequate databases. Based on the idea that biodiversity is not evenly distributed, biologists are trying to conduct rapid assessments in areas where development is proposed. These assessments involve assembling a group of knowledgeable people who can conduct field surveys of plants or animals in a short period of time (around a month). It is hoped that *hotspots* of biodiversity can be identified. If one of these hotspots is identified in the proposed impact areas, conservationists hope to convince the nation to undertake the environmental changes elsewhere. They further propose that these hotspot areas receive some type of protection. If this rapid assessment approach is successful, some tropical areas that are uniquely endowed with plant and animal life may be preserved. So far, hotspots have been identified in Central and South America, Africa, Asia, and many of the islands in the Pacific and Indian Oceans.

TOOLS USED TO MANAGE FOR BIODIVERSITY

The computer has provided people with a tool that generates data quickly and accurately. Input data are assumed to be accurate, but may not always lend themselves to large assumptions.

One system mentioned in Chapter 1 as a useful tool for planning is the geographic information system (GIS). The geographic information system involves the development of a series of maps based on field data. Thus, maps of vegetation, soils, and moistures may be developed individually from field data. These maps can be overlaid a distribution of one or more species of animals or a species diversity map. Correlations between physical factors, and high diversity, large numbers of endangered species, and many other facets can be then identified. Computer programs can assist the correlation process. This system allows us to develop comprehensive biological diversity information that can be used to evaluate changes on a numerical level. While the system is promising as a tool for effective biodiversity management, it also points out where we lack information and must therefore make assumptions. The system is multileveled and encourages the flow of information from level one to upper levels or from the upper to the lower levels.

To be effective in biodiversity management, therefore, a GIS system must involve the clear understanding of the entire process of data flow. There are five components of the system necessary for an effective management tool: (1) encoding, (2) data management, (3) retrieval, (4) manipulation and analysis, and (5) data display.

Encoding begins with the collection of distributive information on an organism. These data can be collected and plotted on a map and then placed into the computer system based on a system of spatial coordinates such as latitude and longitude. Encoding of weather patterns such as rainfall, topographic features such as elevation, and soil type can follow a pattern similar to that of species distribution. This whole process is obviously very tedious and expensive; however, it allows data management.

Data can then be retrieved and range maps drawn for species and other distributional factors. It is possible to overlay distributional maps with climate and topographical factors. Overlays between different groups of species can show relationships between animals and plant communities.

Data on vegetation can be collected from aerial and satellite photography. These techniques should be supplemented by field verification. The data must be put into the system and managed so that it can be retrieved to answer different questions. The system must be designed to allow manipulation of the data (Figure 11.13).

Spatial modeling output involves a series of range maps about the distribution of animals in relation to different components of the habitat or of other animals. Thus, the range maps could go beyond the correlation of species with habitat to show correlation between animals species, and a species map can be drawn. Both the species richness map and the vegetation map could be compared to maps of land use and land suitability in order to determine its extent and potential loss of biodiversity. The result would be the spa-

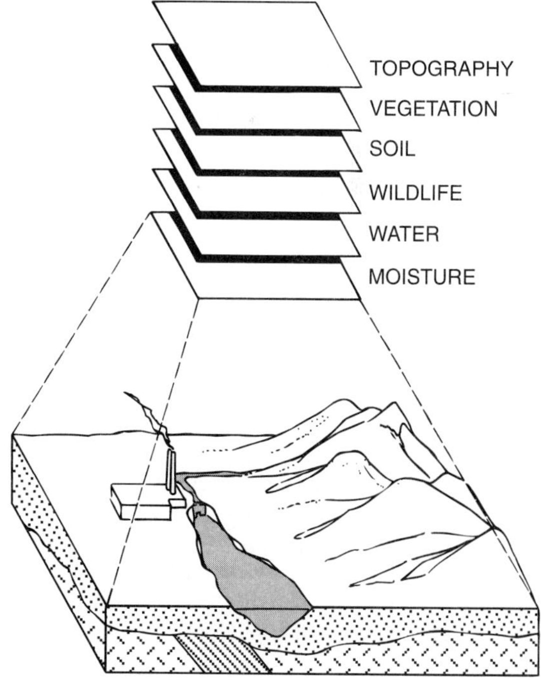

Figure 11.13
The geographic information system.

tial location of the richest and rarest comparison of biodiversity and indications of which are at greatest risk.

Many places lack biological data. GIS systems, if programmed properly, could be used to identify high-priority areas. If the relation is known between species richness and landscape features, then data on one could be used to select high-priority areas for collecting the other data.

Currently, people feel that biological diversity might be preserved on lands set aside for natural reserves. Some sites could be managed for preservation while others are preserved for recreational use. Such potential preserves might be identified by plotting patterns of species richness for all groups of species on which data are available. In the United States this usually applied to vertebrates and butterflies. Species would then be categorized according to their abundance (abundant, common, uncommon, rare) and their occurrence in different habitat types could be plotted (e.g., one habitat type only, two habitat types, etc.). The species that occurs regularly in human-altered habitat would be plotted.

The patterns that emerge could then be evaluated. For example, habitats with large numbers of rare species can be identified. This approach allows us to assess patterns of species richness on a wide scale. The result could be suggestions for managing areas with the highest species diversity, highest number of rare species, or highest number of species subject to human impact. This process is called *GAP analysis*.

A biodiversity information system, if in place, should assist in answering some or all of the following questions:

1. What is the range of species and communities?
2. What locations have different species reached?
3. Where are species at risk located?
4. Are some biological sites at risk?
5. Which ecosystems appear protected and which do not?
6. Are there environmental factors related to areas of greatest biological diversity?
7. Is development suitable with maintenance of species diversity?
8. Can trends of species and biological diversity be predicted over time?
9. Where should other preserves be established?

NATIONAL AND WORLDWIDE CONSIDERATIONS

Priorities

People tend to feel that they are really guests in the natural systems. They obtain all their food, shelter, and protection from the earth's ecosystem, but people place no real economic value on the total or diverse system. True, a stand of trees has economic value. However, this is only in terms of wood products. No value is given the stand for nutrient cycling, wildlife habitat, erosion control, or climate modulation such as wind reduction. In the world, our natural resources are seen for their exploitable value. We must reconsider this attitude in our economic-based world and consider the value of our natural system to our survival. The tropics and their forests are now considered valuable for exploitation. How can we place a value on them for their contribution to the structural and functional aspects of our world?

It is necessary for us as a nation and a world to manage our biological resources in ways that provide sustainable benefits. We must integrate our managed agricultural systems into natural systems. Areas with high hydrological, geological, or species value must be part of preserves. Sensitive areas must be identified and managed. This goes beyond the idea of a wilderness area to examine how plants and animals interact in a stand, watershed, landscape, and even a biome basis. Worldwide legislation and zoning must be established based on regional, national, and worldwide planning. Strategies can be developed

Red Clover Valley Erosion Control Project

Watersheds throughout the world are feeling the impact of intensive human activity. People are realizing that they must plan when they implement projects to enhance the natural diversity of these areas. R. T. Callahan edited a series of case studies in the western United States in Report 22 (1990) of the Wildland Resource Center, Berkeley, California. One case follows.

This project was undertaken on Red Clover Creek, part of the East Branch of the North Fork Feather River watershed on the east slope in the Sierra Nevada Mountains of California. Excessive sedimentation was filling reservoirs. Planning sessions were established between a utility company that managed the reservoir for power, land management agencies, and private landowners. Plans were developed to control grazing and develop bank stabilization programs. Monitoring ran throughout the project.

To stabilize eroding stream banks in Red Clover Creek, a four-phase restoration program was developed. Four loose-rock, check dams were constructed. The dams were designed to reduce water velocity, trap sediment, raise the shallow water table, and stabilize stream banks. Dams and resulting impoundments reduce erosional forces affecting the banks, and encourage establishment of streamside vegetation.

The second phase was to control livestock access to the stream and to protect monitoring equipment. The third phase of work, bank stabilization, involved planting several riparian hardwood species along the banks, seeding herbaceous vegetation in upland areas affected by construction of dams, and installing pine revetment to reinforce an eroding bank.

The technical evaluation is not complete. However, early indications are that well-designed check dams and riparian zone revegetation will effectively stabilize stream banks, reduce erosion, and improve forage, fisheries, wildlife habitat, and aesthetics. The nontechnical objectives of this project have already been met, due mostly to the cooperative management planning process based on addressing local interests. Public interest in watershed improvement has increased, and a method has been developed to implement erosion control projects at locations with multiple interests and agency jurisdictions. The project has resulted in reduced sediment flows. Thus, utility rate payers save. At the same time, biodiversity has been enhanced on the watershed. Perhaps one of the significant outcomes of this study is the fact that private landowners and public and private agencies are working together to achieve environmental goals.

LEGISLATION

As of 1993, legislation to establish conservation of biodiversity as a national goal had not been passed. Conservation organizations have been successful in educating politicians to the need for such legislation, but no real push to pass legislation that has been introduced into Congress has occurred.

Worldwide, several agreements have been established that contribute to biodiversity:

1. *The Convention on Wetlands of International Importance Especially as Waterfowl Habitat (Ramsar, 1971).* A convention recommending that signing parties undertake to use wisely all wetlands under their jurisdiction. Defines wetlands very broadly to include all rivers, lakes, coastal areas, and land around other bodies of water.
2. *The Convention Concerning the Protection of World Cultural and Natural Heritage (Paris, 1972).* Recognizes the obligation of all states to protect unique natural and cultural areas and the obligation of the international community to help pay for the protection.
3. *The Convention on International Trade in Endangered Species of Wild Fauna and Flora (CITES) (Washington, D.C., 1973).* Established list of endangered species for which international trade is restricted or controlled by permit.
4. *The Convention on Conservation of Migratory Species of Wild Animals (Bonn, 1979).* Obligates parties to protect migratory species and to endeavor to establish international agreements for the conservation of migratory species.

that are directed at worldwide biodiversity preservation.

Nations must examine their priorities in terms of land use and national activities. The role of government must be expanded to consider not only economic production but also integration of the nation into the natural system. An expanded role of such organizations as the United Nations, World Bank, and many private organizations could be helpful.

Nations promoting tourism can integrate the option into biodiversity planning. Industrial development, pollution control, and human living can all be factored into biodiversity. We need to develop a new approach to our thinking. We need a new ethic to believe in. These ideas necessitate education of all people, particularly politicians. Still, the future depends on how effectively we manage for total biodiversity.

Wildlife Trade

Commerce in plants and animals poses a major threat to world biodiversity. People see colorful animals and unusual plants and want to take them home. Perhaps of greater concern is the existence worldwide of large groups of poachers and smugglers who illegally remove parts of animals or living plants and animals to sell for large profits.

The Convention on International Trade in Endangered Species of Wild Fauna and Flora (CITES) includes over 100 countries. The goal of the treaty is to prevent trade of threatened species surviving in the wild. Participating countries do not allow import or collecting of listed species. Many items are protected under the treaty, including ivory from elephants, bulbs of some plants, some species of parrots, listed furbearers, primates, and others.

CITES provide a means for controlling trade, however, it is often difficult to enforce. Desire for large profits still drives some people to take and smuggle plants and animals.

In the United States, poaching takes a large toll on our wildlife. Migratory birds are often shot for feathers or fur. Ungulates are taken for food. Some endangered species are removed. Federal law enforcement agents generally enforce migratory birds and endangered species laws. However, there are only about 200 federal law enforcement agents in the country, and they are also responsible for enforcing wildlife importation laws.

State game and fish enforcement offices are responsible for enforcing regulations on fish, ungulates, and many furbearers. They have been assisted in recent years by antipoaching campaigns in which the public can call the 800 toll-free phone number to anonymously report poaching.

Stricter control of all wildlife trade and poaching is going to be necessary to maintain biodiversity. Worldwide education programs are needed, and people are a component in preserving biodiversity.

SUMMARY AND CONCLUSION

Traditionally, people have managed their natural systems for single species. They have tended to examine community interactions only as they affect the selected species. Management for biodiversity provides an alternative to single-species management. Biodiversity, however, requires an understanding of how our natural system functions.

Diversity applies to species diversity, habitat diversity, and genetic diversity. The term has two components, richness and equitability. Today human impact on the natural system is causing high rates of extinction, and therefore a loss of species, genetic, and habitat diversity, collectively referred to as biodiversity.

Generally, the world can be divided into a series of biomes, or major ecosystems, which result from interacting physical conditions. Plants and animals in these biomes have similarities resulting from the physical factors. Biologists examine the interactions in a system numerically. A small gap phase can be shown to have an interactive effect with other gap phases and the surrounding community. The community is often a component of a watershed which can then be viewed in its relationship to the total landscape. The interactions, natural or human, can affect the other component of this hierarchical system. Knowledge of these interactions can assist in planning to reduce adverse impacts throughout the system.

Management for biodiversity sometimes conflicts with other forms of land management practices. Multiple-use practices, for example, may demand that clear-cutting or grazing be part of a system, but these practices result in a reduction of biodiversity. Introduction of exotic species confronts managers with a dilemma. Should exotics that have been part of the system for some time be removed? On the other hand, should introduction of new exotics be allowed?

Identification of areas to be protected for their biodiversity requires time and personnel. GIS systems provide opportunities to manage the large amount of data that is necessary. Through a series of overlay maps, data can be manipulated to show areas of high species concentration, endemics, rare species, and unique habitat. Through GAP analysis, areas can be identified that can serve as preserves for biodiversity. A series of smaller preserves are thought to be as effective in maintaining biodiversity as one large preserve, especially if they are interconnected by corridors. Preserves must be selected to protect biodiversity for long periods into the future.

FURTHER READINGS

Davis, F. W., D. M. Stoms, J. E. Esks, J. Scepan, and J. M. Scott. 1990. An Information System Approach to the Preservation of Biological Diversity. *International Journal of Geographical Information* 4:55–78.

Hunter, M. L. 1990. *Wildlife, Forests, and Forestry: Principles of Managing for Biodiversity.* Englewood Cliffs, NJ: Prentice Hall.

McNeely, J. A., K. R. Miller, W. V. Reid, R. A. Mittermeier, and T. B. Werner. 1990. *Conserving the World's Biological Diversity.* Gland, Switzerland. International Union for Conservation of Nature and Natural Resources, World Resources Institute, Conservation International, World Wildlife Fund–VS, and World Bank.

Noss, R. F. 1990. Indicators for Monitoring Biodiversity: A Hierarchical Approach. *Conservation Biology* 4:355–64.

Soule, M. E., and D. Simberloff. 1986. What Do Genetics and Ecology Tell Us about the Design of Natural Resources? *Biological Conservation* 35:19–40.

Urban, D. L., R. V. O'Neill, and H. H. Shugart. 1987. Landscape Ecology. *BioScience* 37: 119–27.

U.S. Bureau of Land Management. 1990. *Riparian-Wetland Initiative for the 1990's.* Washington, DC.

Westman, W. E. 1990. Managing for Biodiversity. *BioScience* 40:26–33.

Wilson, E. O. (ed.). 1980. *Biodiversity.* Washington, DC: National Academy Press.

STUDY QUESTIONS

1. Define biodiversity.
2. What is GAP analysis?
3. Discuss similarities and differences in management for endangered species and management for biodiversity.
4. How can exotic species be an effective component of management?
5. What is landscape ecology?
6. Can you describe times when it would be better to manage for a single species rather than biodiversity?
7. How does a gap interact with functions of a stand?
8. What are the principal physical factors influencing biome distribution?
9. Why are there more species in a tropical than a temperate system?
10. Why is it important to preserve the diversity of a decomposer community?
11. How can we change the economic-driven attitudes of governments to manage for biodiversity?

SUGGESTED ACTIONS

1. Work as a volunteer with The Nature Conservancy, a federal agency, or state agency to collect data for biodiversity management.
2. Suggest by letter or through attending meetings that biodiversity management be a high priority in federal and state management. Work through local conservation groups such as Audubon Clubs to encourage this form of management.
3. Work with other conservation-minded individuals to establish a series of biological preserves in your area. Organize public support and encourage agency funding of such activity.
4. Develop an education program for private landowners—both urban and rural—to show how to maintain diversity on their parcel of land.

12

FOOD

Because of major crop failures in 1965 and 1966, China, India, Russia, and other countries had to import huge quantities of grain to prevent famine. In 1968 Paul Ehrlich, a population biologist, stated in The Population Bomb *that the world had nine years to find a solution to the food–population crisis. At the same time the book appeared on the bookstands, the "famine" had turned into surpluses. By 1970 countries like the Philippines and Malaysia, formerly on the brink of starvation, were talking of food self-sufficiency in several years. Photos of happy families with full bellies in India were distributed. This dramatic change in a few short years was the product of a major agricultural breakthrough—the Green Revolution. The Green Revolution has been compared to other major developments in history, such as the advent of tool making or the beginning of domestication and cultivation as agricultural practices. It was essentially the development of fast-growing high-yield crops from more than twenty years of agricultural research.*

In 1972, food production in the world fell and food prices rose. Production continued to fall for several years, shattering the hopes many had in the Green Revolution. Wholesale starvation seemed imminent.

During 1977, 1978, and 1979, U.S. farmers staged strikes to obtain greater support for their products. Tractors paraded around Washington, D.C., and protest groups marched on the Capitol steps, while other farmers held back beef and dairy products to increase their incomes. In 1982, United States legislators debated the distribution of surplus cheese to the poor. Meanwhile, consumers balked at the high grocery prices. The public was told that foul weather was to a large extent responsible for the increase in food prices. In some cases, striking migrant workers caused price increases as crops were left to rot. By the late 1980s, some writers claimed a food crisis was far in the future, and biotechnology *became an important buzzword. Yet, during this time, one-third of the people in the world suffered from malnutrition.*

What is a realistic appraisal of the world food situation? Was Ehrlich correct in his judgment, but just a few years off? Is the Green Revolution a success or a failure? How can people apply the principles of energy flow, mineral cycling, and productivity to human food consumption?

HISTORY OF FOOD PRODUCTION

The history of human population growth is tied closely to agricultural development. Early societies consisted of hunters and gatherers who relied entirely on natural plant and animal products to satisfy their needs. With the advent and use of weapons, wild animals were captured. A major breakthrough in the use of natural resources occurred some 15,000 years ago, when people began to cultivate plants and domesticate animals to provide a more reliable source of food. About 10,000 years ago, people became accustomed to village life and began systematically gathering and planting seeds from certain grasses and using herd animals. People have not domesticated a plant or animal of major importance in the last 9000 years.

Modern agricultural practices have resulted from the technological developments, scientific research, and exploration undertaken between 1450 and 1700. It was during the latter part of this period that new foods were introduced into various parts of the world. With this expansion, Europe began to grow three of the world's great food plants—potatoes, maize, and rice. Potatoes and maize came from the Americas and rice from the Orient. The New World also contributed beans, squash, avocados, pineapples, tomatoes, and cocoa to the diets of Europeans. The American Indians gave the world such nonfoods as tobacco, rubber, and certain cottons. To the New World the Europeans brought wheat, onions, grapevines, and sugar cane along with domestic animals which included horses, dogs, pigs, cattle, chickens, sheep, and goats.

The early primitive implements of sticks and rough tools developed into plows pulled by horses and later into mechanized planting and harvesting equipment. Eli Whitney's invention of the cotton gin in 1793, which increased the amount of usable cotton, was one of many technological advances in agriculture.

Food production began to increase in the 1850s as a result of agricultural research. Knowledge of soil nutrients, development of synthetic fertilizers, the discovery of the role of microorganisms (such as nitrogen-fixing bacteria), and the development of irrigation techniques all improved yields. The Morrill Act, passed in 1862, turned out to be one of the biggest advances in U.S. agricultural history. It provided federal funds to establish land grant colleges to promote agricultural development in those states agreeing to provide minimal support. This legislation had a huge impact on the growth of agricultural technology.

In today's world, vast differences exist in agricultural technology and food production. The people in many less developed countries are striving to maintain a living from small plots of land using primitive methods. Family farms in developed countries generally are larger and use modern agricultural practices. In the United States, for instance, the head of the family is often a graduate of the state agricultural college. An increasing amount of the farm and grazing land in developed countries is managed by large corporations. Along with this industrialization come energy subsidies, elaborate chemical control, and hybrid plant and animal species. The main objective of these modern practices of managing our natural resources is to maximize yield.

Earlier agricultural practices influenced only the local ecosystem, but the more recent methods of fertilization, pesticide application, and energy use have affected global ecosystems (see Chapters 2 and 14). Other examples of disruption of ecosystems include overgrazing (Figure 12.1), which often causes top-soil to erode, and doubling the yield, which can require a tenfold increase in fertilizer. Industrialized agricultural practices result in air, land, and water pollution. **Monocultures,** fields of single crops, are more susceptible to pests than the natural communities they replace.

Efforts are currently underway to rectify many of the undesirable effects of modern agriculture. Limiting the number of grazing animals per acre retards soil erosion. Some farmers regularly rotate their crops to avoid depleting soil nutrients and sustain areas of natural vegetation in woodlots and fencerows to maintain natural ecological systems.

Figure 12.1
Soil erosion resulting from heavy cattle grazing. (U.S. Department of Agriculture.)

At a global level, human population growth is the dominant cause of the greater demand for food. Before 1973, the year of a major oil price increase, world food output expanded at just over 3 percent per year. A United Nations study in 1975 concluded that food output must continue at that rate or a little higher to keep pace with population growth. Since 1973, however, annual growth of food production has been just under 2 percent—a struggle to keep up with the additional people. Compounding the problem has been the lack of growth of per capita income; since the late 1970s, many poor cannot afford the available food items.

Food supplies are not evenly distributed throughout the world. North American countries have become the dominant grain exporters (Table 12.1). Imports increased in parts of Africa, the Middle East, and East Asia, at the same time that the USSR and Eastern Europe also expanded imports.

The growing affluence of some people in the world is another major drain on the world food supply. Of all grain produced in the world, about half is consumed directly by people. Food for livestock takes up a considerable share of the remainder. In poor countries, the annual availability of grain per person is about 171 kilograms (400 pounds) per year. Most of this is consumed directly. In the United States and Canada, the per capita grain utilization is approaching 907 kilograms (1 ton) per year. Of this amount, about 68 kilograms (150 pounds) are consumed directly in the form of bread and cereal. The rest is eaten indirectly in such forms as

Table 12.1
Grain Exports in Million Metric Tons[a]

Region	1963	1972	1984
United States	33	40	96
Canada	10	15	28
Western Europe	−22	−17	−11
USSR and Eastern Europe	—	−2	−34
East Asia	−4	−9	−12
South Asia	−2	−2	−5
South and Central America	−1	−3	−3
Middle East	−4	−9	−38
Africa	−1	−2	−9
Other	−1	−6	−16

[a]Numbers with minus values are imports.

meat, milk, and eggs. The agricultural resources—land, water, and fertilizer—needed to support an average person in North America are nearly five times those needed to support people in the world's poorer nations.

Green Revolution

The development of strains of grain useful in many of the world's highly populated areas began in 1944, when four young American plant scientists went to Mexico to help its impoverished agricultural program. Twenty-six years later one of those four scientists, Dr. Norman E. Borlaug, was still working in Mexico when he was awarded the Nobel Peace Prize for his work. During their stay in Mexico, Dr. Borlaug and his associates received strains and varieties of grain crops from different parts of the world. By crossing different varieties, they developed new varieties better adapted to the Mexican environment. Exportation of some of these varieties to Pakistan and India during the late 1960s resulted in remarkable harvest in those nations, postponing serious food shortages.

The introduction of high-yield grain varieties on a commercial scale to many less developed countries in the mid-to late 1960s was called the **Green Revolution.** India, for example, began growing high-yield wheat in 1966 and 1967. In 1967, 11 percent of the wheat crop consisted of these new varieties; by 1978 they accounted for 78 percent of the wheat crop and 54 percent of the total annual grain production. The new varieties of wheat increased total yield by 50 to 100 percent, while new varieties of rice increased yields 10 to 20 percent.

If the Green Revolution was so successful in India, why did world crop production fall in 1972 and the world find many of its inhabitants starving by 1978? As with so many developments involving people and their environment, a whole array of physical factors, such as climate and environmental limitations, as well as political, economic, and cultural factors played a part. More people were eating than ever before, but more people were also starving. Food production simply was not keeping up with the population growth. In addition, a small percentage of the world's population began consuming an increasingly larger proportion of the world's food. Although more synthetic foods such as beef-flavored soy granules, imitation eggs made from seaweed, and sugar substitutes appeared on the market, they took additional energy to manufacture.

Local problems also hampered the progress of the Green Revolution. To produce high-yield crops, large inputs of sometimes unavailable or expensive fertilizers and pesticides plus adequate amounts of water were needed. Each strain of wheat, rice, or corn had to be carefully selected and developed to survive under local conditions, and farmers found that new grain varieties were often highly susceptible to disease. People were frequently unwilling to eat new foods that might look or taste different from the traditional diet. Once the local farmers had been convinced that the new crop variety was something they should grow and the local population was convinced it should be part of their diet, methods of planting, harvesting, storing, and transporting had to be learned. Planting and harvesting equipment was frequently unavailable, and small farmers 30 kilometers (19 miles) from the nearest road in South America had no way of trading surplus for other commodities. Such factors were, in part, responsible for the failure to reach the expected results from the new "miracle grains."

Even though the new agricultural techniques did not achieve the predicted results immediately, the Green Revolution was far from a failure. The new grains continued to be an important agricultural product of many nations. More than 90 percent of the wheat sowed in Mexico thereafter consisted of the high-yield varieties. *Genetic breeding* provided greater yields, *genetic selection* developed strains resistant to pest populations, and the understanding of crop nutrition has contributed to improved agricultural conditions in many parts of the world.

The grain shortage from 1972 to 1974 was caused by several factors—not by a failure of the Green Revolution. A few years before nations like the United States had reduced the quantity of grain in storage; then major crop failures occurred in the Soviet Union and China. These political and environmental factors along with the increased cost of transportation caused by the energy crises in 1973 and 1974 boosted the price of grain.

Even though world grain output showed an overall 2.6-fold gain between 1950 and 1984, recent trends reported by the Worldwatch Institute have not been as encouraging. World grain output fell by 14 percent during the 1985–1989 period. World grain stocks have also diminished, declining sharply between 1987 and 1989. A monsoon failure in India contributed to an 85 million–ton drop in world food output in 1987. In 1988, drought-reduced harvests in North America and China resulted in a 76 million–ton decrease in world grain outputs. Other drought-reduced harvests occurred in 1980 and 1983.

The Green Revolution can continue to provide more food if technical problems are resolved, but it is not a panacea to the population-versus-environment problem. The environment and the balance of nature must be considered when we use these agricultural practices. The Green Revolution cannot indefinitely provide a food supply for an ever-increasing number of people, and perhaps it should not if we want to solve the other problems besides hunger caused by increasing populations.

Biotechnology

Although not new, **biotechnology** is adding new dimensions to the production of world foods. Biotechnology has greater potential than the Green Revolution to change food production. The basis of biotechnology is the ability to manipulate the chemicals of reproduction (Chapter 3). Many genetic engineering companies are engaged in the business of modifying the genetic material of plants and animals to produce faster-growing varieties, disease-resistant plants, bacteria that will remove wastes from garbage, plants that will withstand adverse climatic conditions, and so on. The companies are currently seeking patents on the seeds and plants they have developed.

Biotechnology is the manipulation of genetic material in much the manner of a naturally occurring mutation. In this case, however, people determine what desirable genetic characteristics they wish to attempt to combine into one organism. Biotechnology processes include substitution of chemicals on the DNA material of a gene, replacing some or parts of chromosomes, and the use of a variety of hybrid crosses.

There have been a number of success stories in the field of biotechnology. Cereal crops have been developed in Australia that live and produce in arid areas. In Canada corn and wheat are being tested that can grow in the harsh prairie winters. Faster-growing, more efficient bacteria are now used in the fermenting process, thus requiring fewer steps and less equipment.

All these rapid changes naturally cause concern about the negative aspects of biotechnology—the possibility of developing forms of life that will be harmful to the environment. At the same time, companies will be developing products that they want the public to accept. Biotechnology will thus require public awareness.

Food Additives

Additives have been used for centuries to preserve food and to make it more palatable. Examples include salt brine to preserve fish and smoke to treat meat. All forms of food additives have increased in use since World War II because of the development of large processing and food distribution centers. The use of prepared foods and the large number of meals eaten away from home have also contributed to greater use of food additives.

What are food additives? They include all substances that can become a component of the food or affect any of its characteristics, such as flavor. They are classified as nutrients, preservatives, antioxidants, thickening agents, emulsifiers, drying substances, flavor enhancers, or coloring agents.

Some people are concerned about the use of too many additives. They prefer natural foods with no additives. Such an approach, while admirable, has some pitfalls. Without additives, many foods would start decaying and cause sickness if eaten. In jams, jellies, and cured hams, sugar or table salt are added to inhibit microbial growth. Citric, acetic, and phosphoric acids are added to such foods as mayonnaise and carbonated beverages to stop the growth of bacteria. Synthetic chemicals such as butylated hydroxyanisole (BHA) and butylated hydroxytoluene (BHT) are added to some foods to prevent chemical deterioration.

Because people prefer special flavors, processors try to capture these flavors in food products. Taste buds located on our tongues recognize four basic tastes: sweet, salty, bitter, and sour. Different combinations of chemicals cause different receptor stimulations, so very slight chemical changes in foods can make foods much tastier. Monosodium glutamate is one common flavor enhancer. Another favorite is coffee flavor, which is made up of over 400 different compounds. Companies roast coffee with the combinations of these compounds that they feel produce the best flavor.

Another additive group is artificial colorings. People often associate a specific color with a particular product, so butchers add a coloring agent to make meats red and margarine is colored yellow to look like butter. Florida oranges, which have big green splotches when they are picked, are colored orange to compete with the more orange California ones. If the Florida oranges were held until they turned orange, they would spoil before reaching the consumer.

Less desirable meat cuts are being upgraded and combined with textured vegetable proteins (TVP) to form "engineered foods" in the forms of chucks and fillets. Meat rolls of beef, pork, turkey, and chicken are commonly made this way. TVPs are meat extenders that generally consist of soybean products.

The controversy that surrounds food additives has focused on chemicals such as nitrates and nitrites, sulfur and sulfides, and food colors made from coal tar. Nitrate chemicals are added to prevent spoilage, but can form compounds that cause stomach cancer; in fact, some of the chemicals that form are found in cigarettes. Efforts are underway to ban coal tar dyes from foods.

Sulfurs are used to maintain desirable colors in canned fruits and vegetables and as preservatives in beers, juices, and some seafoods. These chemicals cause allergic reactions in some people and are particularly harmful to asthmatics.

Who controls food additives? The U.S. Food and Drug Administration (FDA) was given early authority in 1906, but the real food regulation provisions were established in the Food and Color Additives Amendment of 1958. At this time, evidence of the safety of food additives was required before the FDA would approve their use. In 1959 the FDA published a list of some 600 substances generally regarded as safe because they had been used safely for many years. In 1969, cyclamates had to be removed from the list because of the Delaney Clause. This clause, which stated that no additive should be considered safe if it induced cancer when ingested by humans or animals, became a provision of the 1958 Food and Color Additives Amendment.

As a result of the amendment, a program was instituted to evaluate the safety of natural and flavoring substances in food. After evaluation and FDA approval, substances that can be safely used are designated as GRAS, a term coming out of the amendment. Between 1965 and 1985, the number of GRAS substances increased from 1130 to 1750. But GRAS substances are not the only additives found in food; nearly 3000 chemicals that don't require FDA approval are used as food additives.

The total number of food additives is immense. They are generally safe when taken in small amounts and as part of a balanced diet. Most problems develop when a substance is not properly tested or when it is eaten in such excessive amounts that it creates physiological disorders. Although most people are concerned about the effects of food additives, few know what to do about them. Eating a balanced diet of a variety of foods and a minimum amount of self-additives (salt and sugar) is a good approach. People should understand how their bodies react to different additives and avoid those that cause problems. Asthmatics, for example, should avoid sulfur additives.

WORLD FOOD SUPPLY

The supply of food available for consumption is determined by a number of controllable and uncontrollable factors. Weather and the amount of arable land have always been factors, and government controls and technology are playing an increasing role in the world food supply. The types of farming equipment and fertilizers used as well as research which produces such major changes as the Green Revolution are important controlling factors.

Present increases in U.S. food prices are largely incurred in steps beyond the farm. Only 5 percent of the retail cost of food in the United States now is attributable to land rent in comparison with 8 percent in the 1930s. Land rent in Great Britain (including buildings and land improvements) accounts for only 7 percent of the value of food, down from 40 percent in 1855.

Weather is an uncontrollable force that affects the world food supply. Major changes in weather patterns can reduce the amount of certain crops, causing prices to rise. A coffee shortage developed after freezing weather in Brazil. A severe winter in Florida sent vegetable and citrus prices soaring, while flooding in California's lettuce-growing regions caused the price of a head of lettuce to double and triple. From these examples we see that weather can directly affect food supplies and prices.

Another controlling power is world governments. Within the United States, many forms of incentive payments are made to farmers for increasing or decreasing their crop production or land use. *Parity* is also a government payment program that influences crop production. In 1978 and 1979, many farmers wanted the government to guarantee a price for their product regardless of the market price. This meant that the U.S. government would have to make up the difference between the price the farmers received and the parity price.

World trade in agricultural products compounds problems caused by internal incentive programs. Import and export quotas are partly responsible for restricted world food supplies; however, even if food were readily available, many people would not have the money to buy it. Expenditures for food take 15 to 25 percent of personal income in developed countries and 50 to 70 percent in less developed countries.

Global food markets have become very complex. While some markets such as Japan and the European Common Market have tried to maintain stable prices by isolating themselves from other world markets, they have merely created new problems because they are not self-sufficient and other nations are unable to supplement their food needs. Likewise, while the United States responds to individual crises, it also prefers to remain isolated from the world food market. However, when U.S. policy dictates food sanctions, widespread repercussions develop. For example, President Carter's grain embargo against the Soviet Union resulted in grain surpluses in the United States. When President Reagan came into office, he removed the embargo, thus altering the supply and demand ratio and causing wide fluctuations in the prices and supplies of American foods.

Because food production in the United States is influenced by food consumption, a decrease in consumption does not mean more food for the world's hungry. For example, if everyone in the United States eliminated one hamburger from his menu each week for a year to send to the hungry, it is unlikely that those nations would get additional food. Rather, the production of beef and grain would probably drop the following year. The only way the beef would benefit others would be through an international agreement to transport it directly to the needy. The hungry nations, then, would have to be able to afford the price, and facilities would have to be available for transporting the food and storing it. Furthermore, we would have to assume that Americans' favorite—the hamburger—would be acceptable to the hungry. Indeed, redistribution of food is not an easy task. Think what that means in terms of the comment "Eat your food because the poor people in Asia are starving!"

Fluctuations in the world food supply were reflected in the price changes that occurred in the late 1960s and early 1970s. At that time the development of new technologies and fertilizers resulted in a surplus of food, which depressed food prices. In 1972, the Soviet Union experienced the first of several severe crop failures. Through agreements with other nations, the surplus grains were shipped to the Soviet Union, thus depleting the world grain surplus. Grain prices rose in response to the change in supply.

A large depression had hit the farm states by the early 1980s. This depression was probably the result of an increase in fuel prices nearly a decade before and the sudden change in artificial price supports. Many farmers went bankrupt, and the severe droughts of 1986 and 1988 further complicated the problem. Meanwhile, many nations were losing the ability to remain self-sufficient for food. African countries were nearly self-sufficient in food in 1970, but by the mid-1980s, they imported nearly 24 million tons of food. The result is that, while the world's total food supply may not be low, political problems cause massive shortages.

AGRICULTURAL LAND

Soil Types

Because of different physical and biological factors, the soil formed in various parts of the world is quite different. The general soil map for the world in Figure 12.2 shows similarities to the biotic regions described in Chapter 2. In tropical **lateritic soil,** high rainfall and high temperatures remove many nutrients through leaching. The mineral nutrients in this ecosystem are tied up primarily in the forest vegetation. If a tree is blown down in the Amazon forest, decomposition is rapid and the raw materials are quickly absorbed by other plants. The soil, particularly the top layer, is composed mainly of iron and aluminum salts. When vegetation is removed in order to farm, the soil in some tropical regions forms **laterite**—a tough, hard surface. This process of **laterization** makes tropical areas unfit for such agricultural practices as the annual production of grain crops. However, tropical soils are excellent for tree crops, making pine tree farms or coffee plantations very productive. Properly used, tropical soils can be valuable agricultural areas if the crops are long-growing species which allow mineral cycles to be maintained.

In northern coniferous forests, including parts of the northeastern United States, trees add little organic matter to the soil and **podzolization** takes place. Water falling on the litter becomes acidic and leaches away many minerals, making the soils unsuitable for agriculture. In the New Jersey pine barrens, leaching removes nutrients from topsoil and leaves sand with a little clay or humus. This podzolic soil does not support very productive crops.

Calcification takes place in the **chernozemic soils** of grasslands, where the rainfall is less than in forested areas. Minerals are not removed from the soil and the annual decay of grasses adds much organic matter to the soil. Calcification produces nutrient-rich soils that are very desirable for agricultural purposes such as grain pro-

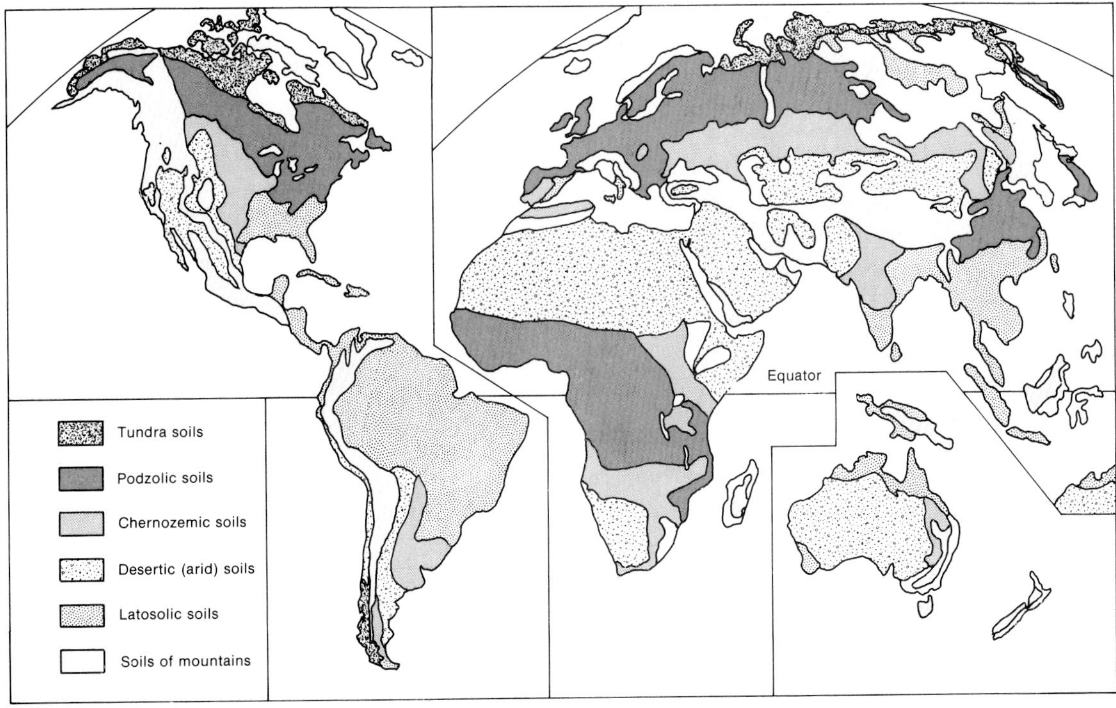

Figure 12.2
Soil types of the world. (U.S. Soil Conservation Service.)

duction (Figure 12.3). Table 12.2 compares the processes of laterization, podzolization, and calcification.

While these soil development processes occur generally throughout the world, local factors influence their rates. Limestone resists podzolization, and granite is slow to show calcification. Desert regions have limited vegetation, organic matter, and nitrogen but they contain high levels of salts because following a rain, the rate of evaporation causes the salts to form an alkaline crust on the soil surface.

Removing natural vegetation from a region by plowing or livestock grazing causes the topsoil to wash away during periods of rainfall. Litter that usually holds the rainwater and allows it to seep into the natural pores of the soil is gone.

Figure 12.3
Grain field in Kansas. (U.S. Department of Agriculture.)

Table 12.2
Summary of Soil Development Processes

	Calcification	Podzolization	Laterization
Precipitation	Low	High	Very high
Temperature	Moderate to high	Low	High
Vegetation	Grassland	Boreal forest	Tropical forest
Soil pH	Basic to neutral	Acid	Acid
TOPSOIL			
Color	Dark brown	Light gray	Red
Texture	Loam	Sandy	Clay
Principal minerals	Various	Silica	Iron, aluminum
Humus	High	Low	Low
Fertility	High	Low	Low
SUBSOIL			
Color	Whitish	Red	Red
Texture	Loam	Heavy clay	Heavy clay
Principal minerals	Calcium	Iron, aluminum	Iron, aluminum
Typical zonal soil groups	Chernozem	Podzol	Lateritic red

Agricultural scientists estimate that on more than 10 percent of the world's cropland erosion is so severe the land is essentially destroyed for further crop cultivation. In poorer countries, millions of acres of unproductive lands are abandoned yearly.

How can erosion be controlled? Basically, the answer is planning and management. Before natural vegetation is disturbed, it is necessary to understand the interrelationships of the particular ecosystem. Planting rapidly growing grasses helps prevent soil erosion. Crops can be planted to follow the land contours in areas of potential erosion problems (Figure 12.4). Leaving rows of the natural vegetation in place is also a useful practice to help avoid soil erosion.

Soil Fertility

The mineral elements essential for plant growth, with the exception of oxygen and carbon dioxide, are taken from the soil. Quantities of these elements in the soil indicate its fertility. Although the nutrient level is usually high in comparison with plant needs, all elements are not in a readily usable form. Soils, then, serve as a medium for bringing nutrient elements into the biological system.

Many factors affect the nutrients that plants can get from soils. The size and arrangement of soil particles determine water flow and storage, air movement, and the soil's ability to release nutrients to plants. Soils deficient in clay and organic matter are structureless and cannot hold water near the surface. For example, undisturbed forest soils can absorb 12.7 centimeters (5 inches) of rainfall each hour for 10 hours, but cultivated soil can absorb no more than 3 centimeters (1 inch) per hour in the same time period.

Organisms also contribute to soil fertility. Some are part of the detritus food chain (Chapter 2) and participate in the breakdown and release of nutrients from dead organic matter. They also play an important part in the formation of *humus*—the decomposed organic material that contributes significantly to soil texture, water-holding capacities, and, in some cases, mineral binding. Some organisms release minerals from soil particles, making them available to plants. They are part of chemical reactions or

Figure 12.4
Contour farming in Wisconsin. (U.S. Department of Agriculture.)

create by-products like acids that change the physical conditions of the soil. Earthworms and other larger organisms are important in maintaining aeration and texture.

The diversity of soil life varies considerably from one ecosystem to another. Soils with the greatest potential for agricultural productivity, such as chernozemic grasslands, have a high diversity of biota, while poorer podzolic soils have a much lower diversity. Unfortunately, many human activities decrease the diversity and thereby reduce fertility. Although the application of chemical fertilizers in standard amounts is not normally detrimental to life, the use of excessive nitrogen and various biocides does destroy groups of organisms.

Fertilizers

Farmers have known for many centuries that the addition of fertilizers in the form of organic materials, including human and animal wastes or animal and plant parts, improved the yield of crops. Massive use of fertilizers, however, began with the manufacture of synthetic fertilizers in the late 1930s. Because plants require large amounts of nitrogen, phosphorus, and potassium—the primary soil micronutrients—these three elements are the most common constituents of manufactured fertilizers. Total consumption varies with the region of the country and especially with the crop.

In the period from 1955 to 1976, use of fertilizer increased 240 percent in the United States because synthetic fertilizers were easily available and because the wornout, overused soil needed it so badly (Figure 12.5). Since 1965, the total acres of corn grown in this country increased by 23 percent, while the amount of fertilizer used for corn increased by 77 percent, as shown in Figure 12.6. In some cases, farmers did not realize that more fertilizer did not always mean greater crop yield. Thus, excess fertilizer only added to ground and surface water pollution.

The growth of fertilizer use was concentrated in the industrial countries during the third quarter of the century; industrial countries used two-thirds of all fertilizers. However, the percentages are now shifting toward the developing countries.

Arable Land

As the human population continues to grow, more and more land is used to raise food. The

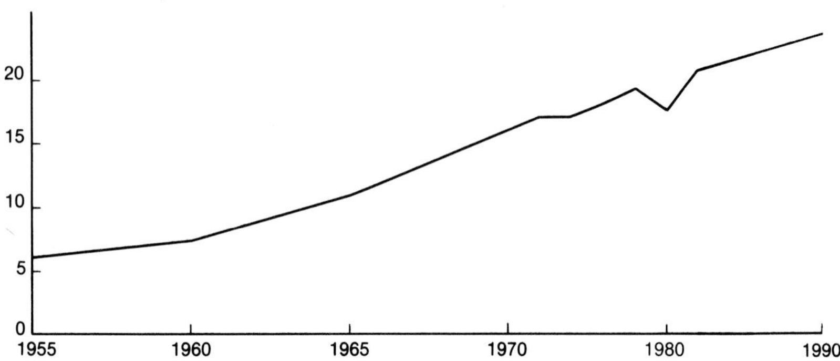

Figure 12.5
Plant nutrient use in the United States. (Tennessee Valley Authority.)

exact amount of land on the earth with soil suitable for agricultural use is difficult to estimate. According to one report no more than 24 percent of the world's land is suitable for plowing. An additional amount can be used for grazing domesticated animals, but about one-fifth of the earth is too cold, one-fifth too dry, one-fifth too rugged, and one-tenth bare or lacking soil to support crops. Irrigating deserts and farming oceans have been suggested, but these methods would take a tremendous amount of energy and would have a major impact on the world's natural cycles of water, air, and minerals. In *The Hungry Planet*, George Borgstrom estimates that some 4.5 billion acres are suitable for agricultural use in the world. This is just slightly less than 1 acre per person. Although estimates of the number of people an acre can support vary and depend on many factors, including the trophic level used for food, most experts visualize no more than five people being supported by 1 acre of land. Estimates for land use patterns in the United States are easier to obtain (Figure 12.7).

Prospects for developing more agricultural land are dim. The tendency over the past 20

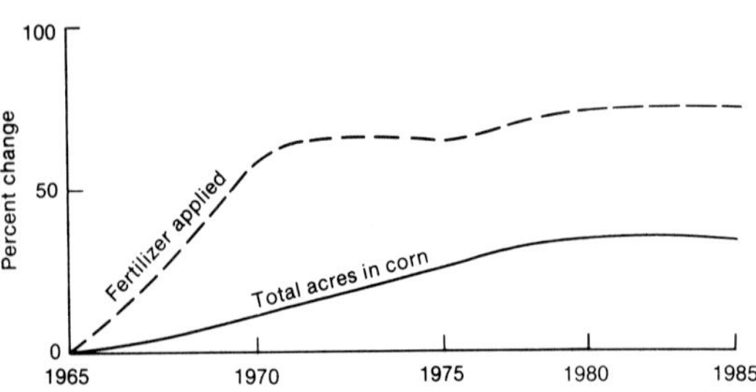

Figure 12.6
Increase in acres of corn planted versus increase in fertilizer use. (Tennessee Valley Authority.)

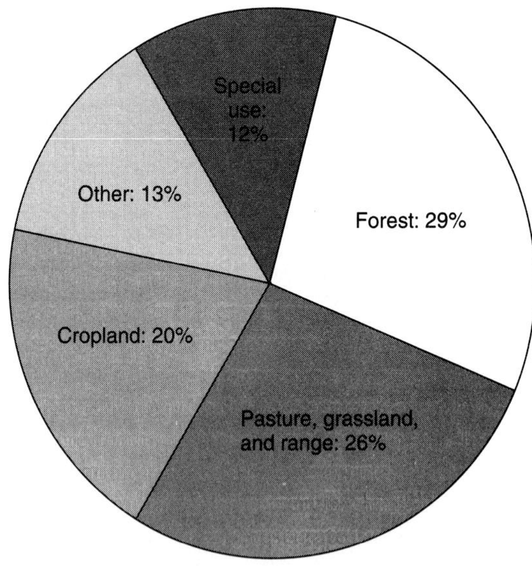

Figure 12.7
Major land use categories for the United States. Special use categories include urban areas, national defense reserves, city parks, and transportation areas such as highways and airports. Others are marshes, swamps, bare rock areas, deserts, and tundra. (Council on Environmental Quality.)

years has been to remove prime agricultural land for housing, industrial development, and road construction (Figure 12.8). Other land has been lost because of poor soil maintenance, due to practices such as overgrazing, failure to restore nutrients to farm soil, erosion, and salt accumulation from irrigation. Roughly one-third of the world population now lives in countries where croplands are shrinking.

In the latter 1950s, world croplands were increasing at about 1 percent per year, thus providing additional food for the world. Because prime agricultural land began to be used for development, cultivated land area increased by only 0.3 percent in the 1970s and 0.2 percent in the 1980s. It is likely that in the 1990s, land for cultivation will increase by an amount that is far below what is needed to maintain food supplies for the projected increase in the world population.

Government programs in the United States influence the amount of cropland. In its effort to reduce government expenditures, the Reagan administration did not idle cropland in 1981 and very little in 1982 by cutting funding programs that paid farmers not to use land. At this time world food supplies were relatively high because of two previous years of bumper harvests in the United States and a worldwide recession that reduced demand. The government apparently overreacted to a farmbelt depression in 1983 by encouraging farmers to divert a large amount of land to nonproductive uses, resulting in the largest diversion in history—over 70 million acres. Then a severe drought occurred, resulting overall, in a 40 percent decline in feedgrain harvest.

During the late 1980s about 11 percent of U.S. cropland was removed from production under the sodbuster and swampbuster provisions of the Food Security Act. These provisions are designed to protect wetlands and land that is highly erodible.

FOOD FROM THE SEA

Oceans, covering some 71 percent of the earth's surface, are an ancient source of food. Today many people mistakenly look to oceans as vast reservoirs of food. Since about 1950, sophisticated technological methods have been used to ensure large catches of fish. Telecommunications, echo sounding, aerial communication, and

Figure 12.8
Housing development in citrus grove in Arizona.

temperature monitoring are used to locate fish, and electronic impulses and light attract them. Extensive information on fish migratory and schooling behavior aids the fishing industry. Modern fishing fleets include processing ships to prepare the catch for the consumer right at sea.

Peru, Japan, Russia, and China have led the world in this expansion into the use of ocean resources (Figure 12.9). Peru's high rank resulted from its coastal anchovy fishing, but that catch decreased in the 1970s. The United States and most western European nations had not expanded greatly in the use of ocean resources in this period. Most poorer nations could not afford to fully utilize marine resources.

The amount of food extracted from the ocean increased from 31 million tons in 1955 to 66 million tons in 1970. By 1987, 79 million tons were removed from the oceans. Starting in the 1970s, many signs of overfishing appeared. Growth in the annual world fish catch slowed from 6 percent per year to less than 1 percent. The collapse of anchovy fishery was another sign. Other seafood items were also declining, including the Alaskan king crab, Atlantic herring, and haddock.

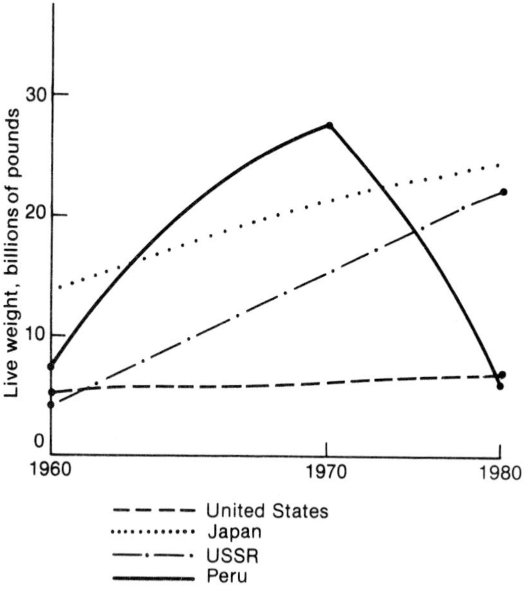

Figure 12.9
Aquatic catch in selected nations from 1960 to 1980. Note the marked increase in catch of Peru until 1970. Japan and the USSR show large increases, while little change occurs in the U.S. catch. (National Oceanic and Atmospheric Administration.)

Experts estimate that oceans could provide 110 million tons of biomass per year without destroying their populations. Nearly half the biomass currently removed is used for purposes other than food, such as manufacturing oil and fish meal which indirectly reach the tables of Western nations. Furthermore, much of the catch is not considered edible. There are over 21,000 species of living fish known, but only about 10 are sought by commercial fisheries. These 10 species are the major sources of food from the sea. They are large or appear in large schools, making them easy to locate and remove. A number of oceanic fishing efforts have led to overfishing, depletion in stock, and an actual decline in catch. Other forms of seafood, such as crab, lobster, and abalone, continue to be used primarily for luxury foods.

People are predators when they remove organisms from the ocean. In order to obtain a continuous supply of biomass from a particular population, the proportion of the total population removed must be considered. *Underfishing* takes fewer fish than the ecosystem can provide, causing other mortality factors to control the population growth. *Overfishing* lowers the reproductive level of the population by removing fish before they mature (Figure 12.10). The goal of the fishing industry should be to establish a *sustained yield.* Closed seasons, catch quotas, nets with larger mesh size, and minimum fish size can help to achieve a sustained yield.

One of the most documented cases of explotation is that of whales. Whales are mammals that generally reproduce only once a year and care for their young for 2 years. Because of thei slow rate of reproduction, it is doubtful that the

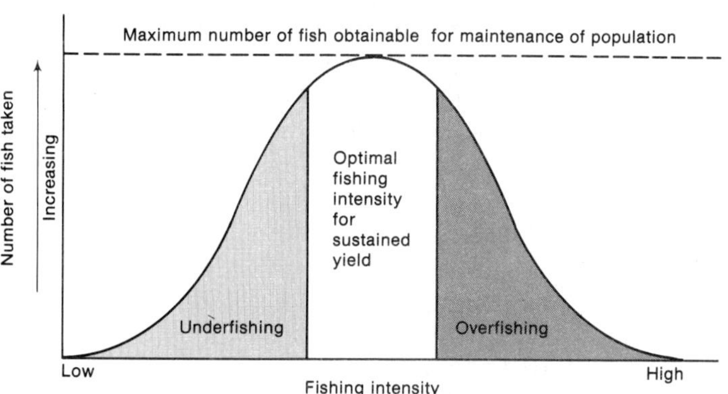

Figure 12.10
Relationship between sustained yield and fishing intensity. Underfishing causes a shift from people to other population mortality factors. Overfishing removes so many fish that changes occur in the species' reproductive behavior.

will become plentiful enough to be used as a resource in the future. The blue whale, the largest creature on earth, appears headed for extinction. The mature animal weighs 83 to 132 tons. Before restrictions were set by the International Whaling Commission, blue whales were caught primarily for their oil and for use in fertilizer and pet food.

How can we prevent further deterioration of the living component in the world's oceans? We must apply the basic principles of ecosystem management: all organisms are part of the natural system; each species has special behavioral attributes that must be understood; and population cannot be depleted below the level at which it can replace itself.

We must also remember that coastal areas, particularly estuaries where the flow of river waters meets the ocean, are vital to marine life because these are the areas where many fish spawn. They are important for anadromous fish that migrate between fresh and salt water and are the habitat for oysters, clams, and some scallops. Estuaries are also important as human recreational areas, and people must not use them as waste dumping grounds.

Aquaculture, a centuries-old practice, is still evident in many parts of the world today. Enclosed lagoons in bays and estuaries in the warmer latitudes provide "farms" for carp, tilapias, gray mullet, and milkfish. Fenced-in coastal areas of Japan provide sources of food for fish fry, shrimp, and prawns. In other areas of the world, clams, oysters, and mussels are placed in screened boxes and lowered into the ocean to feed. Catfish farms are popular in the southern United States (Figure 12.11). The process of aquaculture is important from an energy standpoint because people generally feed on the second or third trophic level—not the fourth or fifth level, as is true for most seafood products. The areas where large amounts of nutrients are

Figure 12.11
Catfish farm ponds in Arkansas. (U.S. Soil Conservation Service.)

available in the oceans are very limited, however. Over 90 percent of the oceans are biological deserts due to lack of nutrients. Coastal areas where aquaculture is possible could provide only a limited amount of additional food for the world.

Through a national education and subsidy program, India has successfully developed aquaculture in rural communities for a variety of fish and shrimp species. This effort has provided jobs as well as low-cost protein to a nation with massive population growth.

Aquaculture methods are being developed to raise fish in freshwater and saltwater ponds. Carp, eel, catfish, tilapias, and milkfish are currently raised in lakes and ponds and some rice paddies in Asia. Using waters for fish cultivation much as we use land to grow livestock could make additional food available.

ENERGY AND FOOD

An important part of the population-food equation is energy production. Today few people in the developed countries come in contact with the soil for food. Neatly packaged food in supermarkets or food at fast-food restaurants are important parts of the American diet. Energy is expended to grow, harvest, process, transport, wholesale, retail, preserve, and cook these foods.

Since 1940, food energy consumed per capita in the United States has increased only slightly. However, during that same period the amount of energy obtained in food per energy input has decreased. If we assume that each bushel of corn can provide 1800 Calories of food energy, then the actual yield in Calories* per input of Calories has decreased 24 percent between 1945 and 1970 (Figure 12.12). This has occurred because total energy use in the food system increased more than 300 percent from 1940 to 1980 (Figure 12.13). All sectors of the food production cycle contributed to this rise.

By looking at the Calorie input versus the Calorie output, we can see clearly that the food system in the United States is receiving an energy subsidy. The United States uses about 10 Calories of fuel for each Calorie of food consumed. Despite recent hikes in energy costs, the United States still has relatively cheap energy. Perhaps this becomes more significant when we consider that 80 percent of the world's annual energy expenditure would be required to feed the world if it had a food system like that in the United States. As the energy costs rise, however, U.S. food prices will continue to go up. Clearly, major changes are necessary, particularly in the developed countries, if the food supply is to con-

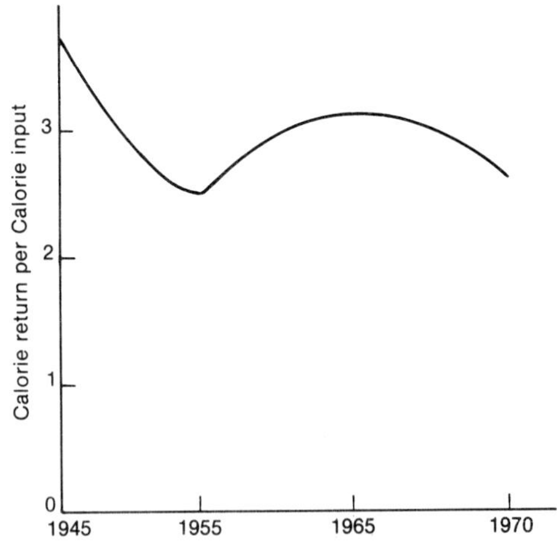

Figure 12.12
Return in Calories per input of Calories for corn in the United States. (Assume 1 bushel of corn provides 1800 Calories of food energy.) (Original data from D. Pimental et al., "Food Production and the Energy Crisis," *Science* 182 [1973]: 443–49. Copyright 1973 by the American Association for the Advancement of Science.)

*A food Calorie is defined as the amount of energy required to raise the temperature of 1 kilogram of water from 14.5° to 15.5°C at constant pressure. This is commonly called the large Calorie with a capital C. In physics, the small calorie (with a lowercase c) is defined as the amount of energy required to raise the temperature of 1 gram of water 1°C. The nutritional Calorie is equal to 1000 physics calories.

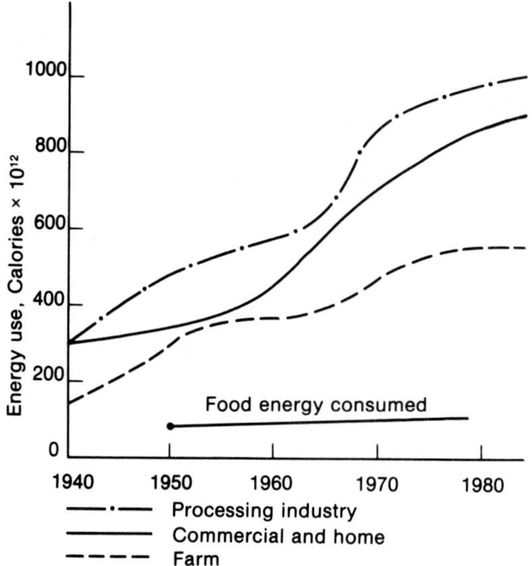

Figure 12.13
Increased energy use in the U.S. food system. Farm energy includes the manufacture of fertilizers, planting, and harvesting. Processing involves preparation, packaging, and transportation. Commercial and home energy use includes refrigeration, cooking, and storage. Food energy consumed increased from approximately 125 to 275 Calories.

tinue to sustain the present population level. Ultimate limits to the world food supply are imposed by land, water, energy, and nutrients.

NUTRITION REQUIREMENTS

More important than food quantity is food quality. Human nutritional needs can be supplied by three categories of food substances:

1. Proteins—body-building substances for growth, repair, and maintenance.
2. Vitamins and minerals—key elements or compounds in many essential body processes.
3. Fats and carbohydrates—energy supply for the operation of the living body.

The prime sources of protein are animal products, including meat, fish, eggs, milk, and cheese (Figure 12.14). In addition some plants such as soybeans contain useful proteins. Proteins are complex chemical chains of carbon, hy-

Figure 12.14
Four major classes of food needed in the daily diet.

DAILY FOOD NEEDS

Requirement	Source
Carbohydrates	Breads and cereals 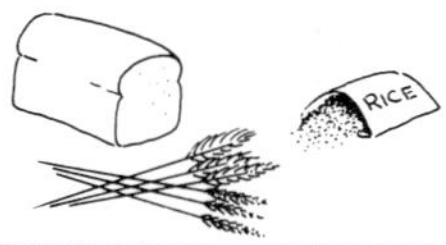
Minerals and vitamins	Fruits and vegetables 
Proteins	Meat, fish, poultry 
Fat-soluble vitamins, some minerals and proteins	Dairy products 

VITAMIN AND PROTEIN SUPPLEMENT FOR LIVESTOCK

The contents of the single yeast or bacterial cell are now being touted as a meal in a pill. Containing 50 to 70 percent protein, rich in vitamins, and high in minerals such as phosphorus, these pills are now sold by several companies as supplemental food for livestock.

Bacteria and yeast are ground and other nutrients dried and encapsulated as a livestock supplement food. In tests at Iowa State University, chickens raised on processed yeast grew as fast as those fed soymeal—the most common protein supplement. Piglets fed on a blend of yeast and soy did better than piglets eating one or the other.

One of the disadvantages of feeding microbes to farm animals is cost. Currently, prices of protein supplements from microorganisms do not compete well with soybean meal, so the processed yeast and bacteria known as single-cell protein may find a limited use. They can be fed to weaning calves, which have trouble digesting soy but can easily handle proteins and yeast.

Other countries in which soybean products are not as readily accessible, as in the Middle East, are experimenting with the use of protein supplement in pills to feed livestock.

For further information, see T. Monmaney, "The Bug Catalog," *Science* 85:38.

drogen, oxygen, nitrogen, and in some cases sulfur and phosphorus. The chemicals are combined into small units called *amino acids*, the building blocks of proteins (Figure 12.15). Of the twenty-two amino acids known to be essential for the operation of the human body, eight cannot be synthesized by the human body and must be obtained from the protein food supply. These eight amino acids must be present simultaneously in correct amounts before the body can use them. Once proteins enter the body, they are broken down into amino acids during digestion and then transported by the blood to sites where they are needed. There they are assembled into the protein required.

A balanced diet generally provides the body with all necessary vitamins and minerals. Some vitamins are water soluble and others fat soluble, so a completely fat-free diet can result in vitamin deficiencies. The major minerals utilized in the body are calcium, phosphorus, magnesium, potassium, sodium, chlorine, and sulfur. Calcium, the most abundant mineral in the body, is found mainly in bones, teeth, and blood. It functions in the regulation of muscle contraction, activation of enzymes, blood clotting, and nerve transmis-

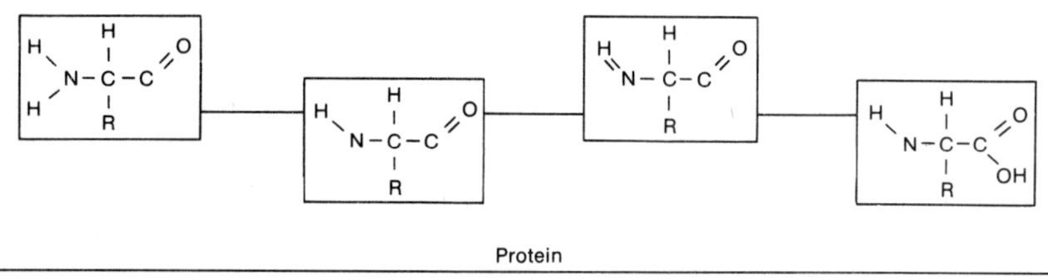

Figure 12.15
Amino acids serve as the building blocks of proteins. Each block represents one amino acid. R depicts different side chains, which vary in different amino acids.

sion. Small or trace amounts of some minerals are essential for good health (Table 12.3). Some trace elements are involved in chemical reactions such as enzyme activation in the body. Others, such as iron and iodine, are essential components of compounds needed by the body. Iron, the most important trace element, is a component of *hemoglobin*—the compound in red blood cells that carries oxygen to the tissues. Deficiencies of vitamins or minerals cause some human diseases. Goiter, for example, is linked to a deficiency of iodine in the diet. Scurvy is found in people who have diets lacking in ascorbic acid (vitamin C).

Fats and carbohydrates are the primary foods providing fuel for the body. The energy contained in food is measured in Calories. We determine the Calorie content of food by taking a known amount of food substance and calculating how much energy (heat) is released when the food is burned. Fats release about 9 Calories per gram, while carbohydrates and proteins each yield 4 Calories per gram. Individual Calorie requirements vary considerably. For example, a very active person such as a lumberjack might need 5000 Calories per day, but an office worker might get by with only 2000.

Table 12.3
Minerals Required in the Human Diet

Major	Trace	
Calcium	Iron	Fluorine
Phosphorus	Copper	Selenium
Magnesium	Iodine	Cesium
Potassium	Manganese	Zinc
Sodium	Molybdenum	Chromium
Sulfur	Cobalt	Vanadium

Source: U.S. Dept. of Agriculture.

Food has a major role in individual health. The term "individual" is used because each person's body responds differently to various foods. Although many people are allergic to some foods, they seldom associate the fact that excess gas or intestinal cramping can be symptoms of food allergies. Evidence has also shown that excessive amounts of sugar can cause antisocial behavior, stress, and depression. People who eat a lot of food high in fat content tend to have more health problems, particularly heart attacks. In order to lose weight, these people require a modification of their behavior patterns. Scientists realize that we have to be concerned with not only how food maintains physical health, but also how it controls human behavior.

Malnutrition

Although infant mortality rates have been lower in recent years, children under the age of 5 are still the major group of people affected by an improper diet. In many parts of the world, infants in particular die or develop poorly because of malnutrition due primarily to the parents' ignorance. An individual who is undernourished might not exhibit a specific dietary deficiency, but could succumb to many bacterial, viral, or parasitic diseases because of a lowered body resistance.

The phrase *protein–Calorie malnutrition* describes a complex syndrome exhibited by young children receiving diets deficient in proteins, Calories, and other essential nutrients. This form of malnutrition results in retarded growth and development as well as an imbalance in many of the body's chemical reactions. Protein- and Calorie-deficient diets most commonly occur together. **Kwashiorkor,** a disease discovered in Africa in 1929, results from a protein-deficient diet in children 1 to 4 years old. The victim slowly loses energy and eats very little, physical growth is retarded, and the hair and skin become reddish (Figure 12.16). Skim milk helps many children with mild cases to recover. In severe cases, the victim's legs, face, and belly become swollen with fluid, hair pulls out at the roots, digestive disorders result, and the child dies. Postmortem examinations show a reduced brain size and a fatty liver. Kwashiorkor commonly occurs when children are weaned because their mother gives birth to another child. The weaned child no longer receives the protein-rich milk from the mother and suffers improper tissue formation because of the protein deficiency. The child is therefore unable to utilize the food supply, so there is no energy available for body functions. Children who survive often have permanent neurological damage.

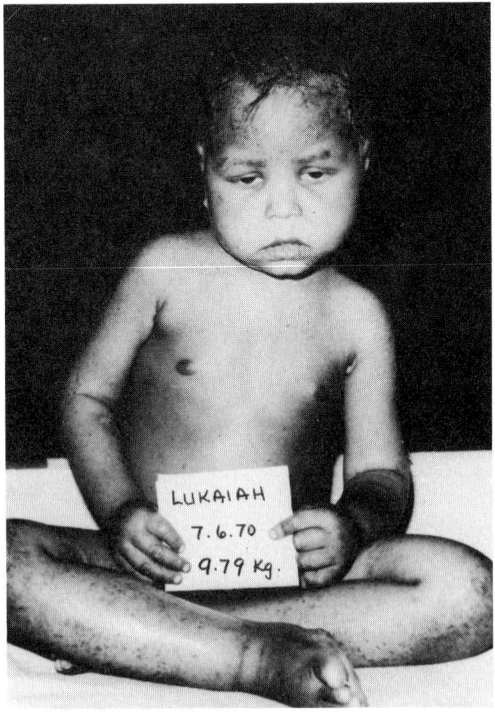

Figure 12.16
Child with kwashiorkor brought to a nutritional education unit in India. After the child was placed on a proper diet for 1 month, the skin and hair lesions and swelling were gone. (Courtesy of the World Health Organization.)

In many parts of the world torn by war and famine, over 50 percent of the infants 1 year old or less have **marasmus,** a disease caused by insufficient food. Marasmus results from a deficiency of many essential nutrients as well as proteins and Calories, causing its victims to suffer from muscle deterioration, loss of subcutaneous fat, and low body weight. This form of malnutrition often occurs in overpopulated city slum areas. Marasmus is increasing in poorer societies where breast-feeding is on the decline and in societies where there is not enough food to feed a weaned child.

People in many parts of the world today have diseases associated with specific forms of malnutrition. Iron deficiency, which can result in anemia, is common in Africa, Asia, and Europe. In the Middle East, 25 to 75 percent of all children are anemic. Even in the United States, iron deficiency occurs in an estimated 20 percent of the population. Current evidence also shows that malnutrition is associated with a decrease in the birth rate. In some countries, the birth rate is declining because undernourished mothers have more miscarriages and fewer children.

World Nutrition

One out of every three people in the world today lives in a country that cannot produce enough food or afford to buy enough from another nation to feed itself. Most of these people live in the four countries of India, Bangladesh, Pakistan, and Indonesia, with the rest in Africa and Latin America. Most major world nutritional deficiencies fall into two categories: inadequate energy and inadequate protein for body-building processes. The average person of 70 kilograms (154 pounds) requires about 70 grams (2.5 ounces) of protein per day—a level many nations do not achieve (Figure 12.17). Much of the protein intake in those nations is supplied by plants that often do not have all the essential amino acids. Although animal products account for about 35 percent of the world protein, they comprise up to 56 percent of the protein intake in the more developed countries. For energy, the average person needs about 2354 Calories per day, a requirement not met in countries with poor food supplies. An encouraging sign is that food energy supplies have been increasing in all regions of the world except sub-Saharan Africa (Figure 12.18). Yet, other nations eat far in excess of this minimum. Some countries, such as Iran and China, do have an adequate supply of food but cannot meet the daily minimum requirement because of problems with distribution and spoilage. The World Resources Institute estimates that 32 percent of the people in sub-Saharan Africa, 22 percent of those in East Asia, and 11 percent of the people in Latin America are undernourished.

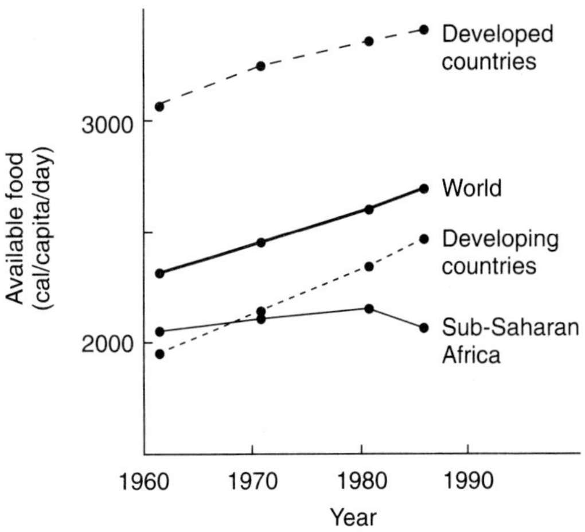

Figure 12.18
Food available for human consumption for the world, the developed countries (North America, Europe, Oceania), developing countries (Asia, Latin America, and Africa), and sub-Saharan Africa, the only region with a declining trend. (From World Resources Institute. 1990. *World Resources 1990–91.* New York: Oxford University Press.)

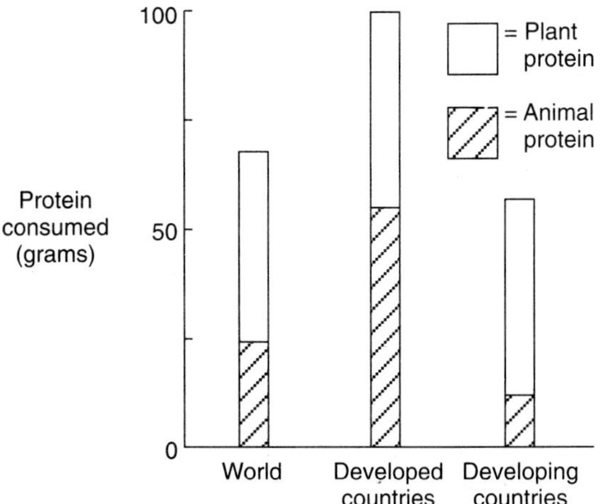

Figure 12.17
Daily per-capita protein consumption for the world, the developed countries (North America, Europe, and Oceania), and the developing countries (Asia, Latin America, and Africa). (From FAO. 1987. *The Fifth World Food Survey.* Rome: Food and Agriculture Organization of the United Nations.)

Just a few developed countries grow more than enough food to feed their own people. World food organizations are trying to undertake massive programs to shift food from the haves to the have-nots and to create world food reserves for nations in time of drought or other crop crises. In addition to the effort and expense of such a program, there must be a major change in the attitude of nations that buy valuable plant proteins from some starving nations unable to afford to buy their own products.

With all its wealth and food, the United States is still unable to provide a balanced diet for everyone in the country. Both financial and educational factors are involved. Food stamp programs help provide food for the poor, but the very poor frequently cannot afford them. In the few instances when they are able to purchase food stamps, the very poor are hindered by a lack of knowledge regarding what constitutes a nutritious diet. Studies of low-income populations reveal that many of these people are suffering from various forms of malnutrition, including some cases of Kwashiorkor and marasmus.

PROSPECTS FOR THE FUTURE

What are the prospects for finding additional food sources or redistributing the present world food supply? Much research is currently being carried out on a number of levels. On the farm and in the sea, new forms of highly mechanized equipment are being developed to produce more food faster. Attempts are being made to bring more land under cultivation and to make better use of the land already cultivated.

Concentrated protein is being developed from meat, fish, soybeans, and peanuts. Fish, higher in protein than beef cattle and potentially one of the richest protein sources, are being raised in confined structures. Different forms of food are being developed from less desirable or less costly products to taste like many popular and well-known ones. The "beef extenders," "bacon chiplets," "turkey-flavored" lunch meat, and imitation pork products on the grocery shelves are all made from soybean and flavor additives. Many forms of microorganisms can also serve as a source of protein in the future. Bacteria, yeasts, algae, and fungi reproduce very rapidly and provide many essential amino acids. Leaf proteins, a source of animal food, may supply over one-half our protein needs by the year 2000. Alfalfa, water hyacinth, and elephant grass can be harvested to provide protein for animals.

Intensive agricultural practices can make arable land support more people; however, this strategy can destroy topsoil more rapidly. Globally, the support ratio of persons per hectare will increase from the current 2.6 to 4 by the year 2000. The ratio will be greater in more densely populated countries like China. Various methods are being employed to achieve greater productivity from limited space. Some chicken farmers keep their chickens in very small cages and use conveyor belts to bring food and carry away eggs. In some areas of Germany, the United States, and the USSR, cattle and hog skyscrapers are highly efficient, although not entirely humane. Plastic domes over fields in the Midwest allow farmers to control the environment of large fields and thereby operate all year. By placing pipes underground, arid land can be irrigated without great water loss. On these farms, there is also a large layer of asphalt several feet below the surface to prevent water loss.

Computer farming is another possibility. By keeping track of farm equipment and the progress of livestock and crops, the computer can inform the farmer when and how much fertilizer or insecticide is needed as well as when the crop is ready to be harvested. Irrigation can be coordinated with current weather control practices.

Agriculture research scientists are encouraging the use of intercropping in many parts of the world. **Intercropping** is the practice of growing more than one crop at the same time in the same field. This practice was actually common in the United States before the 1940s. Intercropping is billed as an alternative to the problems of monoculture, which extracts all of some nutrients from the soil depending on the crop. A combination of factors occur in intercropping. For example, one species may diminish the supply of nutrients but also provide necessary shading for another crop's success. Intercropping requires attention to the time of planting and harvest as well as to ecosystem dynamics.

A growing trend in the United States is to produce fruits and vegetables year-round in family greenhouses, many of which are solar heated. Small city gardens and the use of containers such as pots or window boxes are also contributing to the food supply. These efforts may have to be complemented on a larger scale by programs similar to the World War II victory gardens.

LEGISLATION

Food Security Act (the Farm Bill) of 1985 and 1990. Established the Conservation Reserve Program in which farmers receive annual payments for taking erodible land out of production and planting it in grass or trees for at least 10 years (sodbuster provision). Farmers with highly erodible land must develop government-approved erosion control plans. Similar incentives encourage farmers to protect wetlands, rather than draining them to create new croplands. Farmers may participate in a Wetlands Reserve Program that was added to the Conservation Reserve Program in 1990. Furthermore, those who grow crops on wetlands converted after 1985 can lose their federal benefits (swampbuster provision).

Soil Erosion Act of 1935. Inspired by the problems associated with the Dust Bowl, this law gave the Department of Agriculture the authority to investigate soil erosion and take measures to prevent it. The Soil Conservation Service takes a leadership role in carrying out this mandate.

Even though we may be able to meet the need for large amounts of additional food, new methodologies will have long-term adverse impacts on our natural systems. Furthermore, new approaches to increasing food production raise many social questions, such as what will be produced and who will get the new foods. Our food supply demands are further complicated by the projection that food prices in the United States will rise from 30 to 115 percent between 1979 and 2000.

Until we can implement some of these alternative practices, we face the problem of distributing the food that we have. In 1973 the United States halted soybean shipments to Japan. By 1975 grain shipments were curtailed to the Soviet Union and Poland. Today the United States and Canada produce 80 percent of the export grains. By the year 2000, some food experts feel these countries might be the only ones producing more than they consume. Are these countries to determine who eats and who starves in the world?

SUMMARY AND CONCLUSION

Today we cannot view the world food supply on a nation-by-nation basis. While the Green Revolution has provided a greater supply of fast-growing grains in some nations, the population continues to outgrow the increase in food supply. Biotechnology can provide some alternative foods. Agricultural practices tend to alter natural feedback loops. As we continue to utilize more and more of the world's food energy, we approach the earth's capacity for human population growth. In addition, many interrelated factors control the world food supply. Supply and demand are reflected in food prices, but weather and government controls have major impacts on production. Traditions, income, and local storage and transportation facilities influence people's ability to obtain food.

By improving the nutritive value of the food people presently use, malnutrition can be abated. Energy intake (Calories) must be balanced with the nutritive values of proteins, carbohydrates, fats, minerals, and vitamins in order to reduce suffering from malnutrition.

To properly provide food for the world, we must make the best use of the soil in each region. Soil fertility must be maintained by proper crop rotation and fertilizer use. Better food production results from managing the natural resources of soil, water, minerals, and energy by an application of ecological principles.

No single source, such as the ocean, can be considered a solution to our food problems. Food production can be increased by intensive agriculture, aquaculture, and more efficient use of lands throughout the year. However, we must supplement these methods with conservation measures, such as reducing our use of synthetic and fully prepared foods, in order to use valuable energy more efficiently and productively.

FURTHER READINGS

Altieri, M. A. 1987. *Agroecology. The Scientific Basis of Alternative Agriculture.* Boulder, CO: Westview Press.

Barney, G. O. 1979. *The Global 2000 Report to the President of the United States.* New York: Pergamon Press.

Borgstrom, G. 1972. *The Hungry Planet—The Modern World at the Edge of Famine.* New York: Macmillan.

Brown, L. R. 1989. Reexamining the World Food Prospect. Chap. 3 In L. R. Brown (ed). *State of the World 1989.* New York: W. W. Norton & Company.

Brown, L. R., and J. E. Young. 1990. Feeding the World in the Nineties. Chap. 4 In L. R. Brown (ed). *State of the World 1990.* New York: W. W. Norton & Company.

FAO, 1987. *The Fifth World Food Survey.* Rome: Food and Agriculture Organization of the United Nations.

FAO, 1989. *The State of Food and Agriculture.* FAO Agriculture Series No. 22. Rome: Food and Agriculture Organization of the United Nations.

Hansen, M., L Busch, J. Burkhardt, W. B. Lacy, and L. R. Lacy. 1986. Plant Breeding and Technology. *Bioscience* 36: 29–38.

Horwith, B. 1985. A Role for Intercropping in Modern Agriculture. *Bioscience* 35: 286–91.

Knorr, D., and A. J. Sinskey. 1985. Biotechnology in Food Production and Processing. *Science* 229: 1224–29.

Smith, D. T. 1990. *Americans in Agriculture. Portraits of Diversity. 1990 Yearbook of Agriculture.* U.S. Department of Agriculture.

World Resources Institute. 1990. Food and Agriculture. Chap. 6 In *World Resources 1990–91.* New York: Oxford University Press

STUDY QUESTIONS

1. Explain methods by which the oceans could provide greater amounts of food.
2. Compare the energy expenditure for food derived from the oceans with food derived from the land.
3. Is the term *Green Revolution* appropriate?

4. How would you answer someone who says, "We have no food or population problem because we can put all the population of the United States into two-story homes in California and use the rest of the country to produce food"?
5. People require recreational, urban, and agricultural land. How can the United States avoid using prime recreational or agricultural land for urban sprawl?
6. Discuss the energy required by different methods of food production.
7. What is a balanced diet?
8. How do social factors such as politics, economics, and customs influence food production and availability?
9. What do you see as the ultimate limit to world food production?
10. Modern agricultural practices sometimes have adverse effects on the ecosystem. What are these effects and how can they be avoided?

SUGGESTED ACTIONS

1. Whenever possible purchase food from farmers in your local area. This could encourage diversification of the local food economy. If there is no farmer's market in your area, find out why.
2. Investigate the advantages and disadvantages of organically grown food. If you feel that organic foods are beneficial, encourage your local supermarket to make them available to consumers.
3. Contact agricultural specialists (e.g., Cooperative Extension Agents or the Soil Conservation Service) in your community and find out if local farmers practice conservation tillage and low-input agricultural methods. If they do not, find out why.
4. Visit a farm or ranch and find out what is involved in growing food. Pay particular attention to the decisions that agricultural people have to make as they try to produce food in ways that are economically sound and, at the same time, ecologically sustainable.

13

ENERGY

Did you know that your body and your car use the same form of energy? Both use solar energy converted by green plants into chemical energy. The sun's energy and green plants are essential for sustaining life. Fossil fuels, formed from plants that lived millions of years ago, are storage units of the sun's past energy and are currently the major sources of energy used to maintain our society.

Because readily accessible fossil fuels are now known to be in short supply in the world, we are increasingly extracting energy from other sources such as solar power, nuclear power, and wind power. Why do we not use these sources alone and conserve our fossil fuel supplies? Why do we have an energy problem at all with so much potential solar and nuclear power? What effects do our procurement and use of energy have on the environment? These are some of the questions we are going to explore in this chapter.

WHAT IS ENERGY?

Energy can be defined as the ability to do work, and **work** results when a force is applied through a distance. You do work when you push a rock over a cliff; an engine does work when it moves an automobile; a living cell does work when it divides. Work occurs only if a force causes some object to move. When you push a rock over a cliff, your body uses energy to overcome the forces of friction, inertia, and gravity holding the rock in place. The bigger the rock, the more work is required and therefore the more energy is needed to move the rock. **Power** is the amount of work done in a given time interval. It is measured in either *horsepower*, where 1 horsepower results in the movement of 550 pounds 1 foot in 1 second, or *watts*, in which 746 watts (W) equal 1 horsepower. Electric output is often measured in kilowatts (thousands) or megawatts (millions).

Potential and Kinetic Energy

Energy can be divided into two basic forms: **potential** and **kinetic.** Potential energy, represented by a rock poised at the edge of a cliff or water in a cloud, is energy available to do work. Living organisms store potential energy in the form of high-energy phosphate bonds which, when broken, release energy for use by the living system. Kinetic energy, on the other hand, is associated with movement. A moving car, the wind, falling water, and the earth's movement all have the kinetic energy of motion. You use kinetic energy to place a book on top of a table. The book on the table has potential energy that would become kinetic energy if it were to fall to the floor.

Thermodynamic Laws

Energy, or the accountability of energy, is described by the **first** and **second laws of thermodynamics.** These laws help us understand what happens to energy used to perform work and why energy cannot be recycled like minerals in cans and bottles. The first law, known as the *law of conservation of energy,* states that energy is never created or destroyed but can be transformed from one form to another. The total amount of energy available in a system always remains the same, although it can be distributed in different forms at different times. As gasoline burns, it releases light, thermal radiation, and heat energy which are in part converted to energy to move a car. Although a specific amount of energy in gasoline changes to other forms when burned, energy is neither created nor destroyed.

The second law of thermodynamics states that no reaction involving the transformation of energy from one form to another occurs without some energy changing to a less usable form (heat). No energy transformation is 100 percent efficient. For example, coal is burned to produce steam to turn electric generators, but not all the potential energy of coal is converted to usable heat energy. Heat losses occur in the furnace, some of the steam cools before it can do productive work, and friction in the generator further reduces the amount of usable heat. A young French engineer, N. L. Sadi Carnot (1796–1832), studied the efficiency of steam engines. The difference between the heat received by the engine and the amount rejected or lost by friction equals the amount of heat energy available for doing work. Carnot pointed out that if 100 percent efficiency were to be achieved, no heat could be projected or lost. Therefore, he concluded, steam engines are less than 100 percent efficient. Since most piston steam engines had an efficiency of 10 to 21 percent, those with a high efficiency became known as a Carnot engine because this engine, functioning at high temperature or with a mixture of natural gas and air, operated at 42 percent efficiency.

The sun resupplies the earth with energy that can be converted into potential energy by green plants, making energy available to us in food. From Chapter 2 we know that a high proportion of the energy at each level in the food chain is lost as heat. This is another example of the second law of thermodynamics and it is an important consideration in the balance of energy on earth, the flow of energy from the sun, and energy movement within and between living organisms.

ENERGY DEMANDS AND PRODUCTION

Before industry developed, human demands for energy were primarily associated with food. Energy from the sun provided the food necessary to sustain life. As culture evolved and societies advanced in technology, energy in larger and larger quantities was needed to maintain a new stage in human life. People began to use various forms of energy to counteract the homeostatic, or steady-state mechanisms of natural systems and altered the environment in the process. For example, forests were destroyed as wood was used for fuel. In about 1885, a shift in the major source of industrial and home energy occurred as coal took the place of wood (Figure 13.1). In the early 1940s, petroleum and natural gas became the major sources.

Since 1900, world use and production of energy has increased more than tenfold, but at an irregular rate. During the past 50 years alone, the energy production of the world has increased over fivefold, and on a per capita basis we find it has almost tripled. During the First and Second World Wars and the depression of the 1930s, energy production slackened. It in-

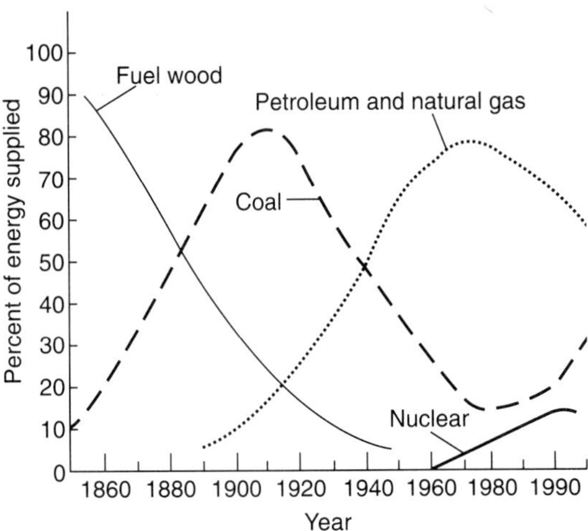

Figure 13.1
Development of new energy sources in the United States. Coal passed wood as the primary fuel source in 1885, and petroleum passed coal in 1944. (U.S. Department of Energy.)

U.S. relations are shaky, at best. In the 7 months prior to Iraq's invasion of Kuwait, over half of the oil in the United States was imported, the highest level ever. Projections show that U.S. oil wells are likely to run dry in large numbers as we enter the twenty-first century. Thus, a greater dependence on foreign oil is likely (Figure 13.2).

From 1985 to 1990, world energy consumption rose about 3 percent per year. Along with increases in the developed nations, many less developed nations are developing a strong dependence on oil. During the 1980s energy consumption increased fastest in the countries of Southeast Asia. In 1988 South Korea increased its consumption by 13 percent and countries in Africa increased their consumption by 4 percent.

Lower energy prices, with a decline in the price of crude oil, have been instrumental in increased again during the post-World War II period.

As the labor force expanded in the 1950s and 1960s, rapid economic growth occurred. Readily available, inexpensive oil and gas were the basic fuels supporting economic expansion. In recent years consumption of these fuels has grown at the rate of 7 to 8 percent per year. Production and use of oil and gas peaked in the early 1970s when major increases in world oil prices signaled the advent of long-term change in the use of world energy supplies.

During the period from 1973 to 1975, a recession and higher prices reduced the demand for petroleum products in the United States. Oil consumption increased again between 1975 and 1979, then declined by about 18 percent between 1979 and the end of 1984. During this same period, demand for coal and nuclear fuels increased.

In 1991 the United States was again confronted with the fact that it imports nearly 40 percent of the oil it consumes. Much of the oil is imported from the Middle East countries where

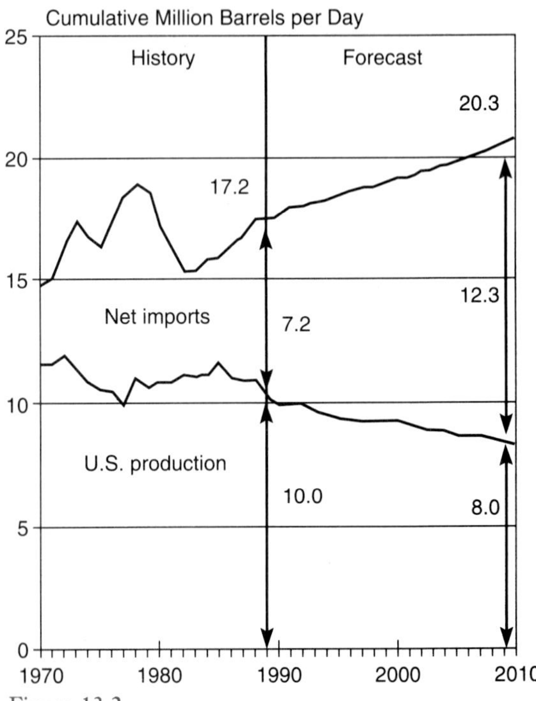

Figure 13.2
Petroleum production and imports in the United States.

creasing energy use. Computerized systems for finding and removing petroleum products has greatly decreased costs of oil and gas. Still, we must remember that these resources are nonrenewable. Conservative estimates indicate that world consumption of oil will increase by 35 percent between 1985 and 2000.

According to some estimates, energy consumption in the next century will not continue at the same dramatic growth rate if recommended conservation methods are followed. This projection is based on an assumed decrease in the population growth rate and a slight decrease in the amount of energy used per individual.

A major problem facing the world today is the inequality of acquiring, distributing, and using energy resources. Institutional and political factors complicate the problem. As industrial nations continue to grow and less developed countries' needs for energy increase, the present supply of energy will be strained and eventually prove insufficient.

Development of New Sources

To investigate new sources to meet our increasing energy demands, long-term research is needed (Figure 13.3). Grants, most of which are awarded by the government, support much of this research. Such grants must be part of a well-planned energy development program that considers both short- and long-term energy needs. Long-term energy requirements seem particularly difficult to address, and political pressures can cut energy research on projects before their worth can be properly evaluated. A federal decision to abort solar energy research in 1970 is one example. Questions about the experimental liquid metal fast breeder reactor, which we shall discuss later in this chapter, will continue into the late 1990s and beyond.

Technology for the use of conventional nuclear power is well developed because the scientists who assembled to develop the atomic bomb during World War II received support to find ways for the peaceful use of the atom. Such sup-

Figure 13.3
New sources of energy being proposed and developed. (Courtesy of Exxon Corporation.)

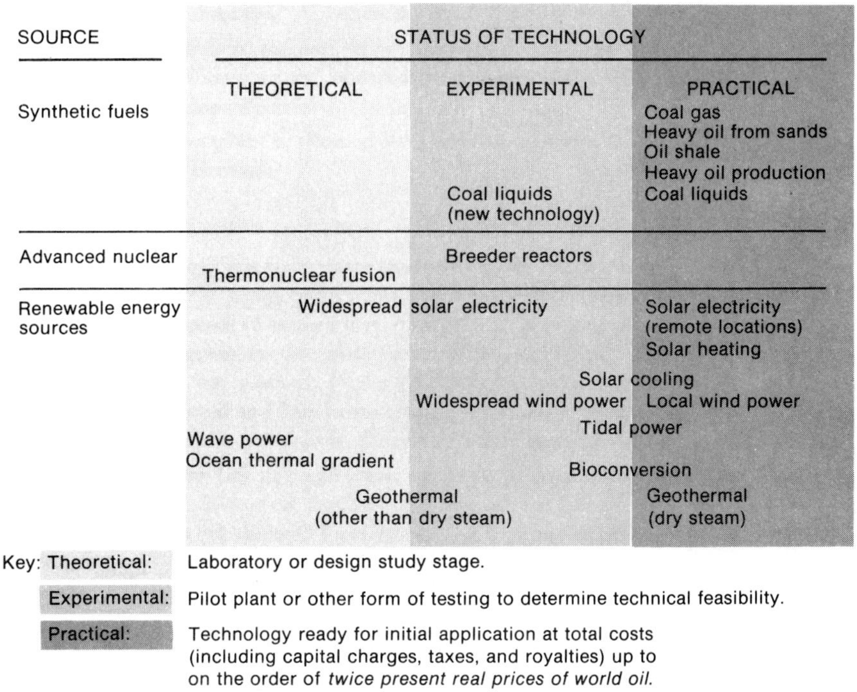

SOURCE	STATUS OF TECHNOLOGY		
	THEORETICAL	EXPERIMENTAL	PRACTICAL
Synthetic fuels		Coal liquids (new technology)	Coal gas Heavy oil from sands Oil shale Heavy oil production Coal liquids
Advanced nuclear	Thermonuclear fusion	Breeder reactors	
Renewable energy sources	Widespread solar electricity Wave power Ocean thermal gradient Geothermal (other than dry steam)	Solar cooling Widespread wind power Tidal power Bioconversion	Solar electricity (remote locations) Solar heating Local wind power Geothermal (dry steam)

Key: Theoretical: Laboratory or design study stage.
Experimental: Pilot plant or other form of testing to determine technical feasibility.
Practical: Technology ready for initial application at total costs (including capital charges, taxes, and royalties) up to on the order of *twice present real prices of world oil.*

HAWAII: SUGAR WASTES PROVIDE ENERGY

Hawaii, which has no fossil-fuel reserve of its own, relies on oil imports for most of its energy needs. In order to lessen the island's 92 percent dependency on imported fuels, officials have developed a plan that calls for alternative energy sources to provide 90 percent of the state's electricity by 2005. The plan is based on producing energy by burning bagasse, a fibrous residue of sugar cane. In the late 1960s when environmental regulations forced the sugar industry to stop disposing of bagasse by dumping it in the ocean, the sugar producers began burning the waste. As oil prices quadrupled in 1973 and 1974, industry owners considered the possibility of turning bagasse into fuel. By burning this waste product, the sugar companies were able to generate more electricity than they needed. During 1980, the Amfac Sugar Company sold 21 million kilowatt-hours of electricity to the Hawaiian Electrical Company. Presently, the largest island in the chain, Hawaii, supplies nearly 22 percent of its electricity needs by burning bagasse. It is estimated that the efficient use of all Hawaii's bagasse would enable the sugar industry to meet all of its own electricity needs while selling enough power to provide 8 to 10 percent of the entire state's requirements.

port did not include the study of the environmental effects of radioactive waste disposal, however. Around 10 percent of the energy used in the United States is produced by the nuclear power process, but scientists project that nuclear power could provide more than half the electrical energy needed by the United States in the year 2020. These projections could change as technologies for other sources are improved.

Denis Hayes, the environmentalist who organized the first Earth Day, argued in 1978 that if we made an immediate and total commitment to develop them, the various forms of solar energy could provide over 80 percent of the world's energy needs by the year 2025. No such commitment has been made, however, nor is it likely to be, but Hayes's proposal points out that policy decisions can affect where our energy will come from and how we will use it.

The price of a particular form of energy determines its development. Costs are incurred from many directions. We are able to produce energy from shale oil and synthetic fuels, but it now takes more energy overall to produce and use shale oil than the oil itself would yield. As the price of fossil fuels soars, however, these sources could become practical. Taxes levied on fuel-producing land also influence costs. Low taxes on land where fossil fuels are found are partly responsible for their low price, but environmental safeguard regulations are causing prices to increase. Reclaiming the devastated land from strip mining (Figure 13.4) should be included in our accounting of fuel cost.

Figure 13.4 Unreclaimed mountainside scars winding through Appalachia show marks of past contour strip mining of coal. (U.S. Bureau of Mines.)

The time it takes from the decision to construct a production facility until energy is commercially produced is called *lead time*. Lead time can vary from six months to two years for oil and gas production in the continental United States to 12 years in more remote areas of Africa and Latin America. Lead time for nuclear energy in the United States is about 12 years; however, government officials are making efforts to reduce that time.

ELECTRICITY

One of the greatest demands for energy is in the form of electricity. Thus, much of our discussion of energy sources will relate to the use of raw materials in the production of electricity.

Electricity is the current resulting from the flow of electrons in a conductor. An electron is a very small, negatively charged subatomic particle, and a conductor is made of copper wire or similar material through which free electrons can move easily. Electric current is relatively easy to produce. A coiled conductor, such as copper wire, is pulled back and forth through a magnetic field. The conductor is loaded with negatively charged free electrons which react to the magnetic field. By forming a loop with the conductor and placing it between the north and south poles of a magnet, an electric current is made to flow through the loop. When the loop is rotated, the current will flow first in one direction and then in the other, creating what we call alternating current (ac). Large generating facilities use an energy source such as water, steam, or wind to turn the blades of a turbine. Steam can be generated by burning coal, oil, or wood, by controlled nuclear reactions, or by solar heating. The conductor loops are attached to the spinning turbine so they move back and forth through a magnetic field (Figure 13.5). The electricity generated is then transmitted via conductors to areas where it is used for lighting, heating, cooling, or other purposes.

The cost of transmitting electricity continues to rise sharply as problems are recognized and treated. Environmental damage from transmission line cuts has resulted in expensive research to achieve the best placement. Throughout their length, some high-voltage electric lines are uninsulated except for the surrounding air. Extremely high voltage (1000 kilovolts) causes lines to spark in the air, creating high levels of ozone

ELECTRIC VEHICLES ARE COMING

Electric vehicles are produced in many parts of the world and may be on the road in large numbers in the 1990s. They are quiet, nonpolluting, and use no oil. They generally have an electric motor, operated by batteries. The vehicles take sixteen to thirty lead acid batteries like the ones in gasoline-powered cars. They take 1000 pounds of lead batteries to produce the same amount of power as 2 gallons of gas. The batteries require frequent charging of 6 to 8 hours. Generally, the car can go 110 to 120 miles on a charge, making them suitable for local driving. Were our world designed for distance travel in this vehicle, service stations could swap battery packs in the same time it takes to fill the gas tank.

While the electric car has been around for a good part of the twentieth century, its evolution has been limited by the battery. Development of a better, more efficient battery would do a lot to make the vehicle attractive.

Recently, hybrid cars have been developed which run on electricity but also contain small gasoline-propelled generators or motors. This extends the range of the car.

Pollution for electric cars is still existent. Production patterns and disposal of batteries all add to pollution. Total hydrocarbon and carbon dioxide pollution is much less than from conventional cars.

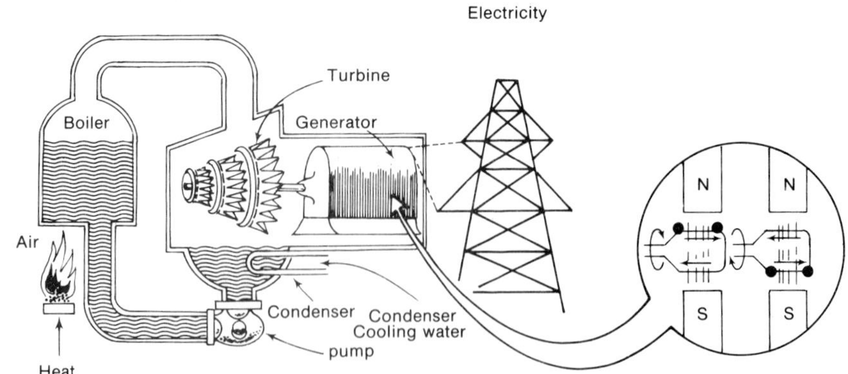

Figure 13.5 Generating electricity requires a power source such as steam. Turbine blades rotate a shaft containing conductor loops, which pass back and forth through magnetic lines of force (insert), producing alternating current. (U.S. Department of Energy.)

in the nearby area—a health hazard. Scientists are developing special forms of liquid and gas for insulating material as well as investigating the possibility of underground transmission. With an anticipated three- to sixfold increase in the use of electric power by the end of the century, solutions to these problems are imperative.

PRIMARY SOURCES OF ENERGY

In this discussion we call energy that requires no conversion process a *primary source* of energy. Sources from which energy can be derived directly include solar radiation, tidal energy from the earth–moon–sun system, and nuclear and thermal energy from the earth. *Secondary sources* include those forms requiring other energy input to convert them to their fuel state. Examples are fossil fuels, wind, water, and hydrogen. The world energy flow is broken down into its major components in Figure 13.6.

Solar Energy

Solar radiation is our greatest energy source, exceeding the sum of other sources we use by a factor of 5000. The direct conversion of solar energy by photosynthesis utilizes less than 0.03 percent of the incoming solar radiation (see Figure 13.6). That energy is used as the foundation of all life on earth and as a future source of energy when stored as fossil fuels. Close to 80 percent of the solar radiation the earth receives is reflected back to the atmosphere or is lost as heat. Some of this sunlight can be collected and converted to electricity by several methods, including *heat conversion* and *direct conversion*. Solar energy stored in the ocean could be used by tak-

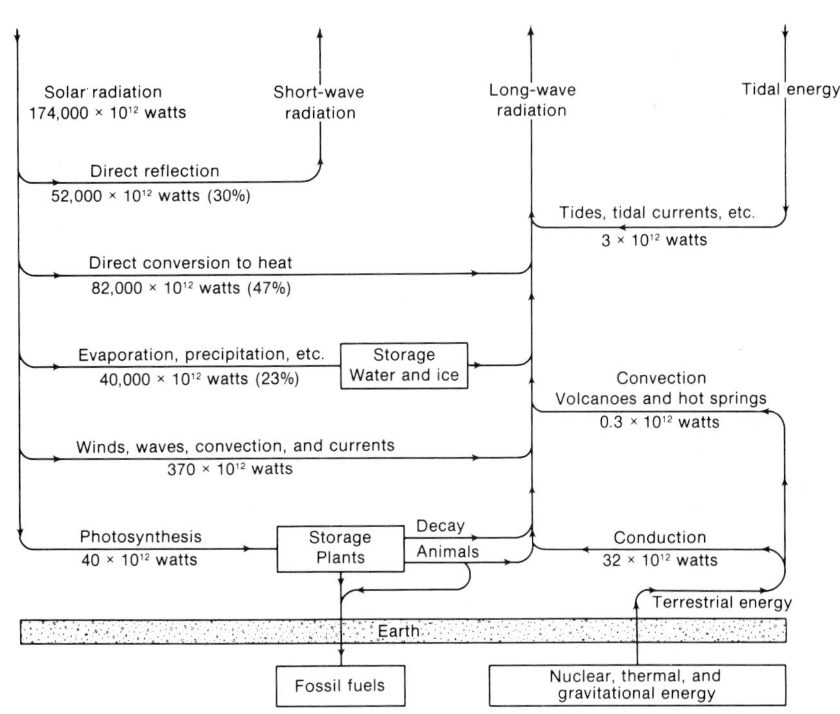

Figure 13.6
The world's energy flow. Note that the principal input is solar radiation. Tidal, nuclear, and thermal energy—all from the earth's surface or interior—constitute a very small part of the earth's energy potential. (From M. King Hubbert, "Survey of World Energy Resources," *Canadian Mining and Metallurgical Bulletin* [July 1973]:38.)

ing advantage of the temperature difference between the sun-heated ocean surface and the colder water under the surface.

The direct use of solar energy is not new. In 1878, A. Mouchot demonstrated a steam engine driven by solar power. In the 1880s, seawater was distilled by solar power in Chile to provide fresh water. A solar collector provided power for a 100-horsepower piston engine in Egypt in 1912. Despite these early starts, economical and technical questions remain unanswered. The Department of Energy believes solar energy technology offers the potential for supplying as much as 25 percent of the energy needs of the United States by the year 2020 if these questions are resolved.

In 1981 about 70 million residential and commercial buildings had the ability to use solar energy for as much as 20 percent of their annual heating and cooling needs. At present, 20 percent of the residential and commercial buildings in this country have *active* or *passive* means of solar heating or cooling. One of the main reasons for this growth in solar usage is federal—and some state—income tax incentives for the installation of solar panels.

Passive solar systems make use of the spontaneous movement of heat by conduction, convection, radiation, or evaporation. They rely heavily on location and architectural design to admit solar radiation in the winter and to block it in the summer. Usually, optically transparent surfaces of glass or plastics on the exterior of buildings allow solar radiation to enter. The radiation is absorbed by a surface, usually dark in color, thereby raising the surface's temperature. The heated surface emits *long-wavelength (infrared)* radiation which is trapped inside the building because glass and plastic prevent the loss of this form of radiation. This is referred to as the *greenhouse effect* (Chapter 8). Some form of *thermal mass* (storage system) is used to retain the heat without pumps or fans.

An active solar system is similar to a passive one except that it contains a distinct collector, heat storage medium, and pumps or fans to circulate the heat (Figure 13.7). In California, commercial solar thermal electric power plants are increasing in number and becoming cost-competitive with conventional power generation. By 1994 about 3 percent of the electric energy use in the Los Angeles area will be from solar facilities. This system uses rows of trough-like mirrors that track the sun with the help of light-sensing instruments. Mirrors focus the sun's light onto coated steel pipes. Oil in the pipes is heated, which passes into a heat exchanger which generates steam to drive turbines. The thermal energy can also be used to operate a compressor-type air conditioner for cooling.

Many homes in the United States have installed solar collectors, liquid-filled tubes under a flat, coated surface. Energy produced by the heated liquid can be used to operate water heaters and heating or cooling systems.

Greenhouses have also become popular as a means of heating homes and growing additional food year-round. A small 3 by 6 meter (10 by 20 foot) greenhouse that is attached to the south side of a 186-square-meter (2000 square-foot) house can provide 30 percent of the winter heating needs of homes in the northern latitudes of the United States.

Hybrid solar wall panels are being tested in Pennsylvania as a means of incorporating solar thermal collection, storage, and distribution into existing buildings. The panel, which uses phase-change material to collect, store, and distribute solar energy, is expected to reduce electrical space heating demands and save energy. Phase

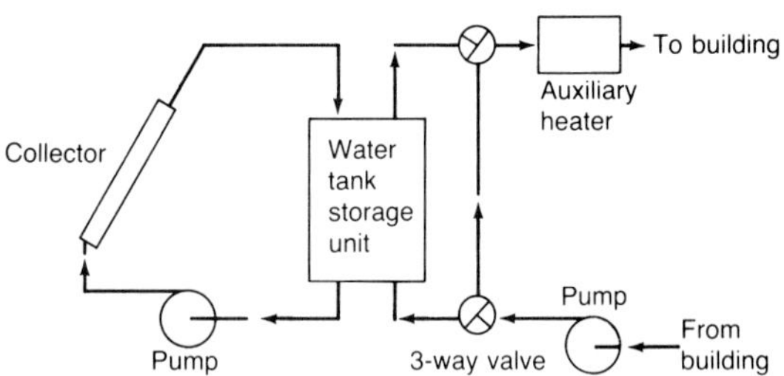

Figure 13.7
An active solar system that uses water to transfer heat to the building.

materials change from a solid to a liquid (change of phase) as they absorb energy, returning to a solid as they release energy. This material can absorb 70 times more energy per pound than water at temperatures between −6° and 55°C (20°F and 130°F) thereby reducing the physical size of the thermal energy storage system.

Since considerable energy is used to heat water in our society, emphasis is being placed on alternatives to gas and electricity for heating water. Presently, active solar hot water heaters are not cost effective; however, passive systems can be cost effective in areas with high utility rates. Do-it-yourself kits that can help keep initial costs down are also available.

Homes in the central latitudes of the United States can have their own solar energy systems for heating and air conditioning (Figure 13.8). As illustrated in Figure 13.9, heat captured in liquids flowing through small solar collectors on roofs provides energy to heat and cool the building. An average three-bedroom house in Connecticut with 90 square meters (1000 square feet) of roof area receives 5.6 million calories daily in January. A roof covered with collectors could provide more than half the 4.8 million calories needed in the winter, even at 50 percent efficiency. The Midwest receives about 2500 kilocalories per square meter daily in January and about 7500 kilocalories in July (Figure 13.10).

The Annual Cycle Energy System (ACES) is being tested in a home in Knoxville, Tennessee. The principal component of the ACES home is an uninsulated tank of water, which serves as an energy storage bin (Figure 13.11). A heat pump draws heat from the bin to provide hot water and to warm the air in winter. Over a period of months the water gradually turns into ice as the heat is removed. In the summer, the chilled water air conditions the building without the heat pump. This action gradually melts the ice and stores heat for winter use.

For well-insulated homes within the applicable zone (the geographic area between Atlanta, Georgia, and Minneapolis, Minnesota), the bin need not exceed 2 cubic feet of water for each square foot of living space. A home with 140 square meters (1500 square feet) of living space would therefore require an 84-square-meter (3000-cubic-foot) tank of water, which could be located in a part of the basement or under a driveway, carport, or patio.

Photovoltaic cells, first developed by E. Becquerel in 1839, convert solar energy directly to electricity. In 1954 Bell Telephone Laboratories developed a silicon photovoltaic solar cell. This device is used in camera light meters to determine the amount of incoming solar energy and as the power source for long-lived satellites. Cells can be placed on buildings to con-

Figure 13.8
Solar collection devices heat and cool George A. Town Elementary School in Atlanta, Georgia. (Courtesy of Westinghouse.)

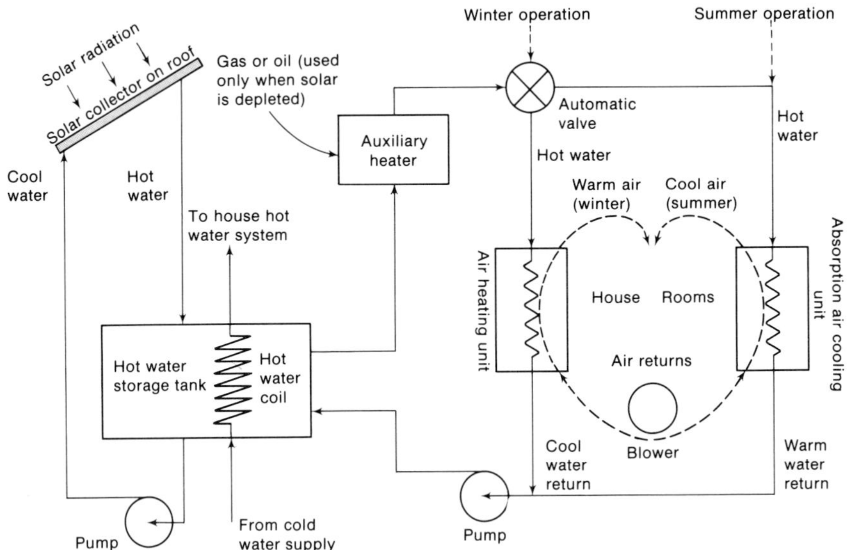

Figure 13.9
Direct use of solar energy to heat and cool a home. Solar energy is collected and stored by heating water in a solar collector. The hot water then heats a second source of water, which is used to operate heating or air conditioning units in the house. (U.S. Department of Energy.)

vert solar energy directly into electricity. The high cost of manufacturing these cells is the primary deterrent to their widespread use. If costs decrease and their efficiency of 13 to 14 percent increases, they could become an unlimited source of energy.

The National Academy of Sciences is encouraging small-scale solar energy technology in developing countries. Although flat plate collectors for space and water heating have been used in some parts of Africa, photovoltaic cells also have potential. Solar energy could create a major change in isolated villages around the world wherever the costs of large central generators with massive transmissions and distribution systems are prohibitive.

Decentralized solar energy units such as water heaters are now used in buildings in Japan, Israel, the Soviet Union, Australia, and the United States. They all operate satisfactorily with little maintenance. These units, enclosed in glass or plastic and mounted on the roof, circulate water in a storage tank. In the United States, the cost of installing such units ($200 to $450

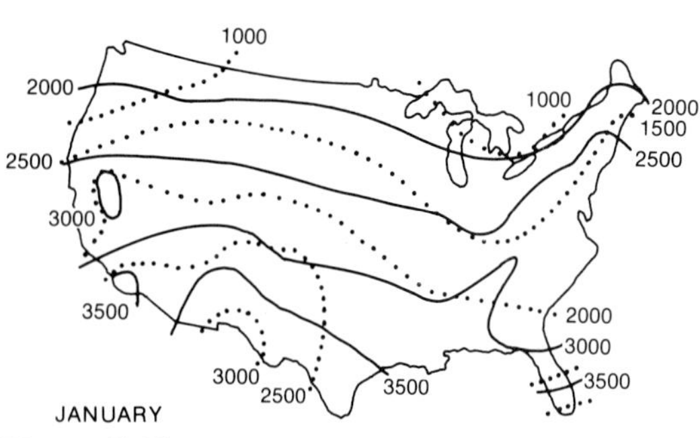

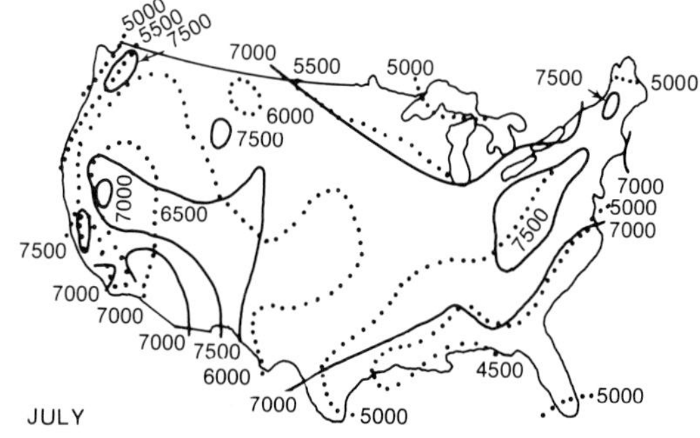

Figure 13.10
Lines of equal total daily solar energy at the ground on cloudless days (solid lines) and on days of average cloudiness (dashed lines) in January and July. Units are kilocalories per square meter per day. The annual average for the United States is 3770 kilocalories per square meter per day, or 58 British thermal units (Btu) per square foot per hour. (U.S. Department of Energy.)

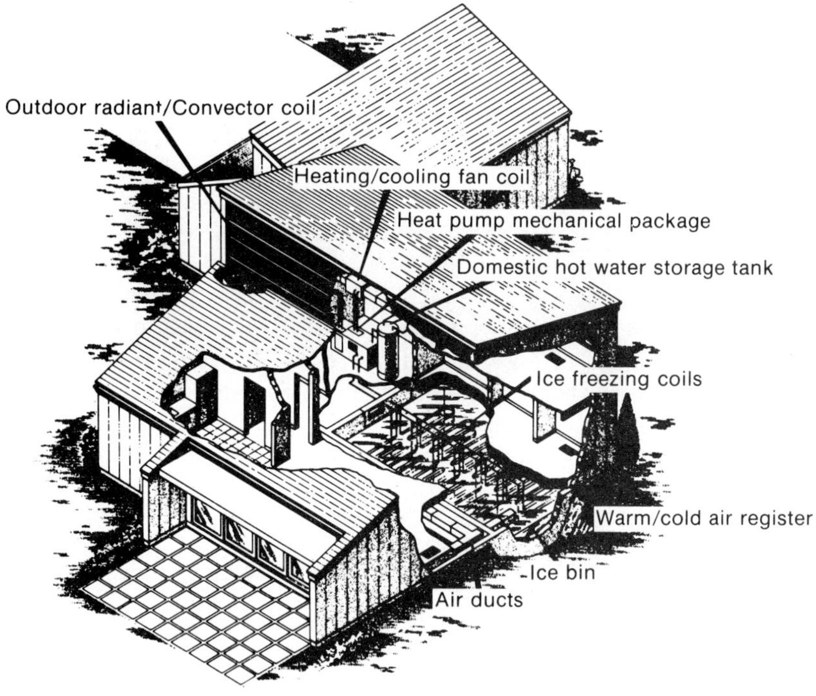

Figure 13.11
Solar ACES (Annual Cycle Energy System) house, showing principal heating and cooling components. (Oak Ridge National Laboratory, Oak Ridge, Tennessee.)

Figure 13.12
Photovoltaic cell on top of a road sign in Arizona turns sunlight into electricity. (Courtesy of Arizona Department of Transportation.)

per single family dwelling) has not been competitive. The Department of Energy estimates that they might be in wider use between 1990 and 1995. Solar devices are also being tested for navigation aids, irrigation pumps, railroad crossing signals, road directional signs, and seawater desalination (Figure 13.12).

Larger solar collectors can gather solar radiation and convert it to electricity for cities. In this case, parabolic reflectors transmit the radiation as light to a central receiver. A liquid, such as sodium or nitrogen, or a gas acts as the receiver and carries the heat to a storage unit, which is often molten salt. Water circulating in pipes through the storage unit is converted to steam to drive turbines (Figure 13.13).

By taking advantage of solar energy falling on the earth, energy from other sources can be made available. A collector 1 meter square provides enough power for a 200-watt bulb. In 24 hours, this is slightly more than 4 kilowatt-hours, or 3770 calories per square meter. Energy equal to 10 barrels of oil falls daily on each acre in the United States. This means the equivalent of 22.4 billion barrels of oil, or about four times our yearly oil consumption, falls each day on the 48 contiguous states.

There are fewer potential environmental problems for solar energy derived from solar collectors than for most other major power sources. However, some problems do exist. Each collection facility with a 1000-megawatt output requires 39 to 52 square kilometers (15–20 square miles) of land. About 15,540 to 20,720 square kilometers (6000 to 8000 square miles) of land area would be required for the present U.S. consumption of 400,000 megawatts. The land, expected to be primarily desert region, would undergo changes in its habitat. The construction of roadways and gradings and filling sites would change the terrain. Water would be in short supply at most sites, and at sites where water is used, the increase in temperature would add heat to the environment as it is released in cooling ponds or towers.

Geothermal Energy

Geothermal resources, generally defined as reserves of heat near the earth's surface, are created when material from the hot interior of the earth protrudes into the cool outer layer. These resources can be hydrothermal reservoirs, geopressured reservoirs, lavas and magmas, or hot, dry rock. The most common and economical are **hydrothermal reservoirs,** where steam is taken from wells reaching 1219 to 1524 meters (4000 to 5000 feet) into the ground. This steam is used to drive turbines on the ground near the wells. The two most highly developed geothermal reservoirs of this type are in Larderello, It-

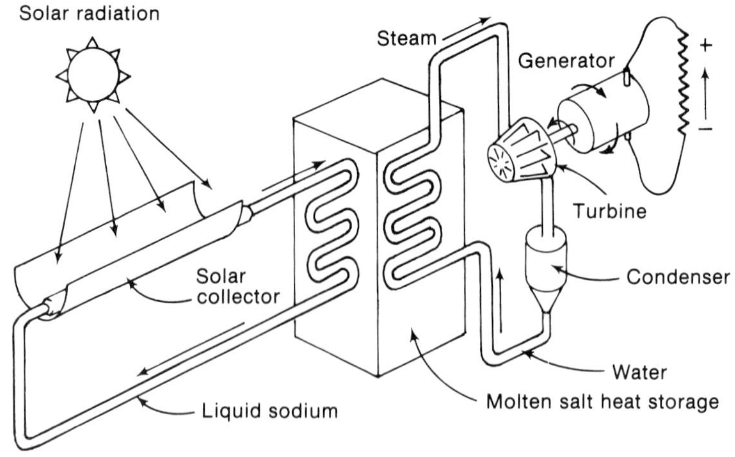

Figure 13.13
The use of solar radiation in the production of electricity. Concentrated solar energy heats liquid sodium, which passes the energy to water (steam). The steam drives the turbine that operates the electric generator. Both sodium and water are recycled through the system. (U.S. Department of Energy.)

Figure 13.14
Geothermal energy generating facility—The Geysers, in California. (Courtesy of Pacific Gas and Electric Company, San Francisco, California.)

aly, and The Geysers in California, shown in Figure 13.14.

Hydrothermal reservoirs are now being used to produce electricity in New Zealand and Mexico. In these areas, hot water deep within the earth releases steam and brine when the reservoir is tapped. The steam is used to produce electricity and the brine is dumped on land or in rivers, causing both mineral and thermal pollution.

In some regions of the world, such as the Gulf Coast from Rio Grande, Texas, to Pearl River, Louisiana, **geopressured reservoirs** exist. Methane-saturated water trapped in layers of sand and shale creates high temperatures and pressures. These zones constitute a potential source of energy if drilling at the necessary depths could be made economical.

Geothermal energy may someday be derived from passing liquid over hot or molten rock near the earth's surface. The Los Alamos Scientific Laboratory in New Mexico is conducting experiments to develop techniques for exposing hot, dry rocks to a liquid that can convey heat to the surface. Another method is being explored by the Nuclear Regulatory Commission. They are attempting to fracture rock by means of controlled nuclear explosions in the earth's interior. They hope to generate electricity by forcing cold water into the fracture and retrieving steam through a second hole.

From an environmental point of view, geothermal energy creates several problems. The highly mineralized water used by a geothermal facility must be discarded. Before it could be discarded, the water would need to be treated in some cases in a manner similar to desalination. Pollution would also result from noxious gas byproducts, especially sulfur dioxide and hydrogen sulfide. In addition, heat released from these plants will have an impact on the surrounding habitat. Once these environmental problems are solved, geothermal energy could provide an alternative power source in the limited areas where the earth's heat is easily accessible for approximately 30 years.

Tidal Energy

Utilization of the tides for energy is not a new idea: there is mention in the Doomsday Book of a tidal mill at the Port of Dover on the English

Channel in 1066. The source of this energy is the gravitational force of the earth–moon–sun system. In the open ocean, tide changes average only about 0.6 meter (2 feet). The physical characteristics of shorelines, estuaries, and bays together with wind conditions greatly amplify these changes. Where amplification is some 50 to 100 times, tides can be used to produce electricity. The greatest release of this energy occurs where the water is forced to flow through dams built in narrow areas. The total amount of energy in ocean tides, if it were accessible, is estimated to be sufficient to provide about the half the energy needs of the entire world. Unfortunately, there are so few sites where harnessing this energy is practical that even if all these locations were used, only from 0.2 to 0.3 percent of the potential amount could be obtained.

One of the largest tidal power stations is located on the Rance estuary on the Channel Island coast of France. It has a power output capacity of 250 megawatts. In the United States, two sites are worthy of consideration. These are the Bay of Fundy on the east coast near Canada and in some bays of Cook Inlet, Alaska.

The effects of tidal energy extraction on marine life could present a major environmental problem, particularly in estuaries (see Chapter 2). Because the velocity of the flowing tide increases near the dam, oyster beds and the habitats of small plants or animals near it would be disrupted.

Nuclear Energy

The transformation of one element to another results in the release of nuclear energy. To understand nuclear power we should know a little about the chemistry of elements.

Atoms of different elements vary from one another in weight and numbers of particles. Atoms are made of three kinds of particles: electrons, protons, and neutrons. **Electrons** have a negative charge, protons a positive charge, and neutrons no charge. **Protons** and **neutrons** are found in the center of the atom while electrons, which are equal in number to protons, orbit the nucleus like satellites orbiting the earth (Figure 13.15).

The sum of the neutrons and protons in a nucleus constitutes the **mass number.** For example, uranium-235 has 92 protons and 143 neutrons, so its mass number is 235. The mass of a proton is about the same as that of a neutron, and each has a mass of about the same as a hydrogen atom. The mass of the electron, however, is only 1/1836 that of a proton or neutron.

The number of protons in an atom is called the **atomic number.** When an element is represented symbolically, the atomic and mass numbers are indicated as $\frac{\text{mass number}}{\text{atomic number}}$ element as in $^{32}_{16}S$ or $^{235}_{92}U$. Normally, only the mass number given at the upper right of the element symbol.

Groups of atoms of the same element with the same atomic numbers but different mass

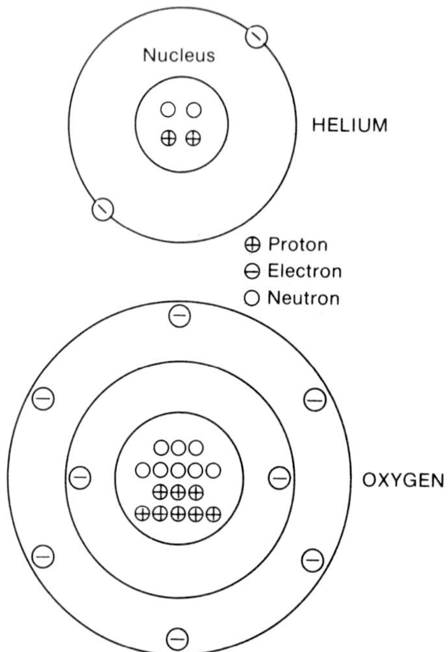

Figure 13.15
Schematic diagrams of helium and oxygen atoms. Electrons orbit the nucleus, which contains the protons and neutrons.

numbers are called **isotopes.** Isotopes of an element have the same number of protons but different numbers of neutrons in their nuclei. Those of the isotopes of uranium are ${}^{234}_{92}U$, ${}^{235}_{92}U$, ${}^{288}_{92}U$. Uranium is a naturally occurring radioactive element consisting of 99.3 percent ${}^{238}U$, 0.7 percent ${}^{235}U$, and a minute fraction of ${}^{234}U$. Because ${}^{235}U$ is readily fissioned, or split into nuclei of lighter elements, it constitutes a major energy source today.

Conventional generating facilities burn fossil fuels such as coal or oil to convert water into steam, which turns a turbine generator to produce electricity. In plants that generate electricity by nuclear power, a fuel such as enriched uranium (${}^{235}U$) is used. The ${}^{235}U$ is contained in fuel rods (Figure 13.16) assembled in the reactor core. When a ${}^{235}U$ atom is bombarded by neutrons, the nucleus of the uranium atom captures a neutron and becomes unstable. These unstable atoms can change in several ways. One possibility is for unstable atoms to **fission,** or split into two or more smaller atoms. The resulting fission products weigh slightly less than the original material. This weight loss represents weight or mass converted into energy. Furthermore, when an atom fissions, several free neutrons are released. These are available to strike other atoms, causing them to fission and thus creating a **chain reaction** (Figure 13.17).

Figure 13.16
Nuclear fuel rods. (Courtesy of Westinghouse.)

If the chain reaction is to continue, there must be enough atoms packed together into a **critical mass.** (Critical mass is the smallest mass of fissionable material needed to maintain a self-sustaining chain reaction.) When several bundles of fuel rods are placed close together in the reactor core, a critical mass is reached. The heat generated within the fuel element where the ${}^{235}U$ is housed is transferred through the fuel element when a coolant flows over the rods. Examples of coolants are water, pressurized water, gas, or liquid metal. The heat now transported in the coolant is used either to turn the turbines directly or to release heat to a secondary coolant which turns the turbines. The rods containing the fuel can be moved in or out of the fuel bundle to increase or decrease the number of neutrons available and thereby control the reaction rate.

Fission Reactors **Light-water reactors** are the primary sources of nuclear energy today (Figure 13.18). Water is used as a *coolant* to remove heat and as a *moderator* to slow down the neutrons so that they fission ${}^{235}U$. Either boiling water or water under pressure serves as the driving force for the turbines. Light-water reactors consume fissionable ${}^{235}U$ and have a low efficiency—about 32 percent. In **boiling-water reactors,** the water passes over the fuel rods and is converted to steam, which drives the turbines (Figure 13.19). In **pressurized-water reactors,** hot water from the core heats water to drive a tur-

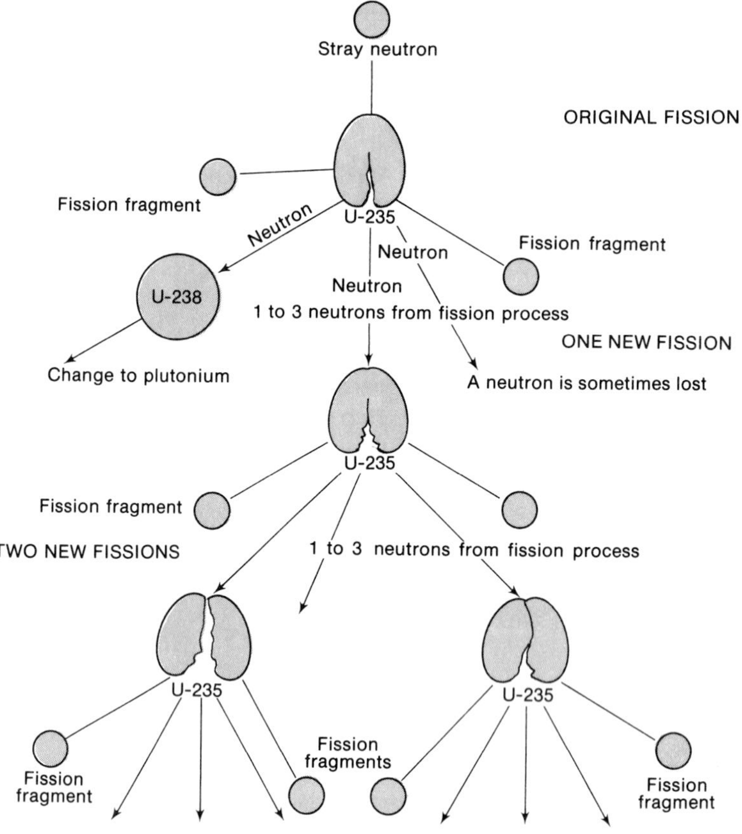

Figure 13.17
An example of a nuclear fission chain reaction. A neutron splits a uranium-235 (^{235}U) atom, which yields some fission fragments and additional neutrons. A free neutron can strike another ^{235}U atom and cause it to split. In some cases, neutrons from a ^{235}U fission can strike two additional ^{235}U atoms, beginning a chain reaction. If uranium-238 (^{238}U) absorbs a neutron, it changes to plutonium-239 (^{239}Pu). (U.S. Department of Energy.)

Figure 13.18
Brown's Ferry nuclear power plant in Athens, Alabama, a light-water nuclear reactor. Its total generating capacity exceeds the output of all thirty TVA hydroelectric dams combined. (Tennessee Valley Authority.)

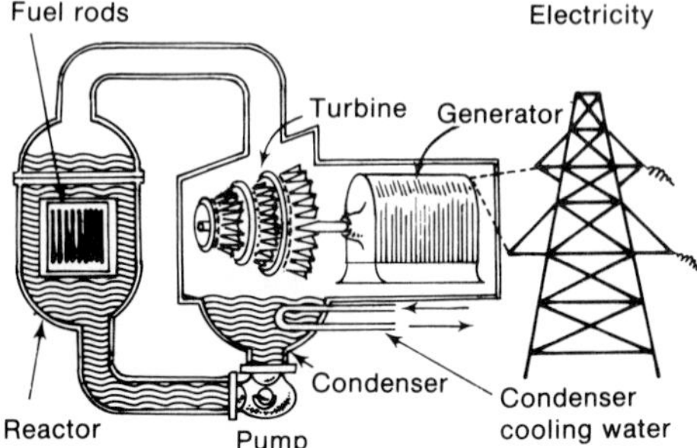

Figure 13.19
Boiling-water nuclear power plant. Fission reaction occurs in the reactor core, made up of bundles of fuel rods containing uranium oxide pellets. The rate of the reactor's operation can be increased, slowed, or stopped by moving the rods. Boiling water is used to heat the air surrounding the turbine. This source turns a turbine and is cooled by an outside water source. (U.S. Department of Energy.)

bine generator. **Gas-cooled reactors** are essentially the same as light-water reactors but use helium gas instead of water. They operate at an efficiency of about 39 percent, which means that they use less than 1 percent of the energy in naturally occurring uranium. They are a drain on our very low supply of ^{235}U.

Breeder reactors, on the other hand, could provide a more economical source of energy because they produce more nuclear fuel than they use. As we learned earlier, when a ^{235}U atom is struck by a neutron, it fissions. Each fission of ^{235}U produces an average of 2.5 neutrons. When ^{238}U, which constitutes more than 99 percent of the uranium supply, is struck by a neutron, it does not fission (see Figure 13.20). Instead, it absorbs the neutron to become a new element, ^{239}Pu (Figure 13.20). Plutonium-239 is a **fertile material**—that is, it undergoes fission to yield energy. Some ^{238}U is converted to ^{239}Pu in light-water reactors, but the process is inefficient. In liquid-metal fast breeder reactors (LMFBR), ^{239}Pu is the fuel.

High-energy neutrons are produced and captured by ^{238}U more efficiently than are the low-energy neutrons of light-water reactors; thus, breeder reactors convert a greater amount of ^{238}U to ^{239}Pu. More ^{239}Pu is actually produced than was initially present (Figure 13.21). About 130 atoms of ^{239}Pu are produced for every 100 consumed. A measure of a breeder reactor's efficiency is its **doubling time,** or the time required to obtain twice the fissionable material originally present. Doubling time of the various proposed reactors ranges from 6 to 20 years.

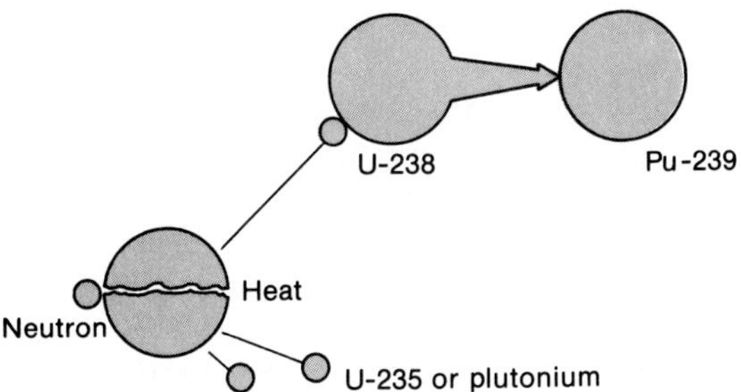

Figure 13.20
The breeding process. A neutron from ^{235}U or ^{239}Pu is captured by ^{238}U, which forms ^{239}Pu.

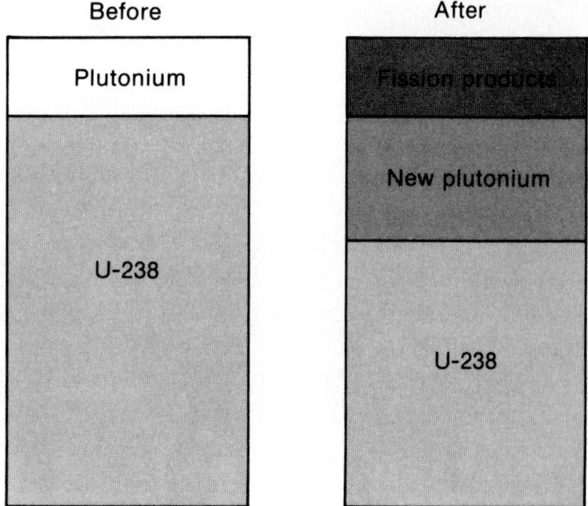

Figure 13.21
A reactor using ^{239}Pu with ^{238}U present will produce more ^{239}Pu than was present before the reaction began.

Currently France, Great Britain, and the Soviet Union are testing and using forms of breeder reactors. Germany, Italy, and Japan also have programs in the developmental stages. There is currently much debate concerning the construction of a LMFBR in Oak Ridge, Tennessee (Figure 13.22). The primary and secondary coolants are sodium, which becomes radioactive (^{24}Na). The potential radioactive contamination from this very corrosive metal requires that the primary coolant system be maintained by remote control equipment. An even greater concern is contamination by and disposal of ^{239}Pu, which has a half-life of 25,000 years (see Chapter 15 for discussion of half-life).

In Canada, the Canada Deuterium Uranium Power Reactors (CANDU) have been in operation since prototype development in 1962. This **heavy-water reactor** operates somewhat differently from reactors operating or being proposed in the United States. Unpressurized deuterium (heavy water) surrounds tubes containing fuel

Figure 13.22
Artist's conception of the liquid-metal fast breeder reactor (LMFBR).

bundles and coolant under pressure. The deuterium acts as a moderator which slows the neutrons but captures very few, allowing the neutrons to create additional fertile material. By using fuels such as thorium isotopes, engineers feel that the CANDU can operate at or near the LMFBR level.

Present Status In 1990 there were 110 operating nuclear generating facilities in the United States and ten reactors either under construction or planned. Total cost for the production of electricity from light-water reactors was estimated by the Department of Energy to be 22.6 mills per kilowatt-hour, while coal-fired steam plants were estimated to produce electricity at a cost of 28.9 mills per kilowatt-hour. (A kilowatt-hour is the amount of work done by a kilowatt in 1 hour.)

Considering the lower costs and greater efficiency of breeder reactors, why do we not shift to nuclear power as our prime source of energy? Aside from the fact that the breeder reactor is only now in the experimental stages, several formidable problems accompany its use. Waste disposal, discussed further in Chapter 15, is a major concern. Heat generated by nuclear plants is a waste product and must be dissipated in the surrounding environment. Environmental changes occur as water taken in from lakes or streams is discharged in a warmer state and as cooling towers release heat, creating local climatic changes (Figure 13.23). In all these ways, nearby life is affected.

Although operating costs would be low, initial construction costs were very high. Between 1975 and 1990, high interest rates combined with other factors to cause the cancellation of 87 planned nuclear reactors. No newly ordered units are projected through 2010 because of the assumption that there will be no changes to existing laws and regulations.

The accident at the Three Mile Island nuclear power plant near Harrisburg, Pennsylva-

Figure 13.23
Cooling towers at the Rancho Seco nuclear generating station. Clay Station, California. (Courtesy of Sacramento Municipal Utility District.)

nia, on March 28, 1979, made the public acutely concerned about the use of nuclear power. Before that time, only a few relatively unpublicized accidents had occurred at nuclear reactors. In 1952 the Chalk River reactor in Canada, an experimental heavy-water reactor, had a block in its liquid coolant system which destroyed the core. In 1957 a plutonium-producing reactor in Windscale, England, released radioactive iodine into the environment. An Army test reactor in Idaho accidentally achieved critical mass in 1961, causing the death of three technicians. Near Monroe, Michigan, in 1966 the sodium coolant was blocked in an experimental breeder reactor and some fuel assemblies were destroyed. The Brown's Ferry nuclear power station in Alabama had a fire in the electric wiring from the control room in 1974. Control of part of the reactor operation was lost as a result.

The Three Mile Island incident showed that a series of systems malfunctions were possible and that design and human fallibility could cause accidents thought to be impossible. The events actually began when auxiliary cooling pumps were shut down several weeks before the accident. A series of mechanical problems occurred on the morning of March 28 which were aggravated by human error, shutting off the emergency core cooling pumps. As a result, the fuel rods were exposed without cooling water, causing a partial meltdown. This damage to the core caused the release of radioactive iodine and other gases into the environment. The hydrogen that appeared later in the reactor was an unexpected event. No one knew if the containment vessel could withstand the force of the potential explosion caused by the pressure. Several studies suggest that a melted core such as the one at Three Mile Island would eventually refreeze even if it was not permanently water-cooled. Damage to buildings and equipment would be high, but a meltdown would not destroy lives, health, or nearby property.

In late April 1986, another major nuclear accident occurred at the Chernobyl Nuclear Reactor Complex in the Soviet Union. An explosion and a fire that lasted for several days released radiation into the atmosphere. More than 100,000 people had to be evacuated from the immediate area, and high radiation levels were detected in other European countries. In July the Soviet government reported that 28 people had died in the accident and 200 others were suffering from radiation sickness. In addition, an estimated 1000 square kilometer area had been contaminated, and economic damage was placed at $2.8 billion. In the aftermath of the accident, several Soviet government officials were charged with gross negligence and fired from their positions.

Many safety questions remain unanswered in the aftermath of the Three Mile Island and the Chernobyl events. The future of nuclear power, according to some people, is now precarious. Releases of radioactivity could occur at points other than the power plant in the nuclear fuel cycle (Figure 13.24): in transit, at the reprocessing plant, or at waste disposal sites.

These nuclear accidents do not in themselves indicate that nuclear power is not a feasible form of energy. Rather, accidents reinforce the fact that we must continue to study and test all possible methods of energy production. Government agencies must establish high standards of safety and performance and provide protection from sabotage of nuclear facilities. If some nuclear developments prove costly in terms of environmental damage or safety hazards, we must see that alternative approaches are sought.

As most of the nuclear reactors in the United States are designed to operate for 40 years, they will need to be dismantled after that approximate period of time. Nine of the reactors in this country have passed their 20-year mark. Now engineers are beginning to think of the process and costs of shutdown. In most cases, parts would not be reusable, thus disposal would be a problem. Nuclear graveyards will be necessary that prevent public access. Utility companies need to begin planning in advance for shutdowns and dismantling and replacing reactors.

Fusion Reactors So far, our discussion of nuclear reactors has centered on fission reactions in which a heavy element like uranium is broken down to release energy. Another type of nuclear reaction is **fusion.** In fusion or thermonuclear reactions, lighter elements are combined into fertile materials, as in the hydrogen bomb. Very high temperatures—hotter than the sun's interior—are required to have a controlled fusion reaction.

Two concepts are currently being considered for fusion reactors: **magnetic confinement** and

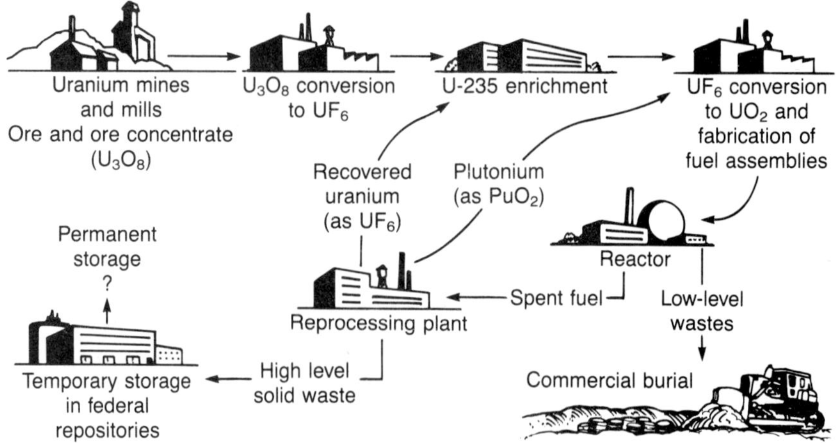

Figure 13.24
Nuclear fuel cycle. (U.S. Department of Energy.)

laser implosion. In the first, hydrogen isotopes are present in a gas or plasma contained in a magnetic field. The magnetic field accelerates the isotopes to high velocities, causing collisions and fusion. In the second concept, concentrated light from lasers compresses a pellet of deuterium and tritium, causing fusion. Overall, there are some thirty possible fusion reactions. Several examples are shown in Figure 13.25. The least demanding technically is the fusion of deuterium and tritium, both isotopes of hydrogen. Deuterium can be easily extracted from seawater, while tritium must be produced by nuclear bombardment of the element lithium.

Because of the low fuel cost and the negligible environmental impact from fuel removal, the fusion reactor could be an important source of electric energy in the early twenty-first century (Figure 13.26). Isotopes released by this reactor have a much shorter half-life than those from fission reactors. Tritium, with a half-life of 12.3 years, is the principal radioactive waste material. People, however, remain concerned about nuclear power, thus, there are social limits to its use.

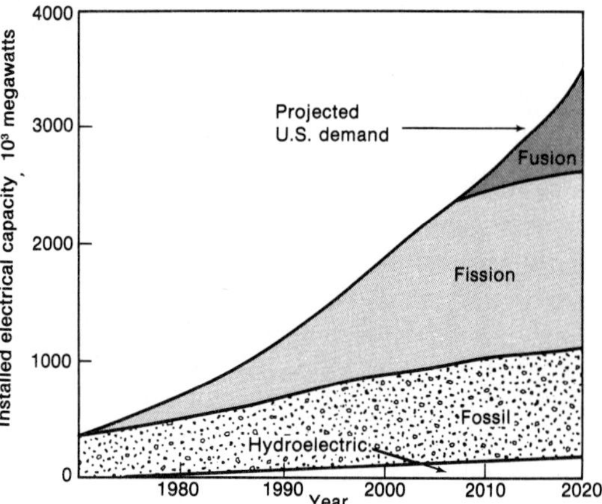

Figure 13.26
Projected use of fusion power as a source of electric energy in the United States (U.S. Department of Energy.)

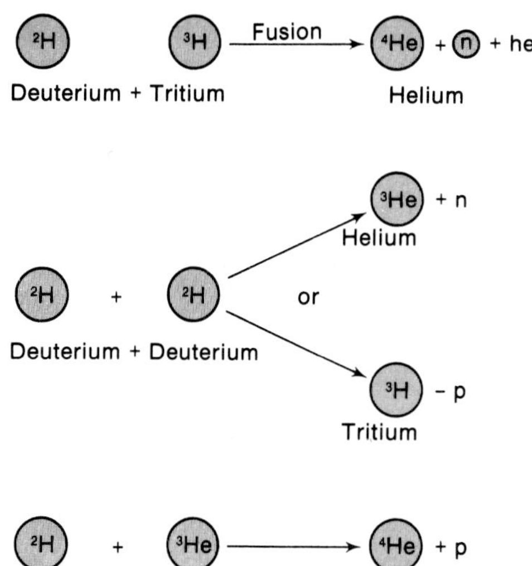

Figure 13.25
Examples of fusion reactions. Deuterium and tritium are isotopes of hydrogen. When these lighter elements combine to form heavier elements, heat is released. This heat, if trapped, could be a source of energy.

SECONDARY ENERGY SOURCES

Most secondary sources involve the release of trapped solar energy. Fossil fuels, organic materials, and wastes generally can release energy trapped by living systems. Wind, rivers, and ocean gradients can be harnessed to release energy.

Fossil Fuels

Coal Coal, oil, and natural gas were formed from plant and a small amount of animal material deposited more than 300 million years ago. Coal developed from deposits in swamps rich in plants that partially decomposed in an oxygen-deficient environment. These plants accumulated in thick layers of peat which were then covered by sand, clay, and silt as the sea level changed. As more sediment was deposited, water and organic gases were squeezed out, increasing the amount of carbon in the deposits. These processes continued until peat became converted into coal. The highest grade was formed as pressure caused a folding of the strata, resulting in a loss of volatiles and concentration of carbon.

Coal occurs in layers 0.6 to 9 meters (2–30 feet) thick throughout the world, with the largest concentrations being found in the United States (Figure 13.27). It is often divided into four classifications according to its percentage of carbon: **lignite** (less than 40 percent fixed carbon); **subbituminous** (40 to 60 percent fixed carbon); **bituminous** (60 to 80 percent fixed carbon); and **anthracite** (80 to 98 percent fixed carbon). Sulfur content is also a means of classification: low (0 to 1 percent); medium (1.1 to 3 percent); and high (more than 3 percent). Much of the United States' supply is low-sulfur subbituminous coal. Anthracite, the least abundant type, is found primarily in Pennsylvania and

Figure 13.27
Coal seam in Ohio County, Kentucky. (Tennessee Valley Authority.)

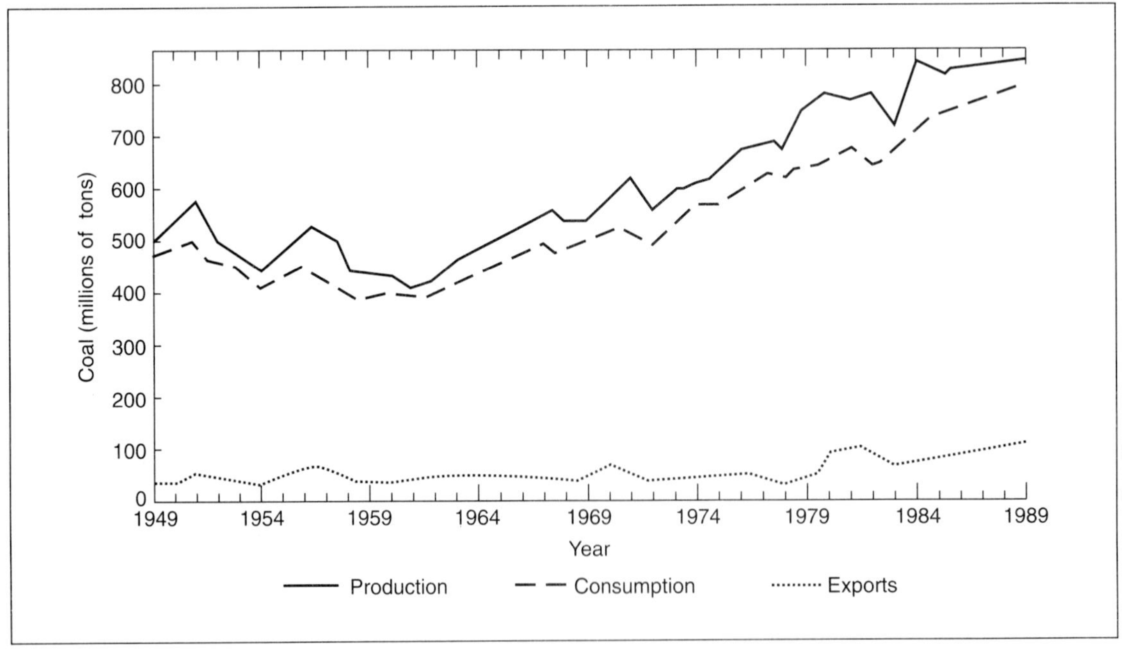

Figure 13.28
Coal supply and disposition, 1949–1989. (U.S. Department of Energy.)

constitutes about 2 percent of the U.S. supply. Bituminous is the most widely used and accounts for about 42 percent of the reserves. The remaining 56 percent is subbituminous and lignite.

Coal accounts for about 18 percent of domestic energy consumption in the United States. This percentage has not changed for a number of years, however, production capacity currently exceeds the demand (Figure 13.28). This trend is expected to change during the next 20 years as demand is expected to increase by 5.1 percent per year in an attempt to ease dependence on imported oil.

Many problems are associated with using coal as an energy source. Coal is a dirty fuel both in terms of recovery and usage. It is difficult to mine because it is found in thin layers or veins between rock formations. Working conditions in mines are often poor, resulting in health problems such as black lung disease. Surface mining, or removing whole layers of hillsides, results in tremendous environmental destruction because, until the land is successfully reclaimed, it is unsuitable for any other purpose (see Figure 13.4). Water leaching from tailings or deposits sends acid runoff into surrounding water. When coal is burned, it produces undesirable gases such as sulfur dioxide and releases them into the environment. A process called *solvent refining* can remove undesirable sulfur and ash from coal, but it is expensive, making low-sulfur coals more desirable.

Coal also contains small quantities of ^{238}U, ^{235}U, ^{232}Th, and other radioactive isotopes. Scientists at Oak Ridge National Laboratory have shown that airborne radioactivity release is greater from coal plants than from nuclear plants, as shown in Table 13.1.

Table 13.1
Radioactivity from Coal-Fired and Nuclear Plants

	Stack Height	Population Dose (relative)[a]
Coal-fired plant:	50 m	1.00
	100 m	0.91
	200 m	0.83
	300 m	0.78
Boiling-water nuclear plant:	20 m	0.56

[a]The population dose is that received by a population within 88.5 kilometers (55 miles) of 1000-megawatt power plants. The dose available from the 50-meter stack was set equal to 1.00 and all other doses were calculated as proportions of that value.

Source: From J. P. McBride et al., "Radiological Impact of Airborne Effluents of Coal and Nuclear Plants," *Science* 202 (1978):1045–50. Copyright 1978 by the American Association for the Advancement of Science.

Oil and Gas The formation of oil and natural gas is not as clearly understood as that of coal. Oil, normally found in pockets in marine sedimentary rock, is thought by geologists to be a product of the decomposition of organic matter that accumulated on the bottom of basins with oxygen-deficient water. Bacteria then decomposed this mixture by removing oxygen and nitrogen, leaving carbon and hydrogen. Pressure and heat converted the material into droplets of liquid oil and minute bubbles of gas when sediments increased. The oil and gas were forced into layers between rocks where the space was large. Oil and gas pockets can be found in a number of areas of the world.

Other Sources Because of the decreasing supply of natural gas, researchers are now considering several methods of converting oil and coal to gas. In **coal gasification,** water is heated to steam and the steam reacts with coal to form a gas containing methane, hydrogen sulfide, and ammonia. The gas is cleared of unwanted constituents and the heat content upgraded to form a product equivalent to natural gas (Figure 13.29). The feasibility of underground coal gasification is being tested by DOE. Inlet wells for the injection stream and one or more outlet wells are drilled to remove the gas produced. Synthetic natural gas could be used to a greater extent in the future as it becomes more economically competitive with natural gas. Much of the potential use of this synthetic fuel depends on the cost associated with refining facilities, future

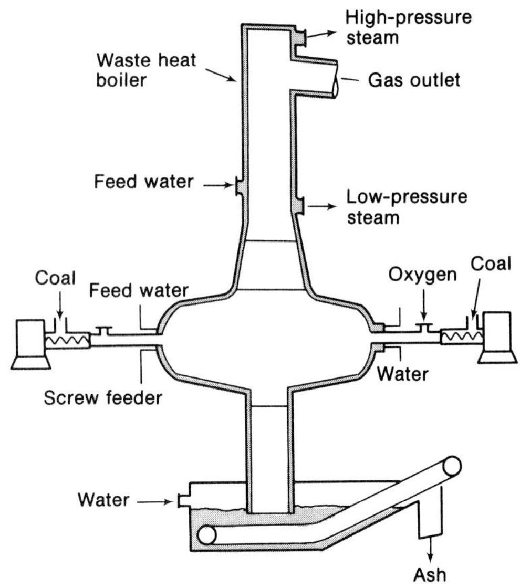

Figure 13.29
One form of the coal gasification process. (U.S. Department of Energy.)

price of other forms of energy, and policies of the United States government.

In the process of **magnetohydrodynamics,** high-temperature combustion of fossil fuels creates gases which are seeded with chemicals to make them electrically conductive, as shown in Figure 13.30. The electrically charged hot gases substitute for the rotating coils of conventional electric generators as they are forced through a magnetic field to produce a flow of current. This process is about 60 percent efficient, or one and one-half times more efficient than a fossil fuel plant, but it has many unsolved engineering problems.

Coal liquefaction involves the conversion of the organic portion of coal to low melting-point solids, liquids, and by-product gases. Direct liquefaction occurs when the hydrogen to carbon ratio is increased by adding hydrogen or removing carbon. During this process, some of the sulfur and most of the ash in coal are removed. When coal is combined with steam and oxygen to form gases of hydrogen and carbon monoxide, indirect liquefaction occurs. The gases are converted to methanol by catalytic action.

If coal liquefaction were used extensively, it would be necessary to contain harmful toxic fumes, prevent water contamination from dissolved solids, and dispose of large volumes of solid wastes.

Other forms of fossil fuels include shale oil and tar sands. **Shale oils** are heavy, viscous crude oils found in shale rock. The largest known deposits, in the Green River formation in Colorado, Utah, and Wyoming (Figure 13.31), con-

Figure 13.30
Magnetohydrodynamics flow diagram. (Courtesy of AVCO Everett Research Laboratory.)

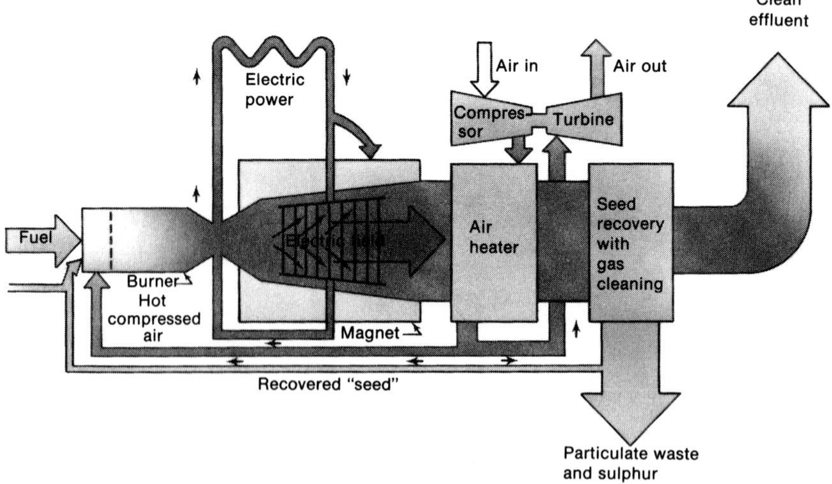

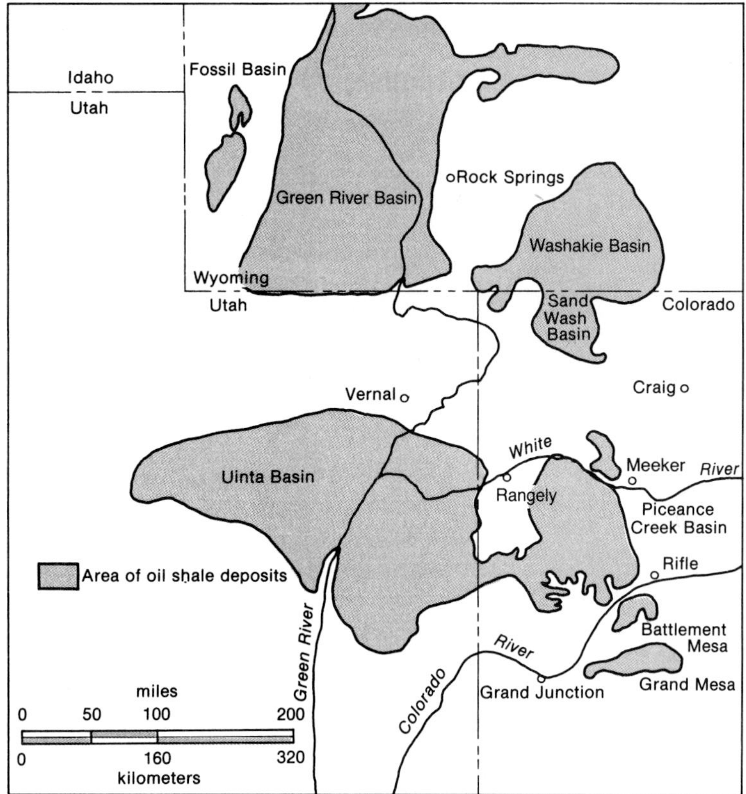

Figure 13.31
Shale oil areas in Colorado, Utah, and Wyoming. (U.S. Department of Energy.)

tain 80 billion barrels of recoverable oil based on present technology. Other minor deposits are scattered from Texas to the Great Lakes. It is not economically feasible to remove oil from shale unless techniques improve to produce more than 0.09 cubic meter (25 gallons) of oil per ton of shale mined. Presently, the most economical method for extracting oil from shale involves heating crushed ore to approximately 482°C (900°F). Oil is then released as vapor. Another extraction method involves breaking up the shale in the ground and heating it in place *(in situ)*.

The environmental problems resulting from shale oil removal are great. Large amounts of shale must be discarded. Engineers estimate that a mine supporting an operation which processes 100,000 barrels of oil per day could produce 82 million kilograms (90,000 tons) of solid wastes each day. Water, a limited resource in many mining areas, is needed for processing, heat dissipation, and waste disposal. Such massive habitat destruction and pollution are detrimental to all surrounding life.

In some ways, shale oil removal is similar to strip mining. Until recently, little attention has been given to the environmental impact of strip mining. The waste and environmental destruction in Pennsylvania, Ohio, and the stripped ridges of Kentucky and Tennessee can be compared to the potential destruction from shale oil removal. Further problems will result from the necessary on-site distillation processing. Commercial shale oil removal will begin in earnest when environmental questions are answered and the price becomes competitive.

Tar sand (deposits of porous rock or sediments containing tarlike hydrocarbon) is being mined in Alberta, Canada. At this time, investigators are studying two methods of fuel recovery. One involves applying various forms of heat to make the tar flow from the parent rock, and the other uses organic solvents to dissolve the tar from the rock. Presently, estimated reserves of 30 billion barrels of tar oil are not economically recoverable in the United States.

Resources How much fossil fuel is available? Table 13.2 shows that an estimated 90 percent of the world's oil supply will be used by the year 2030 and 90 percent of the coal by 2450. As all fossil fuels are limited resources (that is, no appreciable amounts are currently being produced), our energy procurement can be hampered unless alternate sources are found. Fossil fuels are not only finite on the earth, but they are restricted by geographic distribution. A shift to coal as a major energy source in the United States is being discussed because this country has much of the world's recoverable coal but little recoverable petroleum.

The United States became more aware of its dependence on foreign fuel during the Arab oil embargo in 1973. Following the onset of the embargo, national leaders expressed their intention to make the United States "energy independent" by 1985. A task force on energy from the National Academy of Engineering concluded that "wartime" efforts, operating with government intervention and direction, would be required to achieve this in one decade. As Americans waited in gas lines in 1979, they felt that they had a long way to go to achieve energy independence.

In 1979, imports constituted 46.5 percent of the petroleum products supplied in the United States with most of that coming from the Organization of Petroleum Exporting Countries (OPEC). During 1982–1984, petroleum imports were reduced to less than 30 percent, and the proportion imported from Arab OPEC countries dropped from 17.3 percent in 1977 to 5.1 percent in 1984. Low energy prices, however, caused a rise in oil imports in the last half of the 1980s; more than 40 percent of U.S. oil was imported in 1990.

Since 1945 the number of barrels of oil obtained per foot of drilling effort in the continental United States has decreased. This fact undoubtedly represents the decrease in oil reserves (Table 13.2). At the same time, however, costs for drilling and extracting oil have increased. Thus, our future oil supply depends on discoveries in foreign countries such as Mexico and the People's Republic of China.

Biomass Conversion

Stored chemical energy created in vegetation by photosynthesis can be released by **biomass conversion.** This can be done by burning wood,

Table 13.2
Estimated Fossil Fuel Supplies

Fossil Fuel	Estimate of Original Ultimately Recoverable Resources (10^{13} kilojoules)	Year of Peak Production	Year When 90 Percent Is Gone
U.S. petroleum liquids	1.4	1970	2000
U.S. natural gas	1.6	1980	2015
U.S. coal	42.8	2220	2450
World petroleum liquids	12.4	2000	2030
World coal	218.0	2150	2400

Source: Courtesy of M. K. Hubbert, U.S. Geological Survey.

livestock manure, crop remains, food processing wastes, aquatic plants, and other organic waste materials, but not fossil fuels. Although wood-burning electric generating facilities are possible, a 1000-megawatt plant with a 35 percent efficiency requires 3000 20-centimeter (8-inch) diameter trees per hour. This is equivalent to 1.4 million kilograms (1600 tons) per day. Presently a large paper mill uses 2.5 million kilograms (2800 tons) per day. Some farms are obtaining mixed results after building anaerobe digesters which produce fuel gases. The amount of gas produced is small and the costs prohibitive. Interest in converting agricultural crops to intermediate fuels such as alcohol, which can be distilled from grain, is also increasing.

Since the waste materials used in biomass conversion are found in such low concentrations in each region, they must be transported to the conversion facilities, creating additional expense. Some scientists suggest cultivating plants specifically for biomass conversion, but the necessary land is not always available. Existing areas with large quantities of weedy species such as water hyacinth are possible sources. Terrestrial and aquatic energy farms for producing fast-growing vegetation such as eucalyptus trees, sycamore trees, or large kelp could also provide fuel.

Solid Wastes

Solid wastes collected as garbage from homes and industry are another possible source of energy. Over 18 trillion kilograms (200 million tons) of waste per year are generated in the United States, filling up disposal sites quickly and making recycling of materials imminently important. Converting solid wastes into fuels involves some form of chemical process. One way to recover energy is to use the heat from burning wastes for power production. This procedure is the same as biomass conversion. Energy can also be formed through **pyrolysis,** in which wastes are heated in an oxygen-free atmosphere. Fuels such as gas, tarlike oil, and carbon ash are then released and separated.

Bacteria in solid wastes are responsible for the process of **bioconversion,** in which bacteria digest organic wastes and form sludge. After energy is recovered from the released gases and heat of digestion, the sludge leftover can be used for fertilizer. However, this fertilizer must be odorless, treated to bring its acid pH to a level tolerated by the vegetation to be planted, transported without leaking, and applied in a manner that prevents contamination of water supplies and rivers. Current research also shows that sludge can contain toxic metals.

Several cities, including St. Louis, Missouri; Atlanta, Georgia; and Menlo Park, California, burn solid wastes for some energy production. Residues from these processes, such as ash and wastewater, can create undesirable effects. Ash deposits must be carefully handled to prevent harmful levels of leaching by groundwater. Nevertheless, if inexpensive, efficient methods for conversion of solid wastes to energy are found, municipalities will find their garbage an asset—not a headache.

Wind Power

We learned in Chapter 7 that solar radiation is the major cause of winds on the earth's surface. As varying amounts of solar energy fall on the earth, the atmospheric pressure changes, causing winds. Windmills are used in parts of the world today to generate small amounts of electricity. The blades of the windmill, placed so that they rotate when wind strikes their feathered surfaces, are attached directly to an electric generator (Figure 13.32). Wind power is a potential source of energy in the Great Plains and along coastal areas. The Aleutian Islands have a constant high-velocity wind, making them a reasonable source of power.

The potential amount of wind energy is great. Engineers estimate that a wind of 32 kilometers (20 miles) per hour blowing through a rectangle 16 kilometers (10 miles) long and 46 meters (150 feet) high produces about 380,000 kilowatts. Windmills cannot effectively produce commer-

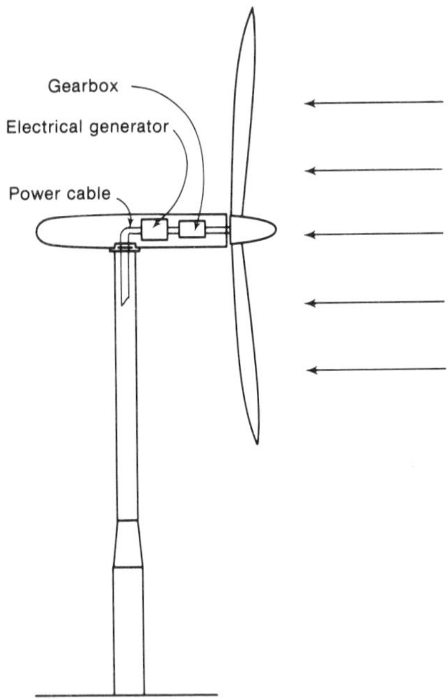

Figure 13.32
Use of a wind rotor system for generation of electricity. The blades of the windmill turn a central shaft which operates an electric generator. (U.S. Department of Energy.)

cial electrical power from wind velocities of less than 20 miles per hour. On the other hand, at velocities above 48 kilometers (30 miles) per hour, the windmill structure ca.ı be damaged. Several midwest power companies derive some electric energy from wind turbines. Wind power extraction causes no air or water pollution thus offers a "clean" method for obtaining electricity.

Ocean Thermal Gradient

In many tropical and subtropical areas of the world, the sun heats the ocean surface to a range of 24 to 30°C (70–86°F). This warm water circulates to the polar regions, where it cools and flows back to the equatorial region along the ocean bottom and on the eastern side of the ocean basin. In these cooler layers 600 meters (1968 feet) below the surface, temperatures range from 2 to 8°C (35–46°F).

In one proposed plant, warm ocean water would enter the upper section of the plant at 25°C (76°F). It would pass over an ammonia boiler, heating ammonia or another fluid with a low temperature of evaporation to 20°C (68°F). The vapor would leave the boiler, expand through a turbine, and then enter a condenser. Ocean water at 5°C (40°F) would condense the vapor (Figure 13.33). Probably the only ocean thermal gradient plant that produced electricity was developed in Cuba by G. Clande in 1930.

Figure 13.33
Platforms utilizing ocean thermal gradients to generate electricity. Water at 25°C (76°F) is taken in at the surface and vaporizes another fluid, which turns a turbine. Water at about 5°C (41°F) is taken in at the bottom and acts as a coolant. (Courtesy of Lockheed Missile and Space Company.)

He was able to produce up to 22 kilowatts of power.

There are practical rather than environmental problems in the use of ocean thermal gradients as an energy source. First, it is difficult to build equipment that can be used in highly corrosive ocean waters. Second, adequate means of transporting electricity to shore must be developed for floating facilities.

Water Power

Hydroelectric energy is an important source of power for the world today, yet only a small part of the estimated worldwide hydroelectric power capabilities is now being used. Since the sun's radiation drives the rain cycles that feed the surface runoff, hydroelectric power uses solar energy indirectly. In fact, it is the only solar energy power used on a widespread basis today. Hydroelectric facilities are operated in a *run-of-the-river mode*, *storage mode*, or both. In the run-of-the-river mode, water runs downstream, passes through turbines and continues in the river. A storage hydroelectric plant accumulates water behind a dam in a reservoir and releases it through the turbine. Shasta Dam, located on the Sacramento River north of Redding, California, is 183 meters high and 1054 meters long (602 by 3460 feet) at the crest and impounds 465 billion cubic meters of water (Figure 13.34). Its five generating units can produce 442,310 kilowatts of electricity.

Hydroelectric power is relatively inexpensive compared to other sources. Africa, South America, Southeast Asia, and some areas of eastern and western Canada have tremendous potential for its development. In the United States most sites are already developed, particularly in the Pacific northwest and southeast (Figure 13.35). Presently the Army Corps of Engineers, which is responsible for planning and constructing hydroelectric energy sites, is being criticized for destroying natural communities and home sites. Because silt builds up behind the dams, some of those built just twenty years ago can no longer function. As the demand for more energy increases, there must be greater efforts to remove silt from behind dams. Although hydroelectric

Figure 13.34
Shasta Dam in Redding, California. Water running through the dam is directed into turbines, which turn the conductor loops in the electric generators. (U.S. Bureau of Reclamation.)

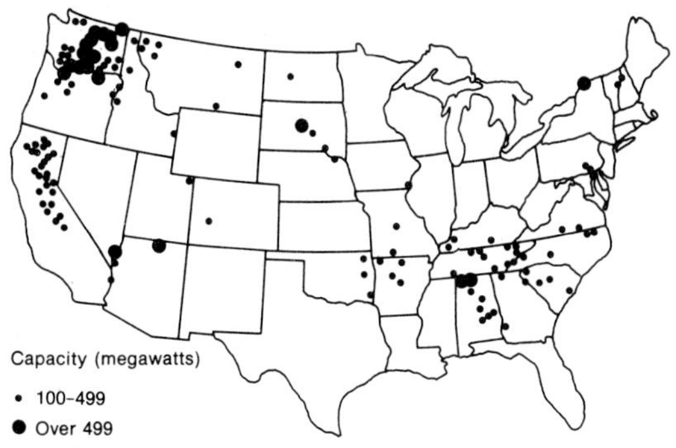

Figure 13.35
Distribution of developed hydroelectric resources in the United States. (Federal Power Commission.)

energy does not cause air pollution and little water pollution, it does block fish migration, disrupt river flow, and flood usable land, particularly when a number of dams are placed along a single river system.

Hydrogen

Hydrogen gas is a potential energy carrier of the future. Currently, hydrogen is produced mainly from methane gas or by steam breakdown processes driven by fossil fuels. Hydrogen can be produced by electrolysis, using electric power from another energy source. As petroleum and natural gas become less available, hydrogen might be used directly as a fuel. Scientists say that hydrogen can be transmitted as a gas in pipelines, stored in holding tanks, and used as fuel in automobiles and other forms of transportation (Figure 13.36).

Present technology is not developed to allow large-scale use of hydrogen. The electrolyzers now in operation emit asbestos particles and use nickel catalysts. Since asbestos is a known carcinogen and nickel is in short supply, health hazards and resource limitations will limit the use of hydrogen.

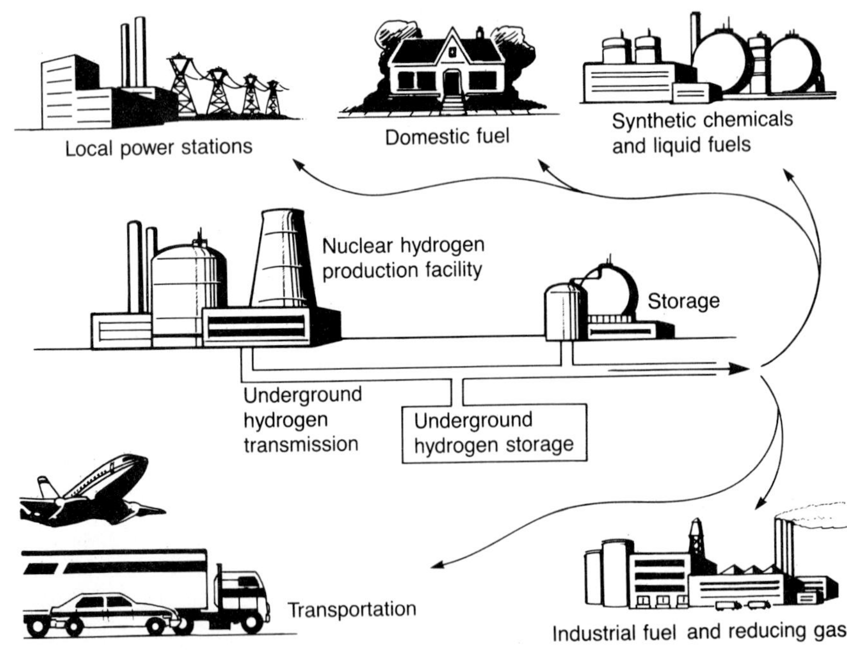

Figure 13.36
Hydrogen economy. Hydrogen could become an important energy source in the future. Converted from water by nuclear, solar, or fossil fuel plants, it could easily be transported to storage facilities near its use site. (From A. L. Hammond, W. D. Metz, and T. H. Maugh II, *Energy and the Future*, AAAS Publication No. 73–7 [1973]: 120. Washington, DC: American Association for the Advancement of Science, 1973. Copyright 1973 by the American Association for the Advancement of Science.)

Research shows that fewer air pollutants result from using hydrogen in internal and external combustion engines than from burning fossil fuels. A major drawback to the use of hydrogen as a fuel is its explosiveness as a pure gas. Many people still recall the newsreel of the hydrogen filled German zeppelin, the *Hindenburg*, exploding after a trip across the Atlantic in the late 1930s. Were a large-scale use of hydrogen to be undertaken, the public would have to be convinced of its safety.

ENERGY AND THE ENVIRONMENT

How can we measure the impact of energy resource development on the environment? The procurement, transportation, use, and disposal of energy wastes disrupt the environment in ways that can prevent a natural balance from being reestablished. In the past, we have not considered the cost of restoring the environment in our procurement of raw materials.

Much is happening to improve this situation. People are devoting more attention to special environmental problems resulting from each power source. Elaborate studies are now being conducted to determine the best methods for removing coal and for returning the surroundings to their natural state after mining. If planned, strip mines can be reclaimed as natural areas or recreational sites. Utility companies now must consider the sites of power plants in their expansion programs because the many miles of transmission line corridors that cut through fields and forests and across swamps and lakes change the natural habitat. In developing regional generic statements, federal agencies are describing local concentrations of wildlife and vegetation as well as areas of historical interest to be avoided when selecting locations of new facilities. Wind direction, type of cooling apparatus, and local climate are all factors in deciding where to place these plants.

Many forms of energy extraction affect the land directly. For example, the impact of coal mining, such as surface disruption, subsidence from underground mining, and waste disposal, pertains specifically to the site. In 1977, when the Soil Conservation Service conducted a major survey, there were 2,308,500 hectares (5.7 million acres) of land disturbed by coal mining, primarily for bituminous coal, of which about 53 percent had been reclaimed. As more coal is extracted, land use will be affected. Because of variations in seam thickness, heating value, and mining techniques, the level of coal production and impact will vary. Wyoming coal is found in 9.2-meter (30-foot) seams, thus 10 hectares (25 acres) of land are disturbed for every 1×10^9 kilograms (1 million tons) produced; whereas 3-meter (10-foot) seams in Arizona, Colorado, and Washington disrupt 28 to 32 hectares (70–80 acres) per 1×10^9 kilograms recovered. In addition, environmental concerns must be weighed against health hazards. In the United States, twelve times as much coal is available from deep mining as opposed to surface mining, however, black lung disease and underground cave-ins are much more frequent in the latter.

Exploration for oil and gas, particularly in the West, is having a major impact on land, vegetation, and wildlife. In many deserts or semideserts where water is scarce and some growing seasons are short, disturbances can destroy wildlife habitat, causing a reduction in populations of species such as golden eagles, antelope, and elk. Oil seepages from off-shore oil and gas drilling upset the balance of the aquatic ecosystem.

America's mobile society demands that oil be available in large amounts to operate its cars, buses, trains, and planes. Nearly two-thirds of the daily consumption of petroleum used in the United States is used to propel us. Not only is this consumption a drain on a nonrenewable resource, but it also is a major environmental pollutant. Since 1950 the number of automobiles in the world has increased faster than the population growth rate. While fuel economy is now emphasized, new technologies are needed to continue to improve the fuel usage and shift to different fuel sources. Research is currently underway to provide efficiency and changes in design to reduce drag.

Currently, compressed natural gas (CNG) is being used in about 700,000 vehicles worldwide. It costs about 70 cents per gallon and leaves little environmental pollution. Crankcase oil lasts about 80,000 kilometers (50,000 miles). The gas, however, must be in cylinders pressurized up to 42 kilograms (3000 pounds) which are four times as big as conventional gas tanks. Since there are no public fueling stations, people must have their own supply of CNG.

Other supplies of fuel are now being used. Alcohols, including methanol (made from natural gas or coal) and ethanol (from corn), have been costly and less efficient.

Technology is improving. With the Clean Air Act of 1990 in place, city drivers will be looking at electric cars as methods of using "clean" fuel. By the early twenty-first century, charging stations in parking lots and along highways may be available.

As new sources of energy are found, new environmentally related problems arise. For example, substantial quantities of biocides would probably be needed to prevent fouling the water flow through boiler passages in installations designed to trap ocean thermal gradients. As the human population redistributes to locations where new forms of solar energy could provide power at low costs, the question of land use would become especially important. Underground coal gasification, although not yet commercially proven, could create a major impact on the environment by destroying wildlife habitats.

With the increased growth of energy industries, the only way we can avoid wholesale destruction of the environment is to develop technologies, such as waste disposal, that can reduce harmful effects on the environment.

ENERGY CONSERVATION

Improving our use of energy can have a substantial impact on total energy consumption. By conserving energy through more efficient use and preventing waste, we can greatly reduce the demand for new energy (Figure 13.37). The total annual energy growth rate required to maintain our nation's standard of living could be reduced from about 3 percent to less than 2 percent, making total consumption 25 percent less than that projected for the year 2000. This reduction can occur without a decrease in services or products if we use energy more efficiently.

If an engine could operate without friction, it would be 100 percent efficient. Through thermodynamic studies we can compare the amount of energy released with the amount of fuel used. The heat loss in any thermodynamic reaction might be used to heat buildings. For example, heat dissipated into the air or water by power plants could be put to such use in industrial sites.

Energy conservation practices require an energy-conscious public. When you go to a

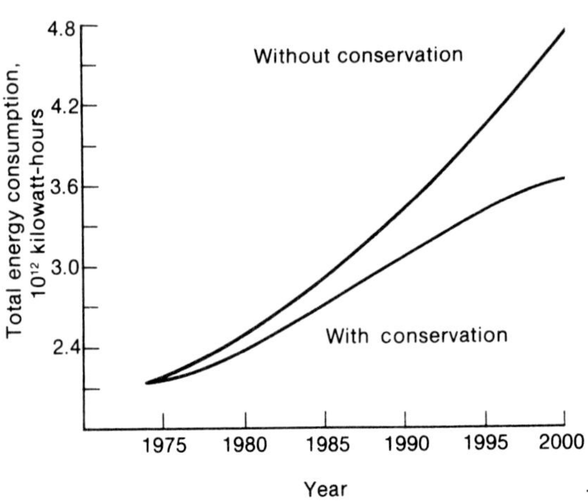

Figure 13.37
Projected total energy consumption with and without conservation measures. (U.S. Department of Energy.)

LEGISLATION

Clean Air Act of 1990. An act aimed at reducing atmospheric pollution. Mandates that gasoline be "reengineered" to reduce harmful pollutants. Requires that starting in 1998 operators of centrally fueled fleets of ten or more vehicles begin to use "clear fuel." Such fuels could be CNG or batteries. Some states such as California have even stronger legislation. In 1998, car builders doing business in that state must offer zero-emission electric vehicles for sale.

drugstore three blocks away, you can take your bike or walk. Upon leaving home, you can turn off the lights and lower the thermostat. Before you leave on an extended vacation, turn off the hot water heater. Installing an inexpensive insulation kit reduces the heat loss of a 66-gallon electric water heater set at 66°C (151°F) by 23 percent. If you lower the thermostat from 66° to 54°C (151° to 129°F), an additional 30 percent reduction in heat loss occurs.

Approximately 30 percent of the energy consumed in the United States is used in the home or in the operation of commercial buildings, and another 40 percent is used in industrial processes. Home uses include energy for heating, cooling, lighting, food processing, and recreation. Heating and cooling of buildings accounts for 18 of the 30 percent. Modern buildings (with very few exceptions) are not designed to conserve energy, and some consume extraordinary amounts. The World Trade Center in New York is an extreme example. About 100,000 kilowatts of power are used in this building—which is more than is used by the 100,000 people in Schenectady, New York.

By using insulation, orienting the building to make the most of natural heating and cooling, and using energy-economic building materials, savings in money and energy can occur. Attic and wall insulation greatly reduces the energy exchange. In the average home, proper attic insulation pays for itself in 2 to 5 years, depending on climate. If glass area is reduced and double-pane glass or storm windows are used, energy is conserved. Weather stripping around windows and doors is also effective. Individuals building their own homes can save fuel by orienting them so that the sun is directed in during the heat of the day in the winter and onto the rooftop during the summer. Overhang, shade trees, and earth banks keep a home cooler in the summer. Through architectural design and use of solar panels, owners of some newer homes address their energy conservation concerns at the planning stage.

Improved industrial processes can be discovered and developed through a systematic evaluation of thermodynamic efficiency. The energy-producing industry, for example, could increase the efficiency of plants using fossil fuel to generate electricity from 33 percent to 40 percent. Better insulated wires and more efficient transmission systems would conserve electric energy. Through price incentives, management can reduce peak electric loads and help reduce energy demands. More efficient heat transfer and storage processes should be developed and used. Energy loss could be reduced by operating factories for 24 hours rather than 8, thus saving the energy needed to warm up the buildings.

Tied closely with industry is transportation, which consumes another 30 percent of our nation's total energy supply. More than half of this is consumed by private autos. Mass transit, carpooling, reduced recreational driving, and better vehicle design are all excellent conservation measures. Electric cars are currently being tested as a fuel and power system.

Secondary or **rechargeable batteries** are effective for energy storage and can be used as a conservation device. Batteries convert and store electric energy by chemical change. To obtain electric energy, the electrochemical reaction is reversed when the battery is discharged. Batteries in electric-generating facilities can store energy in off-peak periods. In the home or industrial plants, electricity generated by wind, water, or the sun can be stored for use when other sources are not producing electricity.

One potential force in energy conservation is the federal government. Cheaper energy prices in the United States are partly the result of national policy to keep prices down through subsidies such as oil depletion allowances, price controls, well-head gas price regulations, and permissiveness with regard to health and environmental risks. Such policies distort the energy market. When the true cost of energy is borne by the consumer, conservation practices will be encouraged.

Energy conservation, therefore, requires a combined effort of individuals, industry, and governmental agencies. With a major effort, the per capita rate of energy use can be significantly reduced.

SUMMARY AND CONCLUSION

All movements or reactions involving work require energy. Energy used in performing work is partially consumed to do the work and partially converted to heat. No energy is ever created or destroyed.

Human energy demands include energy converted by photosynthesis to form food and energy used to maintain society. The latter needs are satisfied by primary energy sources, which include solar, nuclear, geothermal, and tidal energy, and secondary sources that trap energy from one of the primary sources, such as fossil fuels,wind and water power, hydrogen, and solid wastes.

If we are to make a transition from our present use of fossil fuel to new energy sources, it must be done more rapidly than the transition from wood to coal, which took about 80 years. There does not appear to be one best form of energy available for future use—all sources need to be developed to determine the best form for each region. Better planning and management can reduce the problems associated with the development of new energy sources. Placing energy-producing facilities near urban areas reduces transportation costs but sometimes increases health problems. Buffer zones around the industry could help. By evaluating the effects of energy resource development on physical systems and living organisms, we can develop a variety of solutions. We need not one source of energy, not one solution, but a systems approach to the problem.

FURTHER READINGS

Bleviss, D. L. 1988. *The New Oil Crion and Fuel Economy Technologies: Preparing Light Transportation Industry for the 1990s.* Westport, CT: Quorum Books.

Brown, L. R. 1991. *State of the World.* Washington, DC: Worldwatch Institute.

Gibbons, J. H., P. D. Blair, and H. L. Gwin. 1989. Strategies for Energy Use. *Scientific American.*

May, J. 1989. *The Greenpeace Book of the Nuclear Page Age.* New York: Panthean Books.

Pollock, C. 1986. *Decommissioning: Nuclear Power's Missing Link.* Worldwatch Paper 69. Washington, DC: Worldwatch Institute.

Energy for the Planet Earth. 1990. *Scientific American.* Special issue, 263(3).

U.S. Department of Energy. 1989. *Energy Conservation Trends.* Washington, DC.

U.S. Department of Energy. 1990. *Annual Energy Review.* Washington, DC.

World Resources 1990–1991. New York: 1990. Oxford University Press.

STUDY QUESTIONS

1. Compare the environmental impact of light-water nuclear plants and plants using fossil fuels to generate electricity.
2. Did the completion of the Alaskan pipeline increase the oil reserves of the United States? Explain.
3. How can utility company rates and government taxes control energy consumption?
4. What form of population redistribution can you visualize as energy sources change?
5. Using the second law of thermodynamics, explain why the use of hydrogen is less desirable than solar energy for transportation.
6. What is meant by energy conservation?
7. Comment on the following: Because photosynthesis involves the conversion of the sun's energy to chemical energy, a greater supply of chemicals should exist in desert regions of the world where there is an abundant supply of solar energy.
8. What means of energy conservation can you suggest for your home, school, community, and nation?
9. What do you believe the role of government should be in energy research and development? Compare and contrast governments in a highly developed country and a less developed country.
10. A nuclear power plant is planned for a site 1 mile from your home. How do you feel about this decision?

SUGGESTED ACTIONS

1. *Use Cooler Water.* Turn your water heater to lower temperature settings. For every 10° you reduce the temperature, you save 6 percent of the energy required. Also, drain several cupsful of water every 2 months to prevent accumulation of sediment. This improves the life and efficiency of your heater.

   Insulate your water heater. Water heaters store hot water for your use. Many are poorly insulated and lose heat through their sides. Insulation packages are available that can be put around the water heater to reduce heat loss.
2. *Recycle Your Motor Oil.* In some areas, up to 40 percent of the groundwater pollution comes from motor oil that has been put on the ground or in the sewer. You can recycle your oil through local garages or highway departments. If no one does this, encourage it in your community. You thereby reduce pollution and help reuse one of our fast-depleting nonrenewable energy resources. The recycled oil is used for lubrication or ship's fuel.
3. *Use Compact Fluorescent Bulbs.* Compact fluorescent light bulbs are now available. These bulbs produce "normal" light, not light from the standard fluorescent

tubes. These bulbs are great energy savers. They last longer and use about one-fourth of the electric energy of the incandescent bulb. They also generate much less CO_2 than standard bulbs, thereby reducing atmospheric pollution.

4. *Watch Your Tires.* Underinflated tires increase the amount of fuel your car uses. Check the recommended inflation pressure and check your tires monthly to be sure they have the proper pressure. If you buy an inexpensive tire gauge, you can do this at home.

   Radial tires also increase your automobile's gas mileage over conventional tires. Switch to radials and keep them properly inflated.

   Keep your tires in the best possible shape for your safety and that of the environment. Buy long-lasting radial tires. Keep them properly inflated. Rotate as per the manufacturer's suggestions. On older cars, use rethreaded tires to reduce tire pollution in the environment.

PART

FOUR

TOXINS IN THE ENVIRONMENT

14

PESTICIDES

In the fall of 1948, the Nobel Institute in Stockholm, Sweden, announced Swiss chemist Dr. Paul H. Mueller as winner of the Nobel Prize for medicine for his part in the development of DDT. During the late 1950s, however, there was concern about the degradability and mode of dispersal of DDT and related compounds in the environment. In 1962 Rachel Carson discussed food chains, ecology, and persistence of pesticides in Silent Spring, *starting an emotional debate between ecologists and agriculturists that has not been altogether resolved today.*

Today we find that many of the pesticides that were supposed to eliminate insects such as malaria-bearing mosquitoes are no longer effective.

What are these chemicals called pesticides? How do they function? How do they affect people? Can we use other methods to control pest species?

WHAT IS A PEST?

There is no universal biological meaning of the word *pest.* Any living thing that successfully competes with people for food, space, or other essential needs is called a **pest.** The term therefore includes different living things in various parts of the world. Food competitors include birds, rodents, insects, bacteria, mollusca, and nematodes, while weeds, bears, elk, birds, and small mammals are space competitors.

Pests are often animal or plant species displaced by humans from their normal habitat or element. Starlings, rats, and cabbage worms, introduced into North America from Europe; Klamath weed, a native of Europe and Asia now found throughout a large part of the world; and the cottony cushion scale, an insect native to Australia now found in California, are some introduced pests (Figure 14.1). By certain practices, people encourage other pest species. For example, large numbers of mosquitoes begin mating in places where water is impounded and deer become a nuisance when their predators are removed. Sometimes people decide to use areas inhabited by natural populations, which become pests. Malaria and sleeping sickness are examples of diseases that can be transmitted to people and cattle by different vectors.

Most pest species, particularly insect and weed populations, are able to grow and reproduce rapidly. This is common in populations that invade and become established in new areas, as in the case of early successional stages of community development (see Chapter 2). When pest species invade newly plowed land, they follow an

Figure 14.1 Ladybird beetle attacking cottony cushion scale. (Courtesy of University of California, Riverside, Department of Entomology, Division of Biological Control.)

exponential growth curve. Once they have reached a noticeable, rapid growth phase, they are very difficult to control. A pesticide reduces the population where it is applied, but invasions from neighboring areas make repeated applications necessary.

PESTICIDES

Pesticide is a general term denoting substances such as insecticides, herbicides, fungicides, fumigants, algacides, aviacides, and rodenticides used to control objectionable populations. Ecologists use the term **biocide** to indicate all chemicals that destroy pests. In our discussion, we shall emphasize insecticides and herbicides.

Insecticides

Insecticides enter the insect through the stomach wall, body cover, or the respiratory system. Once in the insect, they interfere with the body physiology and kill it. Insecticides are divided into four major classes: inorganic, oil, botanical, and synthetic. Synthetics include organochlorines (chlorinated hydrocarbons), organophosphates, and carbamates.

Inorganics Inorganics have been used for some time, although today they have been largely replaced by more efficient organic compounds. Most inorganics, like lead arsenic used on shrubs and trees to control chewing insects, are stomach poisons. Sulfur derivatives, paris green, and calcium arsenate are some others. This whole group is generally restricted in use because of its toxicity to humans and persistence in the environment.

Botanicals Some plant extracts are used as insecticides. These botanicals are complex chemicals that break down into harmless compounds soon after application. Most are effective in destroying specific insects, usually soft-bodied ones such as aphids and caterpillars. They are relatively safe for people to handle. Pyrethrin, rotenone, nicotine, and dimethrin are a few common ones.

Synthetics By far the most used insecticides today belong to the organic, or synthetic, group.

These synthetics cause much of the debate about the effects of insecticides on the environment. Three general chemical groups of organics are recognized; organochlorines (chlorinated hydrocarbons), organophosphates, and carbamates.

Organochlorines, including DDT, dieldrin, and aldrin, have been used since 1945. Although effective in destroying insect pests, they kill other animals as well. They are fat soluble, persistent, mobile, and very stable. Most of the research has been done on DDT, but all organochlorines have similar patterns of activity with the central nervous system being the main body target. The fatty layer surrounding many nerves appears to absorb organochlorines. Once in the fat, they cause hyperactivity and convulsions by initiating a series of nerve impulses. Scientists believe paralysis and death occur from neural toxicity or the exhaustion of metabolic processes.

Another type of organochlorine with similar structure and chemical properties is **polychlorinated biphenyls,** or PCBs. PCBs are used in manufacturing rubber, plastics, inks, and carbonless reproducing paper. While not a pesticide, PCBs enter our water and atmosphere and pose serious health problems.

Organophosphates are mainly contact insecticides, although some are absorbed through the respiratory system. A variation of Tabun, a nerve gas developed but not used by the Germans during World War II, they interfere to various degrees with the transmission of nerve impulses across nerve synapses. A nerve carrying an impulse releases the chemical *acetylcholine* to stimulate an impulse in the neighboring nerve. The enzyme *cholinesterase* usually breaks down the acetylcholine, but it is inactivated by organophosphates. This results in continuous stimulation of the nerve until the organism goes into spasms and eventually dies. In higher animals, symptoms such as nausea, salivation, muscle spasm, coma, and convulsions occur. This group includes more than forty commercially successful organophosphates, including Azodrin, Parathion, and Malathion.

Carbamates are relatively new and represent a unique, greatly diversified class of insecticides known to act synergistically with several other insecticides. They are absorbed on contact or through the stomach wall and apparently deactivate the enzyme cholinesterase, as do organophosphates. These pesticides are rapidly broken down to less toxic chemicals and eliminated from the insect's body, so they do not accumulate in fat. Carbamates such as Furadan, Lannate, and Sevin are used today to protect a number of agricultural crops.

Herbicides

Both inorganic and organic herbicides are used. Historically, ashes, common salts, and bittern have been used in agriculture, and today the list of inorganic chemicals includes arsenic, copper, and sulfur compounds. In addition, we use more than 100 different organic chemicals as herbicides. Organic herbicides can be placed into three classes according to their effects on plants: contact herbicides, systemic herbicides, and soil sterilants.

Contact herbicides kill plant parts through direct contact with foliage. Generally, they act very rapidly and the plant dies within a matter of a few days. Atrazine, simazine, and pentachlorophenol are examples of contact herbicides in use today.

Some of the best known and most widely used herbicides are systemic. They are absorbed either by the foliage or the roots and can be transferred through the entire plant system. Most **systemic herbicides** have a long-term effect on susceptible plants. Only slightly toxic to animals, the phenoxy herbicides are a group of compounds used for broadleaf weeds and woody plant control. They cause a growth reaction similar to that caused by some plant hormones responsible for controlling the normal size and shape growth patterns. These herbicides create excess hormones, stimulating uncontrollable growth and causing the plant literally to grow itself to death. Included in this group are 2,4-D; 2,4,5-T; silvex; and others.

Another well-known group of systemic herbicides are the so-called substituted ureas. Because they are water soluble, they can be quickly absorbed by plant roots and accumulate in leaves. There they inhibit some of the chemical interactions in the photosynthetic process, preventing the plant from producing the necessary biomass for survival. When used at levels that destroy plants, this group does not appear to be harmful to warm-blooded animals. Examples of some substituted ureas are fenuron, diuron, and norea.

Some herbicides, such as the ureas, fall into two categories. While acting as systemic herbicides, they also remain in the soil and become **soil sterilants** from 48 hours to 24 months. They include such brands as Treflan, Dymid, Aatrex, Dowpon, and Sutan.

IMPACT ON THE ENVIRONMENT

Most pesticides are biologically active chemicals. In some manner they become disruptive to one or more normal physiological functions by combining with, destroying, or interfering with the production of necessary body chemicals. Such characteristics make them harmful to many organisms, not just the target species. Pesticides are used against only some 2000 species, yet pesticide residues are found in over 200,000 non-target species—sometimes in very large amounts. On some occasions birds, fish, and mammals have died in large numbers following pesticide applications (Table 14.1).

Application

Approximately 7.7 million kilograms (1.7 billion pounds) of synthetic organic pesticides are used in the United States each year. By placing this amount uniformly over the 9.3 million square kilometers (3.6 million square miles) of the United States, about 210 kilograms (460 pounds) of active pesticide ingredient would be deposited for every 4 square kilometers (1.7 square mile). Pesticides are applied as aerosols (Figure 14.2), liquids, or solids. In all cases, they move into the air, soil, water, and living organisms. The atmosphere is considered the major route for the widespread distribution of the more persistent pesticides. For example, the DDT levels in samples from the air around Bermuda in 1974 were 100 times greater than the levels sampled 9 years earlier.

Table 14.1
Effects of Organochlorines on Wild Animals

Chemical	Purpose	Location	Effect
Aldrin	Rice seed protection	Texas	Widespread mortality of fulvous tree ducks
DDD	Kill gnats	California	Death of grebes
DDT	Control Dutch elm disease	Maine, Michigan, Wisconsin, New Hampshire	Heavy mortality of songbirds
DDT	Forest protection	Connecticut	Trout kill
Dieldrin	Sandfly larvae control	Florida	Pheasant production reduced
Endrin	Cutworm control	California	Rabbit mortality

From L. F. Stickle, Special Scientific Report 119 (Washington, DC: U.S. Fish and Wildlife Service, 1968).

Figure 14.2
Aerial application of fungicide to a Florida orange grove. (U.S. Department of Agriculture.)

Persistence

Most organochlorines are long-lived and persistent. Fifty percent of the DDT originally applied to a field might be found in the soil after 3 years. When pesticides are applied in the prescribed amounts, 95 percent of the original amount will disappear in the time ranges shown in Table 14.2. Since most pesticides are applied several times a year, buildup in the soil and biota occurs.

In comparison, organophosphates are unstable and most tend to degrade to harmless chemicals after several months. In general, these compounds are not accumulated or stored in living tissue; however, they are toxic to people in some concentrations and can be a special hazard to the worker applying them. Greenhouse workers have been killed in situations where ventilation was poor.

Herbicides generally are removed from the soil by bacteria. The herbicide 2,4-D in small quantities and low concentrations is decomposed in a period of 1 to 4 weeks after application, while 2,4,5-T could take more than a year to decompose. The actual disappearance rate depends on concentration, soil type, moisture, temperature, and aeration. Most herbicides in the group to which 2,4-D belongs are water soluble, so they can be moved via surface water and streams to other areas if applied in high concentrations.

Phenol ureas are quite variable in persistence and solubility. Some are water soluble, and some are oil soluble. A few herbicides in this group, monuron and diuron for instance, resist action by soil microorganisms and so remain active in

Table 14.2
Persistence of Some Organochlorines in Soil

	Average Annual Dose			Time for 95% Disappearance (years)	
Chemical	(lb/acre)	(kg/ha)	Half-life (years)	Range	Average
Aldrin	1–3	1.1–3.4	0.3	1–6	3
Chlordane	1–2	1.1–2.2	1.0	3–5	4
DDT	1–2.5	1.1–2.8	2.8	4–30	10
Dieldrin	1–3	1.1–3.4	2.5	5–25	8
Endrin	1–3	1.1–3.4	2.2	3–20	7
Heptachlor	1–3	1.1–3.4	0.8	3–5	3.5
Lindane	1–2.5	1.1–2.8	1.2	3–10	6.5
Isobenzan	0.25–1	0.3–1.1	0.4	2–7	4

Reprinted with permission from C. A. Edwards, *Persistent Pesticides in the Environment* (Cleveland, OH: CRC Press, 1973). Copyright The Chemical Rubber Co., CRC Press, Inc.

the soil for several years. Soil sterility can result if they are applied in large amounts.

Effects on Food Chains

Studies of bird populations disclose that organochlorines are most highly concentrated in fish-eating birds, followed in decreasing order by raptors (hawks and owls), omnivores, and herbivorous birds. Studies on eggs in museum collections show a decrease in eggshell thickness of bird eggs collected between 1945 and 1947, the period when DDT was first introduced into the ecosystem on a worldwide basis. In the case of the sparrowhawk, for example, DDT disrupts the body mechanism that transports calcium, thus reducing the thickness of the eggshell (Figure 14.3). Some bird populations, such as the bald eagle, peregrine falcon, osprey, and brown pelican, decreased or disappeared from parts of their range because eggshells cracked during incubation.

Studies show that a variety of pesticides and other toxicants are taken up by animals and may interfere with the reproductive process. Wood stork eggs from Florida were found to contain DDE (a DDT derivative), mercury, and PCBs, and hatching success was reduced as amounts of DDE increased. In another project, Loggerhead sea turtles were found to have concentrations of DDE.

Concentrations of pesticides in organisms come from food they ingest. Because many organochlorines are fat soluble, some or all of the pesticide ingested or absorbed is retained and passed on to the consumer. Organochlorines accumulate in each link of the food chain in a process called **biological amplification;** thus, the top consumers receive levels of pesticide that affect their physiology and, in the case of some fish and birds, interfere with their ability to reproduce.

A study at Clear Lake in north central California demonstrated these points. A small midge, attracted to cabins by light, was very irritating to vacationers. In 1949 the organochlorine DDD, a derivative of DDT, was applied to the lake water to reduce the midge population. Initially this program appeared successful, but in 1954 the midge population reached a level

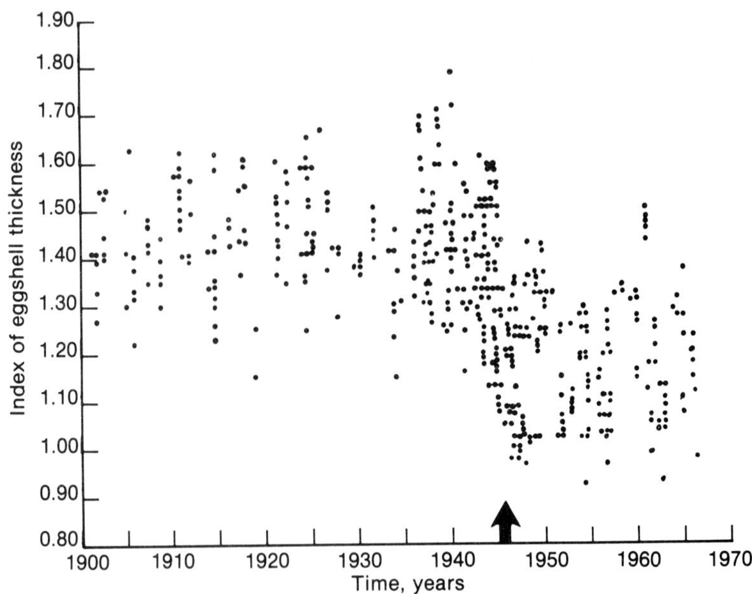

Figure 14.3
Changes in eggshell thickness of the sparrowhawk. The arrow marks the introduction of DDT into the ecosystem. (From D. A. Ratcliffe, "Changes Attributable to Pesticides in Egg Breakage Frequency and Eggshell Thickness in Some British Birds," *Journal of Applied Ecology* 7 [1970]: 67–105. Courtesy of Blackwell Scientific Publications Ltd.)

where respraying with higher levels of pesticide became necessary. In 1957 a third application was required. Studies made at the time showed no traces of DDD in the water. However, following the second spraying, the population of western grebe, a diving bird that fed primarily on fish, failed to reproduce for several years and then died. Studies indicated that the pesticide had been absorbed by many organisms in the lake and concentrated in the food chain leading up to the grebe.

Some scientists believe that top-level consumers such as the grebe absorb insecticides directly from the environment as well as from food. If so, a true case of biological amplification might not be shown. In any case, the pesticide does interfere with population interaction in the ecosystem by destroying top-level consumers. This imbalance in the system can lead only to other effects, possibly detrimental to humans.

In western Oregon, heptachlor is used as a wheat-seed treatment to control wireworm. A number of birds eat the seed directly; others have a diet of small rodents that have consumed the seeds. Since heptachlor has been in use, significant numbers of birds have died off. Biologists have found the pesticide in the tissues of several bird species. Canada geese (Figure 14.4) and American kestrels showed increased adult mortality and reduced natality. This situation illustrates how birds at the primary-consumer level (granivores) and secondary-consumer level (carnivores) can be affected by pesticides in the food chain.

The impact of pesticides on life can be considerable because the stability of an ecosystem is maintained in part by its diversity of organisms (see Chapter 2). When one population level is low, others in the same trophic level utilizing the same food can increase in numbers. When massive destruction of whole groups of population occurs, as can happen with the misuse of pesticides, a less stable system results. This does not mean, however, that because some individuals are harmed by pesticides a whole population is harmed.

Side Effects

Often unwanted side effects nullify any positive effects of pesticides. Populations of insectivorous birds such as the American redstart, parula

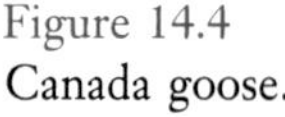

Figure 14.4
Canada goose.

warbler, and red-eyed vireo in a bottomland forest declined 44, 40, and 28 percent, respectively, over a 4-year period. During this period DDT was applied annually at a rate of about 2 kilograms/hectare (2 pounds/acre). This decline was probably due to a reduction in food supply as these birds would normally eat many of the insects the pesticide was meant to destroy.

Sodium monofluoroacetate (1080) was developed during World War II to kill rodents. When it was found to be very effective against the canid species (dog family), bait stations containing impregnated meat designed to kill livestock predators, especially coyotes, were set up. However, studies showed that some nontarget birds and mammals, such as eagles and domestic dogs, ate at the bait stations also. Because of the effect on nontarget organisms, the U.S. Department of Interior restricted the use of 1080. Ranchers, particularly in the West, believed they were being severely restricted in their ability to control coyotes, so the ranchers filed suit in federal court. In 1981 the federal government reversed its position partly due to the development of a sheep collar saturated with 1080. The collars are attached to sacrificed lambs that are placed on the edge of the flock. The coyote attacks the lamb at the neck and receives a lethal dose of 1080.

Because communities are integrated groups of populations, alteration of one or a few populations affects others living there. Herbicide application reduces the vegetation and thus changes the animal composition of the community. Destruction of plants used for food is directly correlated to herbicide application. For example, the herbicide 2,4-D applied to mountain rangeland in Colorado where pocket gophers were abundant reduced the herbaceous growth by 83 percent. Since the gophers were dependent on these herbs for food, their population was reduced by 87 percent.

Many forms of pesticides are used, so their combined effect must be considered. In the analysis of some fish and birds found dead on the coast of Great Britain from 1963 to 1964, traces of dieldrin, heptachlor, BHZ, DDT, and DDE were all found in the body tissue. In most cases, the concentration of each pesticide was below the level that could cause death. At that time scientists were unaware of the combined effects of these pesticides. Were the combined effects additive? Were they synergistic?

Chemicals such as DDT have been thoroughly analyzed, so we have become familiar with their effects on living systems. New chemicals are being developed constantly and are tested by the U.S. Food and Drug Administration for toxicity to some organisms, ability to decompose, and solubility in various solutions. The overall ecological effects of pesticides are extremely variable, however, and little information is available to determine food chain accumulation, movement in the environment, or long-term effects on the ecosystem and on humans. We know little about the mode of action of most pesticides and their physiological interactions in different living organisms. There is evidence that modes of action vary significantly with individual species. For example, DDE is not toxic to many insects but is very dangerous to predatory birds because of its effect on eggshells. At the same time, seed-eating birds like quail are relatively resistant to the effects of DDE.

Carbamates such as carbofuran have become dominant pesticides in agriculture. Despite their being less persistent in the environment than organochlorines, they do cause harm to aquatic life when placed in or on water. Fish, birds, and mammals have been killed by the presence of carbofurans in their food and drinking water.

Tolerance

The properties of DDT as a pesticide were discovered in 1939 by Dr. Paul H. Mueller. The pesticide became a spectacular success in combating typhus and malaria in Italy during World War II by killing human body lice (typhus vectors) and the mosquito vector of malaria. In Egypt, infant mortality was drastically reduced because DDT killed flies that transmitted diarrhea and dysentery. By 1952, the chemical industry was able to produce DDT inexpensively enough to eradicate malaria and sleeping sickness from millions of hectares in Asia, Africa, and South America, making them habitable for humans (Figure 14.5). Malaria is eradicated in 37 of the 149 countries once classified as malarious and is almost eliminated in 16 others. Mosquitoes are now becoming resistant, so a malaria-free world is impossible for the present; in fact, malaria is spreading (Figure 14.6). The development of DDT resistance in mosquito populations is a good example of evolutionary adaptation. Since insects reproduce rapidly, it is possible to see results very quickly. Insects not resistant to the chemical are destroyed, and the resistant ones remain. Without much competition, the resistant organism can reproduce rapidly and occupy the entire area. Reapplication of the pesticide reduces population numbers again, allowing the unlimited opportunity for more resistant varieties to grow. This example of "survival of the fittest" shows how the use of pesticides can have a major impact on the environment. Resistance in insects is usually very specific, so that those able to tolerate DDT or a related chemical form like DDD are not resistant to other organochlorines such as dieldrin or aldrin. However, insects can develop tolerance to more than one insecticide by building up genetic resistance to each. More than 250 of the world's crop pests are resistant to some type of insecticide, and 20 of the worst pests, including the Colorado potato beetle, tobacco budworm, and green peach aphid, are resistant to all types of insecticides. Over a hundred of the insect pests that affect human health are resistant.

Figure 14.5
Twenty-one species of the small tsetse fly transport the blood parasite *Trypanosoma gambiense* between humans and wild animals. People cannot inhabit major areas of Africa because of this parasite, which causes sleeping sickness. Likewise, about 10 million square kilometers (4 million square miles) of grazing land in the world are not used because cattle are infected by the parasite. (Courtesy of Marion Kaplan.)

Plant and higher animal populations can evolve various degrees of tolerance to pesticides. The house mouse populations selected for DDT resistance studies increased their tolerance twofold in ten generations. A mosquito fish population living in a stream near a cotton field developed a 300-fold resistance to Strobane and a 120-fold resistance to endrin. Frog populations are also known to build up a tolerance to DDT. No direct evidence exists to indicate that plants develop resistance to 2,4-D.

PESTICIDES AND HUMAN HEALTH

People ingest pesticides by eating food containing traces of pesticide residue or are exposed to pesticides through house dust, particularly in areas near orchards sprayed with pesticides. Once they enter the human system, pesticides can be detoxified and removed or retained in tissues. In trying to determine the effects of pesticides on people, scientists have found a trace of DDT in

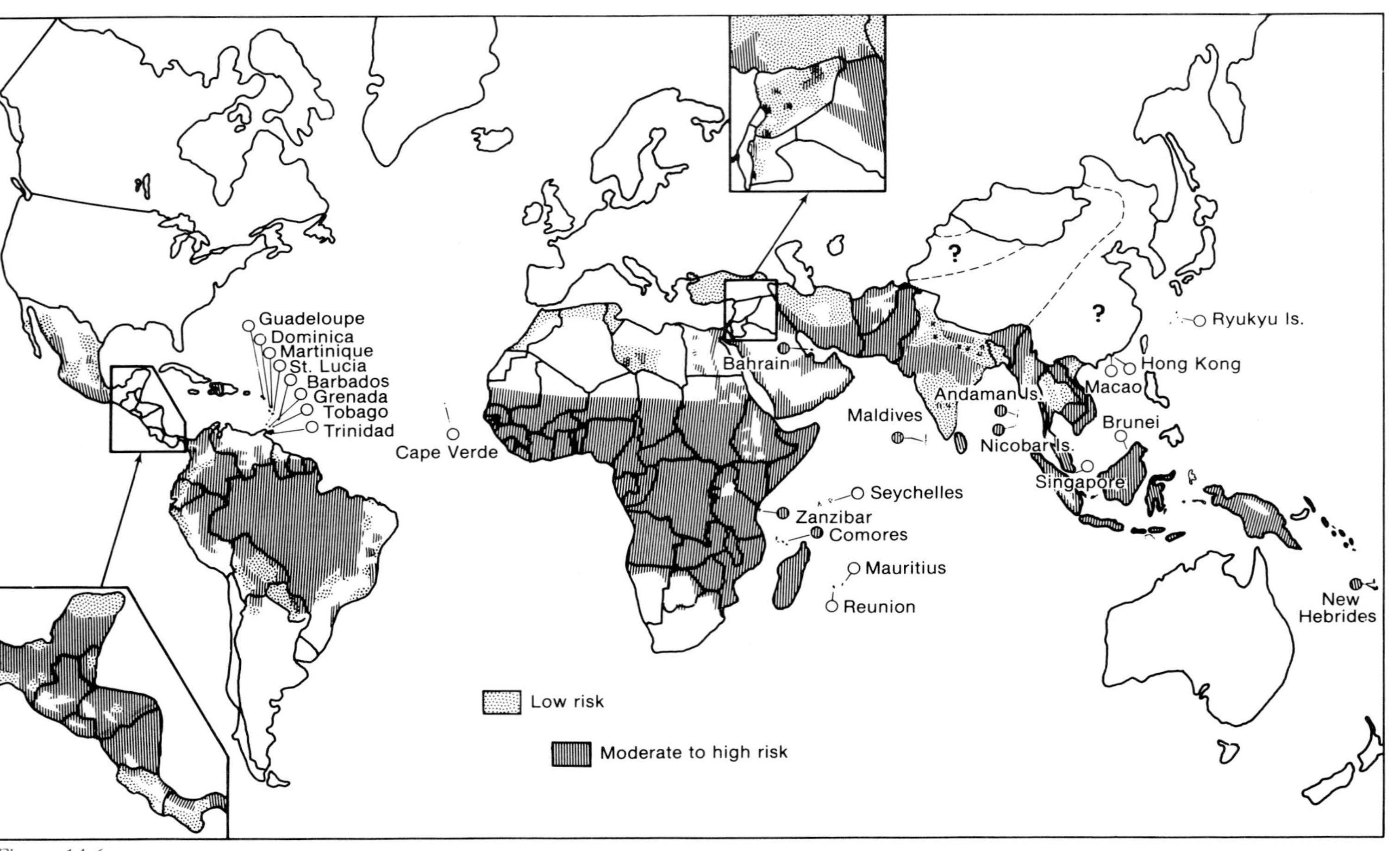

Figure 14.6
Area of risk for malaria transmission. (U.S. Department of Health and Human Services.)

the fatty tissue of every person sampled (Figure 14.7). Eskimos who eat fatty foods from upper-level consumers have very high concentrations in their fatty tissue. This buildup has led the population biologist Paul Ehrlich to suggest that overweight people might not be able to reduce in the future because the release of fat-bound pesticides in their bodies could cause toxic effects.

The effects of pesticides on humans are variable. The World Health Organization maintains that the level of DDT causing death in humans is unknown. When volunteers ingested 0.5 milligram of DDT per kilogram of body weight per day, they developed a tolerance. Other studies have found that symptoms such as nausea, headache, and restlessness occur in people that ingest up to 6 milligrams/kilogram of body weight over a period of weeks.

Several incidents of accidental poisoning by contact with pesticides in production plants have occurred. In one case, employees of a Virginia plant manufacturing Kepone, a chlorinated hydrocarbon, suffered chest pains, peculiar eye movements, and bouts of uncontrollable trembling. When investigators came to the plant, they found a large amount of Kepone dust. Employees were not wearing the proper safety gear and many were sick. Investigators closed the plant and hospitalized thirteen employees who showed severe symptoms of pesticide poisoning.

In 1985 the pesticide *aldicarb* was found on watermelon from the West Coast. People who ate the melons suffered from diarrhea, nausea, vomiting, and blurred vision. Investigators believed the watermelon had been contaminated with aldicarb—normally used on cotton, beans, and potatoes—to take a shortcut to a large yield. Millions of watermelons had to be discarded.

Perhaps no pesticide problem has been as newsworthy as the 1984 disaster in Bhopal, Central India. Leakage from a manufacturing plant of the gas methyl isocyanate, an intermediate for Sevin and other pesticides, caused the deaths of more than 2500 people and uncounted numbers of cattle.

A number of tests show the toxicity of organochlorines to organisms other than humans (Table 14.3). Birds appear to withstand larger doses than mammals.

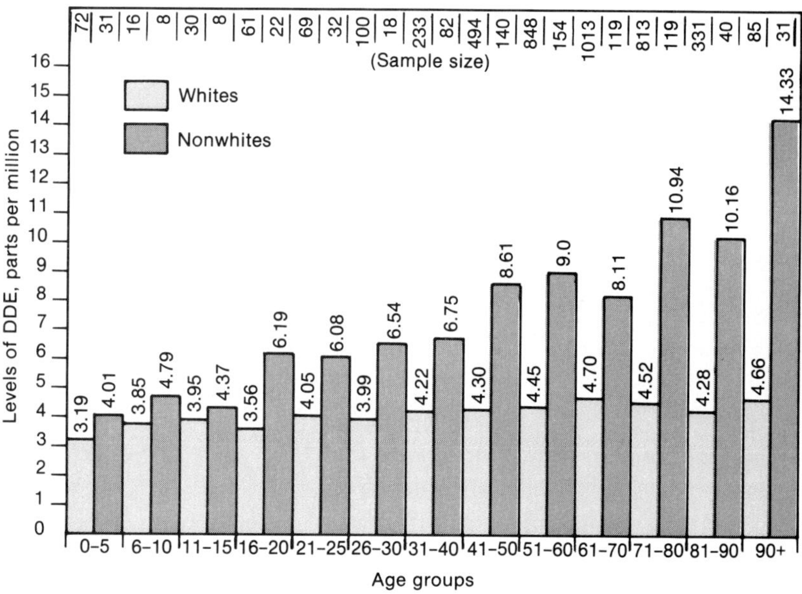

Figure 14.7
Levels of DDT (parts per million) in fat of people in the United States.

Table 14.3
Toxicity of Some Organochlorines

Compound	Rat	Rabbit	Rainbow Trout
	Oral Dose LD_{50} (mg/kg)		LC_{50}* (ppm)
DDT	150–400	250–400	0.042
Dieldrin	40–50	50–80	0.001

*LC_{50} = Lethal concentration for 50 percent of exposed population.
From C. Walker, *Environmental Pollution by Chemicals* (London: Hutchinson Educational Ltd., 1971).

Some forms of pesticides cause mutations in animal and plant species. For example, fumigants containing chlorine or iodines cause mutations in yeast; pesticides containing mercury cause genetic changes in flies; and organophosphates are known to induce mutations in yeast and bacteria. Many other pesticides appear to alter plant and animal chromosomes (Table 14.4). There are some indications that mercury compounds cause chromosome breakage in humans. However, we need more information to ascertain the effects of pesticides on human genetics.

Tests are currently being conducted to determine the effects of pesticides on unborn children. Again, we must rely on the results of animal studies to formulate such tests. Pesticides have been known to cause congenital defects in some animals. Chickens and rabbits exposed to Captan are born with abnormal extremities. Mercury compounds are associated with the development of abnormal limbs in mice. There is evidence that pregnant Japanese women who consume fish and shellfish contaminated by methyl mercury produce offspring with a high incidence of cerebral palsy.

While the evidence for cancer production in many animals is great, there is no direct proof that pesticides cause cancer in humans. For example, these compounds are known to cause cancer in two or more species: aldrin, aramite, chlorobenzilate, DDT, dieldrin, strobane, and heptachlor. Arguments among government agencies and among industrial health groups are precipitated by information from animal studies. Based on such studies, the U.S. Department of Health, Education, and Welfare recommends that groups of pesticides known to cause cancer in animals be eliminated to protect human health. It has also indicated that there are many chemicals about which little is known. It is important to recognize that animal studies generally involve the ingestion of high levels of pesticide. Because the genetics and physiology of test animals are not the same in humans, the extrapolation of data from animals to humans must be viewed with caution.

Although much of our discussion shows the detrimental effects of pesticides on organisms in the environment, immediate benefits to human health are seen in many parts of the world. Pesticides control or eliminate a number of very serious human diseases transmitted by insects. The impressive list includes malaria, Rocky Mountain spotted fever, trypanosomiasis, yaws, tick-transmitted relapsing fever, viral encephalitis, many forms of dysentery, cholera, tularemia, yellow fever, leishmaniasis, sandfly fever, and others. The vectors carrying these diseases or the reservoirs harboring them are destroyed or controlled by use of pesticides.

Sometimes our lack of information on pesticides causes overreaction. During March and April of 1981, endrin was applied in the states of

Table 14.4
Biological Effect of Pesticides on Selected Living Things

Compound	Life Form	Dose	Effect
2,4-D	Onion	25–500 ppm	Chromosome abnormalities
DDT	Mice	105 mg/kg	Mutations
2,4,5-T	Apricot	100 mg/L	Abnormal reproductive cells

Montana, Wyoming, and Colorado to kill army cutworms. In the summer, officials of state and federal governments became concerned about a possibly harmful level of endrin that might be ingested by the public after eating waterfowl from the Central Flyway. Laboratory tests, however, indicated very low levels of endrin—the highest being 0.4 ppm in fat and 0.03 ppm in muscle. Most animals had levels less than 0.1 ppm. Had the public been aware that (1) endrin, being a chlorinated hydrocarbon, had to concentrate in the fat of animals, and (2) most waterfowl have little fat in the breast muscle (the primary food source), unnecessary anxiety could have been avoided.

During the spring and summer of 1981, the Mediterranean fruit fly, also known as the medfly (Figure 14.8), achieved worldwide notoriety as it quickly reproduced and began to overrun California farmland. To help eradicate the medfly, an aerial malathion spraying program was proposed. Because scientific literature was not completely definitive on the use of malathion, a great deal of citizen concern was raised. While there was no evidence of cancer or abnormal births due to malathion use, some people thought more extensive testing should have been done before the pesticide was applied.

Meanwhile the medfly was reproducing and attacking some vegetables and most fleshy fruits. As quarantine officials began checking vehicles moving out of infested areas, and other states and nations started to boycott California fruits and vegetables, farmers in the quarantine area suffered economic loss. Some farmers went to court to block the boycott, but their effort was to no avail.

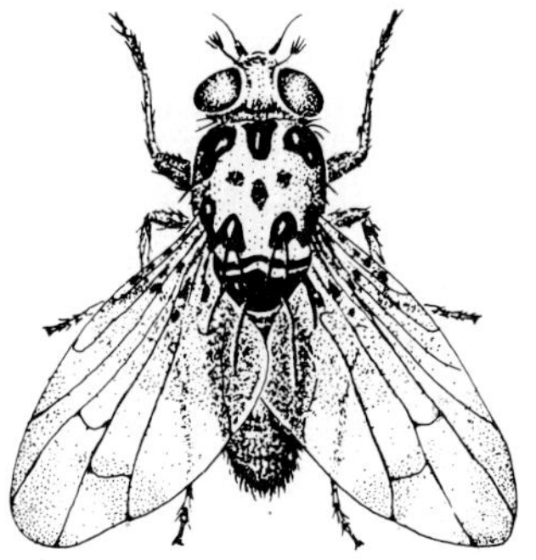

Figure 14.8
Mediterranean fruit fly.

Medflies decreased, then reappeared in California a few years later; meanwhile, the fly appeared in Florida as well. There is no easy solution to the medfly problem; however, we can expect that transporting infested fruit across the country and our ability to modify the natural systems can lead to pest outbreaks.

The health and monetary benefits of pesticides are well documented. It is estimated that in the first decade of DDT use, 5 million lives were saved and 100 million serious illnesses were prevented. In Ceylon, the introduction of DDT reduced the incidence of malaria from over 1 million to 17 cases. During 1968, the spray program was relaxed and 600,000 cases of malaria were reported. A number of developed countries have banned DDT. The United States banned DDT on December 31, 1972, but less developed countries have not been able to afford to give up this precious chemical.

In countries such as Zimbabwe, DDT is used to control tsetse flies, mosquitoes, and other insect pests. Attempts are being made to keep the pesticide at a level that will not harm wildlife, but it is known that DDT threatens a number of nontarget organisms.

Regulations

Between 1970 and 1980, annual pesticide production and application increased by 70 percent—to 771,120,000 kilograms (1.7 billion pounds)—in the United States making regulations mandatory. There are about 35,000 pesticide products on the market that contribute major benefits to society such as increased agricultural productivity, lower domestic food prices, increased exports of agricultural products, control of human disease, and foods that are visually more appealing. On the other hand, pesticides are designed to be injurious to living organisms thus having the potential of harming people and the environment. Their ability to cause damage is heightened because most pesticides are used on human or animal food crops, making human exposure unavoidable.

The Environmental Protection Agency plays a major role in assessing data to determine the safety of pesticides. New products are registered if the EPA finds that the product does not pose unreasonable risks to humans or the environment. By taking into consideration the economic, social, and environmental costs and benefits stemming from the pesticide's use, the agency can issue a registration that allows companies to market the product.

Risks are often quantified in terms of the probability or number of certain health effects in a given population, while benefits are usually stated in dollar values of increased crop yields, lower food costs, reduced chance of disease, or the cost savings from using alternative control measures. Before it is registered, the benefits of a particular pesticide must be shown to exceed the risks. Ecological risk assessments are made. The review and reregistration of all federal and state registered products now on the market are required by the 1972 amendments to the Federal Insecticide, Fungicide, and Rodenticide Act. Most existing products were originally registered before the chronic effects (such as cancer, birth defects, and gene mutations) to toxic chemical exposure were understood. Thus their reregistration requires a more intensive review of test data for both acute and chronic effects. In many cases, the collection of the basic data itself is required.

After the EPA registers products, it conducts enforcement activities in conjunction with the Food and Drug Administration and the Department of Agriculture to ensure compliance with registration decisions. The agencies inspect pesticide product packaging and labeling, manufacturing and formulating plant operations, and actual use of pesticides by consumers. As a result of the 1978 amendments to the act, state governments have acquired the authority to help federal agencies enforce these pesticide laws.

ECONOMIC CONSIDERATIONS

Before 1950, the effectiveness of pesticide applications was measured in terms of application cost versus the cost of crop loss due to pests. Nothing was said about the cost of other forms of pest control or the cost of environmental side effects. People now try to take into consideration all these factors and assess the costs and benefits involved. Since there are many types of pests, each with different means of control, this analysis is not easy. Much of our present research centers around attempts to measure the effects of varying levels of pesticides on farm sales. Frequently such data come from secondary sources such as agricultural surveys.

There is no satisfactory method of evaluating the side effects of pesticides in monetary terms, just as it is difficult to place a dollar value on a recreational resource such as a lake. Dollars cannot be used to measure damage to the ecosystem caused by destruction of nontarget species or pesticide accumulation.

Pesticides are currently marketed very competitively. Salespeople usually recommended pesticides as the best apparent solution to a pest problem. Having no ecological or entomological background, most of them rely on the manufacturers' information when selling. Economic injury from the environmental point of view is rarely considered and analysis of the pest species is often lacking in this information. In some instances, pesticides may be more effective at certain hours, seasons, or temperatures, thereby requiring lower doses than at other times. If a certain level of crop damage is tolerable, it is unnecessary to destroy the pest species completely. Biological information, then, becomes important. If, for example, a person sees a pest, it does not necessarily mean economic damage will occur. On the other hand, it is important to realize that some pest species reproduce rapidly and must be brought under control quickly.

We must recognize that the application of pesticides has helped improve agriculture. In 1913, for example, it took about 35 to 40 hours to produce a bushel of corn. Today, fewer than 15 hours are required. In less developed countries, where every morsel of food counts, improved agriculture based on the application of pesticides becomes a very important consideration. In the Green Revolution, genetic inbreeding has often weakened plant defense mechanisms. In such cases, pesticides are necessary to protect the plant from pests. We must evaluate pesticide use in terms of the amount necessary to accomplish the goal of preserving the crop without contaminating the environment at the same time.

ALTERNATIVE METHODS OF PEST CONTROL

Biological Control

Biological control involves the use of parasites, predators, or pathogens to maintain a pest population at a lower average density than would exist without their use. The purpose of biological control is to suppress the species below the level at which economic injury occurs. This means that some pests must remain, because predators cannot destroy all their prey and survive; but a natural balance develops between the pest and the prey acting as the biological control. Biological control is really an extension of a natural control where a steady state is eventually achieved between the two populations.

The effect of biological control programs on the economics of crop production is shown in Figure 14.9. Economic considerations include the cost of plants, fertilizer, and harvesting and the net profit. The **economic threshold** represents the size of the pest population above which crop damage eliminates net profits. The threshold, or pest size, remains constant with the introduction of a prey. When the pest is reduced below the threshold, a marketable product can be produced for a net profit.

If biological control is so desirable, why are pesticides still used? The history of biological control is full of failures and a few outstanding successes. A successful program involves many years of research and a thorough understanding of pest population dynamics to find a suitable

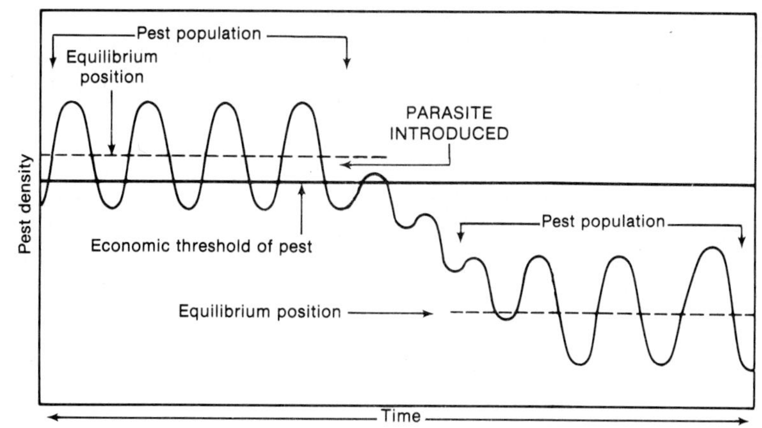

Figure 14.9
Biological control. Pest population density is kept below the population size at which economic damage occurs (economic threshold). (From R. F. Smith and R. van den Bosch, "Integrated Control," in *Pest Control—Biological, Physical and Selected Chemical Methods*, ed. W. W. Kilgore and R. L. Doutt. New York: Academic Press, 1967.)

GYPSY MOTHS: A BIOLOGICAL CONTROL SUCCESS STORY

Gypsy moth caterpillars are large insects with appetites to match. Pennsylvania has been one of their prime feeding grounds throughout the 1970s because of the predominantly oak forests. Heavy infestations have completely defoliated forests, causing many trees to die if repeated and stunting the growth of those that survived.

In 1977, gypsy moths defoliated 526,110 hectares (1.3 million acres) in Pennsylvania. Two years later, only 2428 hectares (6000 acres) were defoliated. This sudden decrease was attributed to the effectiveness of two parasitic flies introduced as biological control agents. One fly was imported from Europe and released in New England in 1908. It spread undetected in Pennsylvania until 1970. The other fly was released as part of a control program of the Division of Forest Pest Management of the Pennsylvania Department of Environmental Resources. One species of fly lays eggs on the gypsy moth caterpillar, and when the eggs hatch, the maggots bore into the caterpillar. The other species lays eggs on leaves that the caterpillars eat. The eggs are then eaten along with the foliage and hatch inside the caterpillar.

The division is trying to establish three exotic wasps and one exotic fly parasite to aid control further. Its goals are to keep damage at low levels, thereby preventing or reducing heavy outbreaks; to bring heavy outbreaks under control the year after damaging levels are reached; and to encourage parasites to spread with the moth as it increases its territory.

control organism. A control organism must be able to grow in the pest habitat and not cause such ecological side effects as the destruction of more desirable populations.

One example of environmental problems caused by biological control is the 1870 introduction of the Indian mongoose to the West Indian Islands to bring about rat control. Initially the mongoose was effective in reducing the rat population; however, the people that introduced the mongoose to the islands did not realize their problem consisted of two rat species—the ground-dwelling brown rat and the tree-dwelling black rat. The nonclimbing mongoose controlled the brown rats; but brown rats competed with black rats, thus the latter increased as the former decreased. When the brown rat population diminished, the mongoose started eating chickens, ground nesting birds, and the ground-inhabiting lizards which previously controlled the sugarcane beetle. The mongoose was also found to be a reservoir and vector for rabies and leptosporosis.

Life tables are often constructed to determine at what stage in its life history a pest is most vulnerable to a controlling agent. Since many pests are introduced from foreign ecosystems, they must be examined in their native habitat to determine the natural controls. Potential controlling agents are subjected to extensive laboratory and enclosed field testing and then to limited field testing if they appear feasible. Finally, if success occurs in all of the preceding steps to control the population, the agent can be introduced in large numbers into the pest's habitat. The whole procedure can take as long as 10 to 20 years, depending on the life cycle of the potential controlling agent. Farmers with an immediate problem are unable to wait for such long periods of time for a possible solution. Still, biological control presents a workable long-term solution to many pest problems.

One of the most noted success stories of early biological control in North America involved the cottony cushion scale (see Figure 14.1), a small scale insect that looks like a bark or leaf discoloration on citrus, pear, acacia, and other plants. It was introduced into California during the last quarter of the nineteenth century. A search in Australia and New Zealand revealed a number of potential parasites and predators, including a ladybird beetle. The beetle fed voraciously on the scale and increased rapidly in numbers. Within 1 year, it controlled the scale by establishing a prey density well below the level of the economic threshold. The balance was disrupted in the early 1950s when DDT destroyed the ladybird beetle population; the scale insect increased rapidly as a result. Reintroduction of the beetle was successful in reducing the scale again.

Control of the oriental fruit fly in Hawaii in the 1940s was another biological control success story. This pest, introduced from Asia, caused great agricultural damage on the islands. A long search led to the discovery of three parasites that attacked and destroyed the fly eggs and larvae. One of the parasites became established and maintained the fly population at a very low level by attacking the larvae.

Bacteria and viruses have been used successfully in biological control programs. The Japanese beetle was controlled in the United States by the dissemination of bacteria spores. In the Central Valley of California, the cabbage looper which fed on cotton was controlled by using a virus from a few infested larvae and spraying the crop with a preparation made from their bodies. Douglas fir tussock moths, tent caterpillars, and cabbage worms have also been controlled by these methods.

Biological control is not confined to insect pests. In about 1894 alligator weed, a South American plant, was introduced into Florida. Since that time it has spread to many waterways in the southern and western United States. The weed grew rapidly, preventing boating, causing floods, blocking drainage canals, and killing fish. Extensive studies in South America revealed that both alligator weed flea beetles and alligator weed thrips were potential control agents (Figure 14.10). After enclosed field testing, both insects were introduced into the infested areas. Alligator weed has been reduced to levels below the economic threshold in areas where the insects have been established.

Agricultural Modification

Generally, the more diverse or the more natural the vegetation remaining in an area, the greater is the opportunity to maintain the system's natural balance and keep pests at a low level. The practice of removing natural vegetation and planting monocultures results in areas dominated by a few species without natural enemies. Since the habitats of natural enemies have been destroyed, farmers can obtain higher plant diversity with a resulting increase in natural control by mixing the crops on adjacent fields or by

Figure 14.10
Twisted, dying alligator weed reveals the presence of thrip on the left. On the right are normal, healthy leaves. (U.S. Department of Agriculture.)

planting alternate strips of wheat and corn or grass and grapes.

Fields interspersed among natural vegetation help reduce pest populations. Farmers are able to utilize natural control species that grow and seek refuge in natural vegetation when they allow such vegetation to grow in fencerows between corn, wheat, and alfalfa fields.

Some crops, like alfalfa, when harvested in strips at different times allow natural enemies of pest species to remain in control of the pests. Removal of whole fields of alfalfa at one time destroys the natural control populations, giving pest species an advantage when regrowth or replanting occurs. The root worms that plague corn crops can be eliminated by planting soybeans or a small grain between the corn rows.

Proper timing of planting and harvesting is also a potential control mechanism. Planting corn early in the North results in lower levels of infestation of army worms, which winter in the South and migrate in the late spring to attack the young plants. Some pests are more destructive if a time lag allows their population to increase in size. Such is the case with the cotton boll weevil. Early harvesting, therefore, reduces economic damage to the cotton crop (Figure 14.11).

Cultivating the soil around some plants helps rid the soil of pests during the growing season. Recultivating in the fall or winter can eliminate some of the pests that remain in the soil during the winter. In other cases, removing weeds along field edges reduces the number of pests entering

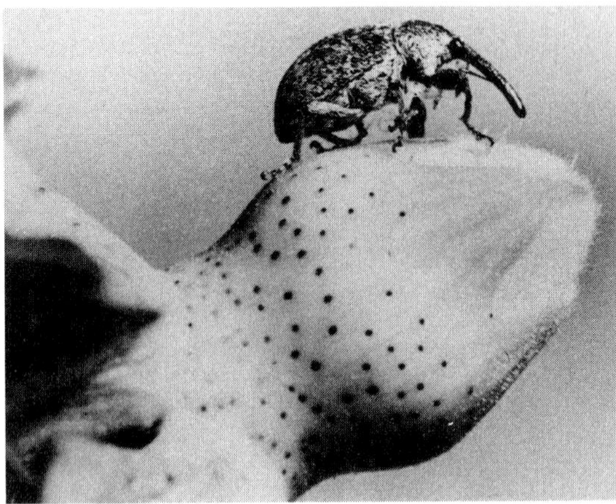

Figure 14.11
Boll weevil on a cotton plant. (U.S. Department of Agriculture.)

cultivated crops. The European corn borer and yellow leaf spot, which both affect corn, can be controlled by secondary tillage. Such practices must be evaluated in terms of additional energy requirements and environmental impact.

Other agricultural techniques include plowing under infected plant parts after harvesting removes the usable portion or using trap crops to divert pests. The latter method involves placing crops preferred by the pests in fields bordering the commercial crop. In Hawaii, for example, corn surrounds fields of melon to attract melon flies.

Genetic Modifications

The science of genetics is a valuable tool in controlling several troublesome insects. The most noted case involves the sterile insect technique used against screwworm flies in the southeastern United States. The screwworm is a serious pest of domestic cattle because it lays its eggs in the animals' skin and the larvae eat the flesh, causing extensive wounds (Figure 14.12). Life history studies reveal that the female mates only once during her lifetime and therefore could be bred to males made sterile by gamma radiation. Large numbers of sterile males were released during 1957 and 1958, resulting in a dramatic decline in the screwworm population. Repeated releases have kept the pest at low levels. In other areas of the world several pest populations, including some species of fruit flies, pink boll worm, and melon flies, are similarly handled. Insects sterilized either by radiation or by chemicals are released into the infested area.

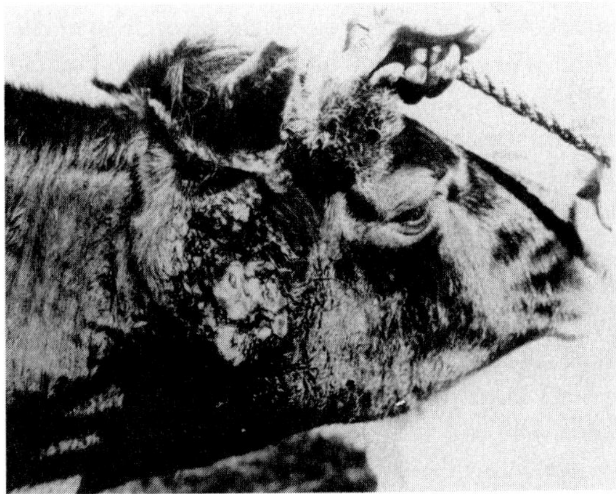

Figure 14.12
Screwworm infestation in a steer's ear. An untreated grown animal can be killed in 10 days by thousands of maggots feeding in a single wound. (U.S. Department of Agriculture.)

Before the sterile insect program can be undertaken, much information must be gathered about the pest population growth patterns as well as the impact of the program on other species in the ecosystem. Because of the extensive background knowledge required, there have been few successful sterile insect programs.

Another form of genetic control is to develop crop varieties that are resistant to damage by several pest species. Resistant varieties of corn, wheat, oat, and barley are now being used, but development of more disease-resistant crops requires a continuing and costly research program. Sometimes sources of resistance cannot be found. In other cases, more virulent races of rusts, fungi, and mildew develop, which means that this form of genetic control must be an ongoing program.

Attractants

Many insects locate their food, mates, and favorable egg-laying sites by built-in automatic responses to chemicals. For such insects, either food attractants, sex attractants, or combination lures can be used to attract them. For example, the boll weevil feeds almost exclusively on the cotton plant. A volatile substance released into the atmosphere by the plant enables the weevil to find it. If this substance could be isolated, it could be placed in some form of insecticide to entice the weevil.

Pheromones are a chemical scent used for communication in some animals. Most phero-

LEGISLATION

Federal Insecticide, Fungicide, and Rodenticide Act (1947, 1977, 1975, 1978). Requires that all chemicals used for pesticides be tested and evaluated. Originally under the control of the Department of Agriculture and transferred to the Environmental Protection Agency (EPA) in 1970, provides a comprehensive regulatory system of chemicals. All the data required for each chemical must be evaluated by industrial and EPA staff before a label registration can be granted. Specific test reviews and data requirements state that, when used in accordance with widespread and commonly recognized practices, it will not generally cause unreasonable adverse effects on the environment.

mones have been studied in insects. For example, ants mark trails with pheromones. Other insects use pheromones to lure the opposite sex. Entomologists find the chemicals are successful in enticing some pest species to concentrations of pesticides. In other cases, the attractant is simply combined and distributed with a pesticide throughout a large area, resulting in a much broader use of the pesticide. The southern pine beetle, pink boll worm, and gypsy moth have been attracted to insecticides by addition of low concentrations of pheromones.

Home gardeners can use a variety of methods to reduce the total amount of pesticide needed. Companion planting is one of the most effective methods. In *Silent Spring*, Rachel Carson mentions the now well-known suggestion of planting marigolds in rose gardens to ward off nematodes. Marigolds apparently secrete a substance through their roots that is toxic to the nematodes.

In other cases, plants can attract beneficial insects to gardens. These insects, in preparing to lay their eggs, seek out hosts for their young. Many times these hosts are also pests. Flowering parsley attracts lacewing flies, which help to control aphids, mites, and other soft insects. Other plants attract tachinids, which attack caterpillars, or pilot bugs, which destroy the corn borer and corn earworm.

Multiple Control

It is important to develop an integrated approach to controlling pests. If we were to use some of the methods we have just described in combination with very low levels of pesticides, the total impact on the natural system would be reduced. For example, if home gardeners have problems with an insect pest, they might consider planting some flowers to repel the insect. In addition, they could introduce some insect species that prey on the pests. Ladybird beetles are good for controlling aphids and scale insects; green lacewings control mites and other soft insects; and the praying mantis controls a large variety of garden pests.

If the insect pest problem in a garden becomes very acute, the gardener can apply small amounts of pesticide. Along with alternate control methods, this single application could successfully reduce the need for a second application. It is important, therefore, to think about combining methods of pest control. Such an approach can be more effective in controlling the pest than simply using a chemical, since chemicals work initially but allow the pest to quickly reappear.

Integrated Pest Management (IPM)

Large farming operations now use IPM to manage pests in ways that are economically and environmentally sound. Crop rotation to avoid buildup of pest population and to improve soil conditions are options. Pest-resistant crops as well as careful monitoring of pest populations to determine the most effective time to apply pesticides are part of IPM.

Pesticides and the Future

Because of the many problems associated with pesticides in the past, extensive research is now underway to find pesticides that are effective yet environmentally sound. Candidate compounds are selected for evaluation after pest problems are identified. These compounds are put through a primary screening process. This process generally involves laboratory tests under controlled conditions. Toxicity tests are part of this process.

Promising compounds are put through a secondary screening process, involving field evaluation. These tests are used to prove that the compound is effective at controlling the pest in the field. The compound is usually applied at different rates and different time periods.

After the tests are completed by companies that want to market the chemicals, the company officials must meet with government officials to establish a specific research program to secure label registration from the government for the chemical. These research programs are usually very costly. Such research must determine safety factors, rates, and methods of application.

Currently a great deal of time and money are going into research. As specific chemicals for various pests are developed, food production can be improved and pests can be reduced. We must always keep in mind that pesticides are influencing the way our ecosystem operates.

SUMMARY AND CONCLUSION

Pests are plant and animal species that compete with people for space or food. Such organisms can be controlled by chemicals called pesticides. Insects are most commonly thought of as pests and are controlled by four groups of chemicals: inorganics, oils, botanicals, and synthetics. Synthetic insecticides are the most commonly used and include organochlorines, organophosphates, and carbamates. Organochlorines are of great concern because of their persistence, stability, mobility, and solubility in living tissue, particularly fat. Herbicides are chemical compounds used to control unwanted plant growth. They are classified as contact, systemic, or soil sterilants.

Because many pesticides interfere with the physiological processes of living organisms and some can be passed from one organism to another in the food chain, many adverse environmental effects occur from their indiscriminate use. For example, fish eggs can be destroyed and bird eggs of upper-level consumers can develop thin shells as pesticides accumulate. Pesticides not only exterminate target and nontarget organisms but also act in many unknown ways throughout the food chain. Long-term effects on the human population are largely unknown because they are difficult to measure.

Biological control is one alternative to the use of pesticides. Agricultural or genetic modifications are also potential means of pest control. The combination of alternative control methods with pesticides requires much smaller amounts of pesticides, thereby reducing their environmental impact.

We need to expand and develop alternative methods of pest control to be used alone or in combination with other means. Pesticides will still be needed in some instances, but in reduced quantities. Management of land—whether a small garden, a city park, or a large area—can be more effective if we consider potential pest problems. Integrating our pest control programs by using certain plants, introducing predators, or tilling can reduce the irritation as well as the economic loss from pest populations while producing more lasting effects.

FURTHER READINGS

Brown, L. R. 1991. *State of the World.* Washington, DC: Worldwatch Institute.

Jacknow, J., J. L. Ludke, and N. C. Coon. 1986. *Monitoring Fish and Wildlife for Environmental Contaminants.* Leaflet 4. Washington, DC: U.S. Fish and Wildlife Service.

Leahey, J. P. (ed). 1985. *The Pyrethroid Insecticides.* Philadelphia: Taylor and Francis.

National Academy Press. 1987. *Regulating Pesticides in Food.* Washington, DC.

Smith, G. J. 1987. *Pesticide Use and Toxicology in Relation to Wildlife: Organophosphorus and Carbamate Compounds.* Resource Publication 70. Washington, DC: Fish and Wildlife Service.

Urban, D. J., and N. V. Cook. 1986. *Ecological Risk Assessment.* EPA-540/9-85-001. Washington, DC: Hazard Evaluation, Standard Evaluation Procedure, Environmental Protection Agency.

STUDY QUESTIONS

1. Compare the movements of energy and DDT along a food chain.
2. How could you convince farmers that alternative pest control measures are better than pesticides?
3. In newspapers today, we often find stories about the ill effects suffered by workers in pesticide plants. How can these effects be documented?
4. How could the best pest control program for a specific pest be established? Who should be responsible for such a program?
5. Should pesticide salespeople have a course in environmental science?
6. Define the word *pest* and give some examples.
7. Describe the mode of action of the major groups of insecticides. Should these facts dictate the use of one group over another?
8. What might be the long-term effects of using herbicides in a transmission line right-of-way or along a roadway?
9. Can you think of ways to make biological control programs more effective?
10. Should all synthetic pesticides be banned from use until their full effects on humans and the environment are known?

SUGGESTED ACTIONS

1. *Manage Your Lawn and Yard Without Pesticides.* Homeowners use about ten times more toxic chemicals per acre of land than farms. These chemicals are harmful to birds, small mammals, natural food chains, and people themselves. Switching to organic pesticides could remove millions of pounds of toxic chemicals from the environment. By talking to local entomologists and extension people, you could reduce pests by introduction of natural predators such as frogs and spiders. By

cutting your lawn high and letting clippings lie, you reduce the need for fertilizer and pesticides. Aeration of lawns also helps maintain a green lawn without the need of chemicals.

2. *Use Organically Grown Food.* By asking that your local stores sell organically grown fruits and vegetables, you assist in reducing the number of foreign chemicals in the environment. Stores will respond to such demands and even advertise organically grown foods.
3. *Don't Use Flea Collars on Your Pets.* Disposal of used flea collars places damaging chemicals in our landfills and water systems. Some of the chemicals can cause paralysis in insects and, in large amounts, are damaging to humans. As substitutes for flea collars, you can use citrus oil sprays on your pets. These are available commercially. You can also make some by running orange or grapefruit skins through a blender and boiling with water to make a flea detractant. After the mixture is cooled, brush your pet's fur with some of the mixture. Use only skins in the mixture, because the fruit material will make the fur sticky. Fleas hate yeast or garlic. If you add some to your pet's food, fleas disappear. Some companies now market biodegradable flea collars or powder.

15

WASTES

Wastes can be considered in three forms: solid, toxic, and radioactive. People generate wastes from all their activities. Typically, individuals determine that products they use in everyday life are no longer repairable; therefore, they discard them. People also discard food wastes, paper wastes, and garden wastes. These items are all called solid wastes. At the same time, industrial operations generate solid wastes—old products as well as processing and manufacturing wastes. Some by-products from industrial processes are hazardous to human health, and home and industrial wastes can be hazardous to the environment by providing breeding grounds for insects that carry human disease.

Society must dispose of a great deal of waste. There are particular problems with disposal of toxic and radioactive wastes. New technologies often result in industrial processes that generate environmentally contaminating products. Unfortunately, disposal processes for the newly generated wastes are not developed at the same time.

A major part of municipal budgets in the United States is spent on the collection and disposal of refuse. In most cases, only public safety and education expenditures exceed the cost of waste disposal. But, aside from costs, we are running out of places to put solid wastes. Few, if any, areas want wastes from another area.

Disposal of wastes has received less attention than air and water pollution. Unfortunately, a "first cost" approach to solid waste disposal has resulted in adverse environmental impacts. Burning wastes contribute to air pollution, and seepage from waste deposits may pollute groundwaters and streams. In turn, techniques to control air and water pollution produce fly ash and sludges that must be disposed of as solid waste.

Why are these wastes such a problem? Do we have to be so wasteful? Why do we not reuse, recycle, or recover materials that we call solid wastes? Are there safe disposal methods available? These questions will be explored in this chapter.

SOLID WASTE

The U.S. Congress, in the 1976 Resource Conservation and Recovery Act, defines **solid wastes** as:

> Any garbage, refuse, sludge from a waste treatment plant, or air pollution control facility and other discarded material, including solid, liquid, semisolid, or contained gaseous material resulting from industrial, commercial, mining, and agricultural operations, and community activities.

Refuse and solid waste are about the same thing. *Garbage* is food waste. *Trash* and *rubbish* are roughly equivalent terms; they contain little or no garbage. Trash frequently refers to grass and shrubbery clippings, paper, glass, cans, and other household wastes. Rubbish is also likely to include demolition materials like brick, broken concrete, and discarded roofing and lumber.

Whether solid waste is readily biodegradable or combustible is important in its handling and disposal. Food wastes are readily biodegraded, but other organic material such as paper and wood are not biodegraded quickly. Plastic is not usually biodegradable. Organic materials are combustible, but excess moisture could retard combustion.

The term *solid waste* includes discarded and abandoned appliances and junk autos. There are special solid wastes from mining and manufacturing, air and water treatment processes, agriculture, lumbering, hospital and research laboratories, and other sources. Some industrial solid wastes are especially toxic or hazardous, including drums of liquid waste chemicals and semiliquid sludges that are disposed of on land.

In many cities, solid wastes of various kinds are collected from private residences by a government agency or a private collector under contract or license. The composition of the wastes collected varies with the economic status and culture of the area, the season of the year, and the regulations and policies of the collection agency. Although we recognize that there will be variation, the composition of typical municipal refuse is shown in Table 15.1. Food wastes constitute about 12 percent and paper items more than 40 percent of municipal solid waste. Garbage contains 72 percent moisture on the average (Figure 15.1).

Table 15.2 shows the relative annual production of solid wastes by several segments of society. Industrial production of solid wastes is increasing at 3 percent per year; most of the increase is due to mandated air and water pollution control improvements that produce more residues. Ten to 15 percent of industrial solid wastes are special hazards to public health and the environment. Furthermore, an EPA survey revealed that 86 percent of a sample of fifty industrial disposal sites showed migration of heavy metals and organics into groundwater. At twenty-six sites, monitoring wells exceeded safe limits for drinking water.

Municipal solid waste in the United States increased 5 percent per year between 1960 and 1970. In the 1970s, the annual growth rate slowed to 2 percent; however, by 1985, the growth rate was over 3 percent. In addition, sludge from municipal wastewater treatment plants is increasing by 4.5 million tons per year.

Impact on Human Health

Early waste disposal efforts began in reaction to the perception that wastes caused diseases. In fact, between 1850 and 1890, some people believed that diseases such as cholera, yellow fever, and typhoid were generated directly from putrefying organic matter. The result was the establishment of laws in England (which carried over to the United States) on waste disposal, cesspools, and privy construction. Industrial wastes were generally neglected.

Diseases are, of course, carried by insects, many of which breed in organic wastes. Flies are a major *disease vector* associated with garbage; that is, they transmit disease organisims from a source to a victim. Rats, with the assistance of rat fleas, are involved in the spread of bubonic plague, and murine typhus fever to humans. Although there has not been an urban epidemic of plague in the United States since one in San Francisco early in this century, it still constitutes a serious health threat because western rodents (squirrels, prairie dogs, rats, rabbits) are infected and the areas where rodents are infected are spreading eastward. During the 1980s an annual average of eighteen cases of plague were reported in the United States, mostly caused by isolated contact with wild rodents.

Table 15.1
Composition of Municipal Solid Waste in the United States

Category	Percentage by Weight of All Refuse	Moisture (percentage by weight)	Caloric Value (Btu/lb)
FOOD WASTE			
Garbage	10.0	72.0	8,484
Fats	2.0	0.0	16,700
	12.0		
RUBBISH			
Paper	42.0	10.2	7,572
Leaves	5.0	50.0	7,096
Grass	4.0	65.0	7,693
Street sweepings	3.0	20.0	6,000
Wood	2.4	20.0	8,613
Brush	1.5	40.0	7,900
Greens	1.5	62.0	7,077
Dirt	1.0	3.2	3,790
Oils, paints	0.8	0.0	13,400
Plastics	0.7	2.0	14,368
Rubber	0.6	1.2	11,330
Rags	0.6	10.0	7,652
Leather	0.3	10.0	8,850
Unclassified	0.6	4.0	3,000
	64.0		
NONCOMBUSTIBLES			
Ashes	10.0	10.0	4,172
Metals	8.0	3.0	124
Glass and ceramics	6.0	2.0	65
	24.0		

Murine typhus fever was prevalent in the southeastern United States before the pesticide DDT was used. DDT was sprinkled in rat runs and enough contacted the rats to kill the fleas. Rats may contaminate food with their urine, transmitting leptospirosis and infectious hepatitis. They can transmit salmonella and other intestinal organisms, bite small children, and destroy more food than they eat.

Rats begin to reproduce at 3 to 4 months of age, spend about 3 weeks in the gestation period, and can mate again 48 hours after giving birth. They can produce 10 to 12 litters per year, with 9 or 10 young in each litter. This phenomenal rate of reproduction means that the only successful rat control measures must include shutting off the rat's food supply. Rats can be denied access to food by placing garbage in metal cans with tight-fitting lids. Rigid plastic cans can be used in many places. Plastic bags, however, are readily ripped open by dogs and rats. If these animals are not roaming the area, plastic bags have the advantages of ease in handling, no noise, and elimination of fly breeding in garbage cans.

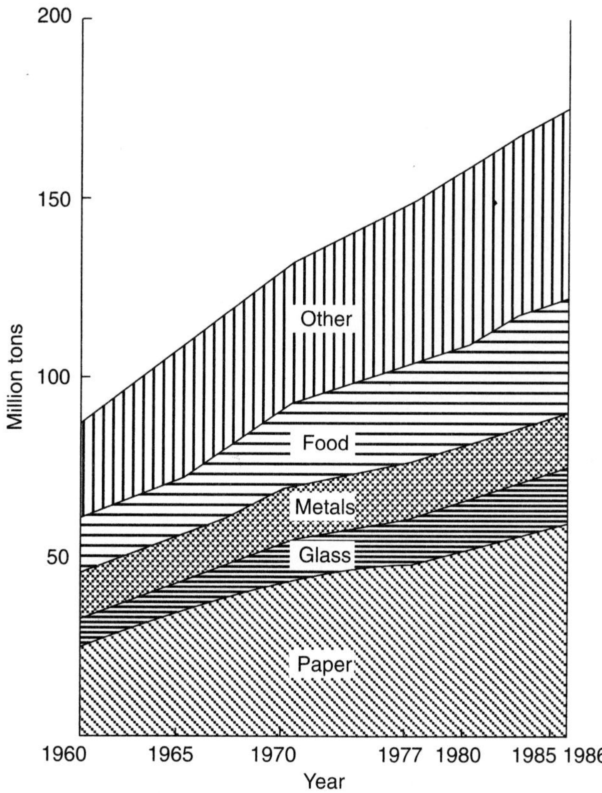

Figure 15.1
Breakdown by material of the estimated generation of residential and commercial solid waste following consumer use indicates a steady increase in waste production. (U.S. Environmental Protection Agency.)

Flies reproduce readily in a very small residue of organic matter. The typical fertilized female housefly, *Musca domestica*, can lay 100 to 150 eggs. Depending on temperature, especially night temperature, several stages of metamorphosis follow. The eggs hatch, producing larvae (maggots) that feed on the garbage or residue for 3 to 7 days. The larvae then migrate to a cool, dry place where they turn into brownish, inactive pupae about the size of a grain of wheat. After 3 to 7 days in the pupa stage, the adult fly emerges. In the summer months these stages take 8 to 20 days to produce a fly. Several generations are produced in a single season, and if all the offspring from one pair survived, there could be millions of flies in a few months.

Table 15.2
Solid Waste Produced in the United States

Sector	Millions of Tons per Year
Municipal trash and garbage[a]	145
Industrial	340
Mining	1700
Agriculture	2300

[a]Does not include sludge from wastewater treatment.

From *Environmental Quality*— 1977 by the Council on Environmental Quality, Washington, D.C., with credit to the U.S. Environmental Protection Agency, Office of Solid Waste.

While a very close-fitting lid could keep a fly from entering a garbage can, the fly can lay eggs in food wastes before they are placed in the can. Also, flies can enter as material is placed in the can. Often some food residue is left in the can when emptied, and if the lid is not placed back on immediately, flies will lay eggs in the residue. Consequently, most garbage cans support fly breeding. In fact, some surveys show that garbage cans are responsible for 80 percent or more of fly breeding. Flies also breed in piles of grass clippings and animal manures.

Fly breeding in garbage cans can be curbed by promptly scrubbing the can after each use to remove the residue or by wrapping garbage before placing it in the can. Installation of garbage grinders will reduce the potential for fly breeding and rat attraction, but there may be an increase in rat habitation where food wastes go to the sewers. Extensive use of garbage grinders also adds to the solids and BOD (biochemical oxygen demand) loads on the sewage treatment plant.

Theoretically, the frequency of garbage collection could influence fly populations in a community. If garbage were removed before the fly larvae migrated from the can to the earth to form pupae, the larvae would be carried off with the garbage. On that basis, collection twice a week would suffice when there are warm days and cool nights. When the nights are also warm, the frequency should be increased to three times a week so there would be no more than 3 days between collections.

The widespread popularity of disposable diapers has added a new hazard to solid wastes handling. Flies and rats in contact with them can spread intestinal diseases of infants, and garbage collectors may be exposed directly in handling.

Garbage compactors, a convenience item for the home, reduce the volume of garbage to be handled. Home incinerators will also reduce volume, but they are not efficient unless they have auxiliary fuel, which makes operation expensive. Home and apartment incinerators frequently produce odors and fly ash.

DISPOSAL

Reduction and Recycling

It would be preferable, of course, to eliminate or reduce solid wastes as much as possible. Some materials or products can be made to last longer, others repaired or reused for other purposes, some recycled, and others replaced by something that will not become a problem later.

If the average life of autos could be extended from 7 to 10 years, there would be fewer autos to dispose of each year, in addition to saving energy and material and avoiding pollution associated with auto production. When states ban nonreturnable containers for beverages or require a deposit on returnable containers, they promote the use of returnable bottles and thus materially reduce the volume of solid waste, save energy, and reduce litter. The ice cream cone is used as a classic example of a case where the container is eaten, thereby eliminating any disposal problem. This might not be a good illustration, however, because packaging is still required for the cones. There would be less solid waste if the ice cream were eaten from a dish that could be washed and reused.

If the generation of solid waste cannot be prevented, the second socially desirable alternative is to recycle or reuse the waste material. In industry, there are companies that specialize in selling one company's waste to another as a raw material. Community refuse usually contains reclaimable materials such as steel (in tin cans), aluminum, and paper. Recovering these materials from a mixture of other material is the main obstacle to recycling. Separation at the source helps.

There is a popular notion that all glass should be recycled, but the energy involved in returning glass to a recycling center and then to the manufacturer for reprocessing is greater than that used in original manufacture. However, reusing glass bottles results in an energy- and cost-efficient operation.

As late as 1950, 99 percent of soft drinks and 70 percent of beer were packaged in returnable, reusable glass bottles. By 1979, only 25 percent of soft drink bottles in use were returnable. The average returnable soft drink bottle is used forty times before being discarded. Some states require a deposit on beverage containers in order to encourage their return.

Persons who have fireplaces in their homes may roll newspapers into tight logs to burn in their fireplaces. There is usually a demand, however, for old newspapers for recycling.

Where garbage alone is collected separately from other refuse, it is sometimes fed to hogs. Raw garbage fed to hogs has transmitted the hog disease *versicula exanthema* and is a means of infecting hogs with trichinae. Thus, persons who eat undercooked pork may contact *trichinosis.* Because of these problems, most states require garbage to be cooked before it is fed to hogs.

Agricultural wastes can be shredded and used in animal feed mixtures or returned to the soil. Animal manure is used to enrich the soil, but at feedlots the disposal of manure is a problem.

Landfill

The simplest method for disposing of refuse for an entire community is to haul it to an area remote from human habitation and dump it. However, these dumps attract scavenger birds and animals, rats, and flies and frequently catch on fire, creating smoke and air pollution. A community can eliminate these features and still have a relatively low-cost operation by using the sanitary landfill method (Figure 15.2). A **sanitary landfill** involves depositing refuse in a low place or a trench excavated for the purpose. The refuse is compacted by a bulldozer and then covered with earth to a depth of 15 centimeters (6 inches). Each day's dumping is completely covered, making a cell of refuse and isolating it from other deposits. The sanitary landfill method of operation excludes flies and rats and retards propagation of fires.

The material placed in the landfill undergoes decomposition initially by aerobic microorganisms. As the oxygen is depleted by microbial action and cannot be replenished because of compaction and cover, the microbial action soon becomes anaerobic, slowing the rate of decomposition. Because the ground settles, the site is not suitable for building anything but light structures, such as horse stables. Some communities grind garbage before dumping it, which reduces volume and speeds decomposition.

Other techniques are under study to enhance the biological processes in landfills. It has been found, for example, that if refuse is compacted in layers 20 to 30 centimeters deep for a year, then piled in large compact fills, much of the biological activity will have occurred.

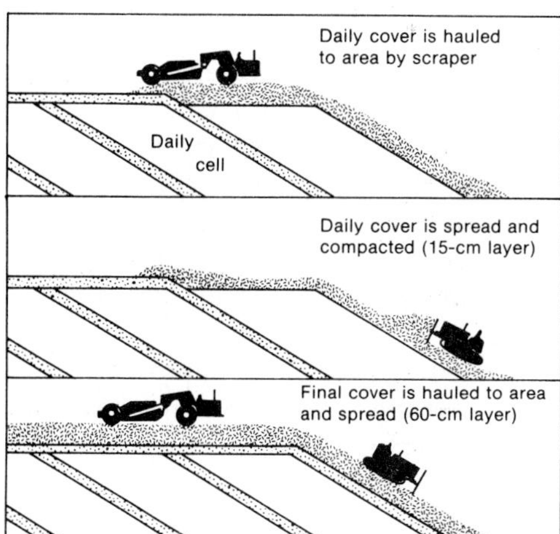

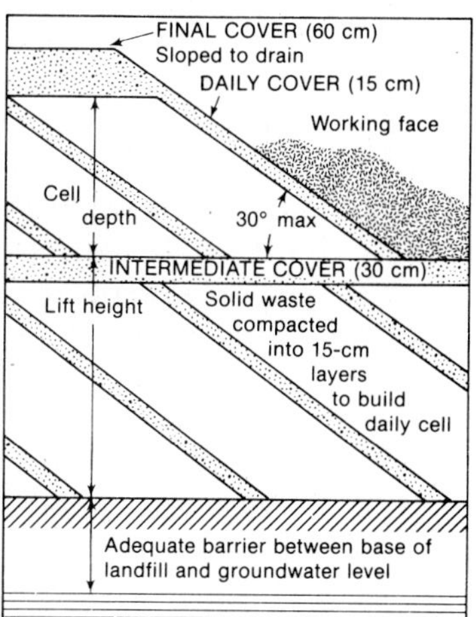

Figure 15.2
Sanitary landfill. (From E. A. Glysson, J. R. Packard, and C. H. Barnes, *The Problem of Solid Waste Disposal.* Ann Arbor: College of Engineering, University of Michigan, 1972.)

GARBAGE ARCHAEOLOGY

In 1973, archaeologist William L. Rathje founded the Garbage Project at the University of Arizona. The project is designed to gain new insights into our society by studying modern garbage, just as other archaeologists learn about ancient societies by examining old refuse. The findings of the Garbage Project have proven to be very interesting and more than a little surprising.

Rathje's project has emphasized the excavation of landfills. As he points out, we need to have a better understanding of landfills and disposal practices because the United States is in a garbage crisis. There are about 5500 landfills across the country, and many of them are reaching capacity. At the same time few new landfills have been approved.

While landfills can be used to reclaim land for other purposes, they take up a lot of space and are potential sources of pollution. For example, New York City's Fresh Kills Landfill on Staten Island covers 1215 hectares (3000 acres) and occupies 67 million cubic meters (2.4 billion cubic feet) of space. This is twenty-five times the volume of the Great Pyramid of Giza.

Discarded newspapers and other paper products represent 50 percent of the volume of refuse in landfills, and Rathje has found that the commonly held belief that paper quickly biodegrades is untrue. Newspapers, books, and even food items have been found intact after decades of burial. Furthermore, Rathje estimates that only 20 to 50 percent of food and yard waste biodegrades in the first 15 years. Plastic, often reported to represent 30 percent of solid waste volume in landfills, actually represents 10 percent in the eleven landfills studied by the Garbage Project.

People have also assumed that polystyrene foam and disposable diapers are a major component (some have estimated as much as 55–85 percent) of landfills. However, these items make up only 0.9 percent and 0.8 percent of the average landfill studied by the Garbage Project.

Rathje's research, and the research of other scientists, have made the choices required of environmentally conscious consumers more complicated. For example, disposable diapers take up landfill space and are a source of pollution, but if you consider such things as transportation energy and water use, cloth diapers from a diaper service may not be a significant improvement. It has also been shown that if factors such as raw materials, energy and water needs, air and water pollution, recycling potential and alternative methods of disposal are taken into account, paper cups are no better as hot drink containers than polystyrene cups (assuming that the foaming process used does not require ozone-depleting chemicals).

It is clear, however, that reusing materials, recycling, and reducing the volume of solid waste are almost always beneficial from an environmental standpoint.

For further information see W. L. Rathje, 1991. Once and Future Landfills. *National Geographic* 179 (5):116–34; and M. B. Hocking, 1991. Paper versus Polystyrene: A Complex Choice. *Science* 251: 504–05.

Because landfills are anaerobic, they produce methane gas, which could migrate to any nearby basements through porous soil. Explosions have blown off manhole covers of some sewer lines running through landfills. However, the energy in methane is useful and experiments are underway to extract the gas from old landfills. In Cinnaminson, New Jersey, the Public Service Electric and Gas Company extracts methane gas and sells it to an iron foundry. Test wells indicate the gas is about 60 percent methane, and burning releases about 60 percent of the heat obtained from natural gas. The landfill is expected to yield 30,000 cubic meters (1 million cubic feet) of this gas per day.

The Pacific Gas and Electric Company built a demonstration facility for processing methane gas produced by a 13-meter (40-foot) landfill in Mountain View, California. The company expects to produce 30,000 cubic meters of raw gas per day, which after cleaning will be 95 percent pure methane and will have a heating value of about 2850 British thermal units* per cubic meter (950 British thermal units/cubic foot).

Rainwater percolating through a landfill dissolves metal salts and organic compounds. This leaching process might not be apparent until several years after filling begins. The **leachate** can seep through the ground to pollute surface or groundwater; thus, landfills should be above the groundwater table. Some states require a minimum of 2 to 3 meters (6 to 10 feet) of earth between the bottom of the fill and the groundwater table. If impervious liners are placed in the area to be filled to contain the leachate, the leachate can then be removed and treated like wastewater before being released. Clay, asphalt, and plastic are used for lining. Plastic membranes may be torn and some solvents dissolve asphalt. In time, even clay may be susceptible to action by some chemicals. Another method of preventing leaching involves diverting rainwater by covering the landfill with an impervious material.

If land is cheap and the haul distance not too far, landfill operations are relatively inexpensive. In some cases, benefits aside from waste disposal are realized. Land can be reclaimed by the landfill method where strip mines have been abandoned. Ravines and canyons can be filled to prevent landslides and the finished site used for parks and similar purposes. It has been proposed to build mountains for skiing by this means. In Los Angeles, solid waste sanitary landfills are referred to as *landfill parks.* People living nearby consider these desirable because the public agency responsible is meticulous about proper engineering and operation to avoid any problems.

Selecting a landfill site compatible with protection of public health and safety and the local ecosystem requires engineering and geological surveys. Landfills are not permitted in wetlands, floodplains, habitats of endangered species, and

*The British thermal unit (Btu) is a measure of heat commonly used in the United States. It is the quantity of heat required to raise the temperature of one pound of water 1°F. One Btu is equal to 252 calories.

recharge zones for local drinking water supplies. To meet EPA standards under the Resource Conservation and Recovery Act, the landfill must have an impermeable barrier on the bottom, a leachate recycling system, provide for gas recovery or venting, cover new refuse with earth daily, and monitor adjacent groundwater for pollution migration.

Some communities are having problems locating landfill sites to replace those that are becoming full. Consequently, alternative disposal methods and ways to reduce solid waste volume have to be developed.

Incineration and Energy Recovery

Where costs of hauling refuse to distant sites or the price of land for the disposal site are too high, refuse may be burned in an incinerator. Large apartment complexes and some commercial establishments burn refuse in incinerators on the premises. However, apartment incinerators are frequently poorly designed and operated, creating air pollution. New York City banned their use completely some years ago, then removed the ban.

In modern designs, refuse is discharged through a hopper to a traveling grate or rotating kiln to be burned. Most conventional incinerators do not completely consume all of the combustible material, but leave a residue that requires landfill disposal. A well-operated incinerator will burn out 85 to 90 percent of the combustibles. If an incinerator is overloaded, telephone books, grapefruit hulls, and other wastes may merely be scorched (Figure 15.3).

Incinerators produce air pollution, so control equipment (cyclones and electrostatic precipitators) is required. In areas where air pollution is already a severe problem, incineration may not be an acceptable means of solid waste disposal. Some persons are concerned that there might be toxic and possibly carcinogenic products resulting from the combustion of plastics. Incinerators also discharge metallic compounds to the air.

With the advent of the energy crisis, many U.S. cities have taken a renewed interest in obtaining heat from refuse. Incinerators can be designed for heat recovery. Where the incinerator is near other buildings, the steam from the incincerator can be piped to heat the buildings. In Europe, municipal solid wastes are viewed as a source of energy. Consequently, heat recovery systems are designed with that end in mind rather than as a by-product of a solid wastes disposal system involving incineration. One estimate is that 350 million barrels of crude oil would not have to be imported annually by the United States if the energy in solid wastes were recovered.

A water wall incinerator produces steam of high quality (that is, high temperature) suitable for generation of electricity, heating, and driving industrial processes. Smaller modular units conduct exhaust gases to waste heat boilers to recover energy. These are less efficient, but capital costs are much less.

Large central incinerators can burn unprocessed refuse (mass burning). The refuse may be shredded and processed to produce refuse derived fuel (RDF). RDF may be stored or shipped to a different location for burning and materials can be salvaged as part of the processing operations.

Worldwide use of refuse energy is shown in Figure 15.4. Denmark leads in energy recovery with 60 percent of its wastes converted to energy. In Switzerland, 40 percent is recovered, and in the Netherlands and Sweden, 30 percent. Presently, the United States converts less than 1 percent.

Examples of European refuse power plants are located in Dusseldorf, Krefeld, and Wuppertal, Germany. The Krefeld plant processes 1100 tons of municipal refuse, 510 tons of sewage sludge cake, and 95 tons of waste oil per day. It produces 5 megawatts of electricity and 200 million British thermal units per hour of steam for district heating. Energy is supplied to the

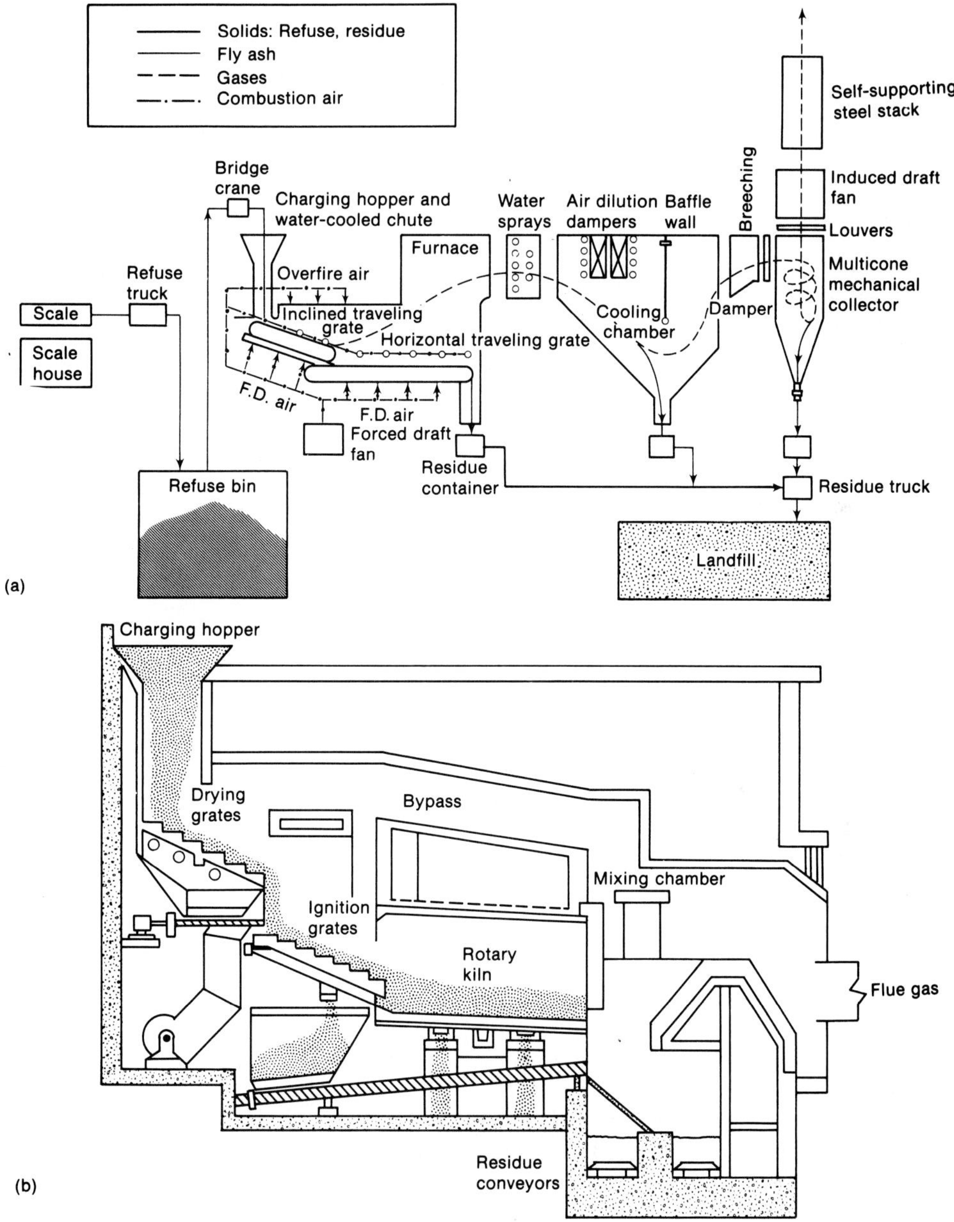

Figure 15.3
Incinerators. (a) Traveling grate. (b) Rotary kiln. (From R. J. Schoenberger and P. W. Purdom, *A Study of Incinerator Residue Analysis of Water-Soluble Components*, EPA-670/2-73-957, NTIS PB-222-458. Washington, DC, 1973.)

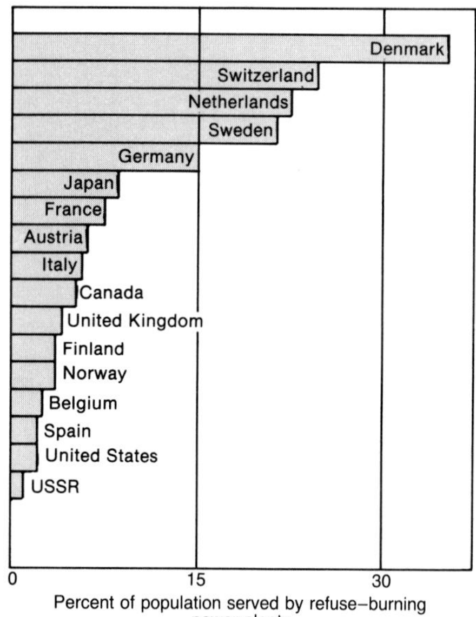

Figure 15.4
Worldwide application of refuse power. Denmark leads the Western world with 32 percent of its population served by district refuse power plants. Of about 250 refuse power plants in operation around the world, only 20 are in the United States. (Courtesy of Battelle Columbus Laboratories, Columbus, Ohio.)

nearby municipal sewage treatment plant and industrial park. Another on-site refuse power plant is located in the center of a modern neighborhood to supply steam to 8000 apartment units, offices, and shops in Nord-West Stadt in the State of Hessia, Germany.

Valuable materials can be recovered from solid waste by a process called **pyrolysis.** Pyrolysis is a method in which waste material is heated to very high temperatures in the absence of oxygen. Gases are driven off, condensed selectively, and recovered to make chemicals. The residue is a char that can be used for fuel. Pyrolysis is accompanied by other salvage operations to recover metals and other nonburnable materials. At present, pyrolysis is not considered economical.

Compost

From time to time, the public becomes fascinated with composting as a means of solid waste disposal. To prepare solid waste for composting, the metals and newspapers are removed. The remaining material (primarily food wastes and lawn trimmings) is turned and agitated frequently to maintain an aerobic condition. Microorganisms break down the degradable material into a powdery material called **compost.** Although it will lighten the soil, adding humus which improves moisture retention, compost has little fertilizing value. Where sewage solids are mixed with garbage, the resulting compost will contain more nutrients to enrich the soil. Metals in sewage could be a problem for vegetation.

When Holland was reclaiming land from the sea, compost was added to the sand to make farmable soil. Under other circumstances, however, it pays to buy fertilizer rather than compost. Most large-scale composting attempts in the United States in the last 30 to 40 years—Oakland, California; Houston, Texas; Tampa, Florida; and others—have failed financially.

Farmers and suburbanites with large lots will sometimes compost leaves, grass clippings, shrubbery trimmings, and vegetable food wastes. A compost pile above ground which is held in place by fencing wire will likely stay in an aerobic condition and give off little odor. If meat scraps are kept out of the compost, the danger of rat attraction will be reduced and odors prevented. Composting of food and other biodegradable wastes in this way results in a material that can improve soil characteristics. Although it has been oversold as to its profitability or fertilizing value, composting is a means of solid waste disposal that enhances the quality of the environment.

In 1978, the city of Philadelphia began a 4-hectare (10-acre) strip mine reclamation demonstration in Somerset County, Pennsylvania. Air-dried, composted, municipal sludge from sewage was spread and disked into the soil; then grass seed was applied. The grass flourished and no adverse effects were observed. Between June 1978 and October 1979, 121 hectares (300 acres) were reclaimed; in 1980, 405 hectares (1000 acres). There are more than 101,175 hectares (250,000 acres) of strip-mined land in Pennsylvania alone that could be reclaimed in the manner demonstrated.

Other Methods of Waste Disposal

The Japanese have experimented with high-pressure compaction of refuse for ocean disposal. The blocks of garbage are so dense that they sink, but the blocks do not maintain their integrity and ocean pollution results. Similar efforts at encasing refuse in some binder, such as asphalt or plastic, and using it for building blocks have not succeeded because the refuse decomposes and is combustible.

Open ocean dumping has been practiced by New York and other cities. Damage to ecological systems on the continental shelf floor has been attributed to this practice, and some materials float and wash up on beaches. The federal government is studying ocean dumping limits. While the ocean ecosystem could absorb a limited amount of biodegradable garbage, preliminary processing is necessary to remove nondegradable and toxic materials.

TOXIC WASTES

In 1978, New York State health officials advised pregnant women living in a middle-class residential section next to an old filled canal (Love Canal) near Niagara Falls to move immediately. The officials were concerned about a rise in rates of miscarriages, birth defects, and liver disease among persons who lived in this area (Figure 15.5). The state bought the homes of 239 families for $10 million, forced the people to move, and fenced in the area.

Studies in the late 1970s and early 1980s assessed birth weight, prematurity, gestational age, and birth defects in 239 children exposed prenatally to Love Canal wastes and compared them to 707 children in a similar area without toxic wastes. Birth weight was lower and there were more birth defects in the exposed group. There was no difference in the length of gestation. The study showed the exposed group exhibited postnatal problems in terms of learning problems, hyperactivity, eye irritation, skin rashes, and abdominal pains.

The Love Canal, dug in the 1890s, was used by the Hooker Chemical and Plastics Corporation as a landfill for chemical wastes in drums from 1947 to 1953. Subsequently, 16 acres of this land were sold to the Niagara Falls Board of Education for one dollar. The Board of Education built a school and playground on the land and sold part of the remaining land to developers. After several years of abnormally heavy rain, wastes began leaking from the drums and rising to the surface. More than eighty chemicals were involved, including suspected carcinogens. The company claims an impervious clay liner and cover were used, but the clay cover may have been disturbed during grading.

In 1988, the U.S. District Court ruled that Occidental Chemical Corporation (the new owner of Hooker Chemical and Plastics Corporation) must pay at least some of the Love Canal cleanup costs, estimated to be $140 million. Occidental Chemical has already paid $20 million in an out-of-court settlement with former residents of Love Canal. In an additional court action against Occidental, the state of New York is seeking $560 million for damages to the environment and for punitive damages.

The Love Canal experience is one of over 400 cases of harmful consequences the EPA has documented as due to inadequate hazardous wastes management. These include surface and groundwater contamination, poisoning from direct contact, air pollution, fires, and explosions. The EPA estimates that 90 percent of hazardous wastes are not adequately protected. Since one-half of the drinking water in the United States is from groundwater, contamination of this source by toxic wastes is a serious threat to public health. The residence time for groundwater (that is, the time it takes for it to be completely replaced by natural circulation processes) may be as long as 200 years.

Of 344 million metric tons of industrial wastes produced annually, 10 to 15 percent is

Figure 15.5
Love Canal development around landfill used for toxic chemicals. Adjacent homes were vacated and abandoned when chemicals were released and permeated the surrounding neighborhood. (Copyright © 1978 by New York State Department of Health. Reproduced by permission.)

hazardous (about 35 million metric tons) and is mainly from industrial sources. Waste production is increasing 3 percent per year. Some of this increase is caused by residues created by methods of air and water pollution control. In addition, there are several hundred million metric tons per year of high-volume, relatively low-risk wastes, such as mining wastes. The EPA estimates that waste from 30,000 sources will require treatment, storage, or disposal permits under the Resource Conservation and Recovery Act of 1976. In this act, hazardous wastes are defined as those that may "(a) cause or significantly contribute to an increase in mortality or an increase in serious irreversible or incapacitating reversible illness, or (b) pose a substantial present or potential hazard to human health or the environment when improperly treated, stored, transported, or disposed of, or otherwise managed."

The EPA established comprehensive regulations to control the handling of hazardous wastes including a manifest system where the generator of wastes fills in a form that accompanies the waste and is kept on file by the ultimate disposer.

Hazardous wastes may have the following characteristics: (1) ignitable, (2) corrosive, (3) reactive, (4) toxic, (5) radioactive, (6) infectious, (7) phytotoxic, and (8) teratogenic. Little attention is actually paid to the health effects of wastes.

Major industrial sources are shown in Table 15.3 and Figure 15.6. Seventy percent of haz-

Table 15.3
Sources of Industrial Hazardous Wastes

Source	Million Metric Tons
Primary metals	8.3
Organic chemicals	6.7
Electroplating	5.3
Inorganic chemicals	3.4
Textiles	1.8
Petroleum refining	1.8
Rubber and plastics	0.8
Seven miscellaneous industries	0.7
	28.8

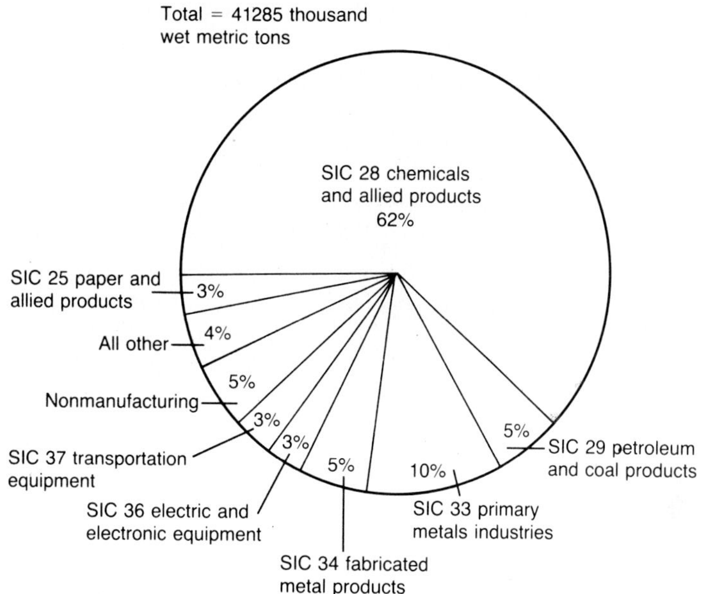

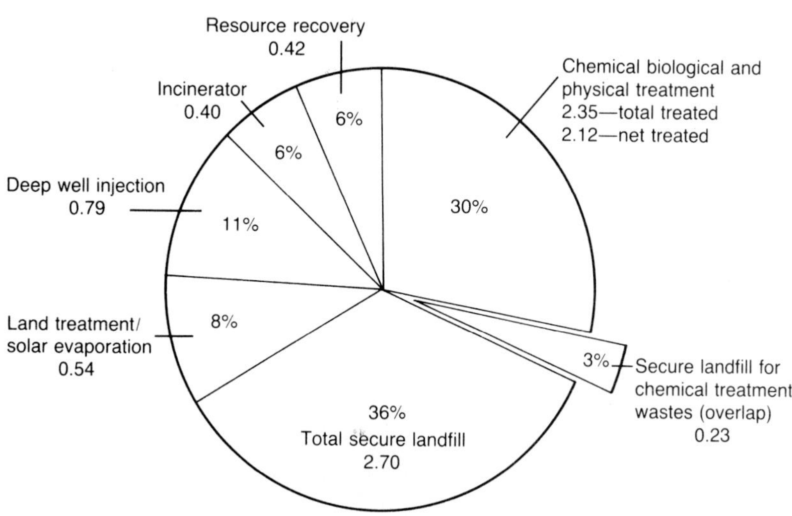

Figure 15.6
(a) Percentage of 1980 hazardous waste generation by standard industrial classification (SIC) code.
(b) Estimated hazardous waste volumes treated and disposed of at commercial off-site facilities by waste management options.
(U.S. Environmental Protection Agency.)

ardous wastes are produced in the "dirty dozen" states (in order, highest first): Texas, Ohio, Pennsylvania, Georgia, Michigan, Indiana, Illinois, Tennessee, West Virginia, California, New York, and New Jersey. The greatest problems are associated with toxic heavy metals that cannot be readily destroyed and organic chemicals, especially those containing chlorine, that resist degradation and accumulate in living organisms.

Management of hazardous wastes involves the following:

- Minimize the amount of waste generated by substitution of raw materials or by modifying manufacturing processes.
- Transfer wastes from one industry for use as raw material by another industry.
- Reprocess wastes to recover energy and to reuse, recycle, or recover materials.
- Separate hazardous wastes from other source material for efficient handling.
- Incinerate or otherwise convert hazardous wastes to a nonhazardous form.
- Dispose of hazardous wastes in a secure landfill or other method to isolate it from the environment and maintain integrity of the containment system for an indefinite time—100 years or more.

Incineration is suitable for some toxic organic wastes. For example, PCBs may be used to supplement the fuel used in cement kilns. In other circumstances, incineration alone may require costly supplemental fuel. Various physical and biological treatment processes may render wastes less hazardous or completely nonhazardous.

There are hazardous wastes that are too low in economic value to recycle, too difficult to degrade, too thick to inject into deep wells, or too contaminated with heavy metals and other noninflammables to incinerate. For these, a **secure landfill** may be the only feasible disposal method. An ordinary landfill may cost $3 to $8 per ton to operate; incineration of hazardous wastes may cost $75 or more per ton; and a secure landfill may cost $50 or more per ton.

There is some question about what constitutes a "secure" landfill. One suggested type consists of cells constructed in dense clay (Figure 15.7). Measuring 150 meters square, the cell would be lined with an impervious reinforced plastic and covered with a meter of clay. Clay barriers between cells would isolate each one. The completed fill would be capped with an impervious liner material and covered with clay to keep out water. Any water that might accidentally intrude would be collected in an underdrain

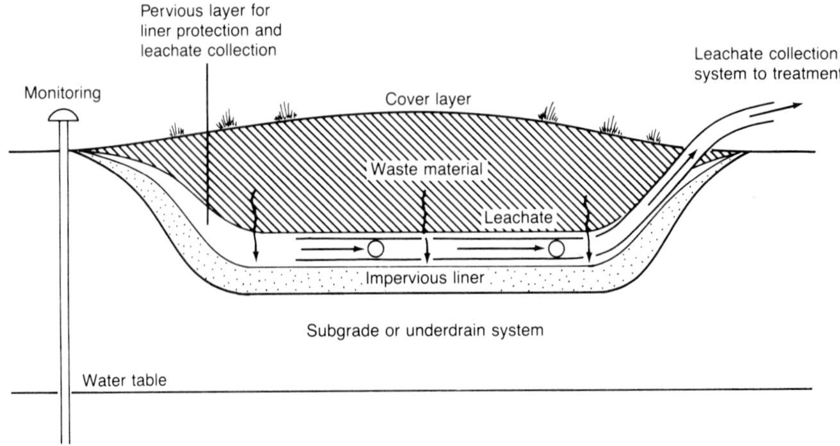

Figure 15.7
Landfill concept for secure leachate management. (From Amir A. Metry et al., *The Handbook of Hazardous Waste Management.* Westport, CT: Technomic Publishing Co., 1980.)

system and pumped to the surface and treated. It has also been suggested that such fills be located 165 meters (500 feet) from a water source and monitored for 20 to 30 years. However, there is no assurance that chemical pollutants will not migrate through the ground more than 165 meters from the source. Also, if the stored chemicals do not degrade readily, they may continue to be toxic after 100 years and may be potentially hazardous.

Some type of chemical solidification (binding in a coherent mass) before burial would probably guard against erosion hazards. Suggested binders include cement-based, pozzolanic or lime-based, thermoplastic, and organic binders.

While new landfills for hazardous wastes can be constructed to be safer, horror stories about old dumps, long forgotten, continue to be revealed. Of some 100 known commercial sites, only about 20 are considered really "secure." Of 32,000 industrial dumps, at least 600 could be potential "time bombs" like the Love Canal. Management of hazardous wastes could be one of the foremost environmental problems of the future.

In December 1980, the U.S. Congress passed the "Superfund" legislation. This law uses contributions from industry and government to clean up spills and abandoned dump sites of hazardous wastes. Funds are generated from a trust fund with contributions from the federal treasury, industrial fees, and fees collected from some chemical manufacturers. This fund, along with stricter enforcement of state and federal laws, led to a major cleanup of some problem disposal sites during the mid-1980s. The legislation that created the superfund also recognized some of the problems of workers exposed to hazardous wastes.

Although 7200 sites are under investigation, the identification of abandoned sites will be a long and difficult process (Figure 15.8). While not a superfund operation, the Rocky Mountain Arsenal dump site in Colorado is an example of a cleanup project. Organic chemicals are being restrained from further migration by installation of trenches filled with clay that serve as a physical barrier. Another method is to pump liquid from wells located at the site boundary at a rate that will lower the water table in the immediate

Figure 15.8
Identified hazardous waste sites in the United States.

vicinity. The liquid from the wells will require treatment if it contains hazardous chemicals.

TOXIC METALS

The release of toxic metals has created some serious environmental problems. Releasing the metals often disrupts the mineral cycle (Chapter 2). For example, bay bottoms serve as a sink for some heavy metals used in industrial processes. When dredging operations were undertaken on Nueces Bay, Corpus Christi, Texas, shorebirds were found to feed in the areas impounded by the dredge materials. Selenium levels in some birds from the dredge pits were within ranges known to cause reproductive impairment in chickens.

Lead poisoning is a major concern for wildlife. Lead shot allows lead to accumulate in the mud of marshes. Bald eagles that eat fish from these contaminated waters show dangerously high concentrations of lead. American kestrel (sparrow hawk) nestlings showed abnormal growth when fed seeds coated with lead. Some hunt areas are experimenting with steel shot, which has generated controversy because many hunters consider it less effective.

Selenium in California's Kesterson Reservoir found its way into the drainwater as irrigation leached it from the soil and built up in the marshes. Waterfowl that fed on small fish, insects, and plants in contact with the drainwater suffered reproductive abnormalities because of selenium bioaccumulation. It was decided to discontinue irrigation of fields from which waters flowed into the marshes because the selenium was believed to be picked up as water flowed through the fields.

During the processing or refining of minerals, air and water contamination can occur. Copper smelters generate about 10 percent of the nation's sulfur oxides, a contributor to acid rain. Phosphate mining permits radiation to emit from some forms of fertilizers that are washed into streams. Geologists estimate there could be a twofold increase in lung cancer among people living in structures built on Florida's reclaimed phosphate lands. Radon gas can also accumulate in homes built over phosphate mines and wastelands.

Waters used in mine waste disposal or percolating through mine sites create additional pollution problems. In Germany, potash (potassium containing salts) mining companies have contaminated the Werra River to such an extent that it is too salty for safe use as a water supply. The famous Rhine River has undergone the same contamination. In North America, Lake Superior is polluted by iron ore wastes which contain microscopic asbestos fibers. These fibers have been found in the drinking water of Duluth, Minnesota, and other cities that take water from the lake.

Planning was an important factor in avoiding environmental contamination from minerals in Urad, Colorado. Before beginning a molybdenum mining operation, a diversion structure and a 2-mile underground pipe were built to allow water to flow around the mine site. The company also provided public recreational access to their holding water reservoir.

To meet the increase in demand for minerals, many alternatives must be considered. Care of the environment must be foremost if we wish to avoid abuses of natural systems.

RADIATION

What Is Radiation?

Radiation was discovered to be useful in 1895, when Wilhelm Conrad Roentgen observed that an electrical discharge traveled a straight line through a simple cathode ray tube and penetrated some material but not others. He found that these rays, called *X rays,* exposed photographic plates and showed the outline of the bones in his hand when he placed it between the cathode tube and the photographic plate. Later, Henri Becquerel accidentally found uranium crystals also exposed photographic plates. Utilizing Becquerel's observations, Marie Curie found that plutonium and radium were **radioactive elements** present in uranium ore (pitchblende). Another radioactive element, thorium, was discovered by Ernest Rutherford in 1905. Rutherford and his associates found that thorium changed to a new element as it emitted radiation. This discovery led to a model of atomic structure.

Radiation is a form of energy and can be compared to light energy absorbed by photosynthetic organisms or chemical energy released in a chemical reaction. It is energy that can be transferred from one body to another across empty space. It takes the form of either subatomic particles (particulate radiation) or waves of energy propelled through space (electromagnetic radiation).

Particulate radiation is caused by subatomic particles released during the spontaneous decay of an atom. Such particles as protons, neutrons, and electrons or products created by their interaction leave the decaying atom to form new elements. Atoms of carbon-14, for example, release electrons to form the stable element nitrogen-14 (Figure 15.9).

X rays and gamma rays are forms of **electromagnetic radiation** with different sites of origin. X rays, which originate outside the nucleus of an atom, occur when electrons are forced at high speed to bombard the nucleus of another element. Some electrons are deflected and decelerated by the positive charge of the nucleus. This deflection and deceleration results in the emission of electromagnetic radiation—in this case, X rays. A simple X-ray tube (Figure 15.10) produces X rays by bombarding a target such as tungsten. Gamma radiation is released by the nucleus during the radioactive decay process.

Particulate or electromagnetic radiation with the ability to dislodge electrons from the orbit of an atom is called **ionizing radiation.** Subatomic particles and electromagnetic waves are the most familiar forms of ionizing radiation. The decay of radioactive isotopes results in the release of three types of ionizing radiation: alpha particles, beta particles, and gamma rays.

Alpha particles consist of two protons and two neutrons emitted from the nucleus. The mass number of the emitter decreases by four units because four nuclear particles are emitted, and the atomic number decreases by two because two protons are released. A new chemical

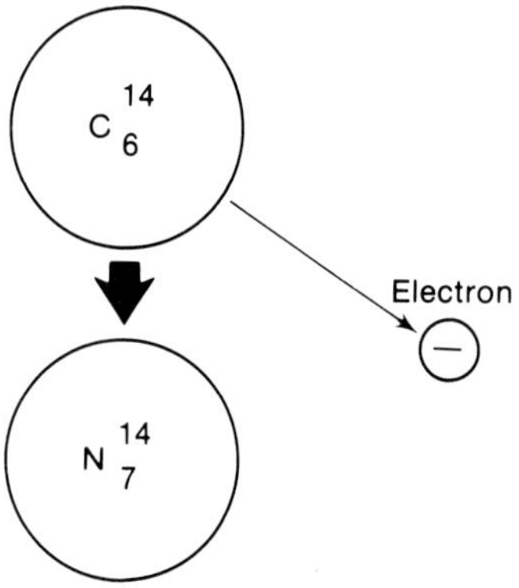

Figure 15.9
An atom of carbon-14 decays to nitrogen-14 by releasing a nonorbital electron (an electron formed in the atom's nucleus) when a neutron changes to a proton. The result is an increase in the number of protons.

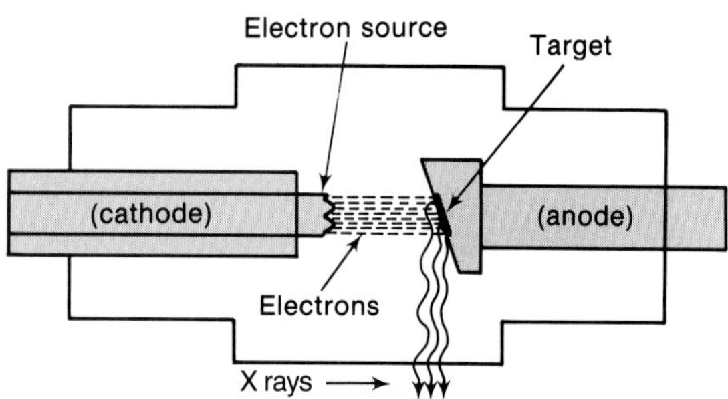

Figure 15.10
The formation of X rays by an X-ray tube. Electrons produced by a cathode (negative terminal) bombard a target on the surface of an anode (positive terminal), producing X rays which can be directed for use in such areas as medicine, physics, and food preservation.

RADIATION: HOW MUCH AND WHAT KINDS OF HUMAN DISEASES DOES IT CAUSE?

In the late 1950s little attention was paid to human exposure during extensive above-ground atomic tests conducted in the desert region of Nevada. Now, as hundreds of radiation-related disability claims flood the Veterans Administration, the U.S. Department of Defense is trying to find the 250,000 military participants who may have been exposed during that time. A follow-up study of 3156 soldiers present at the test called Smokey on August 31, 1957, was made. Instead of the normal leukemia rate of three or four cases, this group had eight cases 20 years after the test. Civilians living even 120 miles from the test site but in the fallout path also showed an increase in leukemia between 1959 and 1967. Following preliminary findings on soldiers from Nevada's atomic tests, the Center for Disease Control indicates that hazards from low-level radiation are either greater than we thought or the monitoring of dose levels at the time of testing was inadequate.

Low-level radiation is present throughout the environment, generated from cosmic rays and building materials but even more from medical and dental X rays. While high doses of radiation can be fatal or cause fatal disease in a few years, we have not been able to prove the short- or long-term effects from low-level radiation. The final conclusions concerning its effects on the human population will be very long in coming.

Interestingly, the expected hereditary influence of the radiation from the atomic bombs dropped on Hiroshima and Nagasaki in 1945 has not materialized. While mental and physical damage was great among unborn exposed infants, it was not among those conceived later by exposed parents. Many of these people are now in their thirties. The rise in leukemia has subsided in the exposed people, but they are experiencing an increase in cancer of the thyroid, breast, lung, and stomach.

On April 25, 1986, the Chernobyl nuclear plant in the Soviet Union exploded, sending a radioactive plume across a large part of the world. The release of radioactive material created great alarm throughout the world. Citizens in the United States were more concerned about ill effects from the Chernobyl accident than they had been about the Three Mile Island accident in 1979. Massive doses of radiation resulted in deaths within days in the Chernobyl accident. Bone-marrow destruction caused additional deaths within several months. Low doses of radiation, however, left the long-term effects of radiation unanswered.

element is thus formed. An example of alpha decay is radon-219 decaying to polonium-215 (Figure 15.11). Compared with other forms of ionizing radiation, alpha particles travel slowly. They are much larger than beta particles and do not readily penetrate most materials (Figure 15.12). A piece of paper or the outer covering of human skin stops alpha particles, so they are not great biological hazards when exposure is external. Most alpha particle emitters are elements of higher atomic weight, such as uranium and thorium.

Electrons ejected by an unstable atomic nucleus which have a negative electric charge are called **beta particles.** They result from the transformation of neutrons or protons in the unstable nucleus and are not orbital. Beta particles have varying amounts of energy, and this energy gradually dissipates when the particles travel through absorbing material. Absorption depends on the amount of energy contained by the particles and the thickness and density of the absorber (see Figure 15.12). Because living organisms are composed of several elements whose isotopes are beta emitters—nitrogen, oxygen, carbon, and sulfur—this form of radiation is very important in the study of life processes.

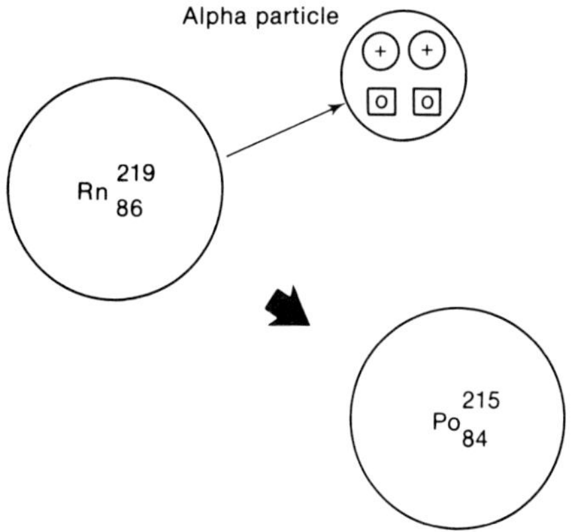

Figure 15.11
Alpha decay by radon-219. An alpha particle, consisting of two protons and two neutrons, is emitted, resulting in polonium-215.

Gamma rays are forms of electromagnetic radiation without associated electrical charge. As shown in Figure 15.12, they travel much further and penetrate material much more readily than either alpha or beta particles. Gamma rays and X rays have similar natures and properties and are readily absorbed by most materials.

Each radioisotope has its own decay scheme. Some decay by the emission of a single form of ionizing radiation; carbon-14, for example, gives off beta particles. Most radioisotopes give off a combination of radiation: bismuth-24 yields alpha, beta, and gamma, and cobalt-60 yields beta and gamma.

Some radioactive elements form unstable decay products. The new element then decays, forming yet another element. This process continues until a stable element is formed. For example, uranium-235 decays to form thorium-231. A decay series follows until the stable element, lead-207, is formed (Figure 15.13). Each element in the decay series has its own half-life and emits its characteristic radiation.

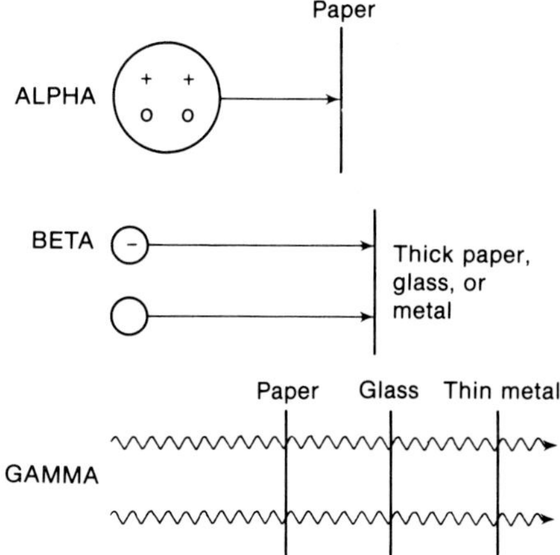

Figure 15.12
Alpha particles are blocked by thin paper. Beta particles can travel a little further but are blocked by thick paper, glass, and thin metal. Because gamma rays penetrate many everyday materials, a lead shield is often used to prevent their escape.

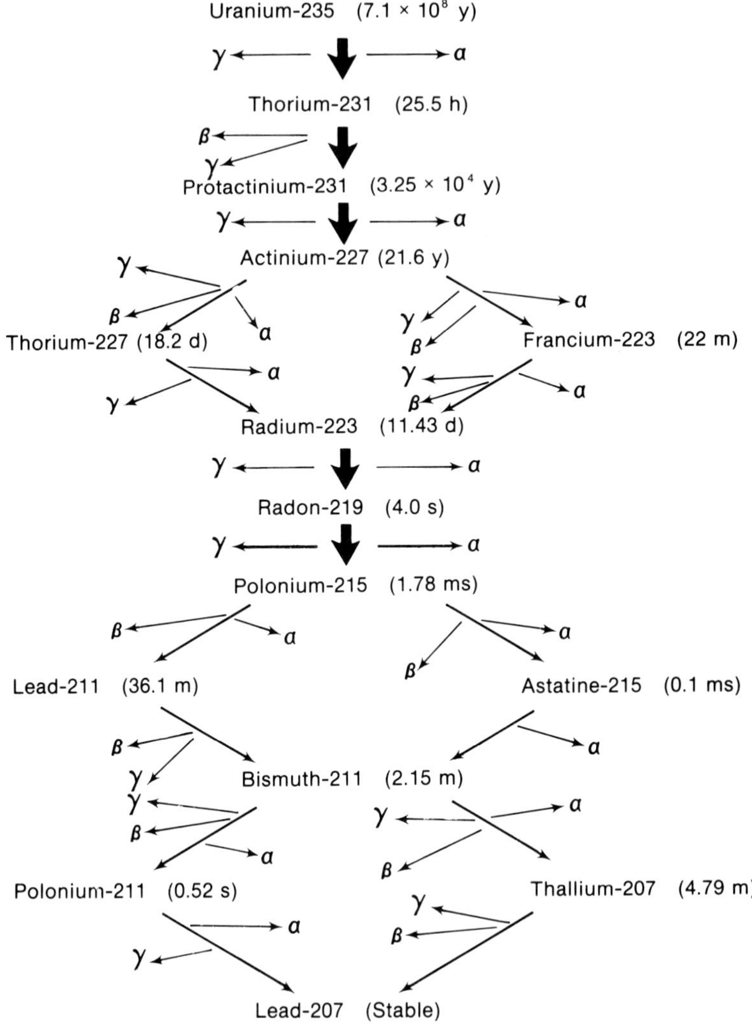

Figure 15.13
The uranium-235 decay series. Uranium-235 decays to a number of intermediate isotopes before a stable one, lead-207, is reached. Each isotope has a characteristic half-life of years (y), days (d), hours (h), minutes (m), seconds (s), or milliseconds (ms). Each isotope also emits characteristic alpha, beta, or gamma radiation as it decays. (From *Radiological Health Handbook.* Washington, DC: U.S. Department of Health, Education, and Welfare, 1970.)

When a large number of atoms of a particular radionuclide are studied, they are seen to decay randomly, making it possible to speak about decay schemes statistically. The statistical concept called **half-life** ($t^{\frac{1}{2}}$) is defined as the time it takes for 50 percent of a radioisotope to decay (see Chapter 4). This value is considered consistent for each radioisotope even though it varies considerably from one to another (see Figure 15.13). We are able to use the concept of half-life because we normally deal with large numbers of atoms. If we were to refer to only one or two atoms of a radioisotope, there would be little value in talking about half-life. The concept is useful in evaluating hazards resulting from the dissemination of radioactive materials in the environment.

Measurement

The radiation energy received by an organism or a particular part of an organism is called the absorbed dose and is based on the amount of energy imparted. Because a number of dose units exist for radiation measurement, a radiation text should be consulted for a complete description. We will mention the most common units here.

Typically, the absorbed dose is expressed in units called **rads.** A human being does not normally show clinical symptoms of radiation exposure until 100 rads or more have been received by a substantial portion of the body. When a group of individuals receives single doses of 350 to 450 rads, 50 percent die—the LD_{50} for humans. Because most people receive smaller doses, radiation is usually measured in **millirads** (1000 millirads are equal to 1 rad).

Since damage to an organism resulting from radiation absorption depends on the type of tissue exposed, the type of ionizing radiation, and the energy associated with the radiation, all three should be considered in the measurement of radiation dose. Special conversion factors based on these three components are used to change the rad into a unit expressing the energy dissipated in tissue and the amount of biological damage resulting from the energy. The rad is multiplied by the qualifying factor to convert it to the roentgen equivalent man, or **rem.** The conversion factor for some common forms of radiation is 1. Since exposures are usually of very low levels, we normally speak of absorption in terms of millirems; 1000 millirems (mrem) equal 1 rem.

In the United States, the average person receives 182 millirems per year (Table 15.4). The highest absorption is from background radiation from cosmic rays and minerals in the earth, followed by radiation from medical facilities. Nuclear energy sources are normally not a source of radiation in the environment. The accident at Three Mile Island did release radiation that gave an average dose of 1.5 millirems to individuals within a 80.5 kilometer (50 mile) radius between March 28 and April 7, 1979. The EPA estimates that by the year 2000 the average yearly human absorption will have increased by less than 1 millirem.

Table 15.4
Estimated Total Annual Whole-Body Doses from Natural Radiation in the United States

Source		Annual Dose (mrem per person)
Environmental		
Natural		102
Global fallout		4
Nuclear power		0.0003
	Subtotal	106.003
Medical		
Diagnostic		72
Radiopharmaceuticals		1
	Subtotal	73
Occupational		0.8
Miscellaneous		2
	Total	181.803

Courtesy of the National Academy of Sciences.

Radiation standards are designed to keep human radiation exposure well below harmful levels. Such levels are difficult to define because it takes a long time to establish the effects of radiation exposure. The National Academy of Sciences estimates that a continuous lifetime exposure to 1 rad per year by 1 million people increases cancer mortality between 3 and 8 percent. This information must be combined with data on dose response characteristics.

Two dose–response curves for living things are drawn in Figure 15.14. The linear curve indicates that response to radiation is directly proportionate to the dose. The other curve indicates that there is a *threshold* below which no injury takes place. Since we are dealing with such low levels of radiation, it is difficult to clearly establish a linear or threshold response. Thus, we should always be cautious when drawing conclusions about radiation. Presently, scientists do not agree on whether or not a threshold curve exists for humans.

Most public and private facilities using radiation have some method of monitoring potential

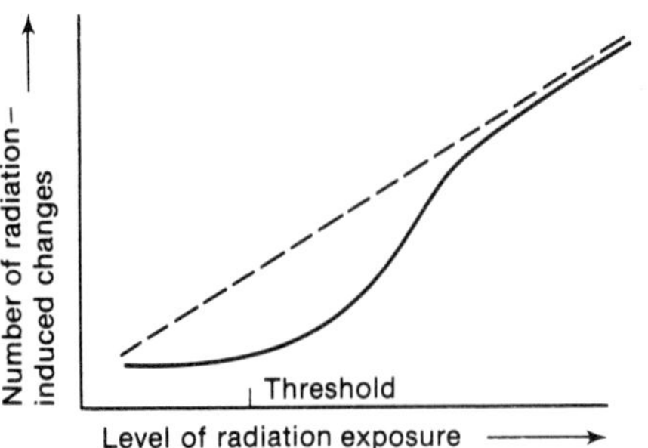

Figure 15.14

Two dose–response curves. If exposure to radiation were to demonstrate a threshold level—a level below which no injury occurs—the curved line would indicate the dose response. If all levels of exposure to radiation were to result in some change, then there would be no threshold level, as indicated by the diagonal line.

releases of radiation. Monitoring instruments generally record radiation in the air, water, and soil continuously or take samples at intervals for analysis. Each instrument must have a detector to indicate the magnitude of the radiation. Typically, detectors pick up flashes of light caused by the interaction of radiation with the gas, liquid, or solid in the detector portion of the instrument. The Geiger–Muller detector, named after the German physicists who invented it, is one of the most common ones. Radiation enters the detector and imparts some energy to the gas inside, causing a countable fluorescence. It is effective in locating sources of beta and gamma radiation.

Effects of Radiation

Acute and Chronic Exposure Effects of whole-body radiation on humans differ with acute (short-term) and chronic (long-term) exposure to ionizing radiation. The human body can normally recover from an acute absorption of up to 250 rems. **Acute absorption** results in damage to bone marrow, the spleen, the gastrointestinal tract lining, and the central nervous system. The blood-producing tissue (bone marrow) is the first to be affected by radiation because rapidly reproducing cells are most easily destroyed. White blood cell counts increase after exposure and then drop. Red blood cell counts begin to drop slightly after exposure, resulting in fatigue, fever, and sore throats. Rapidly reproducing cells which replace cells scraped off the gastrointestinal tract lining as food passes are destroyed, resulting in vomiting and diarrhea.

Chronic exposures to low levels of radiation over periods of months or years are more common than acute exposures. Unfortunately, we know little about the effects of this type of exposure. Some of what we do know is included in the following sections.

Somatic and Genetic Effects Changes that occur in an individual's body as a result of exposure to radiation during his or her lifetime are called **somatic** changes. Ionizing radiation can also induce **genetic** changes by causing a structural change in the chromosomes of sex cells resulting in mutations (Chapter 3). The real danger of ionizing radiation is that members of a population can carry the effects of exposure for many generations before they appear. The National Research Council estimates that, on the average, individuals exposed to a single ionizing radiation dose of 5 rems before or during the child-bearing years would experience an increase of about 0.2 to 2.0 percent in the incidence of genetic disease in their offspring. The increased incidence of genetic disease would not be limited to the offspring of the exposed generation but would extend through many generations. About 20 percent of the total expected increase would be realized in the offspring of the exposed generation. The rest would occur at a declining rate in each succeeding generation. Perhaps 50 percent would be realized by the fourth generation, but it would require over thirty generations to realize 99 percent of the genetic disease expected from this single exposure. Lack of information is a major problem in estimating the effects of radiation-induced changes on the human genetic system. Virtually all data come from studies on other mammal species.

Cellular Effects Radiation-caused structural changes can occur in cells. Water, a principal component of cells, can be broken down into hydrogen and oxygen. These products are free to react with other elements in the cell, producing new compounds that can cause chemical imbalance. Some large molecular structures are affected by different levels of radiation. Proteins, for example, can be changed in some cases and therefore lose the ability to control the rate of the body's chemical reactions. The chemical composition of hemoglobin, the oxygen-carrying pigment of the blood, may be changed so that it does not bind to oxygen and therefore is unable to serve as an oxygen carrier. Other large molecules like sugar and cellulose are broken down so that they cannot serve their usual function. Gross visible damage to cells occurs only with massive doses of radiation. The cell usually dies before these high dose rates are achieved.

Effects on the Organism The exposure of whole organisms to radiation results in generalized changes in their bodies. Each kind of animal and plant develops characteristic symptoms, depending on its age, dose, and number of exposures. Studies indicate an acceleration of the aging process and an increased susceptibility to diseases such as pneumonia in animals exposed to chronic low-level radiation. Species show different sensitivities to radiation. Mammals are more radiosensitive than birds, fish, amphibians, or reptiles, and adult insects can withstand higher doses of radiation than vertebrates. Single-celled organisms, like amoeba, paramecium, and some microorganisms, are least susceptible to radiation exposure. Individual variation also exists within each species. In plants, radiation damage occurs in a similar manner. Irradiated plants often show gross changes in their reproductive structures, particularly the flowers. In some cases, new roots grow from the stem above ground level. Mature leaves tend to show little change when subjected to sublethal doses of radiation.

In humans, radiation can cause a number of changes. Congenital effects of radiation occur in a developing embryo when the mother is exposed to ionizing radiation such as diagnostic X rays or radiation therapy. The type and extent of a defect depend on the developmental stage of the embryo as well as the strength of the dose received. The first trimester is the most critical because the major tissue and organ systems are being formed and tissue division is very rapid. Exposure of the embryo during the latter stages of pregnancy, however, is essentially the same as exposure of a baby. Since not all congenital defects are inheritable, an individual with such defects will not necessarily produce children with the same abnormality. Congenital defects are difficult to attribute to ionizing radiation because abnormalities routinely occur in 5 to 6 percent of all births. However, many women who were pregnant at the time of the Hiroshima and Nagasaki bombings bore children with mental and physical problems.

Long-term effects are often very difficult to trace back to radiation. Statistical comparisons of groups of people exposed either to chronic or acute doses of radiation with groups not exposed provide some insight. Uranium miners, radiologists, and patients treated with radiation suffer more frequently from cancer than do the rest of the population. The value of this kind of statistical study is limited, however, as the comparison does not provide conclusive evidence that uranium or radiation treatment causes cancer. Other genetic and environmental factors might affect both miners and the rest of the population.

Cancer is any condition where cells multiply uncontrollably. A number of *carcinogenic* (cancer-producing) agents are known, including ionizing radiation. Although the mechanism causing cancer is not yet understood, it appears that carcinogenic agents somehow alter the normal reproductive cycle of cells, causing rapid, unchecked growth. Ionizing radiation is paradoxically not only a cause but also a cure.

Although it is difficult to pinpoint the exact cause of most cancers, some scientists interpret statistics to show that people exposed to radiation develop more somatic forms of cancers than people who are not. Cancers thought to be radiation induced are, however, indistinguishable from other forms. Solid tumors have a long latent period, usually appearing 10 years after radiation exposure and sometimes continuing to appear for 30 years thereafter. Leukemia materializes within a few years after exposure but can occur up to 30 years.

Although some body tissues are more susceptible than others, cancers are known to develop in all tissues. Solid tumors induced by whole body radiation occur in the breast, thyroid, lung and some digestive organs. Because radiation seems more likely to induce breast and thyroid cancer, the total cancer risk is greater for women than men.

Age is an important risk factor. Some forms of cancer, particularly leukemias, develop more frequently in exposed children, while women exposed to radiation during their second decade of life have a high risk of developing breast cancer.

There is a growing body of evidence that cancer also can be caused by a variety of chemicals, including some found in cigarettes and the food we eat or drink. Alcohol, which can promote the development of cancer, and cigarettes amplify risk in areas where smoke and alcohol filter through the body.

Cataracts, small tumors in the lens of the eye, are known to occur in humans and other animals exposed to heavy doses of radiation. In a few cases, X-ray doses to the eye of about 200 rems have been found to cause cataracts. The significant dose is between 100 and 600 rems in adults and less in children. Low-level chronic exposure apparently does not result in cataracts in humans, but there is evidence of such exposure producing cataracts in other animals.

Effects on Populations and Communities Radiation can alter the interaction of populations and groups of species in a community. For example, populations exposed to ionizing radiation can be genetically damaged. When the damage is carried for several generations, it can affect the population's ability to survive within that environment. Laboratory studies of fly irradiation show that groups subjected to doses of 5 rads/hour during their lifetime produced an increased number of sterile flies for a period of 140 generations. Similar results were obtained from studies on the long-term effects of radiation on algae. When radioisotopes were added to the nutrient medium, mutant cells immediately built up in the algae and remained at a relatively high level for more than 25 generations.

Background radiation is one ecological factor affecting populations. The effects of higher levels of background radiation, or radiation introduced into the environment over a long period of time, are not completely known. On some islands in the Pacific Ocean where natural radiation is rather high due to the high concentration of radioisotopes in the soil, the local human population has a lower fecundity rate than that of inhabitants of neighboring islands. On the other hand, some studies indicate that microorganisms living in hot springs that contain radioactive material have higher degrees of resistance to radioactive material.

The food chain discussed in Chapter 2 is the primary means of radioisotope transport from soil to plants to animals. This transfer process is moderated by a series of factors that generally reduce concentrations of radioactive elements as they pass through the plant–herbivore–carnivore stages. In vertebrates, the assimilation of most elements is reduced in the intestinal tract. Cesium is an exception to this general scheme in mammals because it concentrates at each transfer in the food chain. Concentration factors for cesium are relatively small. Twofold increases in the concentration of cesium-137 have been measured at each step in the lichen–caribou–wolf food chain, and a ninefold increase has been measured in the plant–deer–cougar chain. Once in the organism, radioactive material can become an internal emitter. Material emitting alpha particles, normally harmless as an external source, can create internal damage when ingested, inhaled, or absorbed through an open cut.

Indirect effects of radiation within natural communities include changes in form and structure, composition, and species diversity, as well as in plant productivity. These can result from alterations in spatial relationships, competition, nutrient supply, and availability of light and moisture.

SOURCES OF RADIATION

In addition to natural radioactivity found in the atmosphere and in some terrestrial deposits, human activities add radioactive materials to the air, water, and soil. We produce them in nuclear reactors or particle accelerators when nonradioactive material is bombarded by radioactive particles. A particle accelerator is any electrical device that increases the velocity and thus the kinetic energy of electrically charged particles (electrons or protons) to a value sufficiently high to convert some of the kinetic energy to electromagnetic radiation (as is the case of X rays). Such a device may also cause the accelerated particle to hit another element and transform that element into a radioactive isotope.

Atmospheric Sources

Radioisotope decay from radioactive material in the atmosphere is a major source of the environment's background of ionizing radiation. Higher elevations generally have higher levels of radiation than that at sea level because the lower atmosphere acts as a shield to block some cosmic radiation.

Cosmic rays, composed of particulate and electromagnetic radiation, are part of the electromagnetic spectrum. As cosmic rays collide with the isotopes of atmospheric gases, a secondary cosmic radiation is formed, increasing the intensity of the incoming radiation. Doses of ionizing radiation from cosmic rays increase with both altitude and latitude (Table 15.5). Whole-body dose rates at sea level from Alaska to Florida vary from 45 to 30 millirems per year. At 45 degrees latitude, the dose equivalent increases from 40 to 200 millirems per year as measured along an elevation gradient from sea level to 2440 meters (8000 feet). The average resident of Hawaii receives a whole-body dose of 30 millirems per year, while the average citizen of Wyoming receives 130 millirems per year. Aircraft crews are exposed to higher levels than the public, but as people fly more often, this form of radiation exposure could become a general concern.

Terrestrial Sources

Naturally occurring radioactive material accounts for some background radiation. The ma-

Table 15.5
Annual Whole-Body Dose to People from Cosmic Rays

State	Average Annual Dose (mrem)
Alaska	45
California	40
Colorado	120
Florida	35
Hawaii	30
Illinois	45
Kansas	50
Maryland	40
Tennessee	45
Wyoming	130

jor sources of external gamma radiation are potassium-140, uranium, thorium, and their decay products. Studies indicate that 90 percent of all areas in the United States produce doses of 15 to 140 millirems per year, with the mean population exposure at 55 millirems per year. From the standpoint of terrestrial whole-body doses (Figure 15.15), the contiguous United States can be divided into three broad areas. Some radioisotopes, such as carbon-14, radium-226, and radium-228, are ingested or inhaled and thereby contribute to the radiation dose. The average annual internal body dose from natural radiation on a worldwide basis is 25 millirems per person per year.

Deposits of uranium ore emit radiation at dose rates about sixty times higher than those found in the environment of U.S. urban areas. These high levels of natural radiation make living near uranium deposits hazardous. Uranium mining is even more harmful for miners because they travel in underground tunnels adjacent to the decaying ore and breathe the dust resulting from the uranium removal processes.

The waste materials from uranium mining operations, known as *tailings*, contain several isotopes harmful to health and the environment. Radium-226 is periodically washed down the Colorado River from such mines and concentrates in high levels in some areas. Tailings are sometimes mixed in concrete, creating built-in radiation hazards in new construction. The Colorado plateau also has an unhealthy level of radon gas because some landfills in that region are made up of uranium mine tailings. A typical uranium mill can generate 1.8 million kilograms (1980 tons) of tailing solids mixed in 2.5 million kilograms (2750 tons) of waste milling solutions per year. Over the lifetime of the mill, 40 to 80 hectares (100 to 200 acres) could be permanently committed to store this material. People living in those regions are exposed to an annual dose rate of about 200 millirems per year.

Radiation Uses

A whole area of nuclear medicine has developed around the use of radioisotopes in the diagnosis

Figure 15.15
Terrestrial dose-equivalent rates in the contiguous United States. (Reproduced from *Effects on Population of Low Level Ionizing Radiation; 1980.* Washington, D.C.: National Academy Press, 1980.)

and treatment of diseases. An X ray is a common method of diagnosing dental cavities, bone fractures, and intestinal disorders. Radiotracers are used to diagnose a number of disorders associated with specific tissues in the body and to determine the amount of blood reaching the extremities, the rate at which materials move from tissue to tissue, the level of some vitamins, and disorders such as brain tumors and liver diseases.

Medical therapy uses radioisotopes to suppress tissue growth or destroy abnormal tissue. World capabilities for generating nuclear power are predicted to rise from 20 to 2000 billion watts of electricity between 1980 and 2000. Consequently, the general public is going to be exposed to potential danger from radiation in all portions of the nuclear fuel cycle. The 1963 Limited Nuclear Test Ban Treaty to halt all nuclear testing in the atmosphere was signed by all nations with nuclear power capabilities at that time. Unfortunately, some countries, like France, have not adhered to the treaty, and countries like the People's Republic of China and India have since developed the ability to produce atomic weapons and have been testing them.

The primary concern about atmospheric nuclear explosives is fallout. **Fallout** is governed by the height and size or power of the explosion. When low-altitude explosions occur, large amounts of soil and water are sucked up by the hot fireball. Radioactive particles condense on this material and form relatively large, heavy particles that usually fall to the earth within a 24-hour period. Little radioactive material is retained in the atmosphere. When the explosion occurs at great heights, little or no dirt or water is sucked up, so only a small amount of local fallout is produced. Instead, small, soluble radioactive particles are produced and are blown by air currents throughout the world. More than 90 percent of this material is distributed as fallout within a 7-month period after the explosion. Low-yield nuclear devices (of a few hundred kilotons or less) exploded near the surface tend to contaminate the troposphere. Most of the resulting fallout is spread around the same latitude as the detonation, and particles are usually washed out of the atmosphere within a month after the explosion. High-yield nuclear explosions (measured in megatons) push fission products into the stratosphere (Figure 15.16). In the stratosphere they move toward the poles, where they descend into the troposphere and are pulled down to the earth.

People in the armed services using certain kinds of detection equipment are also exposed to higher levels of radiation than the general public, as are airport employees who check baggage with similar equipment. The food industry is now using radiation to preserve foods, destroying microorganisms and insects that cause food spoilage by low levels of radiation. Some food products are sterilized by radiation; other foods, like potatoes, are irradiated to reduce sprouting and thereby keep them in a usable form (Figure 15.17).

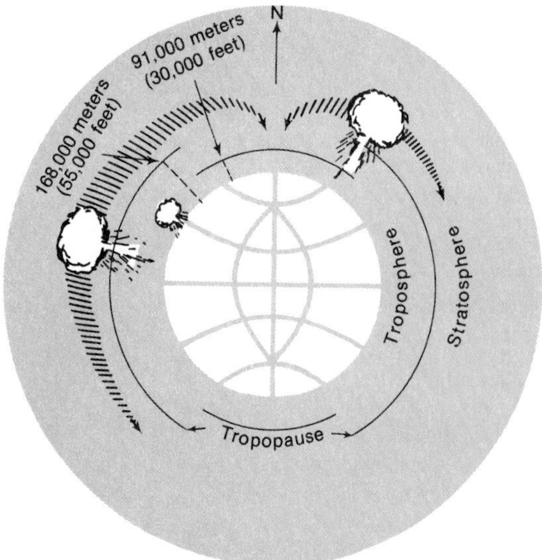

Figure 15.16
Movement of fallout for low- and high-yield nuclear explosions in the troposphere, from west to east. High-yield explosions in the stratosphere cause the fallout to move toward the earth's polar regions, where they descend. (U.S. Department of Energy.)

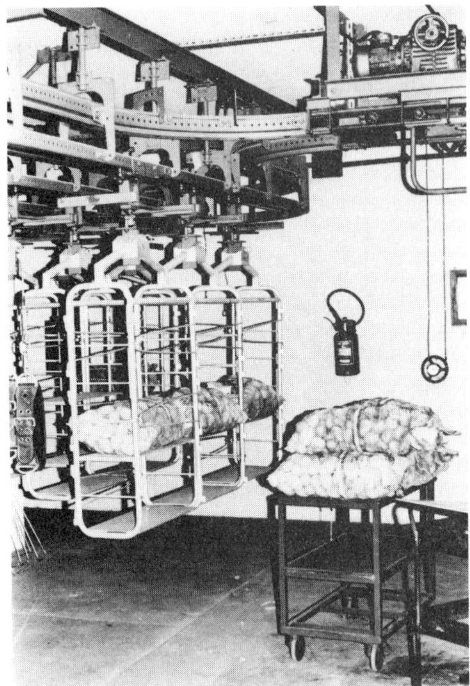

Figure 15.17
A conveyer moves potatoes through a radiation field to reduce sprouting. (Courtesy of the International Project in the Field of Food Irradiation.)

Forms of nonionizing electromagnetic radiation resulting from broadcasting signals, radar, power transmission, and microwave ovens are also of concern to the public. Research is now underway to determine if there are any long-term biological effects from this radiation, which does not dislodge electrons but imparts more energy to objects it strikes.

RADIOACTIVE WASTE DISPOSAL

Radioactive wastes are produced in several ways. Wastes from the nuclear fuel cycle include uranium-238, strontium-90, cesium-137, and others from mining, milling, fuel enrichment, fuel preparation for reactor use, plant operation, and disposal. Plutonium-239 is a by-product of liquid-metal fast breeder and fission reactors. Some industrial processes, as well as nuclear medicine and scientific research, also produce radioactive waste material that must be stored or rendered harmless.

Today, there are three methods for handling radioactive waste material: (1) dilute and disperse; (2) delay and decay; and (3) concentrate and contain. Wastes of low radioactivity can be diluted to permissible levels for dispersal and diluted further in the air, water, and soil. However, such methods risk environmental contamination and accumulation in food chains. State and federal agencies carefully monitor and control such releases.

A radioactive material with a relatively short half-life can be retained in receptacles until its natural decay causes the radioactivity to dissipate. Such "delay and decay" methods are reasonable for radioisotopes with short half-lives; for example, iodine-131 can be safely released into the natural system after 80 days, or after it has gone through ten half-lives.

Wastes with high levels of radiation must be concentrated into small volumes and stored away from wildlife areas and areas of public use. One method involves placing concentrated radioactive wastes in storage tanks. Leakage in one of these underground trench sites on the U.S. Atomic Energy Commission's Hanford reservation in south central Washington occurred between 1950 and 1960. Burrowing mammals apparently exposed dried salt cake containing the radioactive material. Because salt licks were rare in the area, the reservation became a popular spot for wildlife. Field studies conducted in 1972 and 1973 indicated that animals moved the radioactive material, mostly through droppings, up to 3 kilometers (2 miles) from the source in this desertlike region.

At the Oak Ridge National Laboratory in Oak Ridge, Tennessee, a method known as **hydraulic fracturing** is used to dispose of some nuclear waste. As shown in Figure 15.18, liquid wastes are concentrated into layers of soil or

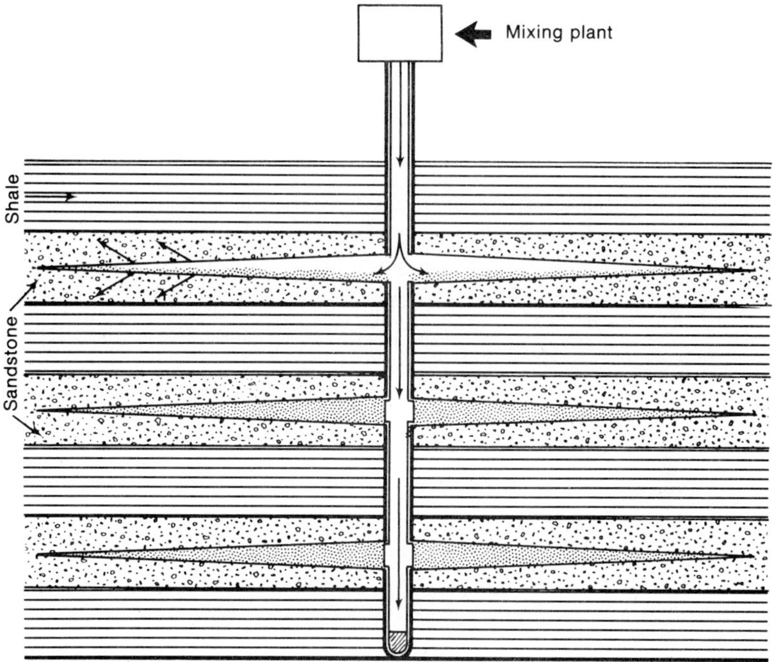

Figure 15.18
Hydraulic fracturing. Liquid wastes are mixed with cement slurry and forced under pressure into seams in the earth.

rock in a watery cement slurry mixture. Layer upon layer of slurry can be forced under pressure into the soil through a well. This method is only effective where the right soil and rock formations exist and where water runoff does not spread the waste material.

The use of salt mines as permanent storage vaults is presently being considered for long-lived nuclear wastes. Salt mines exist in areas where no groundwater runs, providing places where potential disturbances to land are minimized. Once the waste materials are placed in the salt vault, the mine can be sealed (Figure 15.19). A mid-1960 pilot project conducted by the Atomic Energy Commission near Lyons, Kansas, showed salt vaults to be a feasible means of disposing of high-level nuclear wastes. The project was discontinued, however, when water seepage from nearby mines raised concern and local political opposition. Germany is currently using salt mines to store radioactive wastes, and the U.S. Department of Energy is now considering the extensive salt beds of New Mexico as potential disposal sites.

One of several alternative disposal methods is being studied by scientists at Woods Hole Oceanographic Institution in Massachusetts. Be-

Figure 15.19
Preparation of a salt mine for storage of radioactive waste material.

LEGISLATION

Toxic Substance Control Act of 1976. Banned the manufacture of compounds that pose an unreasonable risk to human health or the environment. Also authorized the EPA to monitor chemicals in commercial use and to take enforcement action against those who violate regulations.

Resource Conservation and Recovery Act of 1976 (RCRA). Amended the Solid Waste Disposal Act and authorized the EPA to work with the states to regulate hazardous wastes from their generation to their disposal.

Comprehensive Environmental Response, Compensation and Liability Act of 1980 (CERCLA) and the *Superfund Amendments and Reauthorization Act of 1986 (SARA).* Required parties responsible for sites contaminated by hazardous materials to conduct cleanups under government supervision. Established a superfund to help pay for the cleanup of hazardous waste sites.

Energy Reorganization Act of 1974. Created the Nuclear Regulatory Commission to regulate civilian uses of nuclear material, including nuclear power plants and applications in medicine, industry, and research.

cause of the nature of the clay in deep ocean trenches and the minimal movement of benthic water, they believe that radioactive material could be buried in the ocean floor without harmful consequences (Chapter 4). However, this method involves many unsolved technical and political problems. Another process being investigated by the Department of Energy incorporates high-level waste in ceramics. Small particles could then be handled and disposed of in this form. A method being investigated at Hanford, Washington, involves placing ceramic or vitrified waste pebbles in stainless steel containers and storing them in a ventilated mine of seismically stable granite core. Both salt mines and granite core repositories appear technologically feasible, but solutions to waste disposal problems are often influenced by political concerns because no one wants the wastes stored near them. Research is also being done on the potential conversion of harmful isotopes into harmless ones by means of neutron bombardment.

In 1982, the Nuclear Waste Policy Act established a procedure for building a permanent repository for high-level radioactive wastes from commercial nuclear reactors. A site near Yucca Mountain, Nevada, has been selected and is scheduled to open shortly after the year 2000. Studies are underway to ensure that the volcanic deposits where the waste will be stored are a secure site for the facility. Concerns about safety and potential environmental impacts caused a delay in the opening of the Waste Isolation Pilot Plant (WIPP), a facility designed to store nuclear waste from military uses. Scheduled to be opened in 1988 near Carlsbad, New Mexico,

WIPP would have stored the waste in underground salt beds.

All methods of waste storage or disposal have a risk factor. By producing long-lived isotopes, we are leaving a legacy to future generations. Possible environmental contamination from mishandling is of great concern to many people, so as we continue to use nuclear power, we must emphasize the solution of the problems of waste disposal.

SUMMARY AND CONCLUSION

Our society's wastes are accumulating at an alarming rate. Solid waste, including garbage, refuse, and sludge, pollute air and water. Health problems occur when disease vectors use garbage as a breeding ground. Most solid wastes are disposed of in fills or through incineration; recycling is not common because of costs.

Toxic wastes are frequently included in solid wastes and placed in landfills. Toxic wastes are related to many human health problems, including birth deformities and liver disease. Unfortunately, people have only recently begun to take toxic waste problems seriously. A superfund established by the federal government sets aside funds in trust to assist in cleanup operations when spills occur or when abandoned waste sites are discovered.

Radiation is a form of energy that results in the spontaneous decay of atoms. Particulate radiation such as alpha and beta radiation comes from the release of subatomic particles, while gamma radiation is a form of electromagnetic radiation. The decay scheme of each radioisotope is unique to that isotope, producing alpha, beta, or gamma radiation or combinations thereof.

Radiation is measured in terms of absorbed dose of energy. The average person in the United States receives 182 millirems per year. An acute, or single, absorption of 350 to 450 rems results in the death of 50 percent of the people exposed.

Life is damaged either genetically or somatically by radiation. Genetic damage might not appear for many generations, but somatic effects appear immediately or later in life. Cancer, cataracts, or a shorter life span are some somatic effects of chronic, or long-term, exposure.

The most common sources of human radiation exposure are from cosmic rays, natural earthbound isotopes, and medical diagnostic and treatment procedures. The average person receives less than one millirem per year from energy production; however, we continue to be concerned about the proper handling of raw materials and the use of radioisotopes in energy production because the long-term effects of such exposure are still unknown.

As we increase our use of nuclear energy, we must devote more time and money toward discovering the effects of radiation on people and the environment. Planners must understand these effects and set aside adequate tracts of land to protect the public from waste storage disposal sites and potential sources of radiation. Proper monitoring devices and safety measures must be built in and enforced around facilities using radioactive material.

FURTHER READINGS

Council on Environmental Quality. 1990. *Environmental Quality.* 21st Annual Report. Washington, DC.

Krag, B. L. 1985. Hazardous Wastes and Their Management. *Hazardous Wastes and Hazardous Material* 2:251–303.

Kroesa, R. 1990. *The Greenpeace Guide to Paper.* Vancouver, British Columbia: Greenpeace Books.

Rathje, W. L. 1991. Once and Future Landfills. *National Geographic* 179 (5):116–34.

Tarr, J. A. 1985. Historical Perspective on Hazardous Wastes in the United States. *Waste Management and Research* 3:95–102.

White, A. H., and E. Cromartie. 1985. Bird Use and Heavy Metal Accumulation in Waterbirds at Dredge Disposal Impoundments, Corpus Christi. *Bulletin of Environmental Contamination* 34:295–300.

Young, J. E. 1991. Reducing Waste, Saving Materials. Chap. 3 In L. R. Brown (ed.). *State of the World 1991.* New York: W. W. Norton & Company.

STUDY QUESTIONS

1. How do birds assimilate lead from lead shot during their lifetime?
2. Why is landfilling not a complete answer to solid waste disposal?
3. How long would landfills for toxic substances have to be maintained for safety?
4. What environmental problems are caused by incineration? Landfilling?
5. Should each community be required to dispose of its own wastes?
6. Why do some areas in the country have higher levels of radiation than others?
7. Why do you suppose that toxic and radioactive wastes have a greater impact on unborn organisms?
8. A woman who had a series of X rays during her pregnancy gave birth to a child with a malformed arm. Is this a genetic effect of radiation?
9. How does radioactive pollution differ from chemical pollution?
10. Discuss the future uses of radiation and the implications for the evolution of the human race.

SUGGESTED ACTIONS

(Adapted from *Your Choices Count*, National Wildlife Federation Citizen Action Guide.)

1. Recycle paper, metal, plastics and glass. Recycling centers can be located by consulting:
   a. local sources such as city management departments or telephone directories,
   b. the EPA's RCRA Superfund Hotline (800-424-9346),
   c. the Environmental Defense Fund Hotline (800-225-5333), or
   d. the Reynolds Aluminum Recycling Hotline (800-228-2525).
2. Buy recycled and recyclable products.
3. Dispose of hazardous chemicals correctly and support community programs for safe collection and disposal of household hazardous waste.
4. Take used motor oil to an oil recycling center.
5. Remember the "4 Rs"—reduce (waste volume), reuse, recycle, and reject (products that are overpackaged, toxic, or nondegradable).

FIVE

PEOPLE AND THE ENVIRONMENT

16

HUMAN POPULATION

According to the United Nations, the world population reached 5.3 billion people in 1990. This may not mean much unless you consider that it took from the beginning of time until 1830 for the world to reach its first billion. The second billion took 100 years; the third billion, 30 years; and the fourth billion, 15 years. By the year 2000, the world population will almost certainly exceed 6 billion people. Assuming a decline to zero annual population growth in the early part of the twenty-first century, the world population will still reach a figure of 10 to 16 billion by the year 2100.

Why is the human population growing so fast? Why can we not reach zero population growth immediately if the fertility rate is zero or below? How are people reacting to more people? What methods can be used to slow population growth?

The spectacular population growth of the last two centuries, plus advances in science and industry, have resulted in a human society vastly changed from that of primitive people. These changes are primarily in the form of culture, which dictates how people live and how they use the environment. In short, any form of environmental management or ethic will be greatly influenced by human culture.

CULTURAL EVOLUTION

The total pattern of behavior embodied in actions, artifacts, speech, and thought is human **culture.** The physical characteristics that people developed in adapting to the environment provided the capabilities for subsequent cultural development. Culture itself provides one means of adapting to the environment.

Although culture is defined in many different ways, it is basically our ability to pass on information gained by one generation to members of the next generation. Some see this as the learned portion of human behavior, while others believe that culture includes knowledge, beliefs, art, morals, law, customs, and other acquired capabilities. The key factor is that this information is not only passed on from parents to offspring, as in the case of biological evolution, but is transmitted from all members of society to others. In humans, the unparalleled ability to verbally transmit information has speeded and fostered intricate cultures.

Communication, which began when humans first developed the ability to use language, is but one aspect of cultural evolution. Our modern technology is another product of culture. Technology traces back to the first use of tools. A third aspect of culture, social organization, fostered the cooperation that enabled early man to survive in a hostile environment and has led to the evolution of a highly organized global community.

Cultural evolution, transmitted through symbols, is much more rapid than biologic evolution transmitted through genetic material. Culture can therefore be responsible for faster, more pronounced changes in recent times than **biological evolution.** Still, biological evolution sets the limits for cultural evolution. Most human traits result from both biological and cultural evolution: intelligence, for example, depends both on the limits set by biological evolution and on the cultural attributes available to an individual.

The ability to invent and to accumulate knowledge has allowed people to live in otherwise uninhabitable areas. Eskimos used igloos and clothes from animal furs to make the Arctic habitable, and spaceships and spacesuits enabled people to visit the moon. Overcoming many environmental obstacles, people developed agricultural techniques to provide a reliable source of food and motive power to distribute it. Cultural adaptation led to social organization, which made urban life possible. Technological innovation is another form of cultural evolution that has brought changes in modern industrial societies.

Social organizations and belief systems are also forms of cultural adaptation. **Social organization** provides a structure for society by defining the relationship of one person to another. From these guides, people understand their responsibilities to society and the rights and privileges to which they are entitled if they fulfill these responsibilities. Thus, people expect rewards for behavior that conforms to the rules and punishment for nonconformity. Social organization permits task specialization, mutual protection, and trade. By working together, people accomplish tasks that would be impossible for one person working alone. Modern urban civilization depends upon the support of a complex social structure.

The belief systems by which individuals accept the social dictates of rules, authority, and morals form a basis for social order. Obviously, a great deal of consensus is necessary for a society to function smoothly. Without consensus, chaos and civil strife reign. Beliefs govern all sorts of activities, including dietary taboos and rituals, family ties and responsibilities, the disposal of human remains, property and water rights, and the value of money. The experience of generations is embodied in the dietary taboos and rituals for food preparation and preservation. Those societies that select plants and animals that are safe to eat and prepare and preserve them properly survive to pass on the ritual.

The history of mankind on earth shows a progression of human control over nature. At first man survived by understanding and cooperating with nature. Fundamental changes in the way humans relate to the environment occurred when our ancestors learned to use agriculture and then industrial technology. Raymond Dasmann uses the term *ecosystem people* to describe cultures in which people adjust to the ecological conditions of the environment, rather than try to modify them. Ecosystem people are truly members of the ecosystems they inhabit. Some of these cultures still exist in remote regions of the world. In contrast, most of the world is made up of what Dasmann calls *biosphere people* who are part of a global culture. Biosphere people control nature to a large extent and use the resources of many of the earth's ecosystems. We have learned in recent years, however, that we are still very much a part of nature and that our ability to exert some control over it must be used wisely and creatively.

HISTORY OF POPULATION GROWTH

When humans emerged as a species several million years ago, they were part of a hunting and gathering society. At this time small groups of individuals moved about, obtaining food in different areas. Obviously, life was limited to regions of the world where an ample food supply could be found on a year-round basis. Life was harsh, with whole groups of individuals starving during years when sufficient food was not available.

The first major changes affecting human population growth were the use of fire and the development of tools. Although tools were very primitive at first, stones hurled at wild animals provided a means to obtain more food, thus enabling the population to expand. More tools were developed so that different forms of food could be utilized. Primarily as a result of tool development the human population grew to about 5 million people approximately 100,000 years ago. It then remained at about that level for some 90,000 years. At times, food was not abundant and the population declined; in periods of superabundant food, the population increased.

Eventually, the cultural advantages of tool making and fire caused primitive populations to outgrow game animal populations. The tremendous impact on the environment is referred to as the **Pleistocene overkill.** As a result, the human population could maintain itself only by adopting a more vegetarian diet. This led to the cultivation of food plants.

Approximately 5000 to 10,000 years ago, people began to use tools for agriculture. Grain could be harvested and stored in primitive ways for seasons when food was not as abundant. The agricultural revolution, which occurred over a period of 8000 years, enabled the population density to reach four individuals per square kilometer (0.3861 square mile) of land. Again, the population reached a plateau where over a period of years the birth rate balanced the death rate, even though many fluctuations occurred.

Diseases, often caused by bacteria harbored in wild mammals, generated at least 196 epidemics between 600 B.C. and A.D. 1800. In the fourteenth century the "black plague" epidemic caused the deaths of one-fourth of Europe's population—approximately 25 million people.

Beginning about 1750, population growth began to show the effects of the scientific industrial revolution (Figure 16.1). Development of mechanized methods of food production, transportation, and storage made food available at any time and almost any place on the earth.

In industrial nations, a marked change in population growth began in the late nineteenth

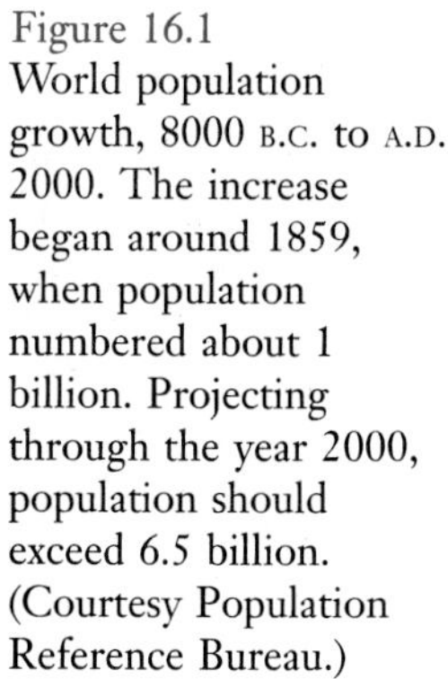

Figure 16.1
World population growth, 8000 B.C. to A.D. 2000. The increase began around 1859, when population numbered about 1 billion. Projecting through the year 2000, population should exceed 6.5 billion. (Courtesy Population Reference Bureau.)

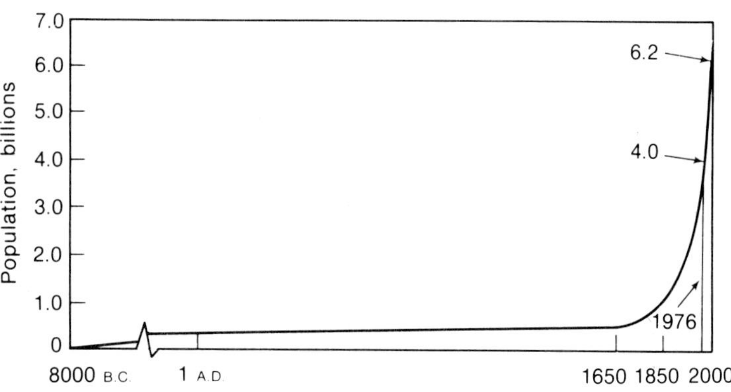

THE GLOBAL 2000 REPORT: ENTERING THE TWENTY-FIRST CENTURY

In 1977, President Jimmy Carter directed the Council on Environmental Quality and the Department of State to make a 1 year study of the probable changes for the remainder of this century in world population, natural resources, and environment. A number of public and private agencies throughout the nation worked together to develop models which could predict these changes. These models served as the basis for government planning as well as projections that might be used by the private sector.

The principal findings of the report, published in 1980, linked an array of population, economic, and environmental conditions. Of primary concern was the profound effect increasing population has on the world. Rapid growth in world population will not be altered by the year 2000, but the population will increase more than 50 percent—from 4 billion in 1975 to 6.35 billion in the year 2000. Although the rate of growth will slow slightly from 1.8 percent per year to 1.7 percent, the population will be growing faster in sheer numbers of individuals. The study's estimates showed 100 million people per year will be added to the population by 2000 compared to 75 million in 1975. It is projected that 90 percent of that growth will occur in the poorest countries of the world.

Along with fuel and mineral shortages, anticipated worldwide water deficiencies become a major concern. These water shortages will be complicated by extensive deforestation. Deforestation will not only result in the loss of fuel wood and building timber, it will diminish the carbon dioxide and oxygen required by different life forms to survive in these areas. By 2000 predictions indicate that about 40 percent of the forest cover in less developed countries will be gone.

Furthermore, data indicate that by 2050 atmospheric concentrations of carbon dioxide and ozone depleting chemicals will intensify at rates that could alter the world's climate and upper atmosphere significantly. Acid rains from increased combustion of fossil fuel, especially coal, will threaten lakes, soil, and crops. Disposal of radioactive wastes and other hazardous materials may well affect a good part of the country.

The conclusions of the Global 2000 Report coincide with a number of other studies completed in recent years. All available evidence indicates that the world faces enormous urgent and complex problems in the decades ahead. Overpopulation is the primary factor influencing diminished supplies of minerals and energy, an increase in the amount of air pollutants, and a general stagnation of the world. Unfortunately, solutions to most of these problems require long lead time and extensive funding. If decisions are delayed until the problems become worse, however, options for effective action will be severely limited.

century as advancements in medical techniques reduced infant mortality and disease. This happened almost simultaneously with the shift from primarily agricultural to technological economies in the scientific industrial revolution. Cities grew as fewer people were needed to run the farms; but the population growth rate declined because children were no longer needed to help support the family and were also more costly and difficult to raise in urban situations. These influences resulted in decreased birth and death rates, causing a **demographic transition** (Figure 16.2). Most nations experiencing this transition had relatively stable governments and public confidence in the government's ability to perform its services.

While the demographic transition was occurring in Europe and North America, the spread of advanced medical techniques reduced the death rate in less developed countries; however, the high birth rate continued (Figure 16.3). Most of these countries were primarily agricultural nations centered on the family unit. The more children a family had, the more food it could produce to sustain the family during the winter. When the death rate was high, the birth rate by necessity had to be high. Even though medical practices introduced new forms of birth control, local customs, religion, and the lack of education often prevented these people from receiving their full benefit. Many people had little or no confidence in the government's ability to provide services, so a large family served as the means of support in old age. This process continues today in less developed countries.

Expanding populations sometimes cause international conflicts. Approximately 17 percent of the people in Honduras came from El Salvador, creating a tension between the two nations that periodically flares into war. Population pressure is partly responsible for the hostilities between some Asiatic and African nations. Illegal immigration from Mexico into the United States causes periodic tension between the two countries.

The distinction between the haves and the have-nots continues to grow: developed countries receive more and more food and material wealth, while less developed countries continue to have more and more children. This disparity does not lend itself to an easy solution. We see the present world growth rate leading to a population of 8 billion people by the twenty-first century, with more people being born in less developed countries that are already overcrowded and unable to support themselves.

In modern industrial countries, other subtle factors affect population fluctuations. The so-called Madison Avenue effect influences the types of lives people lead. By changing expectations and offering new opportunities to spend money, businesses redirect people's goals from raising families to seeking status and material goods.

In 1974, the first World Population Conference was held in Bucharest, Romania. Industrial

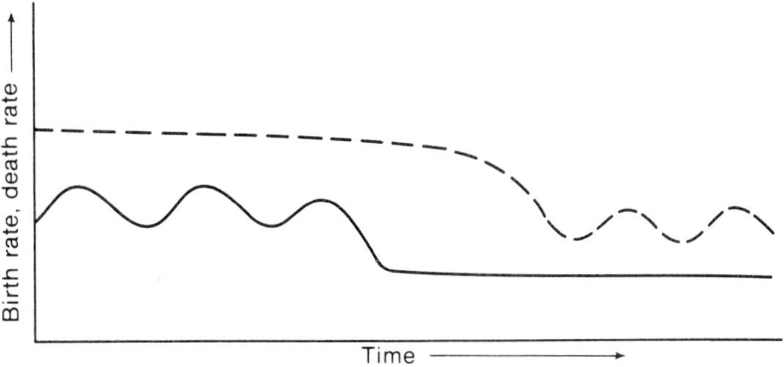

Figure 16.2
The demographic transition: first, death rate (solid line) declined; later, birth rate (dashed line) declined as attitudes about having children changed.

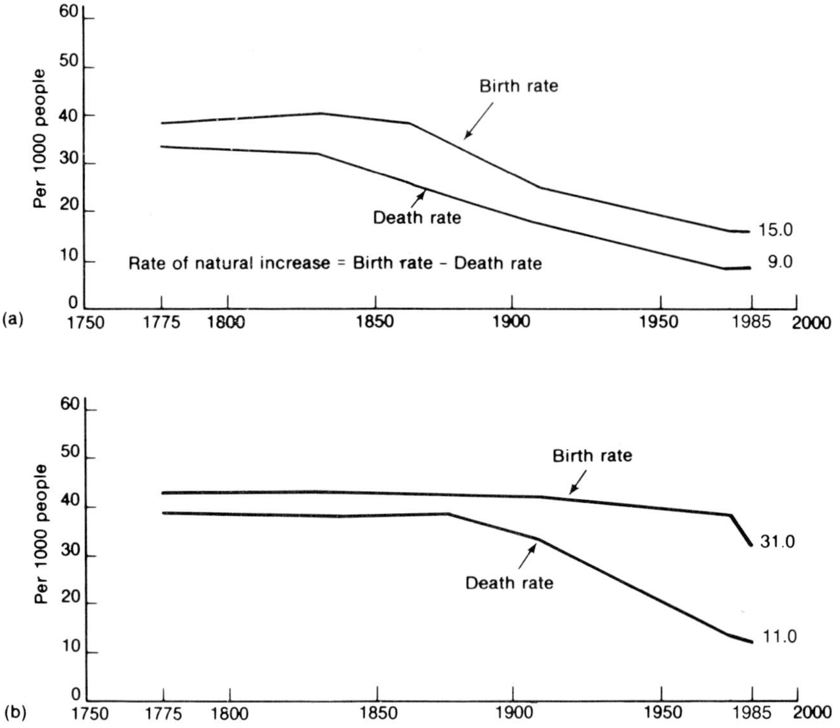

Figure 16.3
World birth and death rates per 1000 people in (a) developed countries in Europe and (b) less developed countries in Asia between 1775 and 1990. (Modified graph. Original graph courtesy of the Population Reference Bureau, Washington, D.C.)

nations pushed for an increase in family planning in less developed countries. Less developed countries felt that social and economic growth were the keys to slowing population growth.

A major change of opinion on the population question surfaced at the U.N. International Conference on Population in Mexico City in 1984. The African nations with their massive hunger problems were trying to institute family planning. China had changed its earlier motion advocating two children per couple to one child per couple. Some less developed countries began to realize they must actively slow population growth. This awareness has been an important step toward population control.

In the United States, however, top political leaders expressed concern about low birth rates in the mid-1980s. They believed a decline in birth rate would result in an economic slowdown. Thus, while less developed countries concentrated on controlling population growth, some people in the United States encouraged it.

HUMAN DEMOGRAPHY

Demography is the science of vital statistics of populations. The data from various demographic studies show us how populations grow and help us determine future trends. Once aware of what to expect, we can predict future population growth patterns.

Doubling Time

The world population growth rate of about 2 percent per year can be deceptive. An examination of **doubling time** gives a better idea of the real impact of growth (Figure 16.4). At a growth rate of 2 percent, the world population doubles every 35 years. However, many developed countries, including Europe, the Soviet Union, Japan, the United States, and Canada, grow at the rate of 1 percent or less (Table 16.1). This means that none of these populations will double in less than 70 years. Most less developed countries have growth rates of 2, 3, and sometimes 4 percent. Projecting the doubling time for these rates can be compared with the methods used for compounding interest in a savings account.

Consider the United States. The 1991 figure for **natural increase** (number of births more than deaths) is 0.8 percent. Most people interpret the figure as almost zero population growth. In reality, 0.8 percent means a doubling time of 88 years.

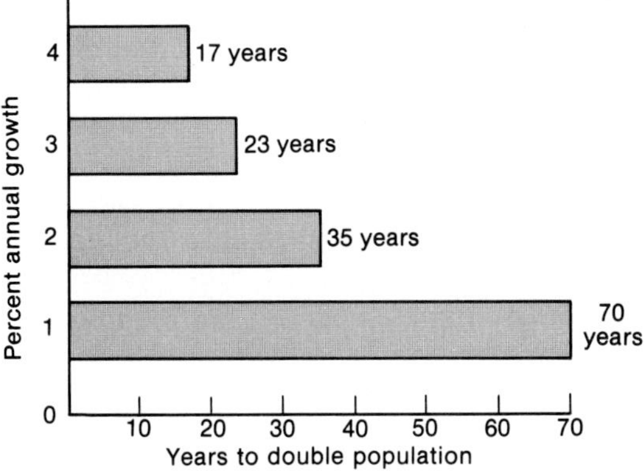

Figure 16.4
A seemingly small increase in the rate of annual population growth can lead to rapid population increase because of its effect on doubling time.

Table 16.1
Selected Annual Population Growth Rates, 1990

Country	Annual Growth Rate (percent)
World	1.8
Algeria	3.1
Bangladesh	2.5
China	1.4
Costa Rica	2.5
Cyprus	1.0
Denmark	0.0
Dominican Republic	2.5
England	0.2
Kuwait	2.5
Laos	2.5
Syria	3.8
United States	0.8
Venezuela	2.3

Fertility Rate

Another way of examining population growth is to compare **fertility rate** (the number of offspring produced) with **fecundity** (the ability to reproduce). Fecundity should be the same for the same number of people anywhere in the world. Fertility, however, differs among the nations of the world and among social classes within nations.

In recent years, the fertility rate in the United States and some other developed countries has decreased in response to changing social attitudes. Fluctuations in the fertility rate reflect the changing pattern of life in any country, making it important to examine population data at intervals of 5, 10, and 25 years.

Too often newspaper accounts view the change in the number of births as a long-term trend, even though it is usually based on 1 year's data. Others hail any brief decline in the fertility rate as the end of the population explosion. When a large number of girls in a population are below the age of 15, as a group they will produce more children when they reach childbearing age, even though individual contributions might be small. Thus, while replacement level fertility (or below) may be reached, a population increase is still likely for some time. It is unreasonable to assume that fertility will remain stable for extended periods as this is an unusual occurrence. Age structure of the population is thus more significant for predicting future population growth than other numbers.

Impetus for population growth comes from the younger members of the population. By calculating the *total fertility*, or the average number of children a woman bears up to age 49, we can determine the number of births required per female for the population to replace itself. A fertility rate of 2.1 births per female is considered the replacement value for the United States. Infant mortality and childless adults are accounted for in this ratio.

Age Distribution Pyramids

Population distribution by age and by sex provides information about the current status of a nation and its potential growth. The age distribution of a country is easily studied through the use of **age–sex pyramids,** which reflect birth and death rates and also dispersal to different political subdivisions.

Three representative pyramids are illustrated in Figure 16.5. As a result of a high birth rate, there are large numbers of children in Mexico's population, giving the pyramid a large base. As the death rate falls, greater numbers of infants and children survive to the reproductive ages, further increasing the base of the pyramid by the number of children they bear. The small number of elderly persons in Mexico reflects the previ-

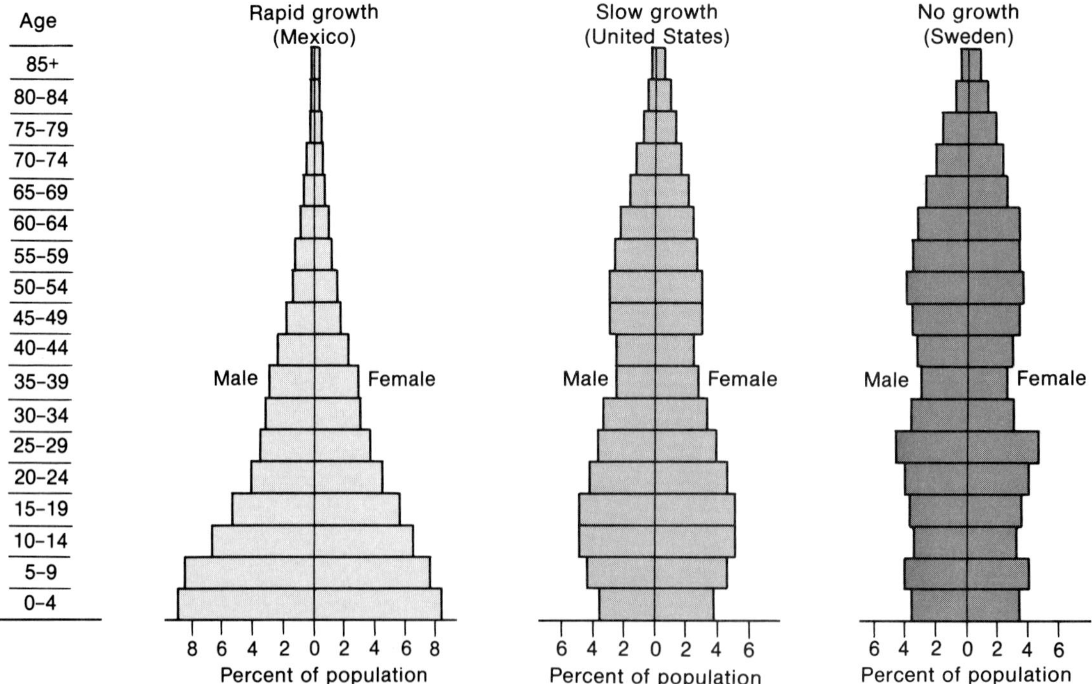

Figure 16.5
Examples of age–sex population pyramids in a rapidly growing, a slow growing, and a no-growth nation. (Data from Population Reference Bureau, Washington, D.C.)

ously high death rate. This pattern of rapid population growth due to high fertility and declining mortality is typical of most developing countries.

The age–sex pyramid for Sweden, in which the population is fairly evenly distributed among all age groups, is representative of a highly industrialized society. Sweden's "old" population reflects an extended period of low birth and death rates. This means that while fewer children are born, most of those born survive to reach old age. This is an example of a country approaching no-growth or zero population growth.

The United States has had relatively high birth rates and has been considered a "young" population compared with most European countries. In the past century, however, the population has been growing "older" due to the long-term downward trend of the birth rate. This trend was interrupted by the post–World War II baby boom, which has significantly affected the nation's age structure. Women of the baby boom generation are now adults in their prime years of childbearing. Assuming there is no significant increase in the number of women of childbearing age, the population is expected to grow moderately for the next 60 to 70 years despite an overall decline in the birth rate.

An excessive number of either children or retired people places an added burden on the working population. In Mexico, 42 percent of the population is under 15 and 4 percent over 65 years of age (Figure 16.6). This means that nearly one-half of the population in Mexico is either too young or too old to work. In Sweden, 64 percent of the population is in the working class.

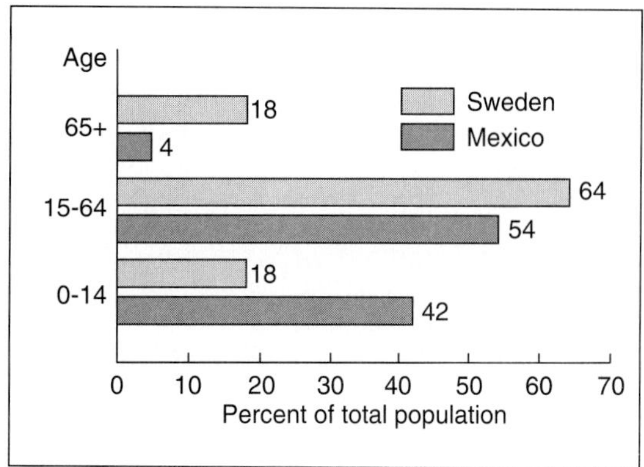

Figure 16.6
Population distribution by age categories in Mexico and Sweden. (Population Reference Bureau.)

Life Tables

Human population growth over the last 200 years is attributed to a decline in the death rate—not to an increase in the birth rate. The average individual in a less developed country is expected to live to the age of 50, while a person living in a developed country has a life expectancy of 73 years.

Death rates are commonly examined by means of a **life table.** A life table starts with a significant number of individuals (usually 1000) and displays their survival pattern throughout the life span of that group. They are used by population biologists to study big game animals, by entomologists to study the life patterns of insects, and by life insurance companies to estimate the risks involved when insuring people. There are two forms of life tables, **static** and **cohort.** The static life table is most commonly used by demographers studying human populations to determine the number of deaths in each age category during a particular year. From such data, they can make predictions about the life expectancy of a population (Table 16.2). Cohort

Table 16.2
Static Life Table for Israel

	Expected Remaining Years of Life	
Age	Males	Females
0	69.55	72.96
25	48.02	50.38
50	25.04	26.67
70	10.73	11.45
85	5.98	4.80

life tables are more useful but are more difficult to compile. Data are gathered by following a specific group of individuals (cohort) from birth to death and noting the survival rate within each age category.

Birth and Death Rates

A **crude birth rate** is obtained by dividing the number of live births by the population size in a particular area. The **crude death rate,** or mortality rate, is computed by dividing the number of deaths by the number of individuals in an area. These figures are generally calculated on the basis of 1000 individuals. The data in Table 16.3 show that Europe and most of North America have birth rates well below 20 per 1000 individuals. The world birth rate is 27 per 1000 people. Latin American countries are generally above 30, while some African and Asian countries are in the high 40s. These figures reflect population growth patterns throughout history (see Figure 16.3).

Table 16.3
Crude Birth and Death Rates for Selected Nations (1990)

Country	Birth Rate	Death Rate
	(per 1000 people)	
World	27	10
Algeria	40	9
Botswana	40	11
China	21	7
Costa Rica	29	4
Cyprus	19	9
Denmark	12	12
Dominican Republic	31	7
England	14	12
Ethiopia	44	24
Iraq	46	7
Syria	45	7
United States	16	9
Venezuela	28	5

Migration

Measurement of population growth depends not only on natality and mortality rates, but also on migration. People tend to move to new areas or nations to seek freedom or a better quality of life. In addition to influencing population size, mass movement more subtly changes the entire character of the population because the young and more adventurous individuals are more likely to emigrate. Single males in their early reproductive years are the most common emigrants.

The United States has traditionally been a nation of immigrants, admitting a greater number than any other nation in the world. This high influx of people continues to contribut significantly to the changes in its population size (Figure 16.7).

In 1970, illegal immigration accounted for 8 percent of the population increase in this country. By 1990, the figure swelled to more than 33 percent (Figure 16.8). While most Americans favor an all-out effort to stop illegal immigration, little has been done to correct the situation. Federal budget cuts and unanticipated influxes of political refugees have often forced the Immigration and Naturalization Service to shift priorities and allocate funds elsewhere. Aside from being a major contributor to population growth, illegal aliens have altered the social structure in some parts of the United States. They have become, in effect, an underground society. By taking jobs usually held by poor Americans, illegal aliens have created tensions among themselves, resident aliens, and U.S. citizens. Furthermore, our country's sporadic enforcement of the law has resulted not only in a disrespect for our regulations, but also has enabled profiteers to exploit illegal aliens who cannot seek protection for fear of exposing their unauthorized presence in this country.

Economic conditions have had a major impact on immigration. The trend to move out of Sweden was reversed in the 1930s as technology provided more jobs for the many people coming

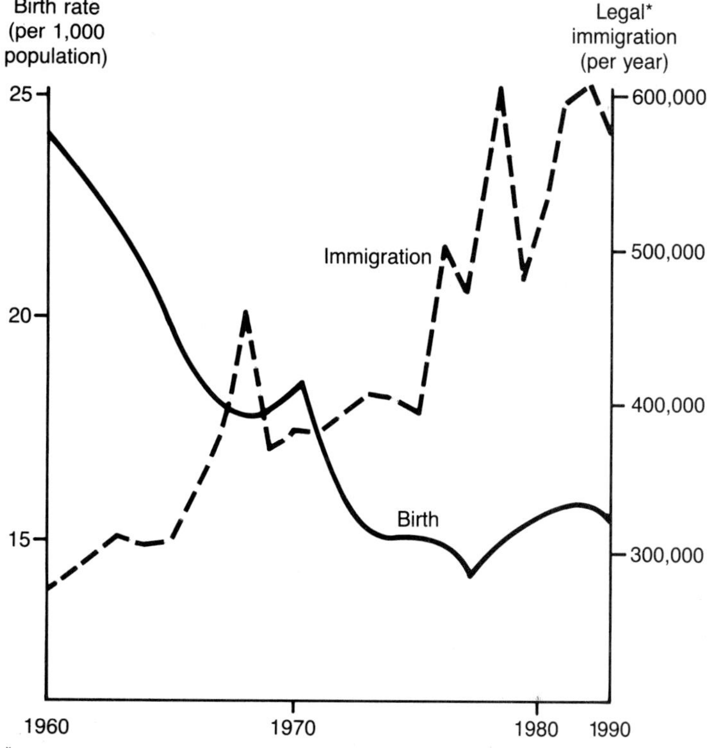

Figure 16.7
Changes in immigration rate now have a greater impact on population growth in the United States.

from rural areas. A lack of food in Ireland caused a mass emigration to the United States between 1851 and 1871. Ireland would otherwise have had close to 12 million people today instead of its present 3 million.

Most nations, with the exception of Israel and Australia, have stopped seeking immigrants, and many nations like the United States have specifically controlled their numbers. Immigration quotas are one measure that some countries have used to alleviate potential overcrowding. Quotas are outdated and often ignored, however, and do not consider illegal immigrants.

Population ebb and flow within a country can result in different living patterns. The 1990 census in the United States showed major population shifts during the preceding decade. Populations in the East and Midwest declined slightly, indicating an exodus from those areas; whereas the southern mountain and western states had large increases.

Young people seeking employment and retired persons are most likely to relocate. This redistribution of people creates a major impact on local economies. Local governments in the East and Midwest are finding it difficult to maintain all established services; the southern and mountain states are incapable of providing schools and other basic services at the rate necessary to meet the increased demand.

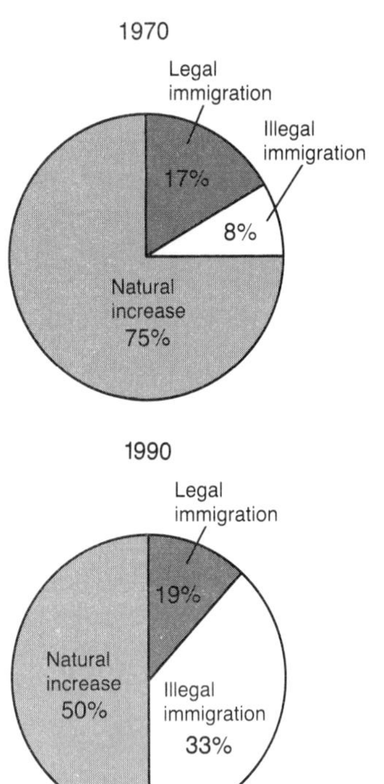

Figure 16.8
A substantial increase in illegal immigration between 1970 and 1990 contributed to population growth in the United States.

STRESS AND DISTRESS IN MODERN SOCIETY

Everyone needs a certain amount of stress to function soundly. In fact, most of us perform best under some stress; but the proper amount varies considerably from person to person. Distress occurs when we feel we cannot cope with our problems. We need to gain insight into some of the ways we are placed under stress, the effects it has on our bodies and minds, some unhealthy ways of handling distress, and ways of relieving distress.

Our performance is governed by our biological past, the environment in which we live, and our internal physiological state. These three factors are often interrelated since change in one causes changes in others. Technology and the pace of life are parts of our environment that greatly modify performance.

Environmental Influences

While past environmental stimuli partly influence performance, many everyday factors such as landscape, color, buildings, and proximity to other people also play important roles. For example, the color green is considered an important environmental stimulus because it is thought to create a relaxed mood. Color together with form, shape, and feeling of elbow room contribute to a pleasant drive in the country. Color, shape, and odor affect people as they walk from a city park into an overcrowded slum of old homes and drab colors. The change in environmental stimuli may be felt in the body as a mild state of depression. Consider the effects on the millions of people who live in slums and ghettos without the benefit of trees, grass, flowers, or shrubs (Figure 16.9).

Some people become depressed when they find that the goals they are seeking cannot be attained. Thus, it is common to find poor people from rural areas moving to the city to find employment and a better life. Instead, they find that jobs are difficult to obtain with their limited skills, and they are forced to live in low-cost housing. This may lead to a mental state of apathy, depression, or a lack of concern about their surroundings and cause them to lose their drive to find work or to escape to alcohol and drugs. The environment, in effect, produces negative stimuli. Psychologists find that they can treat these people by interesting them in projects with people in similar situations. When they begin to help themselves, the apathetic cycle is broken.

Rapid changes also upset people. Returning to a favorite campsite and finding a factory complex instead has a disruptive effect. When a street changes from two-way to one-way traffic, it requires time and energy for us to adapt. The problems of energy limitations and possible de-

Figure 16.9
A city slum section. (U.S. Department of Health and Human Services.)

ficiencies by the twenty-first century are worrisome and difficult to comprehend. When new technology appears, waves of cultural changes occur, altering our relation with the natural environment, our health and educational systems, economics, political institutions, and religion. Consider the effects television would have on a less developed country if all the people would have access to it. How would it affect performance?

The continually changing environment makes the occurrence of new situations routine and creates a constant state of stress and distress. Crime in the streets, speed on the interstate highways, sounds of the city, competition in school, and demands from peers, parents, and society—all create stress.

Population redistribution figures indicate that people are trying to escape the crowded conditions, stress, and pollution of urban areas. As a result, the less populated mountain states are experiencing the greatest growth rate. People are willing to accept a somewhat cooler climate such as that found in Oregon, Idaho, and Alaska in exchange for cleaner air, less crime, and a relief from the stress inherent in urban life.

Physiological Effects

As a result of stress from the environment, many physiological changes take place in the body. The human nervous system responds by increasing the glucose level in the blood, dilating eye pupils, altering the digestive pattern, and so on. The purpose of these reactions in animals is to preserve life through the fight or flight response. If these responses are not completed, serious problems arise. Excessive stress in humans is thought to be at least partially responsible for heart attacks, hypertension, overeating, vitamin deficiencies, and lowered body resistance to some diseases. Inability to respond leads to distress, which lengthens reaction time. For example, if a person is distressed, he or she will need a longer time to brake a car after spotting danger. Distress can also trigger abnormal behavior patterns and mental diseases.

Chemical Crutches Stimuli like drugs, tobacco, and alcohol all create physiological changes which in turn cause performance changes. Most families have a supply of drugs for different ailments. These drugs are used for quick relief of the symptoms of distress such as insomnia, depression, nervousness, and worry. Overuse or improper use of most of these drugs can lead to changes in behavior and health and, in extreme cases, death.

The term *drug* is most commonly associated with chemical products that alter the neurophysiological functions of the body and result in a change in thoughts, feelings, or behavior. The effects of drugs are therefore both biological and psychological. The amount, properties, mode of use, and frequency of use interact with the user's expectations, personality, and physiology.

People taking drugs to cope with problems or for health reasons can develop a tolerance to the chemical. **Tolerance** refers to a state in which the body becomes accustomed to the presence of a drug in given amounts and eventually fails to respond to ordinarily effective doses. Hence, increasingly larger doses are necessary to produce the desired effect. When individuals are psychologically dependent on a drug for a sense of well-being, we say that they have a **drug habit.** Others are physically dependent, or **addicted,** so they cannot function normally without repeatedly using the drug.

The purpose of the Federal Controlled Substance Act is to minimize the amounts of drugs available to persons who are prone to abuse them. The five categories of controlled substances listed in Table 16.4 are based on psychological and pharmacological considerations. Most drugs are taken to change some emotion or drive, and they alter the body's behavior patterns by affecting the nervous system. In effect, drugs alter one's internal environment.

The term **narcotic** originally referred to a variety of substances which induced an altered state of consciousness. In current usage, narcotic means opium, its derivatives, or synthetic substitutes that produce tolerance and dependence. Narcotics reduce pain and frequently result in a state of euphoria. Individuals taking narcotics develop both a physiological and a psychological dependence. Unfortunately, those addicted to these drugs can develop serious diseases because they use unsterile needles or purchase contaminated narcotics.

Depressants include some of the common tranquilizers and sleep-inducing drugs. They produce effects similar to those of alcohol, and the user can become both psychologically and physiologically dependent. In large doses, a temporary state of euphoria along with mood depression and apathy can occur. Alcohol and other depressants such as barbiturates, if taken together, have a synergistic depressant effect that can result in death.

Stimulants are mostly used to suppress appetites, increase alertness, increase tension, and cause arousal of the nervous system. Some stimulants are called "pep" pills because they create a sense of well-being. Most can also cause psychological dependence. Caffeine is also considered a stimulant. Individual body reactions to stimulants like cocaine differ; likewise, one's physiological response can vary from one occasion to another.

Hallucinogens produce varying changes in sensation and mood depending on the dose. People who use hallucinogens are often unable to distinguish between fact and fantasy. Recurrent use produces tolerance, inviting the use of greater amounts. Perhaps the greatest hazard with hallucinogens is their unpredictable effects each time they are taken. Toxic reactions can cause permanent personality changes and even death.

Marijuana is placed in a separate group called *cannabis.* Although we do not know the degree of psychological dependence it produces, people who use marijuana do develop a tolerance. Clinical evidence concerning the effects of this drug is unclear and short-term studies are inconclusive. Scientists feel that more long-term studies are needed to document effects, particularly on young people.

Tobacco, a legacy from the North American Indian culture, contains the chemical **nicotine.** Nicotine causes reactions which mimic normal nervous excitation. The force of the heart contraction, respiration rate, and the levels of fatty acids in the bloodstream increase, all of which put a stress on the heart. When smokers deprive themselves of cigarettes, they experience feelings of drowsiness, headaches, nervousness, and digestive upsets; but addiction to smoking appears to be more psychological than physiological. Programs that approach the problem of quitting smoking by establishing the correct mental setting and encouraging a proper diet are quite successful.

Table 16.4
Selected Controlled Substances and Their Effects

Drugs	Some Brand Names	Possible Effects	Effects of Overdose
NARCOTICS			
Opium	Dover's Powder, Paregoric	Euphoria, drowsiness, respiratory depression, constricted pupils, nausea	Slow and shallow breathing, clammy skin, convulsions, coma, possible death
Morphine	Morphine		
Codeine	Codeine		
Heroin	None		
Meperidine	Demerol, Pethadol		
Methadone	Dolophine, Methadose		
DEPRESSANTS			
Chloral hydrate	Noctec, Somnos	Slurred speech, disorientation, drunken behavior without odor of alcohol	Shallow respiration, cold and clammy skin, dilated pupils, weak and rapid pulse, coma, possible death
Barbiturates	Amytal, Seconal, Tuinal		
Glutethimide	Doriden		
Methaqualone	Parest, Sopor, Quaalude		
Tranquilizers	Librium, Valium, Miltown		
PCP (see Hallucinogens)			
STIMULANTS			
Cocaine	Cocaine	Increased alertness, excitation, euphoria, loss of appetite, increased blood pressure	Agitation, hallucinations, convulsions, possible death
Amphetamines	Benzedrine, Dexedrine		
Phenmetrazine	Preludin		
Methylphenidate	Ritalin		
HALLUCINOGENS			
LSD	None	Illusions and hallucinations (except with MDA), poor perception of time and distance	Longer, more intense "trip" episodes, psychosis, possible death
Mescaline	None		
Psilocybin-Psilocyn	None		
MDA	None		
PCP	Sernylan		
CANNABIS			
Marijuana	None	Euphoria, relaxed inhibitions, increased appetite, disoriented behavior	Fatigue, paranoia, possible psychosis
Hashish	None		
Hashish oil	None		

Tobacco, alcohol, and nonprescription drugs are correlated with abnormal development of the human fetus. Investigations of the relation of smoking to pregnancy show that women who smoke during pregnancy have babies of lower than average weight. **Alcohol** appears to have its greatest influence during the last 3 months of pregnancy. During this period, alcohol concentrates in the fetus up to ten times the level in the mother, primarily in the brain and liver of the baby. Research on the effects of drugs shows conflicting results. Studies on LSD, one of the hallucinogens, show that it causes chromosomal breaks in the fetus of some mothers who use the drug. Some children born to heroin addicts experience withdrawal symptoms.

REPRODUCTION

Human sexual reproduction results from the union of a sperm and an egg to form a zygote.

The elaborate reproductive process involves stimulation, hormone secretion, and most important, the proper environment for production of the reproductive cells and development of the new organism. Birth control measures disrupt the process. In terms of evolution, the reproductive drive has been made a pleasurable process to ensure survival of the species.

Sperm production occurs at a temperature slightly below normal body temperature. Males have two testicles suspended in a sac between the legs to maintain this lower temperature. Holding testes too close to the body wall (for example, by wearing tight pants) can interfere with sperm production by increasing temperature. Each testicle has about 800 tubules which together are about half a mile long. The walls of these tubules have cells with paired chromosomes (see Chapter 3), which divide to form cells that contain only one member of each chromosome pair. These cells mature into sperm. The tissue space between the tubules is the area where **testosterone,** the male sex hormone, is produced. Controlled by other chemicals released by the pituitary gland, testosterone stimulates growth of the testes, penis, larynx, and body and facial hair and generally contributes to "maleness," including the male sex drive. If testosterone is not produced, puberty does not occur; if testosterone declines after puberty, more fat may appear, although the sex drive is usually not lost.

Part of the female reproductive organs are the two ovaries. Each is a mass of connective tissue (*stroma*) surrounded by a thin layer, the *epithelium.* While an unlimited number of sperm are produced in the male, the female is born with about 400,000 immature egg cells located under the epithelium. Usually, fewer than 400 egg cells mature in the female's lifetime. After a woman reaches sexual maturity, one or more of the immature cells are released about every 28 days. The cells undergo reproductive division resulting in an egg with one member of each pair of chromosomes. The egg is attached to the ovary and appears as a small bulge—the *Graafian follicle*—which is surrounded by several layers of cells. The follicle ruptures in the process of *ovulation* and releases the egg. The follicle cells release the female sex hormones **estrogen** and **progesterone** before and after ovulation. The Graafian follicle becomes the *Corpus luteum* after the ovum is released. After the ovum is released from the ovary, it is caught by the ciliated end of one of the fallopian tubes, which lead to the uterus.

Estrogen is the hormone that causes the uterus and other female reproductive parts to mature and function. It produces some "female" characteristics such as development of breasts and subcutaneous fat deposits, and prevents growth of facial hair. After puberty, estrogen secretions oscillate on about a monthly cycle. As the ova mature, estrogen secretions increase, preparing the uterus to accept a fertilized egg. Progesterone secretions increase immediately before release of the ovum, apparently triggering ovulation. If the egg is not fertilized, the hormonal secretion from that group of follicles slows or ceases at the end of the month, causing the sloughing off of the uterine wall (menstrual bleeding).

If the ovum is fertilized, it forms the zygote. The zygote begins to divide and forms an intimate connection with the mother's blood. At this time it is called an *embryo.* During the early few months of pregnancy when cells divide rapidly, they are most susceptible to mutations caused by radiation and toxic chemicals.

The uterine cycles are in response to the ovarian hormone production of estrogen and progesterone (Figure 16.10). Production of these hormones is, in turn, regulated by feedback relations from the pituitary gland. During the early days of the menstrual cycle, **follicle stimulating hormone (FSH)** from the pituitary causes a follicle with a developing egg to grow. (Fertility drugs usually consist of large amounts of FSH.) Estrogen from the follicle then stimulates the uterus to prepare for the ovum. About

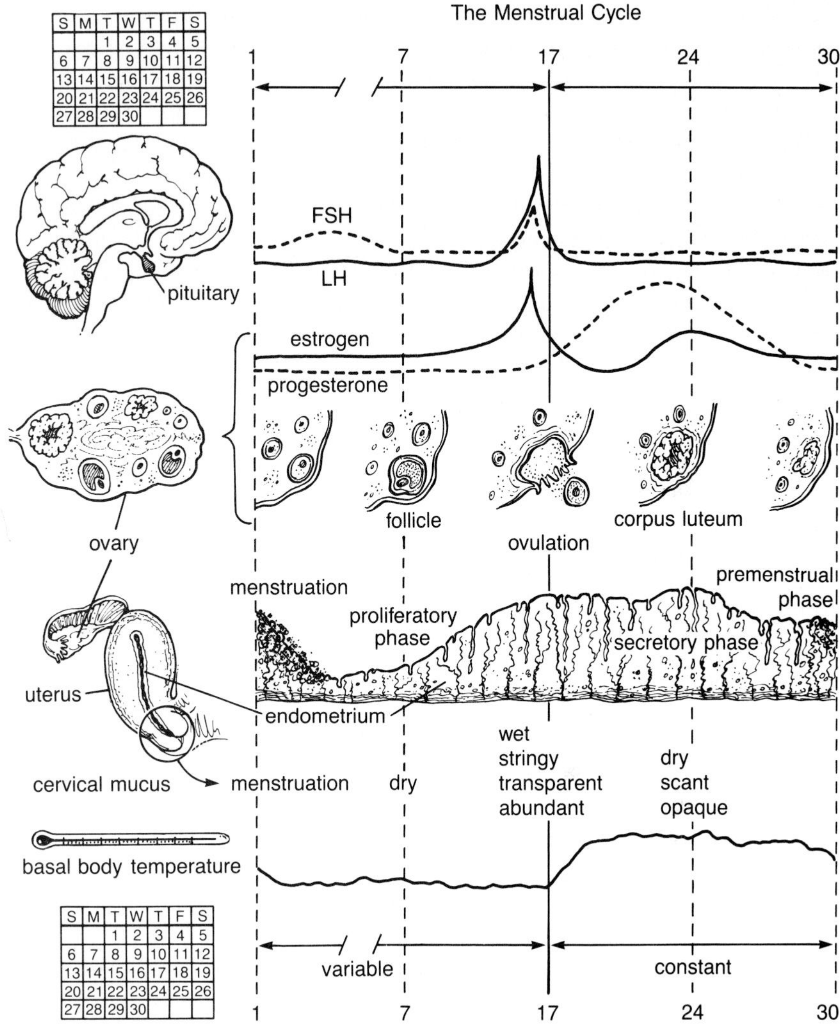

Figure 16.10
Female reproductive cycle.

midway through the average 28-day menstrual cycle, the pituitary releases a surge of **luteinizing hormone (LH).** LH creates ovulation changes in the follicle and release of progesterone, and continues to stimulate the postovulation follicle to develop into the corpus luteum. The corpus luteum secretes estrogen and increasing amounts of progesterone. Progesterone during this time acts to prepare the uterus for implantation of the zygote.

When pregnancy does not occur, the increased level of progesterone causes the LH to decrease, resulting in degeneration of the corpus luteum, whose hormone output then decreases. If pregnancy occurs, the developing embryo produces a hormone that replaces LH and keeps the corpus luteum active. The corpus luteum in turn produces progesterone that keeps the uterine lining intact.

POPULATION CONTROL

Reduction of world population growth must involve either a decrease in the birth rate or an increase in the death rate. In 1798, Reverend Thomas Robert Malthus published *An Essay on the Principle of Population as It Affects the Future Improvement of Society with Remarks on the Speculation of Mr. Godwin M. Condorcet and Other Writers.* In this book, Malthus expressed concern about human population growth and made two claims: "First, food was necessary to the existence of man. Secondly, the passion between the sexes was necessary and would remain nearly in its present state." He then stated, "Assuming my postula as granted, I say that the power of human population is indefinitely greater than the power in the earth to produce substance for man." Malthus claimed that the population increased in a geometric fashion while food production increased in an arithmetic manner. Human population followed a law of nature that, unless deliberately checked, tended to go on expanding until it reached its limit of subsistence. After that, the population was kept in check by starvation, disease, pestilence, and war.

While the ideas of Malthus have been debated for many years and the economic and social systems of the world have changed, we have found some points quite valid. He was the first individual to call attention to the phenomenal potential of human population growth. He concluded that the population cannot continue to grow forever because there is a limit to the supply of food the earth can provide.

Malthus suggested postponement of marriage, assuming intercourse would not occur before marriage. He did not suggest any form of restraint once a couple was married and denounced any physical device for limiting the family. In the latter half of the nineteenth century, a neo-Malthusian movement in England and several other countries in Europe advocated physical means of preventing conception, primarily because children added to a family's economic burden. The movement had become almost worldwide by the end of the nineteenth century. The formal birth control movement was slow to develop in the United States, but became established in 1917 with the founding of the National Birth Control League. There was interest earlier, but the federal Comstock Law passed in 1873 and similar laws subsequently passed in many states prohibited early birth control promoters from distributing contraceptive information through the mails. Some of these laws still exist today.

One of the leading advocates of birth control in the United States was Margaret Sanger, a nurse in the poor section of New York's lower east side. After witnessing the results of hundreds of improperly induced abortions among poor families, Sanger became convinced by 1912 that some sort of birth control had to be made available. In 1916, she opened a birth control clinic and shortly thereafter served a 39-day jail sentence for violating various laws against distribution of birth control literature. As she continued her fight to provide this information to the public, she was constantly harassed by opponents to her birth control methods. At one meeting of the first birth control conference in 1921, an archbishop ordered the New York police to close the lecture hall. As this was done without the approval of the chief of police or the mayor, it created the reverse effect intended by the archbishop, giving birth control free advertisement in the news media.

Methods of Population Control

Family Planning Family planning is considered a form of population control when the family unit is consciously making an effort to adhere to population policies set by a larger social unit like a nation. Such efforts involve methods used by the family to space and limit the number of children. The development of family planning in the United States is frequently hindered by people's overreactions, ignorance, culture, and bias.

The use of birth control devices began in earnest in the less developed countries with the formation of the International Planned Parenthood Federation in the 1950s. The objective of this organization was to disseminate information about birth control methods to the millions in these countries. The World Bank began to aid less developed countries with family planning in the 1960s. Its concern was not with ultimate numbers of people but with the standard of living that could be reached by the people. By 1975, family planning was established in thirty developing countries. Other countries supported private programs but had no national program (Figure 16.11). Because of differences in ethnic backgrounds, religious influence, and local cultures, there was considerable variation in the nature of the programs. Often illiteracy, rural isolation, lack of funds, and local taboos made it difficult to undertake such projects, particularly for outsiders. By 1976 countries such as Taiwan, Korea, and Singapore had developed successful national family planning programs.

In 1990 most countries had some form of family planning. Its approach and success varied considerably. During the early 1970s, several groups advocating a reduction in population growth became established in the United States. The group Zero Population Growth formed a national organization called "The Voluntary Childless," and formed local organizations to encourage couples with similar interests to get together and exchange ideas. Lobbying for revised tax laws and campaigns to change the image of couples with no children are examples of their efforts to change the attitudes of both government and society.

Childbearing Age The age at which members of a society begin sexual activity is an important determinant of population growth. There is a great range of ages at which men and women

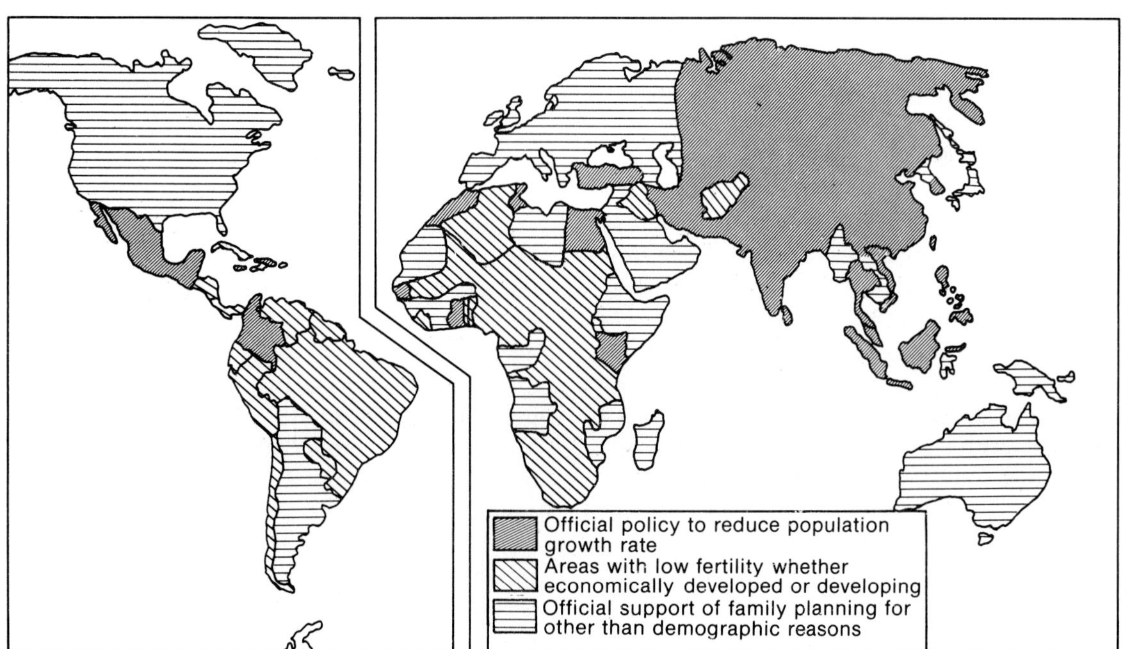

Figure 16.11
Government positions on population growth and family planning among developing countries. (*Population and Family Program.*)

cohabit. The percentage of men and women who live together at the age of 15 to 19 is about 10 percent in Japan, 3 percent in Hong Kong, 5 percent in Singapore, 72 percent in Bangladesh, 55 percent in India, and 6 percent in Nepal. If people want to have children, the availability of modern contraception will not greatly reduce the population growth rate. To raise the age of marriage and the age at which childbearing occurs would require legislation that would be hard to enforce in many societies.

Birth control methods fall into these general categories:

- Abstinence from intercourse
- Suppressing production or release of egg or sperm
- Preventing contact between egg and sperm
- Physically altering reproductive parts to prevent egg or sperm movement (sterilization)
- Changing environment so zygote cannot implant
- Removing an implanted embryo (abortion)

Abstinence Complete abstinence is not generally acceptable, as the sex drive is a strong natural phenomenon. Historically, abstinence was practiced because reproduction was poorly understood and couples did not know how to enjoy sex without risking pregnancy.

Some practice abstinence during the part of the menstrual cycle when the ovum is most likely to be fertilized, usually the several days around ovulation or at the midpoint of the menstrual cycle. This method is not reliable because it is difficult to determine the exact time of ovulation and the length of time the sperm are viable can vary considerably.

Suppression The "pill" is a form of oral contraception that prevents ovulation. Birth control pills contain synthetic estrogen and progesterone; these hormones are released into the body at constant levels and thus suppress the release of FSH and LH. The pill also affects the internal environment by altering the uterine wall in which the zygote would implant. Several other suppressive contraceptives that alter the hormone cycle are available in parts of the world. Injection of long-acting progesterones, used in some 80 countries, has not been approved by the Food and Drug Administration for use in the United States. Pills containing only progesterone (*minipills*) are available but are not as effective as the combined pill in preventing pregnancy.

Oral contraception using the combined pill is the most effective birth control method outside of sterilization or abortion, although not 100 percent effective, as studies in Great Britain have shown. Some women suffer side effects; it is not uncommon for women to take several months to a year to become pregnant after stopping the contraceptive.

Preventing Union of Egg and Sperm Among the methods of preventing egg–sperm contact, withdrawal has been practiced for centuries. In this method, the man withdraws his penis from the woman's vagina before ejaculation. Some people believe that if no sperm are deposited in the vagina, the woman cannot get pregnant; however, sperm deposited on the outer surface of the vagina can wash inward and cause pregnancy.

Mechanical methods include the *condom*, first developed in the late 1800s. Condoms are tubes of rubberized material that fit over the penis and prevent release of sperm into the vagina. Women can be fitted with a diaphragm or rubber dome that is inserted into the vagina and fits over the cervix, the entrance to the uterus. The diaphragm is used with a cream or jelly that kills sperm. Cervical caps are rubber devices placed over the cervix and held in place by suction, and also are used with spermicides.

Sterilization Some couples prefer a form of sterilization when they no longer wish to have children. The male sterilization process is called a *vasectomy* and involves cutting the *vas deferens*, the tube through which the sperm pass from the testes to the penis (Figure 16.12). Vasectomies are relatively easy to perform and can be done in a doctor's office under local anesthesia. Since all of the seminal fluid is present, ejaculation is not impaired, and the sensation experienced in sexual intercourse is not reduced because no muscles, nerves, glands, or hormones are changed. Those men who complain of a change are frequently experiencing a mental adjustment.

Female sterilization by *tubal ligation* involves removing a segment of each *fallopian tube* and cauterizing the ends to prevent the egg from passing into the uterus (Figure 16.13). Tubal ligation by laparascope (a metal tube through which instruments are passed) involves making a tiny abdominal incision and generally takes less than 20 minutes. The woman can frequently leave the hospital 4 to 6 hours after the operation. This method is relatively harmless and inexpensive and therefore increasingly popular.

Environmental Changes An *intrauterine device (IUD)* is placed into the uterus to prevent pregnancy, apparently by causing a slight irritation of the uterine lining that prevents implantation of the zygote. It may also speed movement of the egg so it is not mature when the sperm reaches it.

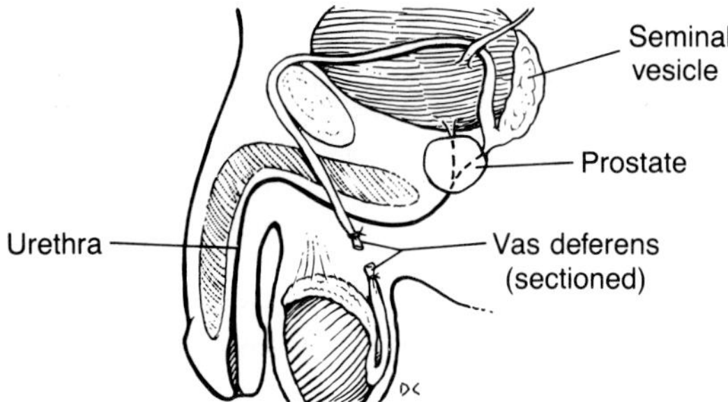

Figure 16.12
Vasectomy is performed by cutting off a small section of the vas deferens (sperm duct) so no sperm can ejaculate during intercourse.

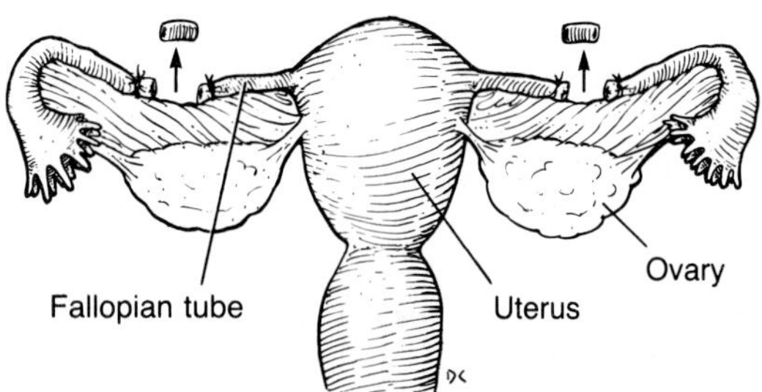

Figure 16.13
Tubal ligation, or female sterilization, involves removing a small portion of each fallopian tube, so the egg cannot reach the uterus.

Abortion Abortion is a method of birth control that removes the fertilized ovum from the site of implantation in the uterus. Much emotional controversy surrounds abortion because many people feel it is a form of murder. Currently, a number of court cases about abortion are pending. The Right to Life groups advocate outlawing abortions. Other individuals feel that each woman should have a right to make the decision.

Historically, abortion has been the world's foremost birth control method. For centuries women have used rocks, sticks, wires, and other objects to perform abortions on themselves. Unfortunately, these methods have frequently caused uncontrolled bleeding and infection, usually resulting in death. Today abortions can be safely performed surgically; however, they are not always available in less developed countries.

New Methods Birth control methods usually have some physiological or emotional drawbacks. Oral contraceptives, for example, are suspected of causing blood clots and cancer and can also cause nausea. Temporary sterility can occur, lasting from months to years after pills are no longer taken. Intrauterine devices (IUDs) can cause excessive bleeding, cramping, and possibly cancer. Because they can become dislodged without the user's knowledge, they need to be checked regularly by medical personnel.

If a particular method is ineffective or improperly used, stress may result from a fear of unwanted pregnancy. Douching and breast feeding are ineffective for most people, and spermicides are unreliable because they are short-acting (Table 16.5). Directions for insertion and removal of the diaphragm with jelly must be followed very carefully as must the guidelines for using temperature rhythm. Other emotional stress may occur from the use of coitus interruptus and calendar rhythm, which requires long periods of abstinence from sexual intercourse.

A variety of new contraceptive methods are currently being developed and tested (Table 16.6). Drugs that can be injected into the body and released over a period of time to prevent pregnancy are now under study. Other research involves the study of drug-loaded microcapsules that can be placed in the vagina and forced to migrate into the uterine cavity. The "morning after" pill is an additional form of contraceptive for women. The "morning after" pill contains high levels of hormonelike chemicals to prevent implantation of the fertilized egg. It is available in many parts of the United States by prescription, but it often causes nausea and discomfort.

Norplant is now available in many parts of the world. A 3 millimeter ($\frac{1}{8}$-inch) incision is made on the upper underarm of women. Six tiny silicone capsules are implanted. Each capsule about 24 millimeters (1 inch) long and about the thickness of a match stick contains synthetic progesterone. The progesterone trickles into the blood stream and finds its way to the pituitary gland, interfering with FSH and LH production. When a woman wants to become pregnant, reverse surgery is necessary to remove the capsules.

Gossypol is now being tested as a male contraceptive in China. Gossypol is a cotton-plant extract that reduces sperm count to nearly 0. Some men have side effects, ranging from nausea to, in rare cases, cardiac arrest. Another problem is that about 10 percent of the men who take gossypol remain sterile after stopping the drug. In Israel, phenoxybenzamine (PBZ) is being tested as an inhibitor of sperm growth. In Denmark a hormone contraceptive that decreases sperm count and suppresses testosterone is being evaluated. An antifertility vaccine for males is also being tested in some countries.

While most of the contraceptive devices mentioned have been available in developed countries, less developed countries could not distribute them uniformly. Education level and local cultures often prevented their proper use. Furthermore, not enough common-sense research was done before their introduction in

Table 16.5
Effectiveness of Contraceptive Methods

Most effective	Oral contraceptives, skin implants
Highly effective	Intrauterine device (IUD); diaphragm with jelly; condom; temperature rhythm
Less effective	Vaginal spermicides; calendar rhythm; coitus interruptus
Least effective	Postcoital douche; breast feeding

Courtesy of the National Institutes of Health, Washington, DC.

Table 16.6
New Forms of Contraceptives

Male Contraceptives	Female Contraceptives
Drugs, both pills and injections	New steroids for injection
Simplified sterilization techniques	Postcoital drugs (for occasional use)
Reversible sterilization techniques	Drugs which regulate the menstrual cycle
	Improved IUDs
	Simplified sterilization techniques

Courtesy of the National Institutes of Health, Washington, DC.

these countries. The Indian government, for example, tried to encourage its people to use the rhythm method in the 1960s. A massive education program was undertaken. Strings of twenty-eight beads, color coded to coincide with the woman's menstrual cycle, were distributed to many families so the couple would know when to abstain from intercourse. Unfortunately, in a land with few electrical lights, couples could not distinguish the bead color in the dark. Other contraceptive techniques were not successful because follow-up treatment was not available. For example, there were frequently no paramedics to check women following the insertion of an intrauterine device. In other cases, men rebelled against the use of condoms because they considered them a challenge to their virility.

ATTITUDES TOWARD BIRTH CONTROL

Many organizations or groups of individuals strongly shape the way we think about population size. Governments, businesses, the medical profession, religious institutions, and social pressures influence our attitudes.

The government of Singapore is successfully reducing the birth rate in its country through many suggestive and coercive techniques. Radio slogans, billboards, and school curriculum all convey the idea that two are enough. The country's tax structure also reinforces the two-child family. The first two children in a family have the privilege of attending the neighborhood school, but additional children are often bused to other schools. Hospital costs increase for each child in that small island country.

In the United States, the government could influence couples' attitudes toward family size by combining education with welfare programs, providing counseling and birth control clinics for low-income families, and using literature depicting happy two-child families. Much of the resistance to government interaction in the area of population control in the United S. caused by private interest groups. Some of feel that the poor are being urged to reduce their family size whereas the weathy can still afford to have larger families. Because more poor people are black, blacks believe their race will be constantly relegated to a minority position, even though most family planning efforts revolve around releasing a family from its poverty status. Opinion polls show that white families desire more children than black families.

The U.S. government subtly encourages large families in a number of ways. Tax deductions for families with children reflect the pioneer society's need to populate new areas. Our free enterprise economy currently depends on an increasing population to promote market growth, so businesses pressure for this through lobbying and legislation. Although the government is often slow to respond to changes in human values and attitudes, recent federal tax laws have eliminated some of the biases favoring married couples.

With the advent of Norplant, more heat is added to the debate over the role of contraception in our society, which is increasingly burdened with costs of accidental pregnancies. The birth rate among teens and poorer people is increasing in the United States. Some groups advocate that birth control devices like Norplant be required after a teenager or sexually active woman has an abortion or baby out of wedlock.

Perhaps the most important influence on human life and history is the church. Many wars and territorial disputes occur in the name of religion, and the everyday and long-term activities of masses of people are shaped by different forms of religion. Because of the tremendous impact the church has on many governments, any major worldwide population control program must consider this important influence.

In the Western world, the Catholic church condemns all "unnatural" methods of birth control. The calendar and temperature rhythm methods are the only ones it considers natural.

In 1968 Pope Paul VI issued the encyclical *Humanae Vitae*, which condemned the use of contraceptives and reinforced the Catholic church's opposition to any form of unnatural birth control. The encyclical resulted from a study by a papal-appointed birth control commission, whose report was reviewed by a commission of sixteen bishops. The scientists on the commission were most adamant in their conclusion that birth control measures must be taken to avoid wholesale starvation, but the bishops overruled their objections. They cited a series of papal decrees beginning with those of Pope Sixtus V in 1588 as the basis for their disapproval. The fact that the papal decrees were issued in the sixteenth century indicated that even at that time people were concerned about limiting the size of their families.

In other religions of the world, we find birth control generally an undiscussed subject. Because many of the Eastern and Near Eastern religions compete with each other for followers, they do not advocate birth control because they view it as a detriment to their own well-being. The more followers each group has, the more power it wields. The Hindu religion in India and the surrounding countries is very loosely organized with many local sects. In this case, it is very difficult to advocate birth control without central channels of communication.

FUTURE POPULATION GROWTH PATTERNS

Today, with the world population growing at a rate of 2 percent per year, the age–sex distribution pyramids have broad bases, indicating that there are always more children in the prereproductive stage moving into the reproductive years. This means that if each woman were to have only two children, the population would continue to grow. It would take at least two generations passing through their reproductive years before the population could stabilize. A stable population occurs when births equal deaths. If the number of births in the United States were held constant beginning in 1980, the fertility rate would begin to drop, but the population would increase at least until the year 2050 (Table 16.7).

The United States Bureau of Census projects the population of the world will grow annually at the rate of 1.8 percent until the year 2000. Problems resulting from population increase are not as easy to ascertain. In nations where people are directly dependent on the land for food and life's other essentials, the impact may be more local; whereas in industrial nations, additional population has a much broader impact.

Projections are still difficult to make. Population growth in the United States will depend on the number of immigrants. If the total fertility rate is lowered to 1.7 (each childbearing woman has 1.7 children in her lifetime), and immigration is held to a half million per year, the population will be 278,000,000 in 2030. Figure 16.14 shows what impact changing the immigration and fertility rates can have.

The Global 2000 Report (see Case Study) states that a small percentage of the world's population lives in industrial nations. These people, however, consume 60 percent of the world's re-

Table 16.7
Projected U.S. Female Population at Constant Birth Rate

Year	Total Female Population (millions)	Fertility Rate per Female to Maintain Constant Birth Rate
1995	117	1.96
2000	120	2.01
2005	123	2.06
2015	129	2.11
2025	133	2.11
2035	137	2.11
2045	138	2.11
2055	137	2.11

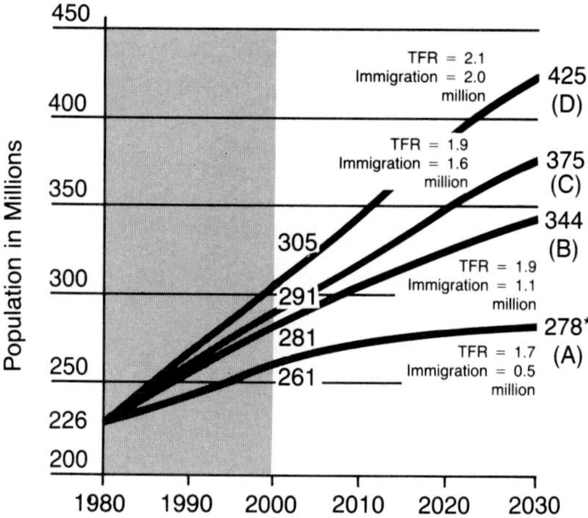

Figure 16.14
U.S. population projections, 1980–2030, based on different immigration numbers and total fertility rate (TFR).

sources. Due to various manufacturing processes, additional resources are lost as waste, producing large landfills and causing soil, water, and air pollution.

The impact of population pressure on people and the environment will continue to increase. Poverty and health problems will become more prevalent and lifestyles will have to change. The only way the human population can maintain equilibrium is to establish a homeostatic relationship with the environment allowing negative feedback to limit human population size just as it limits other life.

In India, an attempt was made to sterilize all males after they had three or more children by providing material incentives, such as transistor radios. This program was not successful partly because medical personnel were not available. Sterilization of all women after their second or third child or the introduction of sterilants into the food and drinking water have been suggested.

Today, the Indian government is experiencing tremendous pressure from the population problem. A program is underway that will deprive all civil servants of some of their benefits, such as housing, if they have more than two children. In other countries, stronger forms of compulsory population control are being enforced. According to Dr. S. Chandrasekhar, the former Indian Minister of State for Health, the People's Republic of China does not allow men to marry until they are 27 and women until they are 25. Jail sentences are imposed on individuals who become sexually active before they marry.

Such actions reduce individual freedom, and some could also alter the population composition by selecting for specific groups of individuals. Other recommendations pose serious social questions. While most of the ideas seem unimaginable, it is necessary to consider the future population and its impact on the environment. As the limits of the earth are approached, we may have few alternatives.

SUMMARY AND CONCLUSION

Major impacts on the human population growth rate have resulted from steps in cultural evolution such as the advent of tools, fire, agriculture, and the scientific industrial revolution. The latter decreased mortality by controlling disease and was followed by an unprecedented growth rate. The demographic transition occurred when the number of births in developed countries was reduced as the nations became urbanized. Less developed countries also had reduced death rates but continued to have high birth rates because the people viewed children as a form of security in old age.

Although no clear approach to population control is emerging, we are exploring various methods. To reduce a country's population growth rate, the birth rate must decrease or the death rate or emigration rate increase. Most approaches seek to reduce the birth rate by employing birth control methods of varying degrees of effectiveness. Sterilization is one possibility for those who do not want more children and for those who wish to remain childless. Changing attitudes can also effectively reduce the birth rate. Industry, the church, the medical profession, and society can learn to view child-bearing from a different perspective. Local educational and medical programs can help introduce workable methods that people could find acceptable.

The exact number of people the world can support is difficult to ascertain. Our models show different numbers based on various input data. However, all models indicate there is a limit to resources, particularly the amount of energy available from the sun and through the food chain, that can be used to support human beings. As the human population continues to grow, personal stress increases, starvation becomes more widespread, and the threat of war seems imminent. Only through sound population control policies can we expect to maintain an environment conducive to society's well-being.

FURTHER READINGS

Barny, G. O. 1979. *The Global 2000 Report to the President of the United States.* New York: Pergamon Press.

Dasmann, R. F. 1984. *Environmental Conservation.* New York: John Wiley.

Jones, E., J. D. Forrest, S. Henshaw, and J. Silverman. 1989. *Pregnancy, Contraception, and Family Planning.* New Haven, CT: Yale Press.

Lele, V., and S. W. Stone. 1989. *Population Pressure, the Environment, and Agricultural Intensification.* Washington, DC: World Bank.

Population Today, a monthly publication of Population Reference Bureau, Washington, DC.

World Resources Institute. 1990. *World Resources.* New York.

STUDY QUESTIONS

1. Explain how marriage postponement can serve as a method of population control.
2. Is family planning synonymous with population control? Explain.
3. Explain the derivation and use of a life table.
4. Discuss various social factors that could decrease the population growth rate.
5. Explain the usefulness of age–sex distribution pyramids.
6. When population control measures are undertaken in one or a few nations without regard to population growth in other nations, what are the possible effects?
7. What are some of the long-term effects on society resulting from the use of chemical crutches?
8. What is your answer to the individual who claims that population control is genocide?
9. What should be the federal government's role in population control? Family planning?
10. Discuss the pros and cons of providing free abortions and/or sterilization to all women who desire them.

SUGGESTED ACTIONS

1. *Doubling Time.* Construct a chart showing how a city with a population of 50,000 will increase in size in the next 25, 50, 100, and 200 years under the following conditions:

Birth Rate (BR)/1000	*Death Rate (DR)/1000*
BR = 2, DR = 1	BR = 4, DR = 1
BR = 3, DR = 1	BR = 7, DR = 1
BR = 6, DR = 1	BR = 8, DR = 1
BR = 9, DR = 1	BR = 10, DR = 1

Now increase the death rate and see what happens in the above examples.

2. *Age of Parents at Birth of First Child.* Draw a chart to show how the country's population might be influenced by the average age of the parents at the birth of their first child. Consider that the average age of death is 73.

   Use the average age of parents when the first child is born as 15, 20, 25, 30, 35, and 40.

17

Where Are We, and Where Do We Go?

In 1750, 3 percent of the world's population lived in urban areas. By 1950, 29 percent of the people were in cities and in 1990 the figure rose to 45 percent. Most current growth is taking place in developing countries. Thus, a figure of 60 percent of the world's people will be living in urban areas by 2025. The level of urbanization varies considerably. In 1990, 72 percent of Latin America was urbanized. In Asia and Africa, the percentage is much lower but is increasing at a very rapid rate.

As our population increases, urbanization will continue. Questions of economics of development, environmental quality, human health, interactions between the natural system and urbanization must be considered. These questions are examined in this final chapter. We then examine how we can plan for the future to achieve an accommodation between humans and our changing world.

DEVELOPMENT OF CITIES

As we look at some of the ancient civilizations, especially in Egypt, Mesopotamia, and China, river valleys and the mouths of rivers (estuaries) seem to have favored the development of human settlements. The estuaries teemed with fish, nearby fertile and level land was available for growing food crops, and the river provided an abundant water supply. In modern times, the early growth of cities such as London, Philadelphia, Baltimore, Charleston, and New Orleans was stimulated by these same features.

The easy grades up the river valley from the coast were suitable for paths, trails, and ultimately roads for access to interior lands. First, the routes were used to distribute the population inland from coastal settlements; later, they became trade routes. The rivers themselves were transportation routes. Construction of canals expanded the inland territory that could be served by waterborne transportation.

Trade

The noted historian Will Durant, realizing the significance of trade to human societies, stated in *The Story of Civilization:*

> Civilization begins in the peasant's hut, but it comes to flower only in towns . . . in that cross fertilization of minds at the crossroads of trade intelligence is sharpened and stimulated to creative powers.

Urban growth and development is indeed stimulated by trade. However, natural resources make some locations more favorable than others. Petroleum deposits, for example, led to the development of oil refineries and petrochemical industries in the Houston, Texas, area. Throughout history gold strikes have created boom towns that became ghost towns when the gold veins were mined out.

Urban Regions

From 1900 to 1970, migration in the United States from rural to urban areas became a consistent trend as farm employment declined and manufacturing jobs in cities increased. There was also movement from the heavily settled East to the sparsely settled West. Not only did the population in previously existing areas increase, but new metropolitan areas became established. By 1920, the peak of the rural-to-urban migration, 50 percent of the population was living in cities. Although the trend has continued, the rate of increase per decade was greater between 1900 and 1930 than in any period since.

The population of the central city as a proportion of the total metropolitan population peaked in the 1930s at 66 percent and declined to 45 percent in the 1980s. In recent years, an outward migration from the center of cities is attributed to such stimuli as electricity and autos and the mobility resulting from these innovations.

In the United States population of nonmetropolitan counties (rural areas and small towns) is growing at a faster rate than that of metropolitan areas. Some metropolitan areas, such as New York, Cleveland, and other areas in the northeast, are losing people. The economy of scale and proximity that favored concentration of industry and business in metropolitan areas may have reached a peak and the economic and environmental costs associated with these aggregations may be outweighing benefits.

The South and West are experiencing the largest growth rate, while the Northeast has negative growth and the north central has minimal growth or decline. In states like Iowa, farm size is increasing as farm properties merge, reducing farm populations. The South and West attracted population as people sought residence in the "sun belt." A large number of illegal aliens are thought to have settled in these areas.

The growth pattern of a particular city shows different phases at different times—a small, almost constant rate in early stages; followed by a rapidly increasing rate of growth; and finally, a declining rate of increase or even an actual decline in numbers. A particular city's population cannot continually expand, however, as there are restraints on unlimited growth. Why do some cities grow and prosper while others stagnate or decline? In the following sections we examine some economic and environmental influences on the growth of cities.

ECONOMIC GROWTH

In developing from a village to a metropolis, urban communities progress through a series of economic stages. We already know some factors leading to settlement and how a village is established to meet the needs of farmers. In isolated and primitive areas, the people on farms were self-sufficient and needed but a few items from the village. But, as cities emerged and offered the special skills of artisans, some activities shifted from the farm to the city: the manufacture of shoes, clothes, plows, fertilizer, and farm equipment, for example. The surplus farm products, in turn, fed the growing populations in cities.

As more of the activities that were performed on farms shifted to the city, more labor was needed in the city and less on the farm. However, farm families still believed that more children meant more labor to tend the farm and more security for the parents in old age—a belief that persists in many less developed countries, contributing to their population growth. Consequently, people living on farms tended to have more children than those in the city. But as the need for human labor on the farm declined as productivity increased, their children sought employment opportunities in the cities.

The recent on-again, off-again pattern of coal development in the Rocky Mountain region has created new cycles of boom and bust in some mining communities. The town of Hanna was established in southern Wyoming in 1899 to provide coal to the Union Pacific Railroad. Hanna's population increased steadily in the early 1900s and temporarily slowed during World War II. In the 1950s railroads shifted from coal to diesel oil, and the mines at Hanna closed. This triggered a bust in Hanna's boom-and-bust economy, and the town's population dropped from 1326 in 1950 to 460 by 1970. As a result of energy shortages in the early 1970s, there were new markets for Hanna coal, and the mines reopened. The economy again began to boom and Hanna grew rapidly; however, a combination of changes in air quality rules and energy markets led to another bust phase as coal mines closed and Hanna's population dropped from 2288 in 1980 to 1433 in 1985.

Farm surpluses produced in the surrounding area can also be the basis for a town's exports and economy, as was the case when wheat was grown in New York State before the middle 1800s. As it became more profitable to grow wheat in the West, the farmers in New York and New England stopped raising wheat. Flour mills then followed wheat farms west. Where a special skill among artisans or the genius of an innovator such as Henry Ford exists, a town might produce specialized products for export. A community might be especially good at making glassware, electronic components, or rubber goods.

However, a community faces certain economic hazards when it becomes the center for manufacturing a single specialized product. As an industry becomes more mature, other communities can acquire the skill or facilities to compete. If the demand for a specialized export decreases, the result is much the same as when a town's mine is exhausted of profitable ore. For example, Lenoir, North Carolina, specializes in furniture manufacture. When times are good in the United States, new families are being formed, and houses are being built, the resulting demand for furniture creates prosperity in Lenoir. But when demand falls, Lenoir's factories are idle, causing unemployment.

As a community develops beyond the manufacture of a specialized product, it enters the stage of growth where it begins to produce a variety of exports. A particularly successful export can lead to the establishment of related satellite industries. For example, iron and steel production might lead to the development of a plant for the fabrication of iron and steel products. If there are flour mills in the community, production of machinery for use in flour mills could begin.

New industries developed in other ways. The plant that makes machinery for flour mills could have the skill and the capacity to make other types of machinery, which might also become export products. Because a variety of machinery can be acquired readily, other new industries may locate in that community.

When a community adds a variety of new industries, it provides both breadth and depth to the local economy and becomes less susceptible to economic obsolescence. If the market for one industry's products is temporarily depressed, the diversity of industry sustains the area's economic vitality. Or, as the market for an old industry's products becomes saturated, the developing new industry keeps the area thriving.

A diversity of goods for export strengthens the economy and provides the impetus for the next stage of growth—the mature city. In the mature city, a wide range of service industries is created to meet the needs of the manufacturing industries. Machines have to be maintained and adjusted, and when they wear out, they need replacement parts. Industries also require utilities: water supply, electricity, and refuse disposal.

Banks, credit companies, and law offices provide financial and other services that are necessary for trade.

Not only do services keep the wheels of industry turning, but they also produce a **multiplier effect;** that is, one new manufacturing job generates consumer demands that lead to the creation of four or five other jobs. For example, the people who work in industries need dry cleaning, shoe repair, medical, grocery, and transportation services. The people in manufacturing and private sector industries who provide these services have additional needs: government services, schools, police and fire protection, street cleaning, and refuse disposal. All new industries and the expanding old ones produce their share of direct and indirect service demands. Many cities approach this mature stage as their populations rise toward 1 million.

When a city develops to the mature stage, its growth is assured. Demands for food and water and similar items continue even in economic slumps, so service industries can help to offset fluctuating business cycles that sometimes occur in manufacturing industries. At the stage where the mature city's population is large enough to provide a market for its industries' products, it might become feasible to produce goods required by local industries and to encourage branches of external industries to locate in the community. The substitution of local products for imported products increases manufacturing activity and creates a need for even more service industries. Because it had reached the mature stage, Los Angeles was able to withstand the curtailment of the aerospace industry in the late 1960s and continued to grow.

A few cities advance beyond the mature stage and begin to function as **regional metropolises.** As such, the city becomes an economic **node,** or the focus of regional economic activity. The neighboring cities that once were rivals now become satellites as the metropolis starts to connect and control local production and economies. Some of these neighboring areas can even be incorporated into the city limits, as were Germantown and other communities in the Philadelphia area. (The earlier identity might persist to designate an area of the city, like Queens and Brooklyn in New York City.) Incorporation of the satellite into the city is not necessary to demonstrate the city's economic dominance. For example, Decatur is very much a part of the Atlanta, Georgia, economy even though it existed as a community before Atlanta. A metropolis's major function is export of services, especially banking, finance and credit, and management services. The export of these services provides an economic stimulus similar to that resulting from the export of goods.

FACTORS INFLUENCING LOCATION

Natural Influences

Natural conditions and sources of raw materials influence a community's location and development. Thus, it seems obvious that arid areas are not apt to be used for agricultural production. But is that so? In the Southwest, irrigation has enabled agricultural production in an otherwise dry region. The area around Los Angeles draws water from the Colorado River for its agricultural and industrial needs, and now the massive Feather River project is bringing water fron northern California to lower portions of the state. From this example we see that while water is essential to regional development, there are ways to overcome a deficiency if an area does not have a readily available source in its own region.

Climate, another natural influence, often favors the development of recreational areas. Areas with warmer weather typically cater to sports such as hiking, fishing, boating, or swimming, but colder areas offering skiing and other winter sports are also attractive spots.

Some natural features of a location could (or should) be a deterrent to development because of certain hazards—flooding, earthquakes, or landslides. In the past, however, natural hazards such as the San Andreas fault in California have not restrained people from building cities. Towns damaged by floods are frequently rebuilt at the same place after the flood waters recede. Today some states are taking a more aggressive approach to controlling development in hazardous locations by adopting zoning laws that regulate land use.

Land use controls also protect some ecologically unique or sensitive natural areas from development. Studies show that even apart from ethical or aesthetic considerations, ecosystems are sometimes more valuable to society when left in their natural state. For example, it was the practice to drain wetlands to make room for urban expansion and subdivision development. But natural wetlands are valuable because of their role in purification of polluted water runoff, temporary storage of potentially damaging floodwaters, erosion control, water supply storage and groundwater recharge, and recreational uses. Wetlands also provide natural products like timber, fish, wildlife, peat, cranberries, blueberries, and wild rice. It has thus become more common to include these ecological considerations in plans for new developments.

The availability of raw materials is an important factor in the initial development of an area, but it can affect its subsequent growth and well-being as well. Plants generating electricity that use oil for fuel and industries that require large amounts of energy or use petroleum as a raw material would be attracted to a location such as the Atlantic seaboard. Many communities in this area are concerned about the future impact from development of deep-water oil ports and exploitation of offshore petroleum resources. While oil spills are a potential pollution problem, a far greater environmental impact could result from industrial development onshore in the vicinity of offshore wells. Petrochemical industries such as plastic manufacturing and oil refining could produce water and air pollution. As other industries locate in the area to secure the advantages of the energy supplies, they will generate additional pollution from autos and from other human activities.

Economic Considerations

If you were the president of a company, where would you locate a new industrial plant? You would locate it where you could be price competitive so you would realize the maximum profits. Of course, the price of raw materials and the cost of labor are significant to the price of the finished product. But if these are roughly equal in two locations, the selection of the site will likely hinge on the cost of transportation. Distance and time are significant factors in transportation costs. Although several cities might have good harbors, railroads, and highways, not all cities are equally accessible to raw materials and markets.

Sometimes we hear of an industry threatening to move if ordered to abate its pollution. The industry will probably find it uneconomical to move unless two conditions exist: (1) the existing plant is obsolete and must be replaced rather than repaired, or (2) transportation, labor, taxes, or other costs favor another location. In most cases, transportation costs will be more significant than any difference in the costs of pollution control.

Transportation also influences decisions about business locations within a region. Where transportation between cities is by water, businesses locate near the waterfront. High-speed rail connections between centers of cities promotes offices in the cities' central areas. Since the center is apt to be the location most readily accessible to the largest number of people, large retail stores also locate there. Nevertheless, if people find it difficult and time consuming to reach the center, satellite shopping areas will develop. Larger cities can have the advantage of cheaper transport and utilities because of the large population served and the economy of large-scale operations. They are also likely to have better education and research facilities and to provide more necessary supportive services for business and industry.

Economic and social inertia favor large cities. That is, where large capital investments are required for new technology or new facilities, the investments already made in buildings and equipment can deter new investment as long as the old are profitably useful. Social inertia is somewhat similar. People do not readily give up an established social status in a community to move to another city. Of course, the quality of life and the environment where they are living would influence this decision.

Business leaders, chambers of commerce, and other business oriented organizations seek ways to broaden industrial activity and stimulate the growth of jobs and payrolls. Persons interested in environmental quality sometimes fear that such growth means degradation of the environment. A wise policy would be to evaluate each new industry to determine its compatibility with the existing economy and the additional burden of public services such as schools, police, hospitals, water, recreation, waste disposal facilities, and other services. Well-planned developments are based on a comprehensive review of the potential impacts.

Limitations to Urban Growth

For several reasons, it is unwise and unrealistic to assume that urban growth will continue indefinitely. First of all, the impetus to move from rural areas to the city is waning. The fertility rate in rural areas has declined to a level equivalent to that in cities, so there is no longer a surplus of labor. Second, labor requirements for farming are apparently approaching a minimum level. In the future we may need to increase the supply of farm labor in order to increase food production.

In the past, cities grew because the farms were able to produce surplus food products and sell them in the city. In the 1930s, when draft animals provided brute power for farm work, about one-sixth of U.S. cropland was dedicated to growing food for horses and mules. Introducing machines to replace animals not only increased output and cut working hours, but it also freed land to produce more food for human consumption. Then, to increase crop yield, farmers expanded their use of fertilizer. Both machines and fertilizer depend on petroleum. As limited petroleum supplies are consumed, how will agricultural production and city growth be affected?

Even if the proportion of population living in urban communities continues to increase, the largest communities may not grow much bigger. Many planners suggest that the most efficient population size in terms of supplying city services is between 50,000 and 100,000. Above that size, management becomes difficult and service costs increase. Another problem plaguing large cities is the lack of managerial resources for administering government services. Managing large population areas requires skills of the highest order, but the status and rewards of municipal governments do not attract the best administrators.

In addition to the impacts of population growth on agriculture and city services, we must consider how environmental quality is affected. When large numbers of people concentrate in one area, the environment is unable to assimilate all the waste materials. This limited capacity could be aided by more innovations in manufacturing processes—changes that would produce less pollution or that would recycle or convert waste materials into usable and salable products.

ENVIRONMENTAL QUALITY

The Dwelling Unit

The dwelling unit is that space under the control of the occupant within the physical structure. A dwelling unit can encompass the whole structure, as in a single-family house, or several units, as in an apartment house.

A dwelling unit should meet all of the physiological and psychological needs of its occupant. The amount and arrangement of space satisfies such needs as privacy, relaxation, sleep, entertainment, study, home recreation, and social life. The New York Tenement House Act of 1901 required a three-room apartment to have one room of 120 square feet and two of 70 square feet, each with 9-foot ceilings. In 1969 the model code recommended for local adoption by the American Public Health Association—U.S. Public Health Service requires 150 square feet for the first occupant and a minimum of 70 square feet in any sleeping room (or a total of about 290 square feet for a three-room apartment) with a 7-foot ceiling. European countries are now allowing flexibility in room size but are retaining overall space requirements. In the Netherlands a single-room dwelling unit for a couple is considered substandard, and in England separate bedrooms are required for children after they reach the age of 10. Most housing officials in the United States believe that enforcing sleeping arrangements is not practical.

For most people facilities for storing, preparing, and serving food in a safe and sanitary manner are needed in the dwelling unit. The dwelling unit should also protect occupants against extremes of heat and cold, satisfy personal hygiene needs, provide sanitation, and safeguard health. Screening and rat proofing are necessary to keep out insects and rodents. In multifamily dwellings, control of rats and other vermin requires a concerted extermination and housekeeping effort by the landlord and all the occupants.

While people's physical needs have long received attention, their psychological needs are just being recognized. For example, provisions for personal privacy are important in avoiding stress. In parts of some cities, personal safety and security are problems. Some of these needs can be met by facilities and space within the unit, and others are met by the unit's external surroundings. The external facilities include those private to the particular unit or shared with neighbors or conveniently located public facilities.

Human Health

The health aspects of housing relate not only to communicable diseases and mental health but also to those features that influence the safety and stability of the structure itself. Although such items as structural stability, explosion hazards, fire hazards, ready escape, and protection from electrical shock, burns, falls, slipping, and tripping are usually covered in local building codes, the ability of local government to enforce these standards varies considerably.

People in urban settings are in close contact; therefore, disease is likely to spread quickly. Vaccination programs are developed to keep people from getting diseases when they come in direct contact with others or with the air that contains disease-causing organisms. Medical research is producing results that may improve the health of people who live in close contact with one another. **Antibodies,** developed by the body's immune system to combat foreign organisms, are being produced in the laboratory to inactivate specific organisms—antibodies for the measles virus, for example, will bind to that virus and nothing else. By injecting foreign substances into mice and isolating the white blood cells that respond, researchers have isolated antibodies against one substance, called *monoclonal antibodies* that can be used to diagnose specific infectious diseases. Such efforts will allow diagnosis and possible cures of many diseases that occur because of close contact, of which the common cold is one.

Improper disposal of human wastes will create conditions favorable to intestinal diseases, including diarrhea, dysentery, and hepatitis. Improper handling of garbage and human wastes will also attract flies, rats, and roaches. Heavy rat infestation in dwellings not only creates the danger of disease transmission, but children have been bitten so severely that they required medical treatment.

Other diseases are transmitted by insects that thrive under certain conditions in warm parts of the world such as Asia, Africa, Latin America, the Pacific Islands, Malaysia, India, and the Caribbean Islands. The mosquito *Aedes aegypti*, which carries yellow fever, breeds in fresh water in vases, old tires, or coconut husks. It is estimated that there have been 200,000 cases of urban filariasis (elephantiasis) transmitted by the mosquito *Culex fatigens*. In Venezuela, homes with dirt floors may have a problem with Chagas disease carried by the *kissing bug*.

A study in Thomasville, Georgia, showed that readily available safe running water for cleansing purposes reduced the prevalence of shigella dystentery. Those homes that had water piped into the kitchen had fewer cases of the disease than those where water was obtainable only from a spigot on the back porch or in the yard, and those with water on the premises had fewer cases than ones where the water was available next door or up the street.

While the use of lead in paints intended for interior use was discontinued in the United States by 1950, there are many old homes that have layers of lead-based paint on walls and woodwork. Children between 1 and 3 years old with a condition known as *pica* have a tendency to chew on various objects, including paint chips. Recurrence of the condition is likely unless the child is moved to another environment or the environment is modified by removing the paint or by putting a shield over painted surfaces. Children living in slum areas have the greatest exposure to old lead paint flaking off the ceiling and walls. In a study of Connecticut inner-city children, 8 percent had lead poisoning. Lead poisoning can cause mental retardation, behavioral difficulties, perceptual disabilities, emotional instability, and distractibility.

Carbon monoxide (CO) is a hazard in homes, particularly in old ones where there is faulty heating equipment and in new ones where the garage is attached to the house. Coal-burning stoves with defective flues and chimneys are responsible for carbon monoxide poisoning and fires. Sometimes where autos are garaged in basements or in attached rooms, carbon monoxide from running motors has seeped into the main body of the house through heating ducts or doorways. In New York City, where apartment houses were built over expressways, carbon monoxide levels in units on lower floors approach hazardous levels.

Some new homes are constructed so tightly to prevent heat loss that the furnace is starved for air, thus increasing carbon monoxide accumulation. A back draft will then put carbon monoxide in the house. This situation is relieved by first warming the outside air with furnace exhaust (optional) and then bringing it in a pipe to the combustion chamber.

Some investigators have concluded that bad housing conditions play an important role in the development of mental illness. While New York City and Malaysia have been the subjects of studies which show mental illness is related to overcrowding, Hong Kong with 2000 persons per acre and a median of 45 square feet per person does not show the same relationship.

INTERACTION BETWEEN URBAN AREAS AND THE NATURAL SYSTEM

Farmland

Land use within an urban area affects the territory around the development as well as the character of the environment within the community. In California, for example, two-thirds of the estuaries have been destroyed by development and 42 percent of the remainder are destined to be destroyed by the year 2000. These estuaries are vital to continued fish production. Near Philadelphia, an area of the Tinicum Marsh has been filled for airport expansion and interstate highway construction; However, the U.S. Army

Corps of Engineers refused to approve further filling for the development of Marco Island, Florida.

Some 72 million cubic meters (60,000 acre-feet) of topsoil are lost annually in California as a result of logging, grading for housing developments and road construction, and forest fires. As the population grew by over 1.5 million between 1950 and 1975, over 100,000 acres of farmland in Orange County were converted to urban uses. In the San Joaquin Valley 407,000 acres of prime farmland is expected to be swallowed by urbanization between 1972 and 2000. One-third of California's prime agricultural land has been lost to development.

In the United States as a whole, more than 3.2 million hectares (8 million acres) were converted to urban development, reservoirs, and other uses from 1967 to 1985. These trends represent a significant loss of food production. Future planning should therefore encompass both rural and urban areas so that the economic and cultural benefits of cities can be developed in ways that will preserve ecosystems and agriculture.

These trends indicate that a major social crisis in the 1990s will be farmland preservation—equal to the energy crisis that became prominent in the 1970s and the hazardous waste crisis that dominated environmental concerns in the 1980s. The issues associated with farmland preservation are complex and not due simply to population growth.

Assessors view farmland located near cities as having value related to development potential—not use as a farm—and raise the appraised value on which property taxes are based. As inflation pushes up the costs of goods and services essential for farming, the farmer's profit margins become depressed. High interest rates make planting and fertilizing costs soar even more. Environmental restrictions prevent adequate pest control while nearby urban air pollution reduces yields. Smog attacks lettuce, turnips, spinach, mustard greens, and artichokes; soybeans are sensitive to acid rain; and lead and other toxic pollutants from auto exhausts contaminate fruits. As urban developments become interspersed with farms, everyday farm activities such as noise and odors become nuisances to the urbanites. Some farmers give up and sell out.

While urban mayors decry the deterioration of the central cities, national tax policies encourage suburban development that consumes farmland. Investment tax credits that cost the public $10 billion per year invite new development. Accelerated depreciation allowances encourage investment in new plants and equipment while tax exempt bonds finance new water and sewer extensions. The 1981 change in tax laws that provides tax credit for improving 30- to 40-year-old homes is one step toward a farmland preservation policy.

Even some planners contribute to land development pressures. Population growth is viewed by many in the business community as a means to increase markets for products and services. Consequently, forecasts of population growth become a political goal. Based on optimistic projections of population growth, an infrastructure of roads and water and sewer lines is built to accommodate the expected increase. With this subsidy, development follows—just as forecast.

An alternative planning policy encourages *compact* versus *leap-frog*, or *spread*, development. This policy will not only save prime farmland, but it will assure other benefits, including lowering the cost of public services and facilities; enhancing the agricultural economic base; and maintaining rural lifestyles. The quality of life will be preserved at a higher level because prime soils require less energy input to farm than poor soils. Also, open space and aquifer recharge areas are preserved.

In England, the government purchases development rights to agricultural lands so that farms remain in production. Farmers also provide open space and, at certain seasons of the year, recreational opportunities. This example shows that if future development of cities were restricted to unproductive areas and prime agricultural land retained for farming, society would receive the benefits of both. Hawaii has adopted policies to preserve the pineapple industry; Massachusetts is experimenting with the purchase of development rights; New Jersey and Pennsylvania have property tax policies for preserving farmland.

Environmental Consciousness in New Development

Taking soil, geology, minerals, topography, hydrology, and biotic resources into consideration, new development can be compatible with the natural environment. For example, 57 to 73 million hectares (40 to 180 million acres) of land (6 to 8 percent of all land) in the United States lies in flood prone areas such as floodplains of rivers and coasts and certain barrier islands. Six and four-tenths million dwelling units plus industrial and commercial facilities have been built in these areas. Rising costs of flood damage reflect this unwise development pattern. In 1966, the cost was $1 billion, while in 1976, the cost had risen to $2.2 billion ($3.8 billion in 1979 dollars). Losses in 1985 were $5 billion (based on 1979 dollars). The ideal use for floodplains is frequently agriculture.

A knowledge of the topography is extremely valuable when selecting building sites and street layout. Unstable slopes are prone to landslides and slumps. A low slope of 1 to 10 percent presents no major obstacles, but indiscriminate removal of vegetation will speed erosion. Gentle slopes are suited to low density residential development (or more dense with caution).

Soil surveys can be taken and maps prepared with the assistance of the U.S. Soil Conservation Service. Prime soils should be reserved for agricultural use, therefore, development should be guided to areas where soils are less than prime. If septic sewage disposal systems are planned, soils should be suited for percolation of liquid. The engineering properties of soils and underlying bedrock will determine if planned structures can be supported at acceptable expense. If valuable minerals or building materials are present, the surface development should be compatible with extraction and restoration.

Development plans should preserve wetlands and aquifer recharge areas. Wetlands are usually low-lying areas saturated with water and overgrown with vegetation. They act as sponges to absorb excess runoff and reduce flooding. Aquifer recharge areas are filters for underground water reservoirs, thus careless development by paving or building over the recharge area diverts water away.

Many people like to see some form of wildlife in cities. As a result, some states have established urban wildlife programs to suggest ways of attracting desirable species such as cardinals and squirrels. Biologists associated with urban wildlife programs are also called upon to suggest structural changes in buildings to prevent house sparrows from nesting and woodpeckers from drilling into wooden siding. Geese in swimming pools and snakes in gardens are also of concern to people.

In several areas of the United States, biologists and planners are working together to restore wildlife habitat along rivers that pass through cities, greenbelts, and city parks. Some features of the city make it possible for certain wildlife to adapt to urban life. For example, young eastern peregrine falcons are being placed on building ledges in Baltimore, Maryland, and Washington, D.C., in the hope that members of this endangered species will nest and reproduce.

Landscaping and preservation of existing trees is being considered more often in large-scale housing developments in the United States as a method of preventing erosion. As we become more concerned with the siltation of streams and lakes resulting from erosion during construction, it is likely that real estate developers will be even more careful to avoid mass denuding of areas with bulldozers. Through careful planning, biotic resources, woodlands, vegetation, and wildlife can be preserved and unique vegetative habitats protected. Proper street layout and building site orientation can allow builders to take advantage of solar energy for both passive and active systems. Evergreen tree plantings can provide wintertime wind breaks for heat conservation and deciduous trees can furnish summertime shade.

LOOKING AHEAD

Today we have many statistics, including data on population size, people moving into and out of urban areas, transportation, and costs associated with supporting people in different regions. From the planning perspective (Chapter 1), we can answer the question "Where are we?" in different ways. We do not have enough data on how people affect the environment or what happens when we lose a plant or animal species. We need to determine some of these answers.

Moving to our next question, "Where do we want to be?," we need to set targets through local, state, national, and even world organizations. We answer the question in many ways. However, we need to be environmentally and socially conscious in considering where we want to be. As we saw in this chapter, we often have planners to guide us in answering the questions about development.

Placement of city parks, development in national parks, construction of interstate highways and other forms of human change in the environment show how important it is for our planning to consider biodiversity, how the natural system can assimilate human by-products, and how humans can fit into the natural system.

Once we know where we want to be we can decide how to get there. Right now we do not have the necessary goals, objectives, and strategies in the "where we want to be" question. To be sure, we have a start. There are plans to look at national weather and vegetation patterns and to inventory our biodiversity. Geographic information systems are providing valuable data. More comprehensive planning needs to be initiated to develop action plans to get where we want to be and to ensure that our actions result in changes that are positive and sustainable.

Right now, our environmental planning, on a national and worldwide level, does not put us in a position to ask the question "did we make it?," although we are beginning to develop local, regional, and global monitoring programs to help us answer this question.

We are just beginning to use planning. Where we want to be is really up to us. Can we decide, and set goals with strategies to get there?

APPENDIXES

1. The Metric System and Conversion Factors
2. Selected Environmental Periodicals
3. Environmental Organizations

1. THE METRIC SYSTEM AND CONVERSION FACTORS

UNITS OF LENGTH

1 micrometer (μm) = 10^{-6} m = 0.000394 in
1 millimeter (mm) = 10^{-3} m = 0.0394 in = 10^{3} μm
1 centimeter (cm) = 10^{-2} m = 0.394 in = 10^{4} μm
1 meter (m) = 10^{2} cm = 39.4 in = 3.28 ft = 1.09 yd = 0.547 fath
1 kilometer (km) = 10^{3} m = 0.621 statute mi = 0.540 nautical mi

UNITS OF VOLUME

1 liter = 10^{3} cm^{3} = 1.0567 liquid qt = 0.264 U.S. gal
1 cubic meter (m^{3}) = 10^{6} cm^{3} = 10^{3} 1 = 35.3 ft^{3} = 264 U.S. gal
1 cubic kilometer (km^{3}) = 10^{9} m^{3} = 10^{15} cm^{3} = 0.24 statute mi^{3}

UNITS OF AREA

1 square centimeter (cm^{2}) = 0.155 in^{2}
1 square meter (m^{2}) = 10.7 ft^{2}
1 square kilometer (km^{2}) = 0.292 nautical mi^{2} = 0.386 statute mi^{2}

UNITS OF TIME

1 day = 8.64×10^{4} s (mean solar day)
1 year = 8765.8 h = 3.156×10^{7} s

UNITS OF MASS

1 gram (g) = 0.035 oz
1 kilogram (kg) = 10^{3} g = 2.205 lb
1 metric ton = 10^{6} g = 2205 lb

UNITS OF SPEED

1 centimeter per second (cm/s) = 0.0328 ft/s
1 meter per second (m/s) = 2.24 statute mi/h = 1.94 kt
1 kilometer per hour (km/h) = 27.8 cm/s = 0.55 kt
1 knot (a nautical mile per hour, kt) = 1.15 statute mi/h = 0.51 m/s

UNITS OF TEMPERATURE

Celsius (°C)	Fahrenheit (°F)	Kelvin (K)	
− 273.2	− 459.7	0	Absolute zero (lowest possible temp.)
0	32	273.2	Freezing point of water
100	212	373.2	Boiling point of water

Conversions:
°C = (°F − 32)/1.8 °F = (1.8 × °C) + 32

UNITS OF FORCE

British: pound (lb)
Metric: dyne

1 dyne = 2.2481×10^{-6} pound (lb)
1 pound = 4.4482×10^{5} dynes
1 newton (N) = 10^{5} dynes

UNITS OF PRESSURE

1 lb/in^{2} = 68.947 mb = 2.0360 in Hg = 5.1715 cm Hg = 68,947 dynes/cm^{2}
1 dyne/cm^{2} = 1.4505×10^{-5} lb/in^{2} = 2.9530×10^{-5} in Hg = 7.506 $\times 10^{-5}$ cm Hg

1 mb = 1000 dynes/cm^2 = 1.4504 × 10^{-2} lb/in^2 = 2.9530 × 10^{-2} in Hg = 7.5006 × 10^{-2} cm Hg
1 cm Hg = 13,332.2 dynes/cm^2 = 0.19337 lb/in^2 = 13.3322 mb
1 in Hg = 0.49116 lb/in^2 = 33,863.9 dynes/cm^2 = 33.8639 mb
1 bar = 10^5 N/m^2

UNITS OF ENERGY

1 gram-calorie [or just "calorie" (cal)]
1 erg = 1 dyne cm = 2.388 × 10^{-8} cal
1 watt-hour = 860 gram-calories (g-cal) = 3.600 × 10^{10} ergs
1 British thermal unit (Btu) = 0.293 watt-hour = 251.98 gram-cal = 1.055 × 10^{10} ergs
1 joule (J) = 10^7 ergs
1 cal = 4.1855 × 10^7 ergs
1 foot-pound = 1.356 × 10^7 ergs
1 horsepower-hour = 2.684 × 10^{13} ergs = 0.6416 × 10^6 cal

UNITS OF POWER

1 watt (W) = 14.3353 cal/min
1 cal/min = 0.06973 watt
1 horsepower = 746 watts
1 Btu/min = 175.25 watts = 252.08 cal/min

2. SELECTED ENVIRONMENTAL PERIODICALS

AMBIO, Royal Swedish Academy of Science, Stockholm.

American City. Publishing Corp., Berkshire Common, Pittsfield, MA 01201.

American Forests. Publication of the American Forestry Association, 1319 18th St., N.W., Washington, DC 20036.

American Scientist. Sigma Xi, 345 Whitney Ave., New Haven, CT 06511.

Audubon. 950 3rd Ave., New York, NY 10022.

BioScience. Publication of the American Institute of Biological Sciences, 1401 Wilson Blvd., Arlington, VA 22209.

Catalyst for Environmental Quality. 274 Madison Ave., New York, NY 10016.

Ceres. Publication of the Food and Agriculture Organization of the United Nations, Via delle Terme di Caracalla, Rome 00100, Italy.

The Conservationist. New York Department of Environmental Conservation, Albany, NY 12233.

Conservation Foundation Letter. The Conservation Foundation, 1717 Massachusetts Ave., N.W., Washington, DC 20036.

Conservation News. Publication of the National Wildlife Federation, 1412 16th St., N.W., Washington, DC 20036.

Design and Environment. 355 Lexington Ave., New York, NY 10017 (a journal for architects, engineers, and city planners).

Discover. Time Inc., 10880 Wilshire Blvd., Los Angeles, CA 90024.

Ecology. Publication of the Ecological Society of America, Dr. Ralph E. Good, Business Manager, Department of Biology, Rutgers University, Camden, NJ 08102.

Ekistics. Athens Center of Ekistics, Box 471, Athens, Greece (a journal on the problems and science of human settlement).

Endangered Species Technical Bulletin. U.S. Fish and Wildlife Service, Washington, DC 20240.

Environment. 4000 Albemarle St., N.W., Washington, DC 20016.

Environment Abstracts. Environment Information Center, Inc., 124 E. 39th St., New York, NY 10016.

Environment Action Bulletin. Rodale Press, Inc., 33 E. Minor St., Emmaus, PA 18049.

Environment and Behavior. Sage Publishing, 225 S. Beverly Dr., Beverly Hills, CA 90212.

Environmental Management. Springer-Verlag, 175 Fifth Ave., New York, NY 10010.

Environmental Science and Technology. Journal of the American Chemical Society, 1155 16th St., N.W., Washington, DC 20036.

Family Planning Perspectives. Periodical of the Planned Parenthood Federation of America, 515 Madison Ave., New York, NY 10022.

Focus. American Geographical Society, 156 5th Ave., New York, NY 10010.

The Futurist. World Future Society, Box 30369, Bethesda Station, Washington, DC 20014 (a journal of forecast and trends of the future).

The Geographical Magazine. Geography Press, Ltd., London, England.

Impact of Science on Society. Box 433, New York, NY 10016.

Journal of the American Public Health Association. 1015 18th St. N.W., Washington, DC 20036.

Journal of the Air Pollution Control Association. 4400 5th Ave., Pittsburgh, PA 15213.

The Journal of Environmental Education. Heldref Publications, 4000 Albemarle St., N.W., Suite 504, Washington, DC 20016.

Journal of the Environmental Engineering Division. Publication of the American Society of Civil Engineers, 345 E. 47th St., New York, NY 10017.
Journal of Environmental Health. Publication of the National Environmental Health Association, 1600 Pennsylvania Ave., Denver, CO 80203.
Journal of the Water Pollution Control Federation. 2626 Pennsylvania Ave., N.W., Washington, DC 20037.
Journal of Wildlife Management. Publication of the Wildlife Society, Suite 611, 7101 Wisconsin Ave., N.W., Washington, DC 20014.
Living Wilderness. Publication of the Wilderness Society, 1901 Pennsylvania Ave., N.W., Washington, DC 20006.
National Geographic. 1600 M Street N.W., Washington, DC 20036.
National Geographic Research. 1600 M Street N.W., Washington, DC 20036.
National Parks and Conservation Magazine. Publication of the National Parks and Conservation Association, 1701 18th St., N.W., Washington, DC 20009.
National Wildlife. Publication of the National Wildlife Federation, Inc., 1412 16th St., N.W., Washington, DC 20036.
Natural History. Publication of the American Museum of Natural History, Box 6000, Des Moines, IA 50340.
Nature. 711 National Press Building, Washington, DC 20045.
Nature and Resources. UNESCO 7 Place de Fontenoy, 75700 Paris, France.
New Scientist. 128 Long Acre, London WC2E 90M, England.
Nuclear Safety. U.S. Department of Energy, Superintendent of Documents, U.S. Government Printing Office, Washington, DC 20402.
Oceans. P.O. Box 10167, Des Moines, IA 50347.
Pollution Abstracts. Data Courier, Inc., 620 S. 5th St., Louisville, KY 40202.
Population and the Environment. Maxwell House, Fairfield Park, Elmsford, NY 10523 (a publication about social demography, ethnic relations, and human ecology).
Population Bulletin. Publication of the Population Reference Bureau, Inc., 1755 Massachusetts Ave., N.W., Washington, DC 20036.
Resources. Publication of Resources for the Future, 1755 Massachusetts Ave., N.W., Washington, DC 20036.
Science. Publication of the American Association for the Advancement of Science, 1515 Massachusetts Ave., N.W., Washington, DC 20005.
Science 1986. 1515 Massachusetts Ave., N.W., Washington, DC 20005.
Smithsonian. Smithsonian Associates, 900 Jefferson Dr., Washington, DC 20560.
Scientific American. 415 Madison Ave., New York, NY 10017.
Sierra Club Bulletin. Publication of the Sierra Club, 530 Bush St., San Francisco, CA 94108.
Solid Waste Report. Business Publishers, Inc., P.O. Box 1067, Blair Station, Silver Spring, MD 20910.
Technology Review. Massachusetts Institute of Technology Alumni Association, Cambridge, MA 02139.

3. ENVIRONMENTAL ORGANIZATIONS

Alliance for Environmental Education, Inc., 10751 Ambassador Dr., Manassas, VA 22110 (works to further formal school and informal public education activities at all levels).

America the Beautiful Fund, 219 Shoreham Bldg., Washington, DC 20005 (stimulates private citizens and community groups to improve environmental quality through recognition, technical support, and small grants).

American Association for the Advancement of Science, 1333 H St., N.W., Washington, DC 20005 (furthers the work of scientists).

American Committee for International Conservation, Inc., care of Thomas B. Stoel, Jr., 917 15th St., N.W., Washington, DC 20009 (promotes research, disseminates information).

American Conservation Association, Inc., 30 Rockefeller Plaza, Room 5415, New York, NY 10020 (a nonmembership, nonprofit, educational and scientific organization formed to advance knowledge).

American Fisheries Society, 5410 Grosvenor Lane, Bethesda, MD 20014 (professional organization to promote conservation).

American Forestry Association, 1516 P St., N.W., Washington, DC 20005 (seeks to advance intelligent management and educate the public regarding all natural resources).

American Geographical Society, 156 Fifth Ave., New York, NY 10010 (sponsors research, holds symposia, publishes scientific and popular books and periodicals).

American Institute of Biological Sciences, Inc., 730 11th St., N.W., Washington, DC 20001 (national organization for biologists).

American Ornithologists' Union, Inc., National Museum of Natural History, Smithsonian Institution, Washington, DC 20560 (aims to advance ornithological science through its publications, meetings, and membership).

American Society of Ichthyologists and Herpetologists, Dept. of Zoology, Arizona State University, Tempe, AZ 85287 (professional society).

American Society of Landscape Architects, 4401 Connecticut Ave., N.W., Washington, DC 20008 (professional society).

American Society of Limnology and Oceanography, Inc., School of Oceanography, University of Washington, Seattle, WA 98195 (professional society).

American Society of Mammalogists, Department of Fisheries and Wildlife, Oregon State University, Corvallis, OR 97331 (professional society).

The Conservation Foundation, 1250 24th St., N.W., Washington, DC 20037 (nonprofit research and communications organization dedicated to wise use of the earth's resources).

Consumer Action Now, Inc., 110 West 34th St., New York, NY 10001 (lobbies on environmental issues).

Cooper Ornithological Society, Department of Biology, University of California, Los Angeles, CA 90074 (professional society).

Defenders of Wildlife, 1244 19th St., N.W., Washington, DC 20036 (promotes preservation of all wildlife through education and research).

Ecological Society of America, Corson Hall, Cornell University, Ithaca, NY 14853 (professional society).

Environmental Action, Inc., 1525 New Hampshire Ave., N.W., Washington, DC 20036 (an action organization interested in political and social change in a broad range of environmental issues).

Environmental Defense Fund, Inc., 257 Park Ave. S., New York, NY 10010 (an organization of lawyers and scientists that serves as the legal action arm for the scientific community).

Environmental Fund, Inc., 1302 18th St., Washington, DC 20036 (educates the public about the need for population control if the environment is to be saved and people are to live in reasonable comfort and dignity).

Environmental Research Institute, Box 156, Moose, WY 83012.

Food and Agriculture Organization of the United Nations, Via delle Terme di Caracalla, Rome 00100, Italy (established to raise people's levels of nutrition and standard of living).

Friends of the Earth, 218 D St., S.E., Washington, DC 20003 (committed to the preservation, restoration, and rational use of the earth).

International Association of Fish and Wildlife Agencies, 444 North Capital St., N.W., Washington, DC 20001 (principal objectives are conservation, protection, and management of wildlife and related natural resources).

Izaak Walton League of America, Inc., 1401 Wilson Blvd., Arlington, VA 22209 (educates public to conserve, maintain, protect, and restore natural resources).

Keep America Beautiful, Inc., 99 Park Ave., New York, NY 10016 (works to combat littering and improper waste handling).

National Audubon Society, 950 3rd Ave., New York, NY 10022 (conservation and education).

National Fish and Wildlife Foundation, 18th and C Streets, N.W., Washington, DC 20240 (encourages private sector to enhance nation's fish and wildlife resources).

National Parks and Conservation Association, 1015 31st St., N.W., Washington, DC 20007 (educational and scientific, nonprofit service organization).

National Science Teachers Association, 1742 Connecticut Ave., N.W., Washington, DC 20009 (educational affiliate of the American Association for the Advancement of Science interested in improving the teaching of science from preschool through college).

National Wildlife Federation, 1400 16th St., N.W., Washington, DC 20036 (conservation education organization).

The Nature Conservancy, 1815 North Lynn St., Arlington, VA 22209 (dedicated to preservation of natural areas; cooperates with colleges and conservation organizations to acquire land for scientific and educational purposes).

The Oceanic Society, Executive Offices, Stamford Marine Center, Magee Ave., Stamford, CT 06902 (fosters informed and sensible management of ocean and coastal resources).

Planned Parenthood Federation of America, Inc., 810 7th Ave., New York, NY 10019 (joins 190 affiliates which operate medically supervised clinics providing family planning services).

The Population Institute, 110 Maryland Ave., N.E., Washington, DC 20002 (works in communications and with key leadership groups to bring population growth into balance with resources).

Population Reference Bureau, Inc., 1875 Connecticut Ave., N.W., Washington, DC 20009 (gathers, interprets, and publishes information on the social, economic, and environmental implications of U.S. and international population dynamics).

Sierra Club, 730 Polk St., San Francisco, CA 94109 (works to protect and conserve the world's natural resources).

Smithsonian Institution, 1000 Jefferson Dr., S.W., Washington, DC 20560 (sponsors a wide variety of programs to educate the public).

Society of American Foresters, 5400 Grosvenor Lane, Bethesda, MD 20814 (professional society).

Soil and Water Conservation Society of America, 7515 N.E. Ankeny Rd., Ankeny, IA 50021 (professional society).

Student Conservation Association, Inc., Box 550, Charlestown, NH 03603 (enlists the voluntary services of conservation-minded high school, college, and graduate students to work and learn during their summer vacations).

The Wildlife Society, 5410 Grosvenor Lane, Bethesda, MD 20814 (professional society).

Wildlife Management Institute, 1101 14th St., N.W., Washington, DC 20005 (promotes better use of natural resources for the welfare of the nation).

The Wilderness Society 900 17th St. N.W., Washington, DC 20006.

Wilson Ornithological Society, National Museum of Natural History, Washington, DC 20560 (professional society).

World Wildlife Fund, 1250 24th St., N.W., Washington, DC 20037.

Zero Population Growth, Inc., 1400 16th St., N.W., Washington, DC 20036 (a citizen's organization formed to stabilize U.S. population growth by voluntary means).

GLOSSARY

Abiotic Nonliving.

Absolute zero The temperature at which no radiant energy is emitted.

Absorption The assimilation and incorporation of substances by solution in a liquid; osmosis by plant roots.

Acid A compound containing hydrogen that upon solution in water produces an excess of hydrogen ions, causing the pH to be lower than 7 (neutral). An acid reacts with a base to form a salt.

Acid deposition Precipitation, fog, or solid particles containing acids produced from a reaction between sulfur and nitrogen oxides and moisture in the air.

Activated carbon The char remaining after wood is heated to a high temperature in the absence of air. The material has many pores and an extensive surface area to which molecules will adhere.

Adaptation Genetically controlled characteristics of a population making it suited to live in its environment.

Adiabatic lapse rate The rate at which a parcel of air loses temperature with elevation if no heat from an external source is added or subtracted.

Adsorption The adherence of molecules to surfaces they contact.

Aeration Agitation of a liquid to allow dissolved gases to escape and to increase concentrations of dissolved oxygen.

Aerobes Organisms that live only under aerobic conditions.

Aerobic Living, acting, or occurring only in the presence of free, uncombined, molecular oxygen either as a gas in air or dissolved in water.

Age distribution pyramids Population distribution by age and sex in a given geographic area.

Alkaline soils Soils in dry regions that contain a large amount of soluble salts, mainly sodium. The salts can appear as a crust during the dry season.

Alluvium Sediments such as sand, gravel, and silt deposited by streams.

Alpha particles Ionizing radiation consisting of two protons and two neutrons emitted from an atom's nucleus.

Amino acids Organic compounds from which proteins are formed.

Anaerobes Organisms that live under anaerobic conditions.

Anaerobic Living, acting, or occurring in the absence of free, uncombined, molecular oxygen either as a gas in air or dissolved in water.

Anthracite coal Coal containing 80 to 98 percent carbon; provides highest heat value of all coal.

Anticline An upfolding of layers of sedimentary deposits with the older rocks in the core of the fold.

Anticyclone A flow of air around the center of a high-pressure area. The flow is clockwise in the Northern Hemisphere.

Aquaculture Farming in salt or fresh water to produce crops such as catfish, shrimp, and oysters.

Aquifer Porous rock or soil saturated with water.

Artesian well A well that penetrates an aquifer confined between two impervious layers so that pressure forces the water to rise to the ground surface.

Asbestos A fibrous mineral used as insulation; a probable carcinogen.

Asthenosphere The flexible layer below the earth's crust.

Atom The smallest particle that exists as an element.

Atomic number The number of protons in an atom's nucleus.

Autotrophs Self-nourishing organisms; green plants and a few bacteria that use the sun's energy to convert inorganic substances to chemical energy by means of photosynthesis.

Bar A deposit of sand offshore in shallow water.

Barrier reef A reef built by coral offshore and roughly parallel to the shore.

Basalt A dense, fine-grained, igneous rock rich in silica and magnesium. Basaltic rocks form the ocean floor and underlie continents.

Base A compound containing hydroxide ions (OH^-) that will react with an acid to form a salt. In solution, the excess hydroxide ions will cause the pH to be higher than 7 (neutral).

Bauxite Rock containing hydrous aluminum oxides; a common ore of aluminum.

Beta particles Ionizing radiation consisting of electrons ejected by an unstable atomic nucleus.

Biochemical oxygen demand (BOD) A measure of the amount of oxygen needed to decompose organic materials in a specific volume of water. Increased organic waste results in a higher demand for oxygen.

Biocide A chemical that destroys pests.

Biogeochemical cycle The cycle of elements essential to life. Stages can be gaseous, sedimentary, and/or hydrological.

Biological amplification A process whereby pesticides or any substances concentrate in each link of the food chain, thus reaching higher levels in each succeeding consumer.

Biological control The use of parasites, predators, or pathogens to maintain a pest population at a lower average density than would occur without their use.

Biological treatment The use of organisms to break down organic compounds, thereby lowering the BOD.

Biomass The total weight of living organisms in a particular area.

Biomass conversion The release of chemical energy stored in animal or plant material that was alive very recently.

Biome A complex of communities that have a distinctive type of vegetation maintained by the region's climatic conditions.

Biosphere The area near the earth's surface where all living organisms are found, including portions of the hydrosphere, atmosphere, and crust.

Biota All living organisms—plant and animal—in a region.

Birth rate The number of births divided by the population size in a given area in a given time.

Bituminous coal Coal containing 60 to 80 percent carbon; provides the second highest heat value.

Black body A perfect absorber of radiant energy.

Blowout A break in a well casing resulting from adverse pressures. It can actually blow part of the casing out of the well, or it may cause only a leak.

BOD *See* Biochemical oxygen demand.

Boiling- water reactor A nuclear reactor that uses steam to drive turbines.

Breeder reactor A nuclear reactor that produces more fertile material than it uses.

Calcification Soil development in chernozemic or grassland soils.

Calorie The amount of energy required to raise the temperature of 1 gram of water 1°C at constant pressure is known as the physics calorie and is spelled with and denoted by a lowercase *c*. Food energy is also measured in Calories, where one Calorie is the amount of energy required to raise 1 kilogram of water from 14.5° to 15.5°C at constant pressure. The food Calorie (or kilocalorie) is spelled with and denoted by a capital *C* and equals 1000 physics calories.

Cannabis The plant from which marijuana is produced.

Carbamates A major chemical group of insecticides absorbed through the stomach. They deactivate nerve enzymes.

Carbonate rocks Mineral deposits containing the carbonate ion (CO_3^{--}), commonly limestone and dolomite.

Carcinogen Any substance capable of producing cancer.

Carnivores Animals that eat animal flesh.

Carrying capacity The number of individuals of a population that a habitat can support at a given time.

Catalyst A substance that facilitates and speeds up a chemical reaction without being involved in the reaction.

Celsius (°C) A temperature scale divided so that the temperatures at which water freezes and boils are 0 and 100, respectively, at sea level; also called Centigrade.

Chain reaction A reaction that stimulates its own repetition. In fission, an atom's nucleus absorbs a neutron and splits, releasing additional neutrons. These in turn can be absorbed by other fissionable nuclei, releasing more neutrons.

Chemical compound. A combination of two, or more molecules of different elements.

Chernozemic soil Soil with a high mineral content; usually found in grasslands.

Chloroplast A plant organelle; the site of photosynthesis.

Chromosome A filamentous structure in the cell nucleus that contains information for reproduction (genes).

Cistern A tank or vessel in which rainfall is stored for future use.

Clear-cutting Removing all trees from an area.

Climax community A stable system that results from succession in each environment.

Closed system A system in which there is no exchange of matter with surroundings.

Coal gasification The reaction of steam with coal to form a gas containing methane, hydrogen, sulfide, and ammonia. The gases are used to generate electricity and burned to operate a steam cycle.

Coitus interruptus Withdrawal of penis before ejaculation.

Commensalism The relationship in which two organisms live together with one receiving some benefit and the other no benefit or harm.

Community All populations in a defined area.

Competition The interactions between two or more organisms or populations that are striving for the same thing.

Compost The residue resulting from composting. It contains humus material that will retain moisture and can have a high metallic content, but usually has little nutritive value.

Composting The process whereby food and agricultural wastes and sewage sludge are partially decomposed by biological action.

Condensation The change from gaseous state to liquid state. For water, it is the dew point.

Condom A sheath of material, usually rubber, worn over the penis to prevent sperm from entering the vagina.

Conductor A material, such as copper wire, through which free electrons can move easily. The flow of these electrons results in an electric current.

Cone of depression The lowering of the water table caused by withdrawing water faster than it is being replaced.

Conservation Wise use of resources.

Continental shelf The portion of the continent flooded by the ocean.

Continental slope The steep slope between the edge of the continental shelf and the ocean floor.

Convection The transfer and transport of heat by a gas or fluid that becomes less dense upon warming and rises with respect to the surrounding colder, denser gas or fluid.

Core The innermost and densest portion of the earth. Its interior is solid (due to pressure), while the outer portion is molten (due to heat).

Coriolis effect The deflecting force acting on a body in motion (e.g., airplane, wind, ocean current) due to the rotation of the earth. This force is clockwise in the Northern Hemisphere.

Critical mass The smallest mass of fissionable material needed to maintain a self-sustaining chain reaction.

Cultural evolution Adaptation through modification of human behavior and alteration of the environment rather than physiological change controlled by genetics.

Culture The information passed from one individual to another or one group to another through the spoken or written word and art, resulting in availability of all knowledge to successive generations.

Curie A measure of radioactivity that describes the amount of a radioisotope that undergoes 3.700×10^{10} disintegrations per second.

Cybernetics The science of controls or self-regulating feedback mechanisms.

Cyclone A flow of air around the center of a low-pressure area. This flow is counterclockwise in the Northern Hemisphere.

Death rate The number of deaths divided by the population size in a given area during a given time.

Decomposers Animals, bacteria, and fungi that cause chemical disintegration of organic material.

Demographic transition Changes in a maturing population that cause a decline in the death rate, then a decline in the birth rate.

Demography The science of vital statistics of populations, usually human populations.

Denitrification The conversion of ammonia and nitrates into free nitrogen by microorganisms.

Density The weight of a substance compared to the volume it occupies. Water, usually used as a reference, is given a density value of one.

Desalination The process of removing or reducing the concentration of salts in water.

Detritus food chain The breakdown of living tissue by decomposers, which results in a release of energy and minerals.

Deuterium An isotope of hydrogen (H^2 or D) with one neutron and one proton, making it approximately twice as heavy as normal hydrogen; also called *heavy water*.

Dew point The temperature at which air is saturated with water vapor that condenses to form water.

Diaphragm A flexible rubber dome inserted into the vagina to prevent conception.

Digestion (sewage sludge) The decomposition of organic matter in sludge by microorganisms.

Disinfection A procedure used to kill pathogenic organisms. In water treatment, chlorination is frequently used.

Dissolved oxygen (DO) Free, uncombined oxygen molecules dissolved in water.

Doubling time (nuclear reactor) A measure of the time a nuclear reactor requires to produce twice the fissionable material originally present.

Doubling time (population) The time it takes for the population in a geographic area to double in size.

Douche A flushing of the vagina with a liquid; sometimes used as a contraceptive method.

Drug addiction A physical dependence on a drug to the extent that physiological changes occur in the body when it is deprived of the drug.

Drug tolerance The state at which the body becomes accustomed to drugs in given amounts and fails to respond to ordinarily effective doses.

Ecological efficiency The rate obtained by dividing the energy received by the total amount available at each trophic level.

Ecology The study of the relationships of an organism or a group of organisms to its environment.

Economic threshold The population size of a pest species above which economic injury occurs.

Ecosystem The living and nonliving components of the environment functioning together.

Ecotone The boundary or transition between two or more diverse communities.

Edge effect The tendency toward an increased variety and density of organisms in ecotone areas.

Electricity The current resulting from a flow of electrons in a conductor.

Electromagnetic radiation Energy emitted from a radiating body (temperature above absolute zero) in wave form and transmitted through space. The wavelength is related to temperature: the hotter the source, the shorter the wavelength. Electromagnetic radiation includes visible light, gamma and ultraviolet (short wavelengths) radiation, and infrared and radio (long wavelengths) radiation.

Electrons Negatively charged subatomic particles that normally orbit the nucleus of an atom.

Electrostatic precipitation The process of removing particles suspended in air by imparting an electrical charge to them. The particles then move in the direction of rods carrying the opposite charge.

Elements Substances that are composed of one kind of atom and cannot be broken down further by chemical change.

Eminent domain The power of government to take private property for public use provided the individual is compensated.

Energy The ability to do work.

Environmental resistance Biotic and abiotic factors that limit population growth.

Environmental science Application of knowledge of environmental principles to sound environmental management.

Enzyme A protein that acts as a catalyst or speeds up chemical reactions.

Epilimnion The upper layer of water in a stratified lake where there is usually circulation induced by wind action.

Estuary A partially enclosed body of water that contains a mixture of fresh water from land drainage and tidal sea water, such as a river mouth or a coastal bay.

Eutrophic lake A lake with abundant nutrients.

Eutrophication The process of increasing productivity or fertility in a lake, accomplished through natural succession or human disruption of the surrounding ecosystem.

Evolution Change in the genetic makeup of populations through natural selection, which occurs in the course of successive generations.

Exponential growth Population growth that is slow at first, then rapidly accelerates beyond the carrying capacity of the habitat, and stops suddenly as environmental resistance or other limits become effective.

Facultative anaerobes Organisms that live under anaerobic conditions but also tolerate aerobic conditions.

Fahrenheit (°F) A temperature scale divided so that the temperatures at which water freezes and boils are 32 and 212, respectively, at sea level.

Fault A fracture in a rock mass where opposite sides of the rock have moved independently.

Fecundity The number of offspring a female could possibly have in her lifetime.

Feedback The process of returning part of the output to a system as input.

Fermentation Anaerobic respiration in which oxygen is supplied by organic compounds.

Fertile material Material that undergoes fission to yield energy.

Fertility rate Actual number of offspring produced by a female during her lifetime; number of offspring in a population per unit of time.

First law of thermodynamics The law which states that energy is neither created nor destroyed but can be transformed from one state to another.

Fission The splitting of an atom into two or more smaller atoms with release of energy and neutrons.

Floc A mass of particles suspended in liquid, produced by chemical or biological action, with other particles adhering upon collision.

Fly ash The fine residue suspended in exhaust gases that results from incomplete combustion.

Folds Bends developing in stratified rocks as a result of tectonic movement.

Food additives Substances added to food that become components of foods or affect their characteristics (includes nutrients, preservatives, antioxidants, thickening agents, emulsifiers, drying substances, flavors, and colors).

Food chain The energy flow from green plants (autotrophs) through consumer organisms in each trophic level. There are two broad forms—grazing and detritus.

Food web Food chains interrelated within a community.

Fossil fuel Substances, such as coal, oil, and natural gas, that are formed by pressure on plant and animal material.

Fusion The formation of heavier nuclei from lighter ones with the release of energy.

Gamma rays Electromagnetic radiation with a wavelength shorter than 10^{-4} micrometer released by the nucleus during the radioactive decay process.

Garbage Refuse containing a large proportion of food wastes.

Gas-cooled reactor A nuclear reactor that uses helium as a coolant.

Genetic changes Structural changes in chromosomes that can be passed on to future generations.

Genetic selection Selection for specific characteristics controlled by genetic material.

Geothermal energy The conversion to electricity of natural heat from the earth's interior.

Granite A fine- to coarse-grained igneous rock rich in silica and aluminum.

Grazing food chain The movement of energy from green plants to herbivores to carnivores (excludes decomposers).

Green Revolution The development and use of high-yield grain varieties on a commercial scale.

Greenhouse effect The selective energy absorption characteristics of carbon dioxide that allow short-wavelength electromagnetic radiation to pass through but absorb longer wavelengths.

Groin An artificial barrier to protect shorelines in littoral drift areas, usually built at right angles to the beach edge and extending offshore.

Gross primary productivity The rate at which plants convert solar energy to chemical energy; the total amount of solar energy converted to chemical energy per unit area per time period, including that used by plants for respiration.

Guano Animal droppings that contain large amounts of phosphate.

Habitat The area wherein all the needs of a population are satisfied.

Half-life A statistical concept of the time it takes for 50 percent of a radioisotope to decay.

Hazard *See* Natural hazard.

Heavy-water reactor A nuclear reactor that uses deuterium as a moderator.

Herbicide A substance used to destroy plant populations.

Herbivores Animals that eat green plants.

Heterotrophs Organisms that depend on autotrophs directly or indirectly for food energy.

Homeostasis The tendency for biological systems to resist change and remain in a state of equilibrium.

Hormones Chemicals secreted in living organisms that control the rate of biochemical processes.

Horsepower A measurement of work required for movement; for example, one horsepower moves 550 pounds one foot in one second.

Host The living plant or animal that supports a parasite.

Humanae Vitae The encyclical issued by Pope Paul VI in 1968 condemning the use of artificial birth control.

Humus A complex mixture of organic and inorganic compounds in soils that is a major source of plant nutrients.

Hydrocarbon A compound containing hydrogen and carbon.

Hydrological cycle The movement of water between bodies of water on or in the earth and the atmosphere, including the passage through living organisms.

Hypolimnion The lower layer of water in a stratified lake where the water is usually stagnant.

Igneous rock Rock formed by solidification of molten magma.

Infiltration The seepage of a portion of rainfall and snowmelt into the ground.

Inorganic compound A compound containing no hydrogen or carbon.

Insecticide A substance used to destroy insect populations.

Intrauterine device (IUD) A small plastic or metal device placed in the uterus to prevent conception.

Inversion A condition in the troposphere where warmer air overlays colder air. It is the reverse of normal conditions.

Ionization The process of forming ion pairs.

Ionizing radiation Particulate or electromagnetic radiation having the ability to dislodge electrons from the orbit of an atom.

Isostasy The concept that describes how continents "float" on crustal plates due to a difference in their densities.

Isotope Atoms of the same element with the same atomic numbers but different mass numbers.

Jet stream A long, narrow, meandering, high-speed air flow or stream at high altitudes, generally near the tropopause.

Juvenile water Water from the earth's interior that has not previously existed as atmospheric or surface water.

Karst topography An area abounding in sinkholes and caverns, underlain with carbonate rock.

Kelvin The scale of absolute temperature in which zero is approximately $-273°C$.

Kinetic energy Energy of movement, or energy doing work.

Kwashiorkor A disease caused by a shortage of proteins that can be accompanied by a Calorie deficiency.

Laparascope A metal tube through which instruments are passed; used to perform tubal ligations.

Laterite A tough, hard covering of iron and aluminum formed by lateritic soils when vegetation is removed.

Lateritic soil Tropical soil where nutrients are found mostly in vegetation; contains insoluble iron and aluminum.

Laterization A soil development process in lateritic soils.

Law of the minimum The law which states that under stable conditions the growth of a plant is dependent on a minimum amount of nutrients (sometimes expanded to include animals).

LD_{50} Lethal dose, the level of radiation at which 50 percent of the exposed population dies.

Leaching A downward movement of nutrients and solids through soil.

Lead time The time it takes from the decision to construct an energy-producing facility until energy is commercially produced.

Legumes Plants of the pea family (e.g., beans, clover, alfalfa), many of which live in symbiotic relationships with nitrogen-fixing bacteria.

Levee A natural mound at the edge of a river formed by deposits left when the river overflowed.

Life table The survival pattern of a group of individuals in a population. *Cohort life tables* account for a group of individuals until all are dead. *Static life tables* determine the number of deaths in an age class by examining a representative number of individuals in that age class.

Light-water reactor A nuclear reactor that uses water as the coolant.

Lignite coal Coal containing less than 40 percent carbon.

Limit The ultimate capacity of a system or its components to function.

Limiting factor Anything (e.g., light, water, nutrients) present in insufficient amounts so that an organism's habitat for survival and reproduction is restricted.

Lithosphere The solidified material forming the earth's crust.

Littoral Having to do with nearshore and beach environments.

Littoral drift The transport of sand by the littoral current; erodes the leading edges of offshore islands and deposits sand at the trailing edge.

Logarithm The power to which a number must be raised (or the number of times it is multiplied by itself) to produce a given number. Thus, for $10^2 = 10 \times 10 = 100$, the logarithm is 2.

Magma A naturally occurring liquid rock, usually composed of silica.

Magnetohydrodynamics The seeding of gases produced from fossil fuels at high temperatures with chemicals to make the gases electrically conductive. Gases are forced through magnetic fields to produce electric current.

Management The process of directing or controlling production or use of resources.

Mantle The flexible material surrounding the earth's core and under the solid crust.

Marasmus A disease caused by a combined shortage of proteins and Calories.

Mass number The number of protons and neutrons in an atom's nucleus.

Mesosphere The layer of the atmosphere above the stratosphere. Temperature declines with height in this layer.

Metamorphic rock Rock formed when extremely high temperature and pressure cause changes in a rock's mineral composition and texture.

Methyl mercury A deadly form of mercury (CH_3Hg^+) formed by microorganisms from elemental mercury and mercurial salts.

Mineral A naturally occurring inorganic crystalline substance with a unique chemical structure.

Mineral reserves Identified deposits of rock from which minerals can be extracted profitably.

Mineral resource Potentially usable concentrations of materials in a particular location on or in the earth, including reserves.

Mining water Extraction of groundwater at a rate exceeding recharge.

Mitochondrion A plant organelle; the site of aerobic respiration.

Model A representation, through structure or actions, for describing the real system or predicting the effect of certain actions on the real system.

Moderator A material (e.g., water, deuterium, graphite) used in a nuclear reactor to slow the velocity of neutrons, thereby increasing the likelihood of further fission.

Molecule A chemical term indicating two or more atoms bound together.

Monoculture The growing of a single crop (e.g., corn, wheat, cotton).

Mortality rate *See Death rate.*

Mutualism The relationship in which two organisms live together to the benefit of both.

Narcotic Opium, its derivatives, or synthetic substitutes that produce tolerance or dependence.

Natality rate *See Birth rate.*

National Environmental Policy Act (NEPA) The federal law that requires environmental impact statements before federal agencies can permit or finance projects under their control. This act established the Council on Environmental Quality.

Natural hazard A natural material, event, or process that is physically or economically harmful to humans.

Natural resource A natural material or process that is beneficial or useful to humans.

Natural selection The dominance of a new genetic form as a result of environmental change.

Negative feedback A command originating from output to change the system in the opposite direction.

Net primary productivity. Gross primary productivity minus energy used for plant respiration. This energy is available for plant growth and consumption by animals.

Neutrons Subatomic particles with no charge found in the nuclei of atoms.

Niche The physical space and functional role of an organism in a community.

Nitrification production by soil bacteria; a two-stage process involving conversion of ammonia ions to nitrite and then nitrite to nitrate. Nitrate is picked up by plants.

Nitrogen fixation The changing of atmospheric nitrogen by microorganisms into a form usable by other organisms (nitrates).

Nonrenewable resource A resource whose amount cannot be increased.

Nuclear energy The energy liberated by nuclear reactions (fission or fusion) or by radioactive decay.

Obligate anaerobes Organisms that cannot tolerate oxygen.

Oligotrophic lake A lake with few nutrients.

Omnivores Animals that feed on both plants and animals.

Oral contraceptive Synthetic hormones taken to prevent conception.

Ore A deposit in which the mineral concentration is great enough to make recovery economical.

Organic compound A compound containing carbon and hydrogen. It can also contain other elements.

Organochlorines Chlorinated hydrocarbons; synthetic pesticides (including DDT, dieldrin, and aldrin) that alter nerve impulses in the body. They are very persistent and mobile in the environment.

Organophosphates Synthetic contact pesticides affecting nerve transmission.

Osmosis The movement of a solvent (usually water) in one direction through a semipermeable membrane resulting in a higher concentration on one side than the other.

Ozone A molecule composed of three atoms of oxygen (O_3). In the upper atmosphere, it forms a thin layer that absorbs very-short-wavelength ionizing radiation. On the earth's surface, it is a principal component of smog, is chemically active, and irritates the eyes and nose.

PAN Chemicals known as peroxyacetylnitrates found in photochemical smog.

Pangea The ancient landmass that at one time comprised all the continents.

Panthalessa The sea that surrounded Pangea.

Parasite An organism living in or on another and obtaining part or all of its nutrients from its host.

Parasitism The relationship between two organisms in which one benefits at the expense of the other.

Particulate radiation Radiation made up of particles (e.g., alpha or beta particles).

PCB Polychlorinated biphenyl; a chemical similar in structure and properties to organochlorines used in rubber, plastics, and ink. It has become a major environmental pollutant.

Permafrost Permanently frozen ground.

Pest Any living thing that successfully competes with people for food, space, or other essential need.

Pesticide A substance used to kill or control a pest population.

pH A measure of a liquid's acidity or alkalinity. It reflects the hydrogen ion concentration: a pH of 7 is neutral, less is acid, and more is alkaline.

Photochemical smog A complex mixture of air pollutants produced in the atmosphere by reactions of hydrocarbons and nitrogen oxides with sunlight.

Photoperiod The length of daylight. This stimulus is used by some organisms to time their activities.

Photosynthesis Autotrophic transformation of sunlight into chemical energy usable by life.

Photovoltaic cell A device that converts solar energy directly to electricity by means of chemicals.

Phytoplankton Microscopic plants floating in aquatic ecosystems; the basic organism in aquatic food chains.

Plate tectonics The theory which states that the earth's crust is separated into large plates that are moved by convective forces below the crust, creating rifts.

Plume The path followed by the particles, liquids, or gases discharged from an exhaust stack into the air or from a pipe into a body of water.

Podzolic soil Soil formed when the acids from coniferous litter are leached out along with soil nutrients during heavy rains, leaving nutrient-poor, sandy soil. Such soil occurs in coniferous forests in cold, damp climates.

Podzolization The soil formation process in podzolic soils.

Pollution The addition of matter to air, water, or soil until an undesirable level of this material is reached.

Population Members of one species of organism occupying a particular place at a particular time.

Positive feedback A command originating from output to change the system in the same direction.

Potential energy Energy available to do work; stored work.

Power The amount of work done in a given period of time.

Predation The feeding of free-living organisms on other organisms.

Preservation Maintaining a resource intact.

Pressurized water reactor A nuclear reactor that uses water under pressure to drive turbines.

Primary consumers Organisms that feed on producers or green plants.

Primary producers Organisms such as plants that synthesize their own organic substances from inorganic substances.

Primary treatment The removal of settleable solids from sewage.

Proprietary function An activity not absolutely essential to the protection of public health and safety.

Protein–calorie malnutrition A syndrome exhibited by young children or others who have diets deficient in proteins, calories, and other nutrients.

Proteins Amino acids linked end to end in a specific order. They are supplied by meat, fish, eggs, milk, cheese, and some vegetables.

Protons Positively charged subatomic particles in the nuclei of atoms.

Pyrolysis The process of heating a substance to a high temperature in the absence of sufficient oxygen to cause combustion. Volatile substances are driven off and recovered.

Rad Radiation absorbed dose; the unit of power of radiation absorbed by a target. It is expressed as energy per gram of absorbing material (erg).

Radiant energy Electromagnetic energy transmitted through space.

Radioactivity The spontaneous disintegration of atoms of unstable elements, emitting radiant energy (gamma radiation) and possibly atomic particles (alpha, beta, neutrons).

Radioisotope The isotope of an element with a nucleus capable of spontaneously emitting radiation in the form of alpha or beta particles or gamma rays.

Recharge area Ground where the soil and geology permit rain to infiltrate, thus recharging the groundwater.

Recycling Using a resource more than once, either in the same or another form.

Refuse Discarded, unwanted materials, including combustible and noncombustible articles.

Rem Roentgen equivalent man; radiation dosage equivalent to rads absorbed multiplied by the biological effect of radiation.

Resource *See Natural resource.*

Respiration The process whereby energy converted from the sun by photosynthesis is made available to life; the taking in of oxygen and releasing of carbon dioxide in breathing.

Rhythm method Birth control method involving abstention from sexual intercourse during a woman's fertile period.

Rift The place where upwelling convective forces in the earth's mantle cause a rise and separation of the earth's crust into plates.

Riparian rights The right of people who own property downstream to use the stream's water. The people upstream cannot divert the water, foul it, or flood the downstream users.

Saltwater intrusion The penetration of salty seawater into an aquifer. This occurs when excessive pumping lowers the water table.

Savanna A grassland with scattered trees or clumps of trees. It is an intermediate community between a grassland and a forest.

Scrubbing The process in which an airstream containing a pollutant is brought into intimate contact with a liquid. The pollutant may be washed out if it is a solid particle or dissolved if it is a gas.

Second law of thermodynamics The law which states that no reaction involving the transformation of energy from one state to another occurs without some energy being changed to a less usable form; that is, no reaction is 100 percent efficient.

Secondary treatment The removal or reduction of BOD from sewage.

Sedimentary rock Rock formed from cementing and pressure in sediments deposited in unconsolidated layers.

Sedimentation The settling of solids suspended in air or water.

Sewage The water carrying wastes from homes. In most municipalities other wastes, such as industrial wastes, may be present.

Shale oil Heavy, viscous crude oils found in rock shale.

Sigmoid growth Population growth that is slow at first and then increases rapidly. As environmental resistance is met, growth slows gradually until an equilibrium is established and maintained around the environment's carrying capacity.

Sinkhole A ground surface depression caused by a collapse over an underground hole or the dissolution of supporting carbonate rock.

Sludge The settleable solids removed by sedimentation; the residue after digestion.

Snags Standing dead trees.

Solar energy The energy produced by fusion reactions occurring on the sun which reaches the earth as radiant energy. This energy must be converted by physical devices into heat or electricity.

Solid wastes Refuse, sludge, and junk items (e.g., autos, refrigerators).

Somatic changes Changes to an individual's tissue during its lifetime (not including changes to reproductive cells).

Species diversity The ratio between the number of species and the number of individuals in each community.

Stimulus An environmental change detected by a receptor.

Stratosphere The layer of atmosphere above the troposphere. Temperature increases in this layer because the ozone absorbs short-wave-length electromagnetic radiation.

Subbituminous coal Coal containing 40 to 60 percent carbon.

Subsidence The lowering of the ground surface because support has been removed by extraction of minerals, oil, or water or by solution of underlying carbonate rocks by percolating water.

Succession Predictable changes in species composition and community organization brought about by organisms during a period of time in a natural community.

Sunshine laws Laws that require agencies to hold hearings in public and permit public access to official records.

Sustained yield The number of individuals that can be removed from a population without destroying it. This term is frequently applied to forestry and fishery practices.

Symbiosis A state in which dissimilar species live together (sometimes used in the same sense as *mutualism*).

Syncline A downfolding of layers of sedimentary deposits with the younger rocks in the core of the fold.

Synergism The combined, simultaneous action of two or more substances such that the total effect is greater than the sum of the effects if each substance is used individually.

System A group of interacting and interdependent components functioning together.

Tar sand Tarlike hydrocarbons found in porous rock or sediment.

Tectonic Pertaining to rock deformation.

Tertiary treatment The third in a series of sewage treatment processes, usually a special process such as chemical treatment to remove chemicals or disease-causing bacteria.

Thermocline The boundary between layers of water in a stratified lake.

Threshold The minimum level or concentration at which an effect first occurs.

Tidal energy Electricity generated by tidal power.

Trace element An element present in small amounts in the geochemical environment.

Transform fault The boundary between tectonic plates where there is motion in opposite directions.

Transpiration The evaporation of water on the surface of and within plant leaves.

Trophic level The level at which energy (food) is transferred from one organism to another.

Tropopause The transition zone between the troposphere and stratosphere.

Troposphere The lowest layer of the atmosphere, next to the earth's surface. Generally, temperature declines with height in this layer.

Tubal ligation Sterilization by surgically removing a segment of each fallopian tube.

Turbidity Suspended particles in water that reflect or interfere with light transmission.

Turbine The part of an electric power facility that captures energy from a source such as water, which turns curved vanes on a rotating spindle. The turbine then spins conductor loops, which generate electricity.

Tundra The biome of the far north characterized by a short growing season and low temperatures. It is essentially a wet arctic grassland.

Vasectomy Sterilization by surgically cutting the sperm-carrying tubes (vas deferens).

Volt The work required to bring a positive charge to one point from another.

Water table The level below which all ground is saturated with water.
Watershed The whole region contributing to the supply of a river or lake.
Watt A unit of power, or the rate at which work is done; for example, 746 watts equal one horsepower.
Wavelength The distance between successive crests or troughs in electromagnetic radiation or sound pressure waves.
Work The movement of an object by force.
X ray Electromagnetic radiation originating outside the nucleus of an atom.

Zone of stress The upper and lower levels of a zone of tolerance where the population can survive but cannot reproduce.
Zone of tolerance The range of environmental factors (e.g., temperature) in which a population can survive.
Zooplankton Microscopic animals floating in aquatic ecosystems.

INDEX

Abortion, 441
Acid, 95
Acid deposition, 210–214
Activated sludge, 163
Activity pattern, 60
Adaptation, 70
Aerobes, 24
Aerobic, 24
Age Distribution Pyramid, 428–429
Air currents, 46
Air pollutants
  carbon monoxide, 204
  hydrocarbons, 205
  lead, 206
  nitrogen oxide, 205
  ozone, 205
  standards, 200–210
  sulfur oxide, 203
  suspended particles, 201
Air pollution, 191–214
  human health, 192–200
  plants, effects on, 197–200, 212
  sources of, 201–206
Air quality, 206–210
Alcohol, 435
Allele, 64
Alluvial fans, 104
Alpha particles, 403
Amino acid, 316, 317
Anadromous fish, 271
Anaerobes, 24, 393
Anaerobic, 24
Animal Welfare Act, 75
Annual cycle energy system (ACES), 335
Annual rainfall, 116
Antarctica, 14–15
Antibodies, 194, 390
Anticyclones, 183
Aquaculture, 29, 313–314
Aquifer, 117
Arable land, 309–311
Artesian well, 117
Atmosphere
  air circulation, 182–185
  composition, 171–176
  earth's heat balance, 172–174
  origin, 171, 172
  pressure, 48
  storms, 185–188
  weather modification, 182
  weather prediction, 178–179
  weather systems, 182–185
Atom, 94
Atomic number, 338
Automobile emission standards, 200
Autotrophs, 24, 26

Bacteria, 143, 144, 153, 154
Bag houses, 202, 203
Barrier island, 110
Barrier reef, 109
Basalt, 80
Base, 95
Beaches, 108
Beta particles, 405
Big Bang theory, 3
Big game, 259, 260
Bioaccumulation, 402
Biochemical oxygen demand (BOD), 150, 151, 162–164, 390

Biocide, 364
Biodiversity, 274–278, 283–286
Biological amplification, 368
Biological control, 377–379
Biomass, 27
Biomass conversion, 350–351
Biomes, 278–284
Biosphere, 5, 20, 422
Biosphere II, 37–38
Biotechnology, 303
Biotic potential, 54
Biotic regions, 278–283
Birth control, 437–443
  attitudes toward, 443
  methods, 437–442
Birth rate, 430
Black body, 174
Black smokers, 84
Blood type, 64
BOD. *See* Biochemical oxygen demand
Boreal forest, 280
Borgstrom, G., 310
Borlaug, N., 302
Breeder reactor, 341
British thermal unit (BTU), 393

Cadmium cycle, 39–40
Caldera-Cladera, 86
California Ballot Initiative to Close Cougar Hunting, 75
California Water Project, 133
Calorie (food), 314, 318
Calorie (physics), 314, 318, 393
Canada Deuterium Uranium Power Reactor (CANDU), 341
Cancer, 409–410
Carbamates, 365
Carbohydrates, 315, 316
Carbon cycle, 31
Carlson, R., 363
Carnivores, 25
Carrying capacity, 53, 54, 248
Catalyst, 205
Catalytic converter, 205
Cataract, 410
Cavity nesters, 252
Celsius, 173
Chain reaction, 339
Channelization, 120, 121
Chaparral, 282
Chemical bonds, 24
Chernobyl nuclear plant, 343
Chernozemic soils, 306–307
Chlorinated hydrocarbon, 365
Chloroplast, 23–24
Cholera, 143, 144
Chromium cycle, 40
Chromosome, 63, 66, 67
Cistern, 157, 158
Cities, development of, 449–455
  location, 451, 455–457
  mobility, 449, 451, 452
  power, 452
  trade, 448
Civilian Conservation Corps (CCC), 223–224
Clean Air Act, 217, 357
Clear-cutting, 127
Climate, 176–188
Climax community, 41
Cloud seeding, 182
Clouds, 182, 184
Coal, 345–347
  anthracite, 346
  bituminous, 346
  lignite, 346
  pollution, 203–204
  subbituminous, 346
Coal gasification, 347
Coastal Zone Management Act, 112
Cold front, 183
Coliform bacteria, 153
Commensalism, 59
Community, 21–23, 61
  pattern, 49
  species diversity, 49, 50
  stratification, 48, 49
  succession, 41–43
Competition, 54, 56, 242
  interspecific, 56
  intraspecific, 54
Compost, 396
Comprehensive Environmental Response Act (Superfund), 416
Comstock Law, 438
Conservation, 9
Consumers (food chain), 25
Continental drift, 82–86
Continental shelf, 107, 108, 122
Continental slope, 107
Convention on International Trade in Endangered Species of Wild Fauna and Flora, 296
Convention Concerning the Protection of World Cultural and Natural Heritage, 296
Convention on Wetlands of International Importance Especially as Waterfowl Habitat, 296
Core, (structure of the earth), 80
Coriolis effect, 183
Cosmic rays, 411

Critical link species, 284
Critical mass, 339
Crown of Thorns, 69
Crust, 80
Cultural evolution, 421–423
Culture, 421
Cybernetics, 6
Cyclone (equipment), 201, 202
Cyclone (wind), 183

Dams, 130
Dark reaction, 24
Dasmann, R., 422
Data bank, 12
Day length (photoperiod), 46
DDT, 71, 152, 275, 365, 366–376, 388
Death rate, 430
Decay series, 340
Deciduous forest, 226, 280, 281
Decomposers, 25–27
  macrodecomposers, 27
  microdecomposers, 27
Deforestation, 238
Delaney Clause, 304
Demographic transition, 425
Demography, 426
Denitrification, 33
Density-dependent factors, 59
Density-independent factors, 60
Deoxyribonucleic acid (DNA), 2, 63, 274, 303
Depressants, 434
Desalination, 132, 161
Desertification, 136
Deserts, 281, 282
Detritus food chain, 29
Detritus food web, 29
Disaster Relief Act, 189
Dissolved oxygen (DO), 150, 151
DNA, *See* Deoxyribonucleic acid
Dominant gene, 65
Doubling time (nuclear), 341
Doubling time (population), 427
Drugs, 433–435
Durant, W., 448

Earth
  asthenosphere, 80
  core, 80
  lithosphere, 80
  mantle, 80
Earth Day, 224
Earthquake, 86–94
  prediction of, 88, 89
Ecological efficiency, 29–30
Ecologist, 5
Ecology, 5
Economic growth, 449–451, 452
Economic threshold, 377
Ecosystem, 5, 20, 92, 93, 225, 240, 397, 422
  energy conversion, 27–30
  energy flow, 23
  stress, 50
Ecotones, 50
Edge, 249
Edge effect, 50
Electrical vehicles, 330
Electricity, 329, 331
Electromagnetic radiation, 173, 174
Electromagnetic spectrum, 173, 174
Electromagnetic waves, 173, 174
Electron, 338
electrostatic precipitator, 202
Element, 2–3, 94
Endangered species, 131, 261–264
Endangered Species Act, 265, 266
Energy, 7, 324–329
  conservation, 356–358
  conversion, 27–30
  flow, 25, 26
  food, 314, 315
  kinetic, 324
  movement in ecosystem, 23, 25, 26
  potential, 324
  world energy flow, 331
Energy production, 325–327
  lead time, 329
  new sources, 327–329
Energy Reorganization Act, 416
Energy sources
  biomass conversion, 350–351
  electricity, 329–331
  fossil fuels, 345–348. *See also* Fossil fuel
  geothermal, 336–337
  hydrogen, 354–355
  nuclear, 338–345
  ocean thermal gradients, 352–353
  primary, 331–345
  secondary, 345–355
  solar, 331–336
  sugar cane, 328
  tidal, 337–338
  water, 353–354
  wind, 351–352
Environmental management, 8–13
Environmental modification, 5–6
Environmental quality, 453, 454
Environmental science, 1, 5
Epilimnion, 123
Erosion, 103, 104, 295, 301, 308
Estrogen, 436

Estuary, 122
Euphotic zone, 48
Eutrophic, 44
Eutrophication, 43–45, 151
Evapotranspiration, 118
Evolution, 68–72
  biological, 68–72, 422
  cultural, 421–423
Exotic species, 288–290
Exponential growth, 53, 54
Extinction, 73–75

Facultative anaerobe, 24
Facultative sludge lagoon, 154
Fahrenheit, 173
Fallout, 413
Family planning, 438, 439
Fats (food), 315, 316
Featured species, 288
Fecundity, 427
Federal Funding Legislation, 270
Federal Insecticide, Fungicide and Rodenticide Act, 382
Federal Land Policy and Management Act, 243
Federal Water Pollution Control Act, 168
Feedback, 7
  negative, 7
  positive, 7
Fermentation, 24
Fertile material, 341
Fertility rate, 427
Fertilizer, 309–310
Fire, 236, 237
First Law of Thermodynamics, 325
Fish culture, 269
Fish ladders, 131, 271
Fisheries, 266–272
  anadromous, 271, 272
  freshwater, 267–269
  saltwater, 270, 271
Fission, 339–340
Fission reactors, 339–340
  boiling water, 339
  breeder, 341–342
  gas-cooled, 341
  light-water, 339
  pressurized water, 339
Fitness, 69
Flood Control Acts, 137
Flooding, 129–132
Floodplains, 119
Follicle stimulating hormone (FSH), 436, 437, 440
Food additives, 303–305
Food and Color Additives Amendment, 304
Food chains, 25, 27, 368, 369
Food Security Act (the Farm Bill), 321
Food supply, 305
  future, 320, 321
  from oceans, 311–314
  world, 305–306
Food web, 25, 26
  detritus, 25
  grazing, 25
Forest management, 225–239
  clear-cutting, 232
  fire, 229, 236
  multiple use, 225, 229, 230
  new perspectives, 236
  pest control, 237
  reforestation, 236
  selective cutting, 232
  sustained yield, 225, 230
  tree farming, 237
Forestry, 231–239
Fossil fuel, 345–350
  coal, 345–347
  coal gasification, 347
  coal liquefaction, 348
  magnetohydrodynamics, 348
  natural gas, 347
  oil, 347
  resources, 350
  shale oil, 348
  tar sands, 349
Fossils, 81
Fusion, 344–345
Fusion reactors, 344–345
  laser implosion, 345
  magnetic confinement, 345

Gamma radiation, 405
Gamma rays, 405
Gap analysis, 294
Gap phase, 285
Garbage, 387, 392
Gene, 63
Gene pool, 63, 64, 71
General Mining Law, 112
Genetic breeding, 63–68, 302
Genetic counseling, 72, 73
Genetic engineering, 72, 73
Genetic inbreeding, 68
Genetics, population, 62
Genotype, 62, 63, 67, 69
Geographic Information Systems (GIS), 12, 293
Geological time, 79–80, 177
Geopressured reservoir, 337
Geothermal energy, 336–337
Giardiasis, 144
Glaciers, 119–123

Global 2000 Report, 424
Gossypol, 442
Grasslands, 239–244, 280, 281
Grazing, 242
Grazing food web, 25
Great Lakes, 152
Green Revolution, 302, 303
Greenhouse effect, 177, 332
Groins (beaches), 108
Gross primary productivity, 27, 28
Groundwater, 118, 155, 156
Guano, 34

Habitat, 74, 267, 268
Half-life (radioisotope), 79, 80, 406
Hallucinogens, 434–435
Hardy-Weinberg Law, 66, 67
Hayes, D., 328
Headland, 108
Heat islands, 181
Heavy-water reactor, 341
Hemoglobin, 194–197
Hemophilia, 66
Herbicides, 365–366
Herbivores, 25
Hertz, 173
Heterotrophs, 25, 26
Heterozygous, 65
Homeostasis, 6–7
Home range, 60
Homozygous, 65
Hoover, Herbert, 223
Horsepower, 324
Hot spot, 86, 89, 90
Humanae Vitae, 444
Humus, 102
Hunting, 58, 59
Hurricanes, 185, 186
Hybrids, 71, 72
Hydraulic fracturing, 414, 415
Hydrogen power, 354–355
Hydrograph, 128
Hydrologic cycle, 115–123
Hydrothermal reservoir, 336
Hypolimnion, 123

Imhoff tank, 162
Immigration, 431
Incineration, 394
Infrared radiation, 174
Insecticides, 364–365
Intercropping, 320
Intrauterine Device (IUD), 441
Inversion, 191, 192
Ionizing radiation, 408–411
Irrigation, 134
Isostasy, 82
Isotope, 339
International Union for the Conservation of Nature and Natural Resources (IUCN), 6

Jet streams, 184, 185
Juvenile water, 115

Karst topography, 118
Kelvin, 173
Kennedy, J. F., 224
Keystone species, 284
Kinetic energy, 324
Krill, 15
Kwashiorkor, 318

Lakes, 123–126
  epilimnion, 123
  eutrophic, 44
  eutrophication, 43–45
  hypolimnion, 123
  oligotrophic, 44
  spring turnover, 123
  stratification, 123
  thermocline, 123
Lamprey, 152, 271
Land classification, 12
Land and Water Conservation Fund Act, 243
Landfill, 155, 156, 391, 393, 400, 401
Landslides, 91, 104
Laser implosion, 345
Laterite, 306, 307
Laterization, 306, 307
Law of the minimum, 45
LD_{50}, 194
Leaching, 393
Lead cycle, 36, 206
Lead poisoning, 35, 402
Lead time (energy production), 329
Legislation, 16
Leopold, A., 9
Levees, 120, 121
Life, 2
Life tables, 429
Light reactions, 24
Limestone, 118
Limit, 7–8
Limiting factors, 45
Liquid metal fast breeder reactor (LMFBR), 341, 342
Lithification, 306
Lithosphere, 81
Littoral currents, 108, 110
Littoral drift, 108
Livestock management, 243

Logarithm, 95
Love Canal, 397
Luteinizing hormone (LH), 436, 437, 440

Macrodecomposers, 27
Madison Avenue effect, 433
Magnetic confinement, 345
Malaria, 372, 373
Malnutrition, 317–318
  kwashiorkor, 318
  marasmus, 318
  protein-Calorie malnutrition, 317, 318
Malthus, T., 438
Management
  fisheries, 266–272
  forest, 224–239
  grassland (range), 239–244
  wildlife, 247–266
Mantle, 80
Marasmus, 318
Marijuana, 434, 435
Mass number, 338
Mediterranean fruit fly, 375
Mercury
  cycle, 35–36, 39
  metallic, 36
  methyl, 36
Mesosphere, 172
Microdecomposers, 27
Migration, 60, 430, 431
  genetic, 68
Migratory Bird Treaties, 270
Mineral, 94–97
  cycles, 30
  imports, 97, 98, 99
  recycling, 100, 101
  reserves, 100, 101
  resources, 97
  role in economy, 97–100
  supply and demand, 100, 101
  use, 98, 99
Minerals (food), 315–317
Mitochondrion, 24, 25
Model, 12, 293
Molecule, 94
Monoculture, 300
Montreal Protocol on Substances that Deplete the Ozone Layer, 189
Morrill Act, 300
Mortality rate. *See* Death rate
Mount St. Helens, 91, 92, 93
Mudflow, 91
Multiple gene, 65
Multiple use, 225, 229, 230, 288
Multiple Use-Sustained Yield Act, 224, 225, 243
Mussel Watch Program, 152
Mutation, 68, 71
Mutualism, 59

Narcotic, 334, 335
Natality rate. *See* Birth rate
National Environmental Policy Act (NEPA), 51
National Forest Management Act, 243
National Wild and Scenic Rivers System Act, 16
National Wilderness Preservation System Act, 16
Natural selection, 68, 69, 71
Net primary productivity, 26
Neutrons, 338
New Jersey Pinelands, 47
Niche, 61
Nicotine, 334–335
Nitrification, 33
Nitrogen cycle, 32
Nitrogen fixation, 32
Noise, 214–218
Noise Control Act, 217
Nongame, 260
Norplant, 442
Nuclear accidents, 342–344
Nuclear energy, 338–345
Nuclear fuel cycle, 344
Nuclear reactors, 339–345
Nuclear waste disposal. *See* Radioactive waste disposal
Nutrition (human), 315–320

Obligate anaerobes, 24
Ocean thermal gradients (OTEC), 351–352
Oil formation, 347
Old growth, 228–239
Oligotrophic, 44
Omnivores, 25
Oparin, 3
Ore, 94–97
Organochlorines, 365
Organophosphates, 365
Orographic rain, 185
Oxygen cycle, 31
Oxygen depletion, 150, 151
Ozone, 174, 175

Pangea, 84
Panthalessa, 84
Parasitism, 59
Passenger pigeon, 73
Pattern (community organization), 49
PCB, 152, 156, 365, 400
Perception of environment, 5
Permafrost, 280
Pest control, 364–383
  argricultural modification, 379–381

Pest control, (*Continued*)
  attractants, 381–382
  biological, 377–379
  genetic modification, 381
  integrated pest management (IPM), 383
  multiple control, 382–383
  pesticides, 363–383
  range, 243
Pesticides, 363–383
  and economics, 376, 377
  environmental effects of, 366–371
  and human health, 371–375
  types, 364–366
Pests, 244, 363
pH, 95, 212
Phenotype, 62, 69
Pheromone, 381–382
Phosphorus cycle, 32–33
Photoperiod, 46
Photosynthesis, 23
  dark reactions, 24
  light reactions, 24
Photovoltaic cell, 333
Phytoplankton, 25
Pinchot, 9, 223, 231
Planning, 9–13, 258, 259, 294, 458
Planning cycle, 10, 458
Plate tectonics, 82, 94
Pleistocene overkill, 423
Podzolization, 306, 307
Pollution
  air, 191–214
  chemical, 151–152
  noise, 214–218
  nonpoint sources, 147, 148
  point sources, 147, 148
  thermal, 166, 167
  water, 143–153
Population, 23, 53
  cycles, 59, 60
  distribution, 60–62, 428
  dynamics, 53–59
  genetics, 62–68
  growth, 53, 54, 423–426
  human, 423–432, 435–446
  interactions, 54–59
Potential energy, 324
Power. *See* Energy, 324
Prairie, 280, 281
Precipitation, 124, 125
Predation, 58, 59
Predator, 58, 265, 266
Predator-prey relationships, 58, 59
Preservation, 9
Preserves, 291, 292
Privy, 158
Producers, 25
Productivity, 26–27, 30, 240
Progesterone, 436
Protein, 33, 315, 318, 319
Protein-Calorie malnutrition, 317, 318
Protons, 338
Public domain, 224
Pyramids, age-sex distribution, 428, 429
Pyrolysis, 351, 395, 396

Radiation, 402–414
  background, 410, 411, 412
  effects, 408–411
  electromagnetic, 172, 173, 403
  fallout, 413
  infrared, 174
  ionizing, 403
  measurement, 406–408
  particulate, 403
  sources, 411, 412
  ultraviolet, 174, 175
  uses, 412–414
Radiation measurement
  Geiger-Mueller detector, 408
  rads, 407
  rem, 407
Radioactive decay, 79
Radioactive waste disposal, 414–417
Radon, 209
Rads, 407
Range management, 240–244
Recessive gene, 65
Reclamation Act, 137
Recycling, 390
Rem, 407
Reproduction, human, 435–443
Resource Conservation and Recovery Act (RCRA), 112, 416
Resources, 8
  energy, 324
  fisheries, 266
  food, 299
  forest, 231
  land, 102, 306
  mineral, 94–101
  nonrenewable, 13, 100, 101
  renewable, 13, 100, 101
  reserves, 94–101
  water, 353
  wildlife, 247
Respiration, 24, 25
Response, 7
Restoration ecology, 51
Ribonucleic acid (RNA), 2

Rift, 83, 84
Riparian rights, 132
RNA. *See* ribonucleic acid
Rocks, 80
  basalt, 80
  cycle, 101
  dating, 79, 80
  formation, 79, 80
  granite, 101
  igneous, 101
  limestone, 101, 118
  marble, 101
  metamorphic, 101
  sedimentary, 101
Roosevelt, F. D., 223
Roosevelt, T., 223, 231
Runoff, 127, 128

Safe Drinking Water Act, 168
Salmonellosis, 143, 144
San Andreas Fault, 85, 86
Sanger, M., 438
Schistosomiasis, 144
Scrubbers, 203, 213, 214
Seamount, 89
Second Law of Thermodynamics, 325
Selenium, 402
Septic tank, 158
Sex-linked characteristics, 65, 66
Shale oil, 348
Shigellosis, 114
Sickle-cell anemia, 70
Sigmoid growth, 53, 54
Sinkhole, 118
Sludge, 165
Smog, 191, 199
Snag, 227
Snail darter, 130–131
Soil Erosion Act, 321
Soil types, 102, 126, 306–308
  chernozemic, 306, 307
  latosolic, 306, 307
  podzolic, 306, 307
Soils, 45, 102, 107
  classification, 306, 307, 308
  composition, 102
  development, 308
  fertility, 308, 309
  formation, 102
  horizons, 102
Solar energy, 24, 331–336
Solar radiation, 45, 172, 173, 331–334
Solid waste, energy conversion from, 350, 351
  bioconversion, 350, 351
  incineration, 394
  pyrolysis, 395
Solid waste disposal, 386–397
  compost, 396
  incineration, 394
  landfill, 391
  recycling, 390
  reduction, 390
Speciation, 71
Species diversity, 48–49, 248, 274, 275, 286
Springs, 157, 158
Sterilization (human), 440, 441
Stimulants, 434, 435
Stimulus, 7
Strategic Plan for Comprehensive Management of Wildlife, 11
Stratification (community organization), 47–48
Stratosphere, 172
Streams, 119–123
Stress (human), 432–435
Stress, zone of, 61
Subduction zone, 83, 90, 91
Subsidence, 105, 106
Succession, 9, 41–43, 54, 92, 93, 229, 248
  primary, 41, 42
  secondary, 41, 42
Sulfur cycle, 34–35
Super fund, 416
Surface Mining Control and Reclamation Act, 112
Surface water, 156
Sustainable development, 6, 13
Sustained yield, 13, 312
Symbiosis, 8, 56
Synergism, 194
System, 8
  closed, 8
  open, 8

Taiga, 280
Tailings, 412
Tay-Sachs disease, 72
Taylor Grazing Act, 224, 240
Tectonic plates, 83
Temperature conversion, 173
Temporal activity, 60
Teratogenic changes, 83
Testosterone, 436
Thermal profile, 172
Thermocline, 123
Thermodynamics, 325
  laws of, 325
  law of conservation of energy, 325
Threatened species, 262
Three Mile Island, 343
Threshold level (physiological), 194, 407, 408
Tidal energy, 337–338

Tobacco, 434, 435
Tolerance, 434
  drug, 434
  pesticide, 370, 371
Tolerance, zone of, 61
Topographic divide, 127, 128
Tornadoes, 186–188
Toxic pollutants, 194–197, 402
Toxic Substance Control Act (TSCA), 416
Toxic wastes, 397–402
Trace elements, 40
Transpiration, 116, 226
Tree farming, 237
Trickling filter, 163
Trophic level, 25, 26
Tropical forests, 233–234, 282, 283
Tropopause, 172
Troposphere, 172
Tsetse fly, 371
Tubal ligation, 441
Tundra, 278–280
Typhoid, 143

Ultraviolet radiation, 174, 175
United Nations Environmental Program (UNEP), 6
Urban forests, 230–231
Urban growth, 453
Urban land use, 455–457
Urban regions, 449
Urban wildlife, 264, 265

Vasectomy, 441
Vitamins, 315, 316
Visibility, 198–199
Volcano, 86–92

Warm front, 183
Wastes
  human, 158, 159
  radioactive, 414–417
  solid, 386–397
  toxic, 397–402
Wastewater treatment, 159–166
  activated sludge, 163
  aeration, 160
  facultative sludge lagoon, 154
  filtration, 160
  Imhoff tank, 162
  primary treatment, 162
  sanitary pit, 158
  secondary treatment, 162
  sedimentation, 160, 162
  sludge digestion, 162
  tertiary treatment, 164
  trickling filter, 163
Water, 44, 115–123
  conservation, 136
  cycle, 44, 124–132
  pollution, 143–153
  power, 353–354
  quality, 143–159
  recycling, 167
  rights, 132
  storage basin, 130
  supply, 132, 134, 137
  table, 117
  treatment, 159–166
  watershed, 128
Waterfowl, 255, 256, 260
Water purification
  aeration, 160
  chemical precipitation, 160
  chlorination, 160, 161, 164
  filtration, 160
  sedimentation, 160, 162
Watts, 324
Wave frequency, 173, 174
Wavelength, 173, 174
Weather, 178–179, 182–188
Wells, 157, 158
White House Conferences, 9, 223
Wild and Scenic Rivers Act, 137
Wildlife management, 247–266
  fire, 248–249, 250–251
  habitat manipulation, 249–253
  hunting, 253–254
  single-species management, 248, 287, 288
  snags, 227, 252
  water impoundments, 252–253
  wildlife refuge, 256, 257
  wildlife trade, 296, 297
Wind power, 351–352
Work, 324
World Conservation Strategy, 6
World Population Conferences, 444
World Wildlife Fund, 6

X-rays, 402

Zero Population Growth (ZPG), 439
Zoning, 456, 457
Zooplankton, 27
Zygote, 63, 64